International Scientific and Practical Conference on Innovations in Engineering and Technology (ISPCIET'2018)

AA006660

IOP Conference Series: Materials Science and Engineering Volume 441

Veliky Novgorod, Russia
28 – 29 June 2018

Editor:

Sergey B. Sapozhkov

ISBN: 978-1-5108-7575-3
ISSN: 1757-8981

Printed from e-media with permission by:

Curran Associates, Inc.
57 Morehouse Lane
Red Hook, NY 12571

proceedings.com

Some format issues inherent in the e-media version may also appear in this print version.

Printed by Curran Associates, Inc. (2019)

For permission requests, please contact the Institute of Physics at the address below.

Institute of Physics
Dirac House, Temple Back
Bristol BS1 6BE UK

Phone: 44 1 17 929 7481
Fax: 44 1 17 920 0979

techtracking@iop.org

Additional copies of this publication are available from:

Curran Associates, Inc.
57 Morehouse Lane
Red Hook, NY 12571 USA
Phone: 845-758-0400
Fax: 845-758-2633
Email: curran@proceedings.com
Web: www.proceedings.com

TABLE OF CONTENTS

IOP Conf. Series: Materials Science and Engineering **441** (2018) 011001 doi:10.1088/1757-899X/441/1/011001

Abstract

International Scientific and Practical Conference on Innovations in Engineering and Technology (ISPCIET'2018)

PREFACE

The conference serves as a platform for discussion of advanced and innovative technical solutions in various areas of modern science and technology. The aim of the conference is to present the innovative approaches to developing convergent solutions in engineering and technology, and also in interdisciplinary scientific researches. The conference's objective is to analyze and appraise innovative solutions in latest scientific researches in terms of applicability, performance, and prospects for the development. Key topics of the conference cover priority research areas of technical and related sciences and take into account global trends in development.

ISPCIET'2018 topics are built on the fundamentals of sustainability, with vigorous attention to all forms of engineering professions and enabling research tools including the following topics.

EDITOR

Sergey B. Sapozhkov

Polytechnic Institute of Novgorod State University - Federal State Budgetary Educational Institution of Higher Education "Yaroslav-the-Wise Novgorod State University" (NovSU)

E-mail: **iit novsu@mail.ru**

IOP Conf. Series: Materials Science and Engineering **441** (2018) 011002 doi:10.1088/1757-899X/441/1/011002

Peer review statement

All papers published in this volume of *IOP Conference Series: Materials Science and Engineering* have been peer reviewed through processes administered by the proceedings Editors. Reviews were conducted by expert referees to the professional and scientific standards expected of a proceedings journal published by IOP Publishing.

IOP Conf. Series: Materials Science and Engineering **441** (2018) 012001 doi:10.1088/1757-899X/441/1/012001

Adaptive neural network control of hexapod for aerospace application

Y A Zhukov[1], E B Korotkov[1], A V Moroz[1], V V Zhukova and A M Abramov[3]

[1] Baltic State Technical University «VOENMEH»
[2]Peter the Great St.Petersburg Polytechnic University
[3]Novgorod State University, ul. B. St. Petersburgskaya, 41 173003 Veliky Novgorod, Russia

Abstract. This research is a part of the work implemented by BSTU "Voenmeh" under the financial support of the Ministry of Education and Science of the Russian Federation for design and development of a precision mechanism with the parallel kinematics called "Hexapod". Based on the Stewart platform Hexapod as a positioning system with the parallel kinematics has many advantages in comparison with the serial kinematics mechanism. In engineering applications, very precise motion of parallel manipulators to track the desired trajectory has received a great deal of attention. The inverse kinematics problem of the hexapod is described, on the basis of which the desired trajectories are formed. The simulation model of hexapod dynamics implemented in the Matlab Simulink environment is shown. In this paper provides adaptive neural networks algorithm of the hexapods control with on-line tuning synaptic weights of network. Our results show an advantage of applying of the investigated algorithm in the problem of tracking control of the space applications hexapod.

1. Introduction.

In BSTU "Voenmeh" and JSC "ISS" named after M.F. Reshetnev is currently working to create a number of multi-step mechanisms with parallel kinematics (MIC) to ensure the precise positioning and orientation of spaceborne instruments and devices [1]. Most of the ultra-precise positioning and orientation systems based on the IPC created in modern times is a modification of the Guy-Stewart platform [2].

In this paper, the object of the study is an IPC ("Hexapod") of space based on six linear drives with stepping motors, depicted in Figure 1. The hexapod consists of a fixed base and a movable platform controlled by six identical linear actuators - legs (bars, posts). Each leg consists of two half-stems connected by hinges to the base and platform.

The main task of the hexapod control system is the development of the position of the mobile platform in the Cartesian coordinates relative to the base with an accuracy of ± 10 μm and orientation with an accuracy of ± 30 arc.sec. The working load of the hexapod is large-sized transformable objects with large mass-inertia characteristics and the center of mass remote from the moving platform.

Hexapod functions in the conditions of open space [3] - an extreme working environment, which determines the stringent requirements for the implementation of hardware. The control system is developed on the basis of a domestic radiation-resistant microcontroller [4]. Therefore, actual and important tasks are to assess the feasibility and improve the quality of the developed hexapod control algorithms. The presence of parametric and functional uncertainties of the hexapod under extreme operating conditions causes the application of adaptive approaches to control.

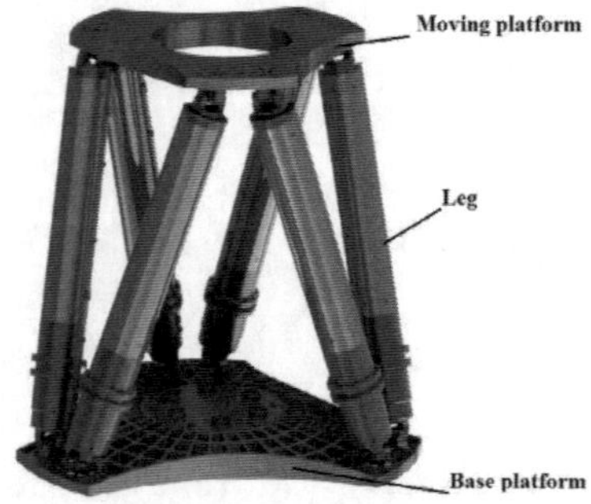

Figure 1. Hexapod sketch.

The exceptional approximating properties of artificial neural networks ensured the uninterrupted interest of the research engineers in their application in a variety of control tasks for mechatronic objects. Multilayer neural networks are used in identification problems, as nonlinear regulators or for forecasting the state of control objects. The main disadvantage of neural networks is the complexity of synthesis and learning in the problems of precise control of dynamic objects.

Nevertheless, in practice the methods of direct adaptive control based on trained on-line "light" neural networks have proved themselves well. In robotics, the approach [5, 6] of the application of an adaptive neural network with sigmoidal activation functions for controlling a manipulation robot with unknown dynamic parameters is known. Works [7-9] reflect the synthesis and implementation of the algorithm in the real-time operating system. In [10], a method of direct adaptive control based on a neural network with RBF - radially basic - activation functions is described. In [11], the RBF-based algorithm is applied in the problem of vibration isolation of the control object using the Stewart platform [2].

In the existing works, the questions of neural network control of the hexapode are not fully disclosed. Thus, the main goal of this work is to investigate the algorithms of direct adaptive neural network control and to evaluate its application in the problem of contour control of the hexapod of space purpose.

2. Hexapod Kinematic

The hexapod control is based on solving the inverse problem of the hexapod kinematics. A hexapod with a symmetrical arrangement of joints to the base and platform is considered, as shown in Figure 2. The pairs of hinges of the base and platform on the dividing circles form an angle of 120° with respect to the center. The hinges of the platform are shifted by an angle of 60° relative to the hinges of the base.

To solve kinematics problems, we introduce coordinate systems, as shown in Figure 2. The fixed coordinate system OXYZ is connected to the base, the mobile coordinate system O'X'Y'Z 'is connected to the platform. We define the initial "zero" position of the symmetrical hexapod, in which the legs have the same elongation. Thus, in this position, the coordinate system O'X'Y'Z 'relative to the OXYZ coordinate system is shifted along the OZ axis by the parameter h_0.

The numerical solution of kinematics problems will depend on the design parameters of the hexapod:

R_b, R_p — radii of dividing circles of the base and platform, respectively, on which the hinges are placed;

C_b, C_p — the distance between adjacent pairs (1-2, 3-4, 5-6) of the base joints and pairs (2-3, 4-5, 6-1) of the platform hinges, respectively.

The length of the legs in the "zero" position is defined as L_0 — "zero length of legs". Enter the leg numbers, as shown in Figure 2.

During the design of the hexapod design, the following parameters are selected:

$$R_b = 0.175 \text{ м}, R_p = 0.15 \text{ м}, C_b = 0.045 \text{ м}, C_p = 0.045 \text{ м}, h_0 = 0.4 \text{ м}.$$

The position of the center O 'of the platform with respect to the fixed coordinate system is specified by the Cartesian coordinates X, Y, Z. To determine the orientation of the platform, the Euler angles [3] φ, θ, ψ. Thus, the linear position and the angular orientation of the platform are given by the vector $\mathbf{q} = [X, Y, Z, \varphi, \theta, \psi]^T$.

The set working range of the hexapod is: with respect to the X coordinate ± 100mm, the Y coordinate ± 100mm, the Z coordinate ± 25mm, the angular coordinates ± 7 deg.

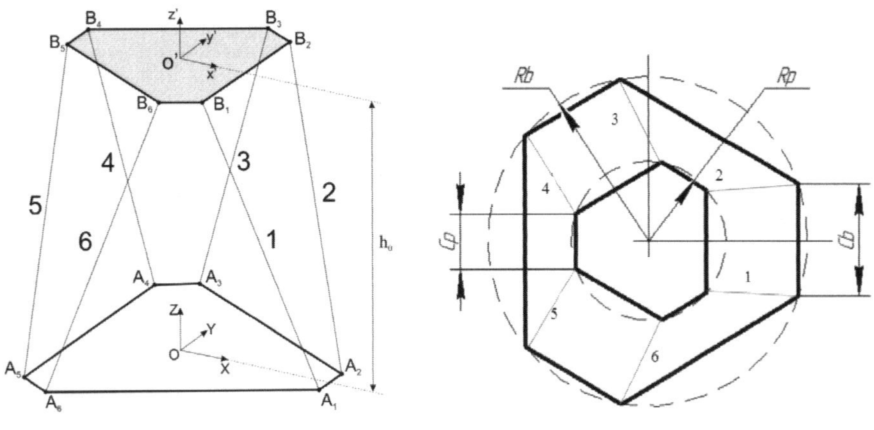

Figure 2. Kinematic scheme and design parameters of the hexapod.

For the connection between a fixed and a moving coordinate system, we use the matrix transformation in homogeneous coordinates [12]:

$$\mathbf{r} = \mathbf{T} \times \mathbf{r}', \tag{1}$$

where $\mathbf{r} = [X, Y, Z, 1]^T$ – vector of homogeneous coordinates of some point in the coordinate system OXYZ, T is the index denoting transposition of the vector, $\mathbf{r}'$ – is the vector of homogeneous coordinates of the same point relative to the moving coordinate system In the transformation (1), the extended matrix T is defined by $\mathbf{p} = [X', Y', Z']^T$ - the coordinates of the beginning of the mobile system relatively fixed - and the rotation matrix **R**

$$\mathbf{T} = \begin{vmatrix} & \mathbf{R} & & \mathbf{p} \\ 0 & 0 & 0 & 1 \end{vmatrix}, \tag{2}$$

Rotation matrix for choosing Euler angles – expression

$$\mathbf{R} = \begin{vmatrix} \cos\varphi\cos\psi + \sin\varphi\sin\theta\sin\psi & \sin\varphi\sin\theta\cos\psi - \cos\varphi\sin\psi & \sin\varphi\cos\theta \\ \cos\theta\sin\psi & \cos\theta\cos\psi & -\sin\theta \\ \cos\varphi\sin\theta\sin\psi - \sin\varphi\cos\psi & \sin\varphi\sin\psi + \cos\varphi\sin\theta\cos\psi & \cos\varphi\cos\theta \end{vmatrix}, \quad (3)$$

Calculating from (1-3) the transformation matrix T for a given vector $\mathbf{q}$ and knowing the coordinates of the points of attachment of the hinges of the legs to the base ($\mathbf{r}_{A_i}$ - vector of homogeneous coordinates of the point A_i в in the OXYZ system) and to the platform ($\mathbf{r}'_{B_i}$ -vector of homogeneous coordinates of a point B_i relatively O'X'Y'Z'), the solution of the inverse kinematics problem is easily found-the length of the i-th leg of the hexapod

$$L_i = \left\| \mathbf{r}_{A_i} - \mathbf{T} \times \mathbf{r}'_{B_i} \right\|, \quad (4)$$

As a rule, when the hexapode is contoured along a given trajectory $\mathbf{q}_r(t)$ in the Cartesian space of coordinates, on the basis of expression (4), the leg lengths defining in space $\mathbf{L}_r(t)$ are determined, and the hexapod control system must work out the trajectory.

3. Hexapod Neural Network Controller

In the structure of the adaptive neural network control algorithm for the hexapode shown in Figure 3, an artificial neural network (ANN) together with a nonlinear controller by error of control forms a force acting on linear drives - the legs of a hexapod.

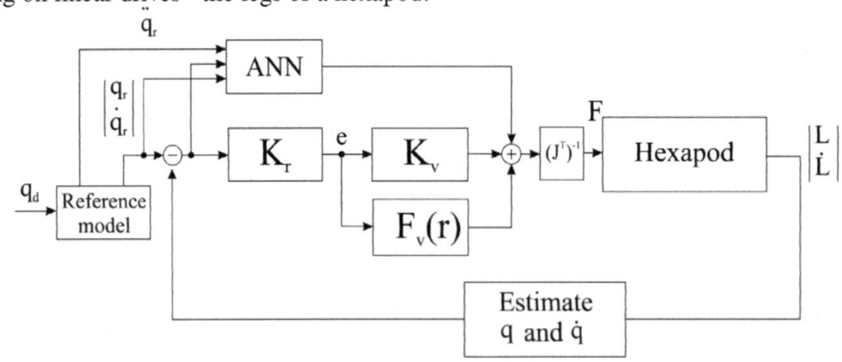

Figure 3. Block diagram of hexapod control controller.

The calculated forces of hexapode control are formed from the expression [5,6]

$$\mathbf{F} = (\mathbf{J}^T)^{-1} \cdot (\mathbf{W}^T \sigma(\mathbf{V}^T \mathbf{x}) + \mathbf{K}_v \mathbf{e} + \mathbf{F}_v(\mathbf{e})), \quad (5)$$

where $\mathbf{K}_v$ - diagonal positive definite coefficient matrix, $\mathbf{x} = \left| \mathbf{q}_\mathbf{r}^T \quad \dot{\mathbf{q}}_\mathbf{r}^T \quad \ddot{\mathbf{q}}_\mathbf{r}^T \quad (\mathbf{q}_\mathbf{r}^T - \mathbf{q}^T) \quad (\dot{\mathbf{q}}_\mathbf{r}^T - \dot{\mathbf{q}}^T) \right|^T$ - entrance of neural network, $\mathbf{W}^T$ и $\mathbf{V}^T$ - matrices of neural network weights, whose values are calculated in a parallel adaptation loop based on the integration of differential equations [5]

$$\begin{cases} \dot{\mathbf{W}} = \mathbf{\Gamma}\sigma(\mathbf{V}^T\mathbf{x})\mathbf{e}^T - \mathbf{\Gamma}\sigma'(\mathbf{V}^T\mathbf{x})\mathbf{V}^T\mathbf{x}\mathbf{e}^T - k\mathbf{G}\|\mathbf{e}\|\mathbf{W} \\ \dot{\mathbf{V}} = \mathbf{G}\mathbf{x}(\sigma'(\mathbf{V}^T\mathbf{x})\mathbf{W}\mathbf{e})^T\mathbf{e}^T - k\mathbf{G}\|\mathbf{e}\|\mathbf{V} \end{cases}, \quad (6)$$

where $\mathbf{G}, \mathbf{\Gamma}$ - positive definite matrices of the algorithm adjustment coefficients, k is the scalar adjustment coefficient, $\mathbf{e}$ - filtered control error, σ - sigmoid activation function, σ' - derivative of the activation function, calculated from expressions

$$\mathbf{e} = \Lambda(\dot{\mathbf{q}}_\mathbf{r} - \dot{\mathbf{q}}) + (\mathbf{q}_\mathbf{r} - \mathbf{q}), \sigma(z) = (1 - e^{-z})^{-1}, \sigma'(z) = \sigma(z)(1 - \sigma(z)), \tag{7}$$

$\ddot{\mathbf{q}}_\mathbf{r}, \dot{\mathbf{q}}_\mathbf{r}, \mathbf{q}_\mathbf{r}$ - speed, and positions in Cartesian coordinates obtained from the reference model based on the signal $\mathbf{q}_\mathrm{d}$ - desired trajectory, $\dot{\mathbf{q}}, \mathbf{q}$ - real velocities and positions in Cartesian coordinates, calculated on the basis of the solution of the direct kinematics problem [14-16] by measurements of positions and velocities of lengths of legs, Λ - diagonal positive-definite matrix of filter coefficients, $\mathbf{J}$ - the Jacobi matrix, whose calculation algorithm is described in detail in [16], $\mathbf{F}_v(\mathbf{e})$ - the robustness rule ensures the stability of the algorithm with a variation in the weights of the neural network and is found from expression

$$\mathbf{F}_v(\mathbf{e}) = \mathbf{K}_z(\|Z\|_F + \mathbf{Z}_m)\mathbf{e}, Z = \begin{vmatrix} \mathbf{W} & 0 \\ 0 & \mathbf{V} \end{vmatrix}, \tag{8}$$

where $\|Z\|_F$ - Frobenius norm from the block matrix Z, $\mathbf{K}_z$ - positive definite matrix of adjustment factors, $\mathbf{Z}_m$ - the maximum value of the Frobenius norm of unknown target weights for the matrix Z.

The procedure for synthesizing the adaptive algorithm consists in choosing the adjustment coefficients $\mathbf{K}_v, \Lambda, k, \mathbf{G}, \mathbf{\Gamma}, \mathbf{K}_z$ and $\mathbf{Z}_m$. We note that the pair of coefficients $\mathbf{K}_v$ and Λ determine the proportional-differential (PD) algorithm for controlling the hexapode, so it is rational to sequentially synthesize the parameters of the PD controller, and then adjust the coefficients of the adaptive neural network controller.

4. Hexapod Control System Model

To assess the quality of hexapode control in the SimMechanics mathematical simulation package of the MatlabSimulink system, a hexapod control system model was created, shown in Figure 4.

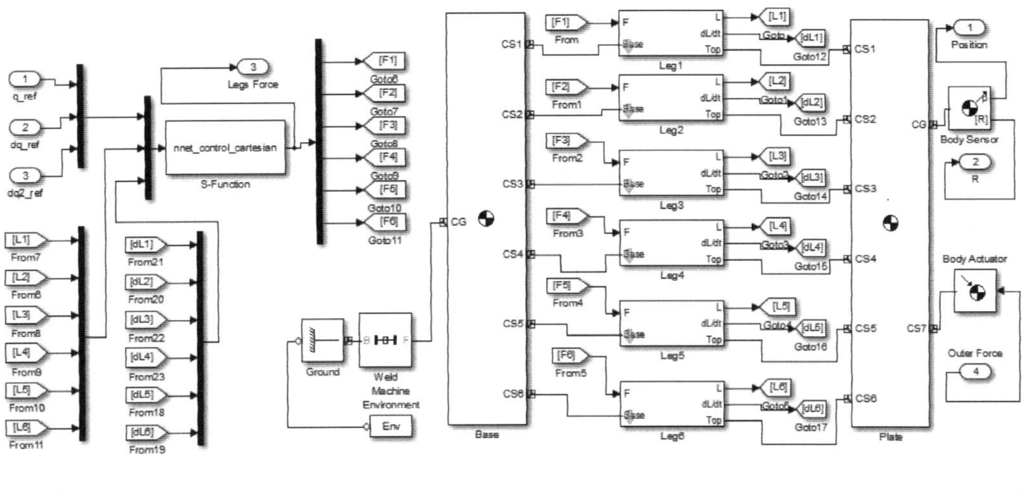

Figure 4. Hexapod control system model «Hexapod».

The hexapod control is simulated under conditions of zero gravity. The hexapod control system generates leg efforts based on the regulator (5-8) implemented in the S-Function "nnet_control_cartesian". The input of the S-Function unit receives signals: set points, speeds, acceleration of leg lengths (q_ref, dq_ref, d2q_ref) and feedback signals - the position and speed of leg lengths. The reference signals in Cartesian coordinates are calculated from the desired desired position of the platform based on the solution of the task of planning the trajectories. The hexapod dynamics model consists of the following elements:

Base – a base, Plate – a platform, (Leg1-Leg6) – models of the dynamics of the hexapod legs, Machine Environment – general environment parameters of the simulation system (zero acceleration of free fall, accuracy of assembly of the mechanics model), Ground – the basis of the absolute frame of reference, Weld – base spike with an absolute coordinate system. The model allows you to calculate the positioning and orientation errors of the platform by the signal from the absolute position sensor BodySensor. On the input of the OuterForce, signals from external forces are applied to the center of mass of the platform. The hexapod's leg model, shown in figure figure5, corresponds to the kinematic structure [17] *RRPRR* - consists of two bodies of semi-stems BaseCylinder and PlateCylinder, connected by a cylindrical hinge of LegCylindrical, two-axis hinges of base and platform, represented by objects 2d jointbase и 2d jointplate respectively. Block JointSensor – element of the information system - generates in the model the signals of the linear position and speed of the drive, JointActuator transmits to the mechanics model the resulting force that determines the motion of the bodies in the linear actuator. The model takes into account the viscous frictional force, which depends linearly on the rate of change in leg length. The parameters of the mechanics of the hexapodshexapap for cosmic purposes are given: the mass of the platform with an inertia load of 100 kg, the main moments of inertia of the platform $J_{xx} = 4900\ \text{kg} \cdot m^2, J_{yy} = 4900\ \text{kg} \cdot m^2, J_{zz} = 6300\ \text{kg} \cdot m^2$.

The mass-inertial parameters of the legs are found from the solid hexapod model created in the SolidWorks environment [18].

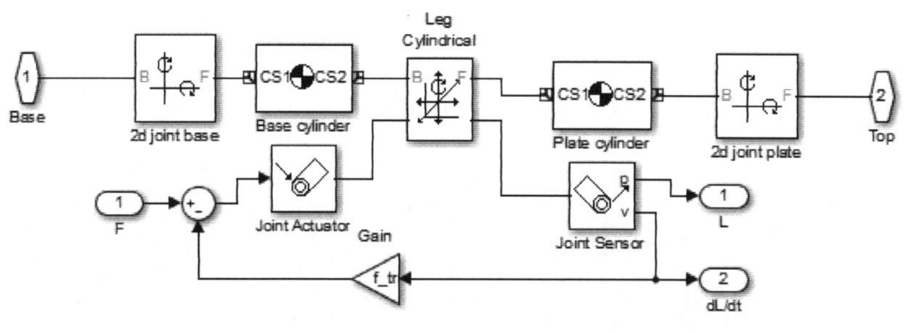

Figure5. The model of the hexapod leg «Leg».

5. Simulation of hexapod control

During the simulation of the hexapod control system, the parameters of the regulator PD were synthesized - the parameters of the matrices $\mathbf{K}_v$ and Λ. Synthesis was carried out in the mode of position control when input to the input of the planned trajectories generated in the reference model. After the synthesis procedure of the PD controller, the parameters of the artificial neural network are determined for the successful adaptation of the regulator. The influence of the number of neurons on the control error has been studied. On figure 6 a) the positional errors of working off the X coordinate in the test jump mode with constant dynamics parameters and the absence of external forces are shown.

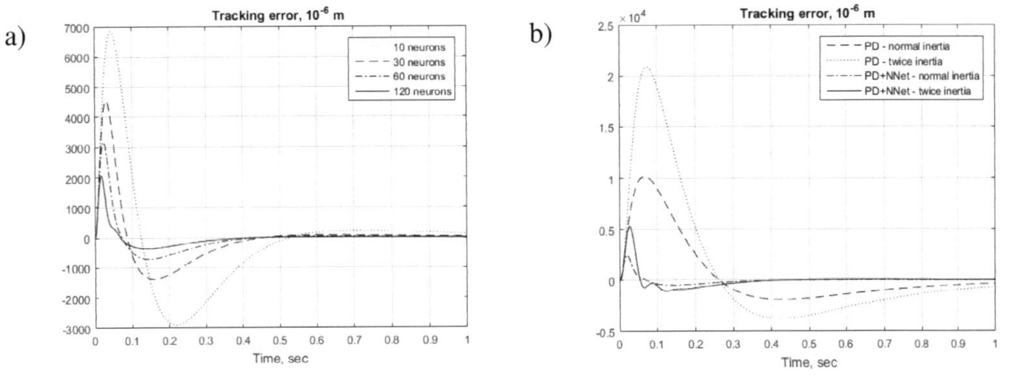

Figure 6. Control error: a) when neurons vary in the network, b) when the inertial parameters change.

The simulation results demonstrate a significant dependence of the quality of approximation on the number of neurons in the hidden layer: with an increase in the number of neurons, the error decreases. However, note that an excessive increase in the number of neurons leads to poor convergence of the adaptation contour, therefore, when evaluating the regulator under the conditions of parametric and functional uncertainties, we will take the number of neurons equal to 90. The influence of parametric uncertainty on the quality of control is also investigated. Figure 6 b) demonstrates the effect of a two-fold change in inertial parameters on the control error. Note that even with a two-fold change in parameters, the transient time for neural network control hardly changes.

To assess the quality of adaptation, comparative studies of synthesized PD and adaptive regulator with harmonic input action, double increase of mass inertial characteristics, as well as action of harmonic external forces on the platform

$$F_x(t) = F_0 \sin(\omega t), F_y(t) = F_0 \sin(2\omega t), F_z(t) = F_0 \sin(3\omega t), \tag{9}$$

where F_0 - amplitude, a ω - frequency of exposure. The simulation results are presented in Figure 7.

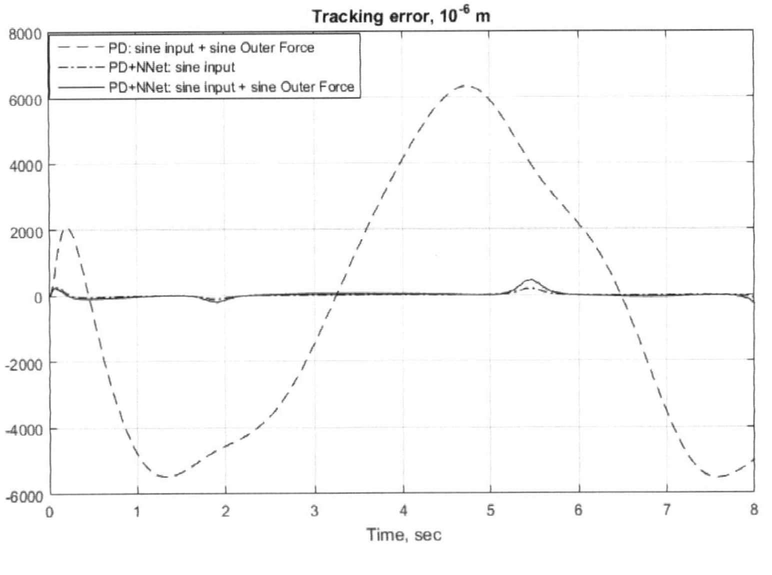

Figure7. Errors of positional control of a hexapode in a harmonic mode under the action of an external force.

Analysis of control errors shows that the adaptive neural network controller successfully copes with variations of the driving influences and external forces, improving the quality of hexapode control.

Conclusion

In the course of this work, Matlab-Simulink implements a hexapod control system model based on the adaptive neural network controller [5, 6]. Modeling of a hexapod control system with a workload under conditions of parametric and functional uncertainties is carried out. The performed investigations showed the efficiency of the regulator in the problem of contouring the hexapod. Adapting to the variable parameters and functional disturbances, the controller successfully parries the error, is easy to configure and robust. It does not require the calculation of the dynamics parameters of the control object. We note the simplicity of the structure of the artificial neural network of the regulator. The neural network has one layer, sigmoid activation functions, is adjusted during the control, allowing the initial state to use zero synaptic weights. To solve the problems of adaptation, there are 60 to 100 neurons. The errors of hexapod control are obtained, which demonstrate the advantages of the adaptive approach with respect to the proportional differential. The results of the research make it possible to give recommendations for the use of an adaptive neural network controller in control systems of mechanisms with parallel kinematics. In the future, it is planned to implement a regulator in the control system based on the domestic radiation-resistant microcontroller [3], as well as to solve the problems of automated tuning and synthesis of the parameters of the adaptive circuit.

References

[1] Korotkov E B, Matveev S A and Jakovenko N G 2016 The paths of increase quality rating of control system of machine with strut-type structure as hexapod and tripod on based Russian and global available electronic component *Questions of radio electronics* **8** pp 85–91

[2] Stewart D A 1965 platform with six degrees of freedom. Proc. of the Institution of mechanical engineers **180** pp 371–85

[3] Zhukov Yu A, Korotkov E B and Slobodzjan N S 2017 Radiation resistant microcontroller with Cortex-M4F core based control system of parallel kinematics mechanism designed for spacecraft's onboard devices movements *Questions of radio electronics* **7** pp 48-54

[4] Microprocessor 1986BE8T [Official web site of MILANDR GROUP] 2018 Available at: http://ic.milandr.ru/products/mikrokontrollery_i_protsessory/ (accessed: 28.02.2018).

[5] Lewis F, Yegildirek A and Kai L 1996 Multilayer neural-net robot controller with guaranteed tracking performance *IEEE Transactions on Neural Networks* **7** pp 388–99

[6] Lewis F, Dawson D and Abdallah C 2004 *Robot Manipulator Control Theory and Practice* (New-York: Marcel Dekker) 614 p

[7] Zhukov Yu 2007 Research of robot-manipulator drive control algorithms *Mechatronics, Automation, Control* **5** pp 21-5

[8] Nemtsev E and Zhukov Yu 2009 Adaptive neural network control of mechatronics objects *ActaMechanica et Automatica* **2** (Bialystok: Bialystok Technical Universityvol2) 81p

[9] Nemtsev E 2011 QNX Platform Application for Discrete Time Adaptive Neural Network Control of PUMA-560 Actuators *Mekhatronika, avtomatizatsiya, upravlenie* [Mechatronics, Automation, Control] pp 17-21

[10] Shuzhi S Ge, Hang C and Woon L C 1997 Adaptive neural network control of robot manipulators in task space *IEEE Transactions on Industrial Electronics* **44** Issue 6 pp 746 – 52.

[11] Jia Ma, Tao Yang, Zeng-Guang Hou and Min Tan 2009 Adaptive neural network controller of a stewart platform with unknown dynamics for active vibration isolation *IEEE International Conf. on Robotics and Biomimetics* pp 1631–36
[12] Fu K S, Gonzalez C S and Lee G 1987 *Robotics: Control, Sensing, Vision, and Intelligence* (Published by Mcgraw-Hill Book Company) 580 p
[13] Gavrilenko V A, Zhukov Yu A and Moroz A V 2016 Kinematics Solutions on the ARM microprocessor for mechatronic control systems of hexapods *Questions of radio electronics* **8** pp 92–8
[14] Cardona M N 2015 A New Approach for the Forward Kinematics of General Stewart-Gough Platforms *Proc. of the 2015 IEEE thirty fifth Central American and Panama convention* pp 1-6
[15] Campa R, Bernal J and Soto I 2016 Kinematic Modeling and Control of the Hexapod Parallel Robot *Proc. of American Control Conference* pp 1203– 08
[16] Liu K, Fitzgerald J and Lewis F 1993 Kinematic Analysis of a Stewart Platform Manipulator *IEEE Transactions on Industrial Electronics* **40** pp282-93.
[17] Lung-Wen Tsai 1999 *Robot Analysis, the Mechanics of Serial and Parallel Manipulators* (New York: Wiley) 520 p
[18] Zhukov Yu A, Gorbunov A V and Lychagin Yu V 2017 Create the simulation model of hexapod in Solidworks and Matlab *Automated design in engineering* **5** pp 98-103

Acknowledgments
This work was carried out in accordance with the RF Government Decree of 09.04.2010 number 218 (PROJECT 218) within NIOKTR carried out with the financial support of the Ministry of Education and Science of the Russian Federation (the contract from 01.12.2015 № 02.G25.31.0160). The work is carried out in the organization of the head performer NIOCTD FGBOU in BSTU "VOENMEKh" named after D. F. Ustinov.

A control of hexapod based on the neural networks solution for forward kinematics

Y A Zhukov[1], E B Korotkov[1], A V Moroz[1], V V Zhukova and A M Abramov[3]

[1] Baltic State Technical University «VOENMEH»
[2] Peter the Great St.Petersburg Polytechnic University
[3] Novgorod State University, ul. B. St. Petersburgskaya, 41, 173003 Veliky Novgorod, Russia

Abstract. This research is a part of the work implemented by BSTU "Voenmeh" under the financial support of the Ministry of Education and Science of the Russian Federation for design and development of a precision mechanism with the parallel kinematics called "Hexapod". In the paper the inverse kinematics problem of Hexapod is described. The solution of the forward kinematics problem is solved by the iterative Newton-Raphson method. A solution of the forward kinematics problem and approximation of the Jacobi matrix of the hexapod based on the apparatus of artificial neural networks is proposed. The results of synthesis of neural networks are presented. An algorithm for the application of neural network kinematics in the hexapode control system is described. A model of a digital control system based on the neural network algorithm under investigation is implemented. The estimation of control accuracy in the positional mode is obtained.

Introduction

BSTU "Voenmeh" and JSC ISS-Reshetnev Company are working together on creating a number of multi-step mechanisms of parallel kinematics [1, 2] (MPK) to ensure the precise positioning and orientation of spaceborne instruments and devices [3]. The object of our research is the MPK ("Hexapod") based on six linear drives with stepper motors, depicted in Figure 1. The hexapod, constructed according to the Stewart platform scheme [4], consists of a fixed base and a movable platform controlled by six identical linear actuators - legs (bars, posts). Each leg consists of two half-stems connected with hinges to the base and platform.

The task of the hexapod control system is to develop the positions of the mobile platform given in the Cartesian coordinates with the respect to the base with an accuracy of ± 10 μm and orientation with an accuracy of ± 30 arc sec.

The main difficulty of controlling the MPK is that when regulating in Cartesian coordinates, it is required to generate forces in linear drives-the legs of the hexapod. The most simple and popular approach for solving this problem is to implement a separate control of the lengths of the hexapod's legs [5], in which the control system is divided into six regulators by each leg. The input of each regulator is fed with a signal of the required length of the leg, calculated on the basis of the solution of the inverse kinematics problem [6], control is formed on the basis of the signal from the linear actuator feedback sensor. Disadvantages of separate control are especially acute when positioning and orienting a large object with large moments of inertia and a remote center of mass.

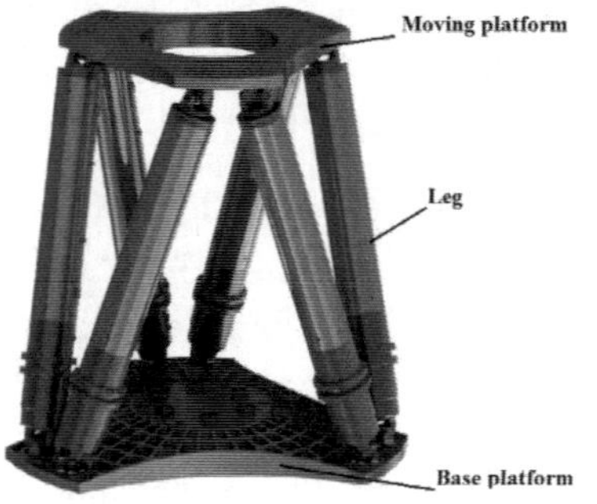

Figure 1. Sketch of hexapod.

In our work, we study the approach to controlling based on the solution of the forward kinematics problem [6], in which regulation with the legs of the hexapod is based on a control error in Cartesian coordinates. The main difficulties of this approach are connected with obtaining a fast solution of the forward kinematics problem, as well as with the transformation of regulation signals from the space of Cartesian coordinates to the space of legs lengths. In [6-8], a forward kinematics problem was solved by iterative numerical methods, but an approach based on artificial neural networks (ANN) is of particular interest.

At present, many researchers are considering with great interest the possibilities of applying ANN theory to solving the problems of controlling mechanisms of parallel kinematics. In known works on the application of ANN for the forward kinematics problem of the Stuart platform, only a part of the problems are solved. In [9-12] the possibility of successful application of ANN direct propagation is shown. In [13-14], the possibilities of recurrent networks (RNN), and in work [15] - nonlinear autoregressive exogenous neural network (NLARX). Essential shortcomings of the research are their poor orientation to the hardware implementation and the insufficient validity of the ANN device application in comparison with the traditional approaches to solving of the forward problem of kinematics.

Thus, the main goal of the present paper is to obtain a neural network solution of the forward kinematics problem and to study its application in the control problem in Cartesian coordinates with a hexapod of space purpose.

Kinematics of hexapod

When managing the hexapod, two kinematics problems are solved [2, 6]:

1) inverse problem of kinematics (IPK) - determination of the length of the legs (half-stems) according to the given position and orientation of the mobile platform;

2) forward problem of kinematics (FPK) - determination of the position and orientation of the platform for the given length of the legs.

To solve kinematics problems, we introduce coordinate systems, as shown in Figure 2. The fixed coordinate system OXYZ is connected to the base, the mobile coordinate system O'X'Y'Z 'is connected to the platform. We define the initial "zero" position of the symmetrical hexapod, in which the legs have the same elongation. Thus, in this position, the coordinate system O'X'Y'Z 'relative to the OXYZ coordinate system is shifted along the axis OZ by the parameter h_0.

The numerical solution of the kinematics problems will depend on the design parameters of the hexapod shown in Figure 2:

R_b,R_p are the radii of the pitch circles of the base and platform, respectively, on which the hinges are placed;

C_b,C_p is the distance between adjacent pairs (1-2, 3-4, 5-6) of the base joints and pairs (2-3, 4-5, 6-1) of the platform hinges, respectively. The length of the legs in the "zero" position is defined as L_0 - "zero leg lengths". Let's enter the leg numbers, as shown in Figure 2.

During the design of the hexapod construction, the following parameters are selected: h_0 = 0.4 m; R_b = 0.175 m; R_p = 0.15 m; C_b = 0.045 m; Cp = 0.045 m.

The position of the center O' platform with respect to the fixed coordinate system is specified with the Cartesian coordinates X, Y, Z. The Euler angles are used to determine the orientation of the platform [16] $\varphi,\ \theta,\ \psi$. Thus, the linear position and the angular orientation of the platform are given with the vector $\mathbf{q}=[X,\ Y,\ Z,\ \varphi,\ \theta,\ \psi]^T$.

The set working range of the hexapod is: with respect to the X coordinate ± 100mm, the Y coordinate ± 100mm, the Z coordinate ± 25mm, the angular coordinates ± 7 deg.

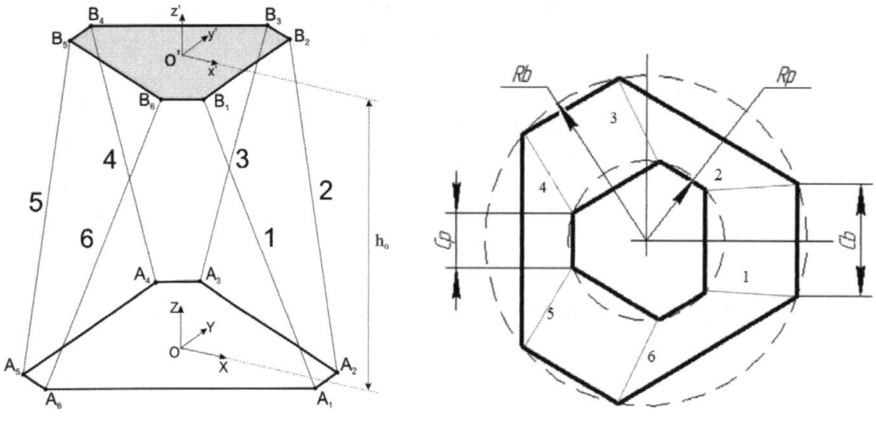

Figure 2. Kinematic scheme and design parameters of the hexapod.

For the connection between a fixed and a moving coordinate system, we use the matrix transformation in homogeneous coordinates [16]:

$$\mathbf{r}=\mathbf{T}\times\mathbf{r}', \tag{1}$$

where $\mathbf{r}=[X,\ Y,\ Z,\ 1]^T$ is the vector of homogeneous coordinates of some point in the OXYZ coordinate system, T is the index denoting the transpose of the vector, $\mathbf{r}'$ is the vector of homogeneous coordinates of the same point relative to the mobile coordinate system.

In transformation (1), the expanded matrix $\mathbf{T}$ is defined through $\mathbf{p}=[X',\ Y',\ Z']^T$ - the coordinates of the origin of the mobile system with respect to the fixed - and the rotation matrix $\mathbf{R}$

$$\mathbf{T}=\begin{vmatrix} & \mathbf{R} & & \mathbf{p} \\ 0 & 0 & 0 & 1 \end{vmatrix}, \tag{2}$$

Rotation matrix for choosing Euler angles - expression

$$\mathbf{R}=\begin{vmatrix} \cos\varphi\cos\psi+\sin\varphi\sin\theta\sin\psi & \sin\varphi\sin\theta\cos\psi-\cos\varphi\sin\psi & \sin\varphi\cos\theta \\ \cos\theta\sin\psi & \cos\theta\cos\psi & -\sin\theta \\ \cos\varphi\sin\theta\sin\psi-\sin\varphi\cos\psi & \sin\varphi\sin\psi+\cos\varphi\sin\theta\cos\psi & \cos\varphi\cos\theta \end{vmatrix}, \tag{3}$$

Calculating from (1-3) the transformation matrix **T** for a given vector **q** and knowing the coordinates of the attachment points of the hinges of the legs to the base ($\mathbf{r}_{A_i}$ - the vector of homogeneous coordinates of the point A_i in the OXYZ system) and to the platform ($\mathbf{r}'_{B_i}$ - the vector of homogeneous coordinates of the point B_i with respect to O'X'Y'Z ') , it is easy to solve the inverse kinematics problem-the length of the *i*-th leg of the hexapod

$$L_i = \left\| \mathbf{r}_{A_i} - \mathbf{T} \times \mathbf{r}'_{B_i} \right\|, \tag{4}$$

Essentially, the solution of the inverse kinematics problem is represented by a set of nonlinear algebraic expressions

$$\mathbf{L} = H(\mathbf{q}), \tag{5}$$

The solution of the forward kinematics problem can be represented by finding the inverse function

$$\mathbf{q} = H^{-1}(\mathbf{L}). \tag{6}$$

Since an analytical solution of problem (6) is difficult, in practice, numerical iterative Newton-Raphson methods [6] are used to obtain vector estimates $\mathbf{q}_{k+1}$ from a given vector $\mathbf{L}_r$

$$\mathbf{q}_{k+1} = \mathbf{q}_k + \mathbf{J}^{-1}(\mathbf{L}_r - H(\mathbf{q}_k)), \tag{7}$$

where **J** is the Jacobi matrix of the Stewart platform, an algorithm for computing which is given in [17].

Neural network solution of kinematics problems

In the present paper, the apparatus of artificial neural networks is used for the approximation solution (6). The investigations have shown that the separate neural network approximation of the components of the vector **q,** the scheme of which is shown in Figure 3, is the best way to use ANN.

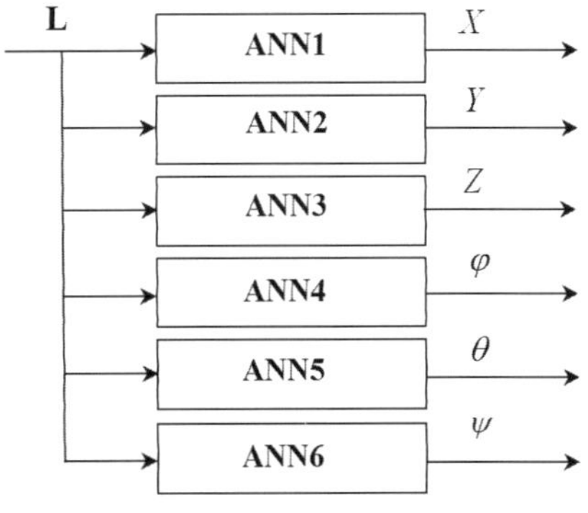

Figure 3. The scheme of the neural network solution of the forward kinematics problem.

The application of this approach makes it possible to accelerate and simplify the process of learning and synthesizing ANN for the solution of a FPK with a specified accuracy, during which the following are determined: algorithm for forming a training sample, the network architecture (network type, number of hidden layers, number of neurons in each layer), learning algorithm, matrix network implementation.

It is convenient to create and train a neural network suitable for realizing the problems of approximation of multidimensional nonlinear functions using modern mathematical modeling environments, which include the Matlab Neural Network Toolbox expansion package [18]. The toolkit of this package was used in the design of neural networks for solving the FPK.

The research has shown that the architecture of the three-layer cascade ANN of direct propagation with two hidden layers, the architecture of which is shown in Figure 4, satisfies the given working range and the required accuracy of the FPK.

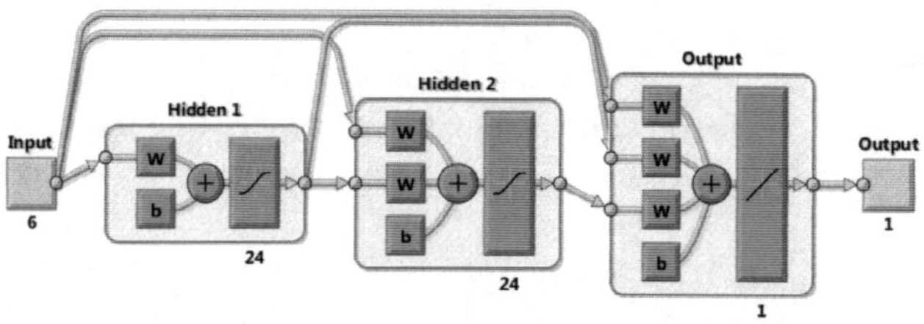

Figure 4. Structure of a cascade neural network of direct propagation.

The synthesis procedure is performed for six ANNs approximating the components of the vector **q**, in this paper the results for the ANN1 are demonstrated. On the basis of the solution of the inverse problem of kinematics (5), a training sample is formed as a linearly distributed array for the values of the working range of the hexapod. The dimension of the array for the training sample was 6 × P6 (P = 5). A three-layer network with two hidden layers was selected, 24 neurons were specified in the hidden layers, and a hyperbolic tangent was used as an activation function. In the output layer there were linear neurons. In the course of ANN training the Levenberg-Marquardt method was used. Learning outcomes for 1000 epochs are shown in the figure 5.

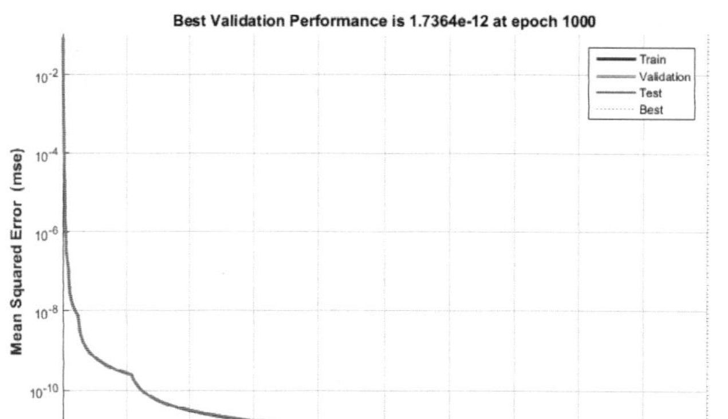

Figure 5. Learning outcomes of the neural network

To test the accuracy of the solution, a test sample of dimension $6 \times P^6$ (P = 10) was formed, the maximum positional error of the solution of FPK for the coordinate X $-\Delta_{max} = 1.4\ \mu m$ was calculated, which satisfies the requirements of the permissible error of hexapod control.

For each synthesized ANN, a matrix realization, defined with expression

$$\begin{cases} \mathbf{y}_1 = \mathbf{K}_{inp}(\mathbf{x} - \mathbf{x}_0) + \mathbf{x}_{\min} \\ \mathbf{y}_2 = F_{th}(\mathbf{W}_{11}\mathbf{y}_1 + \mathbf{b}_1) \\ \mathbf{y}_3 = F_{th}(\mathbf{W}_{21}\mathbf{y}_1 + \mathbf{W}_{22}\mathbf{y}_2 + \mathbf{b}_2) \\ \mathbf{y}_{out} = \mathbf{W}_{31}\mathbf{y}_1 + \mathbf{W}_{32}\mathbf{y}_2 + \mathbf{W}_{33}\mathbf{y}_3 + \mathbf{b}_3 \\ \mathbf{y} = \mathbf{K}_{out}(\mathbf{y}_{out} - \mathbf{y}_{\min}) + \mathbf{y}_o \end{cases}, \quad (8)$$

where the matrices $\mathbf{W}_{11}, \mathbf{W}_{21}, \mathbf{W}_{31}, \mathbf{W}_{22}, \mathbf{W}_{32}, \mathbf{W}_{33}$ - set the synaptic weights in the layers of the neural network, and $\mathbf{b}_1, \mathbf{b}_2, \mathbf{b}_3$- the displacements in each layer, their values were obtained during the training of the ANN; $\mathbf{K}_{inp}, \mathbf{x}_0, \mathbf{x}_{\min}$ - matrices of coefficients of valuation of input values (values of the length vector of legs **L**); $\mathbf{K}_{out}, \mathbf{y}_0, \mathbf{y}_{\min}$ - matrices of recovery coefficients of the output of the network **y** are determined in the normalization of training samples; F_{th} - the activation function of the hyperbolic tangent, is found from expression

$$F_{th}(z) = \frac{2}{1 + e^{-2z}} - 1. \quad (9)$$

A similar approach was used to synthesize the neural network of approximation of the inverse Jacobi matrix in expression (7). The three-layer network was trained; its structure is shown in Figure 6. Learning outcomes are shown in Figure 7.

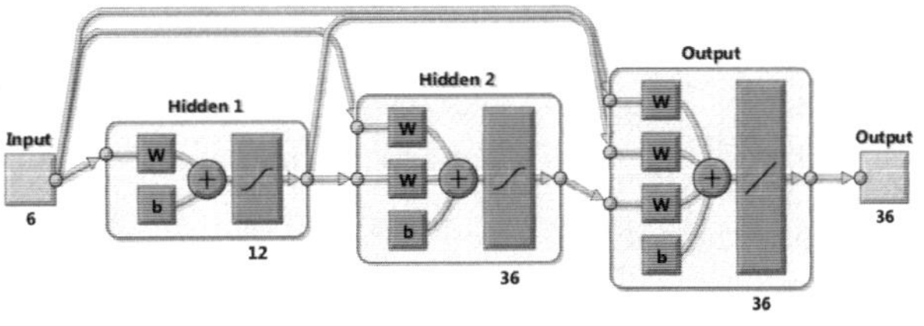

Figure 6. The structure of the cascade network of approximating the inverse Jacobi matrix.

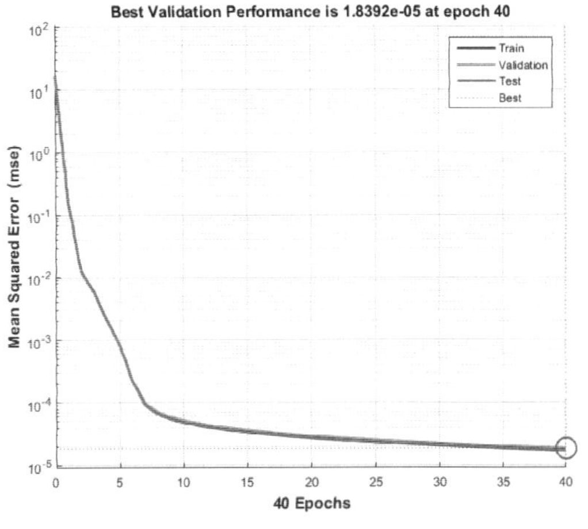

Figure 7. The results of the approximation of the inverse Jacobi matrix.

The maximum absolute error of approximation of the inverse Jacobi matrix on the test sample for the synthesized network was $\Delta_{\max} = 0.062$ rad/m.

Hexapod Management System Model

An estimate of the quality of hexapod control is realized for the algorithm of the stabilizing proportional-differential control in Cartesian coordinates with the calculation of the control forces based on the solution of the inverse dynamic problem of Lagrange's equation of the second kind for the mobile platform [19]

$$\mathbf{M}(\mathbf{q})\ddot{\mathbf{q}} + \mathbf{N}(\mathbf{q},\dot{\mathbf{q}}) = \mathbf{J}^T \cdot \mathbf{F}, \tag{10}$$

where $\mathbf{M}(\mathbf{q})$ is the matrix of inertia of the platform, $\mathbf{N}(\mathbf{q},\dot{\mathbf{q}})$ is the vector of centrifugal, gravitational and Coriolis forces. The algorithm for calculating the components of expression (10) is described in detail in [17, 19].

When implementing the control method on the basis of the estimated moment (called "computed-torque control") of the hexapod control forces $\mathbf{F}$ are derived from expressions

$$\begin{cases} \mathbf{F} = (\mathbf{J}^T)^{-1} \cdot (\mathbf{M}(\mathbf{q})\ddot{\mathbf{q}}_d + \mathbf{N}(\mathbf{q},\dot{\mathbf{q}}) + \mathbf{K}_p\mathbf{e} + \mathbf{K}_d\dot{\mathbf{e}}) \\ \mathbf{e} = \mathbf{q}_d - H^{-1}(\mathbf{L}) \\ \dot{\mathbf{e}} = \dot{\mathbf{q}}_d - \mathbf{J}^{-1} \cdot \dot{\mathbf{L}} \end{cases}, \tag{11}$$

where $\mathbf{K}_p, \mathbf{K}_d$ - the diagonal matrices of the adjustment factors of the PD controller, the $\mathbf{e}$-vector of the control errors $,\ddot{\mathbf{q}}_d, \dot{\mathbf{q}}_d, \mathbf{q}_d,$ - the accelerating signals, velocities and positions in Cartesian coordinates obtained from the trajectory planner, $\dot{\mathbf{L}}, \mathbf{L}$ - measured with the speed sensors and the length of the hexapod's legs, $H^{-1}(\mathbf{L})$ - the iterative solution of the forward problem kinematics (7).

$F_{ann}(\mathbf{L})$ - matrix function of the implementation of the neural network solution of the FPK, obtained on the basis of (8) and (9) for the ANN structure shown in Figure 3.

Applying the neural network approximation for the components of the kinematic problem in (11), we obtain a control algorithm based on the estimated moment with a neural network solution of the FPK and approximation of the inverse Jacobi matrices, which is determined by the relations

$$\begin{cases} \mathbf{F} = F_{ann2}(\mathbf{L}) \cdot (\mathbf{M}(\mathbf{q})\ddot{\mathbf{q}}_d + \mathbf{N}(\mathbf{q},\dot{\mathbf{q}}) + \mathbf{K}_p\mathbf{e} + \mathbf{K}_d\dot{\mathbf{e}}) \\ \mathbf{e} = \mathbf{q}_d - F_{ann}(\mathbf{L}) \\ \dot{\mathbf{e}} = \dot{\mathbf{q}}_d - F_{ann1}(\mathbf{L}) \cdot \dot{\mathbf{L}} \end{cases}, \tag{12}$$

where $F_{ann}(\mathbf{L})$ is the matrix function of the implementation of the neural network solution of the FPK, obtained on the basis of (8) and (9), $F_{ann2}(\mathbf{L})$ - neural network approximation of the reciprocal transposed Jacobi matrix, $F_{ann1}(\mathbf{L})$ - neural network approximation of the inverse Jacobi matrix.

Based on the structure of the regulator (12) and the results of the ANN synthesis in the Matlab Simulink mathematical modeling environment, a digital hexapod control system model was created, shown in Figure 8.

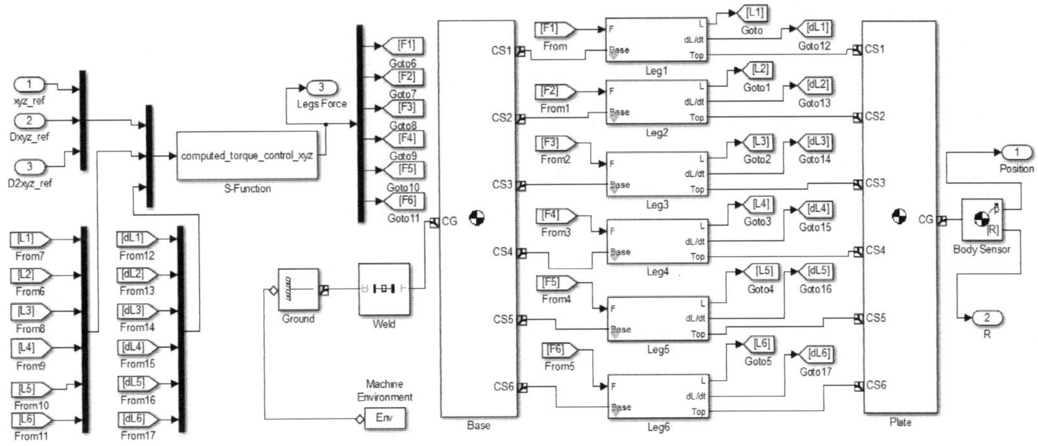

Figure 8. Model of Hexapod Control System.

The hexapod control is simulated under conditions of zero gravity. The hexapod control system generates foot forces on the basis of the regulator (12) realized in the S-Function "computed torque control xyz". The input of the S-Function block includes: the signals of position, velocity, acceleration in Cartesian coordinates ("xyz_ref", "Dxyz_ref", "D2xyz_ref") and feedback signals - position and speed of leg lengths. Commanders in Cartesian coordinates are calculated according to the desired position of the platform on the basis of solving the problem of planning trajectories.

The hexapod dynamics model consists of the following elements: "Base" - base, "Plate" - platform, ("Leg1" – "Leg6") - models of the hexapod's legs dynamics, "Machine Environment"- general environment parameters of the simulation system (zero acceleration of free fall , the accuracy of assembly of the model of mechanics), "Ground" - the basis of the absolute reference frame, "Weld" - the base soldering with the absolute coordinate system. The model allows you to calculate the positioning and orientation errors of the platform by the signal from the absolute position sensor "Body Sensor". On the input "Outer Force" signals of external forces are applied to the center of mass of the platform. The same elements "Go to" and "From" determine the connections in the model. The model of the hexapod leg, shown in Figure 9, consists of two bodies of semi-stems "Base Cylinder" and "Plate Cylinder", connected by a cylindrical hinge "Leg Cylindrical", two-axis hinges of the base and platform represented by the objects "2d joint base" and "2d joint plate" respectively. The "Joint Sensor" block - an element of the information system - generates signals in the model for the linear position and speed of the drive, "Joint Actuator" transmits to the mechanics model the resulting force that determines the motion of bodies in the linear drive. The model takes into account the viscous friction force, linearly dependent on the rate of change in leg length.

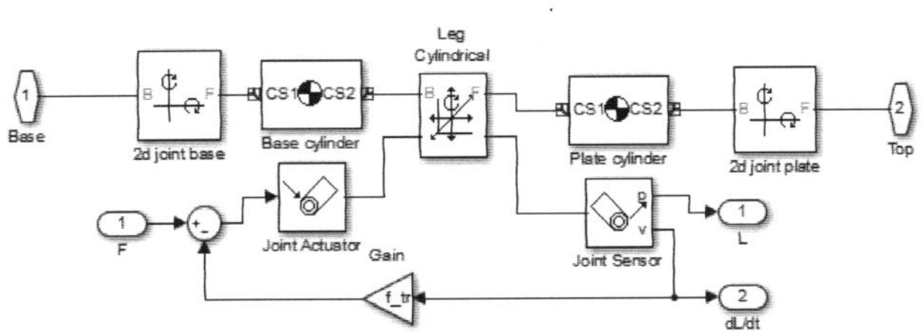

Figure 9. Leg model of the hexapod "Leg".

Modeling of the control system

During the simulation of the hexapod control system, the parameters of the regulator PD - the matrix values $\mathbf{K}_p, \mathbf{K}_d$ - were synthesized. Synthesis was carried out in the mode of position control that is transfer of the system from the initial position $\mathbf{q}_0 = [\ 0,\ 0,\ 0.35,\ 0,\ 0,\ 0]^T$ to the final position $\mathbf{q}_f = [\ 0.1,\ -0.1,\ 0.375,\ -0.122,\ 0.122,\ -0.122]^T$. The parameters of the hexapod mechanics are specified: the mass of the platform with the inertial load is 100 kg, the main moments of inertia of the platform $J_{xx} = 4900\ \text{kg}\cdot\text{m}^2, J_{yy} = 4900\ \text{kg}\cdot\text{m}^2, J_{zz} = 6300\ \text{kg}\cdot\text{m}^2$. For the digital realization of the algorithm (12), the time-quantization period T = 0.001 sec is determined. An estimate of the hexapod tracking errors, which are shown in Figure 10, is obtained.

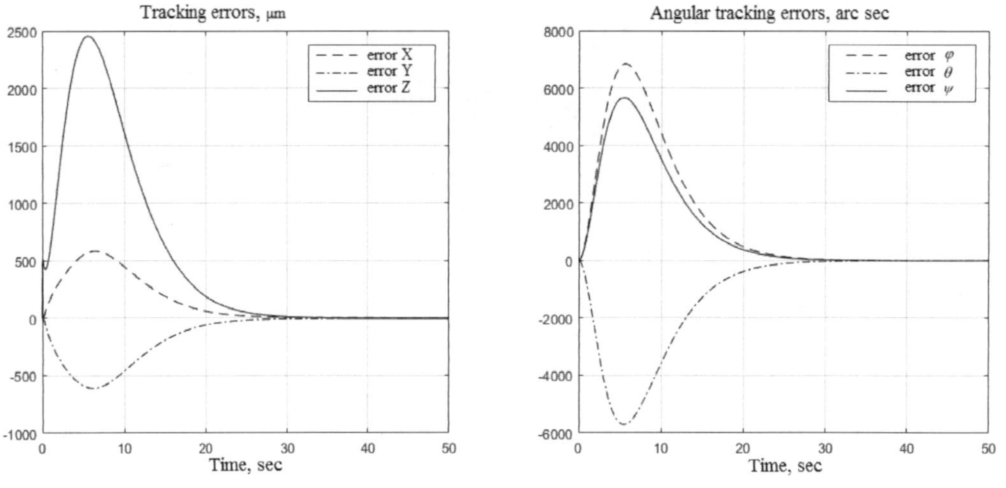

Figure 10. Hexapod control errors.

As can be seen from the figures, the use of the neural network approximation of the hexapod kinematics provides a stable control; the steady-state control error is within the required range.

Conclusion

In the course of this research, algorithms for managing a hexapod based on the neural network solution of the forward kinematics problem and approximation of the Jacobi matrices are developed. A model of a digital hexapod control system has been implemented. Modeling of the hexapod control system with a workload in the position control mode was carried out. The performed studies showed the effectiveness of the application of neural network approximation of kinematic problems, as well as their successful implementation in the hexapod control system. When implementing neural network kinematics problems in parallel computing systems, the algorithms under study will get a higher speed than traditional ones based on iterative solutions.

References

[1] Lung-Wen T. 1999 *Robot Analysis. The Mechanics of Serial and Parallel Manipulators* (New York: Wiley) 520 p.
[2] Merlet J. P. 2006 *Parallel Robots* (Springer) 420 p.
[3] Korotkov E. B., Matveev S. A., Jakovenko N. G. The paths of increase quality rating of control system of machine with strut-type structure as hexapod and tripod on based russian and global available electronic component 2016 *Questions of radio electronics* **8** pp 85–91.

[4] Stewart D.A. platform with six degrees of freedom. Proc. of the Institution of mechanical engineers 1965. 180 pp. 371–85.
[5] Zhukov Yu. A., Korotkov E. B., Slobodzjan N. S. Radiation resistant microcontroller with Cortex-M4F core based control system of parallel kinematics mechanism designed for spacecraft's onboard devices movements 2017 *Questions of radio electronics* **7** pp. 48-54.
[6] Gavrilenko V. A., Zhukov Yu. A., Moroz A. V. Kinematics Solutions on the ARM microprocessor for mechatronic control systems of hexapods 2016 *Questions of radio electronics* **8** pp. 92–8.
[7] Campa R, Bernal J, Soto I Kinematic modeling and control of the hexapod parallel robot 2016 *Proc. of American Control Conference* pp. 1203– 08.
[8] Cardona M. N. A new Approach for the Forward Kinematics of General Stewart-Gough Platforms 2015*Proc. of the 2015 IEEE thirty fifth central american and panama convention* pp. 1-6.
[9] Geng Z, Haynes L Neural network solution for the forward kinematics problem of a Stewart platform 1991 *Proceedings of IEEE International Conference on Robotics and Automation* **3** pp. 2650 – 2655.
[10] Choon S. Y., Kahbin L. Forward kinematics solution of Stewart platform using neural networks 1997 *Neurocomputing* 16 Issue **4** pp. 333-349.
[11] Lee H. S., Myung C. H. The estimation for forward kinematic solution of Stewart platform using the neural network 1999 *Proceedings of IEEE/RSJ International Conference. Intelligent Robots and Systems* **1** pp. 501 – 506.
[12] Parikh P J, Lam S Y A Hybrid Strategy to Solve the Forward KinematicsProblem in Parallel Manipulators 2005 *IEEE Transactions on robotics* 21 No**1** pp. 18-25.
[13] Xinxin Guo, Guixi Ke, Fengwu Zheng, Lijie Zhang. Forward Kinematics Analysis of the Stewart Parallel Platform Based on the Elman Recurrent Network 2013*Proceedings of 5th International Conference on Intelligent Human-Machine Systems and Cybernetics* **2** pp. 175 – 177.
[14] Mohammed A., Li S. Dynamic Neural Networks for Kinematic Redundancy Resolution of Parallel Stewart Platforms. 2016. *IEEE Transactions on cybernetics* 46 Issue **7** pp. 1538 – 1550.
[15] Ramesh K. P., Bandyopadhyay B. The forward kinematic modeling of a Stewart platform using NLARX model with wavelet network 2013 *Proceedings of 11th IEEE International Conference on Industrial Informatics (INDIN)* pp. 343 – 348.
[16] FuKS, GonzalezCS, Lee G *Robotics: Control, Sensing, Vision, and Intelligence* 1987 (Published by Mcgraw-Hill Book Company) 580 p.
[17] Liu K, Fitzgerald J, Lewis F Kinematic Analysis of a Stewart platform manipulator 1993 *IEEE Transactions on Industrial Electronics* 40 pp. 282-93.
[18] Neural Network Toolbox Matlab [Official web site of Math Works distributor in Russia and the CIS] 2018 Available at: https://matlab.ru/products/neural-network-toolbox/ (accessed: 22.06.2018) (In Russ.)
[19] Lebret G, Liu K, Lewis F Dynamic analysis and control of a Stewart platform manipulator 1993 *Journal of Robotic System* 10 Issue **5** pp. 629-655.

Acknowledgments
The present work was carried out in accordance with the Resolution of the Government of the Russian Federation No. 218 of 09.04.2010 (DRAFT 218) within the framework of R & D carried out with the financial support of the Ministry of Education and Science of the Russian Federation (01.12.2015 agreement No. 02.G25.31.0160). The work is carried out in the organization of the head performer R & D of BALTIC STATE TECHNICAL UNIVERSITY «VOENMEH» named after D.F. Ustinov.

The investigation of microstructure and mechanical properties of tool steel produced by selective laser melting technology

A V Agapovichev[1, 3], A V Sotov[1], R R Kyarimov[1], V P Alexeev[1], V G Smelov[1], V S Sufiiarov[2], D V Masaylo[2]

[1] Samara University, 34, Moskovskoe sh., Samara, 443086 Russia
[2] National Technology Initiative Center of Excellence in Advanced Manufacturing Technologies at Peter the Great St. Petersburg Polytechnic University, 29, Polytechnicheskaya, St. Petersburg, 195251 Russia

[3] E-mail: Agapovichev5@mail.ru

Abstract. This paper presents study of microstructure of the material (tool steel 150Cr14) produced by the selective laser melting (SLM) technology. The chemical composition analysis and grain size measurements of the initial powder tool steel 150Cr14 were performed. The bulk density and flowability of the powder were investigated. Specimens were built under various building parameters such as laser scanning speed and laser power and tested on uniaxial tension. In this study, the effect of the SLM parameters on the mechanical properties of tool steel is investigated. Due to less surface quality limitations of SLM technology compared with conventional production techniques specimens were subjected to set of tests including measurement of surface roughness, inspection of inner structure which are influenced from building parameters.

1. Introduction

The implementation of new technological processes in blanking and basic production gives an amazing possibility of material and time resources saving for design, development and manufacture of gas turbine engines components [1, 2].

Selective laser melting (SLM) is one of the new additive production methods using high power lasers (typically ytterbium fiber lasers) to create three-dimensional physical objects by fusing metallic powders [3, 4].

Technology of selective laser melting makes it possible to manufacture parts from stainless steels, tool steels, aluminum and titanium alloys, superalloys, cobalt-chromium metal powders [5, 6]. SLM technology uses a high power laser to melt a thin layer of powder in accordance with the 3D model [7, 8]. The SLM method is a promising method for manufacturing parts from materials, the processing of which by traditional methods is complex and very costly. Currently, SLM is a unique technology for manufacturing objects of complex shapes, including the grid structure [9, 10].

The cost of a modern metal stamp often exceeds the cost of the part made with it, and the labor intensity of designing and manufacturing of technological equipment makes up to 80% of the total labor intensity of new product technological preproduction [11]. In the face of increasing global competition in the manufacture of gas turbine engines components, manufacturing time is becoming a key factor in the choice of production technology. Such world tools manufactures as Sandvik Coromant, Kennametal, IMC Group, etc., have created scientific divisions that study the possibilities of additive technologies implementation in the production of cutting tools and complex tooling. This article is shown the influence of the layer construction strategy during powder melting on the strain values. Moreover, the manufacturing time and the amount of consumed materials are determined. Also, results of optimization of internal structure of melting materials are resulted, the best strength was shown by samples with cellular internal structure.

To obtain the required mechanical characteristics of the stamp, it is necessary to obtain a dense internal structure by selecting the optimal technological parameters and layers constructing strategy. Thus [12] this paper deals with establishing of building parameters for tool steel H13 processed using SLM technology with layer thickness of 50 and 30 μm with different part building orientation. In the article [13] the questions of the effect of layer thickness at SLM.

The main benefits of SLM technology using in the manufacture of stamps is the possibility of obtaining complex cooling channels. The paper [14] presents the results of CAD / CAE modelling and the production of dies cooling channels from tool steel H13 by the EBM method and tool steel P20 by the SLS method.

In the article [15] questions of manufacturing of samples of dies with cooling channels from tool steel H13 were discussed. The heat treatment processes for residual stresses elimination and properties modelling were investigated.

It is important to design and evaluate the residual stresses in the dies cooling channels. The authors of the work [16] carried out numerical experiments to determine the influence of the SLM parameters and the lattice designed structure on the surface roughness of the cooled channels; coefficient of stress concentration and uniformity of cooling; structural and thermal insulation channels properties.

Thus, one of the promising areas of SLM technology application is manufacturing of tooling. In this case, the cost of an enterprise is eliminated the cost of expensive serial tooling, since the SLM technologies make is possible to make an individual light alloy dies with less resources. One key aspect of SLM implementation is the elimination of as many intermediate processing steps as possible.

In the manufacture of technological equipment, tool steels with special properties are used, the steels which are normally used in engineering.

The most important requirement for all types of die steels is the combination of hardness with high viscosity [17]. Die steels should also have special technological properties, which include:

- good workability, i.e. ability to accept cutting and cold and hot pressure treatment well;
- good hardenability, i.e, the possibility of obtaining a high and uniform hardness, a uniform fine-grained structure to a great depth;
- low sensitivity to overheating, i.e, the possibility of quenching with heating in a sufficiently wide temperature range; - slight sensitivity to decarburization during heating, which reduces the hardness of the surface (working) metal layer;
- high temperature strength, high temperature toughness and, thermal conductivity;
- good grinding, which determines the high quality of polished and polished surface.

To determine the possibility of SLM technology using for tooling manufacturing, a complex of tool steel 150Cr14 structure and mechanical properties studies were carried out.

2. Research methods and equipment

During the experiments 150Cr14 tool steel powder and SLM produced specimens were used. The experiments aimed to investigate the microstructure and mechanical properties of a selected steel powders at various process parameters.

The test specimens were manufactured using the SLM 280HL. The equipment is designed for single and small batch production of various parts, and has a building camera with a size of 280x280x350 mm. The SLM 280 HL is equipped with an ytterbium fiber laser (wavelength of about 1 mm) with a power of 400 W. The laser has a Gaussian intensity distribution with respect to the center of the laser radiation, also denoted by the transverse electromagnetic mode TEM_00.

Investigation of powder particles morphology and chemical composition was made by a scanning electron microscope Tescan Vega.

The flowability and bulk density of the 150Cr14 tool steel powder investigation was determined with a volumeter.

The hardness of the samples was measured on a hardness modifier ITBRW-187.5-M. In order to investigate the effect of the process parameters on hardness, macro and micro scale hardness measurements were conducted on the side cross-sections of specimens produced with different settings of scan speed and layer thickness.

Tensile testing was performed at the INSTRON 8802 machine.

Study of microstructure of samples was carried out using an optical metallographic microscope METAM LV-32.

The surface roughness of the as-processed samples was measured on horizontal top surfaces using a Hommel-Etamic W55 profilometer. Samples for the study of roughness were made at different sections of the laser radiation power, the scanning speed of the displacement of the scanning contour.

3. Research methods and equipment

The results of the 150Cr14 powder elemental composition X-ray spectral analysis are presented in table 1.

Table 1. Results of the tool steel powder elemental composition analysis 150Cr14

—	Fe	C	Si	Mn	S	P	Cr	Ni	Mo	Ti	O	Cu
Research results	88.77	-	1.44	0.54	-	-	1.14	0.40	-	-	4.08	
DIN 150Cr14	Base	0,95	1.60	0.60	0.030	0.030	1.25	0.4	0.2	0.030	-	0.3

Figure 1 shows scanning electron microscope (SEM) images of the150Cr14 powder.

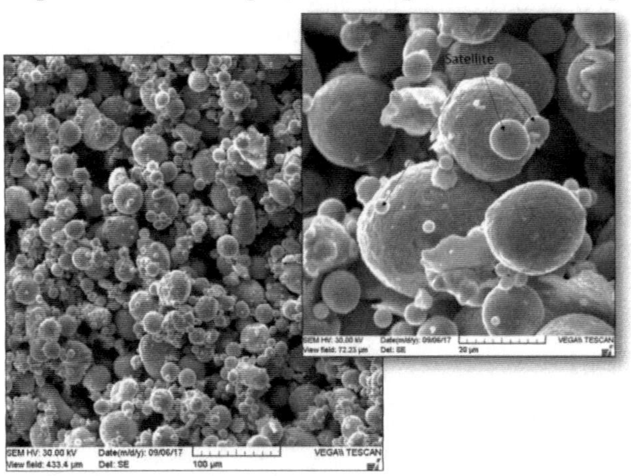

Figure 1. SEM-images of 150Cr14 tool steel powder particles.

The results of particle size distribution measurement show that the predominant number of particles has a shape close to spherical. Some particles are covered with a large number of small satellite particles which are specific for powders obtained by gas atomization [18].

One of the most important parameters of the metal powder used in SLM technology is the particle size distribution. The layer thickness of SLM is changing between 20 and 100 μm. Consequently, the maximum size of the used powder particles should not exceed the minimal layer thickness [19]. The grain size distribution of 150Cr14 powder particles is shown in figure 2.

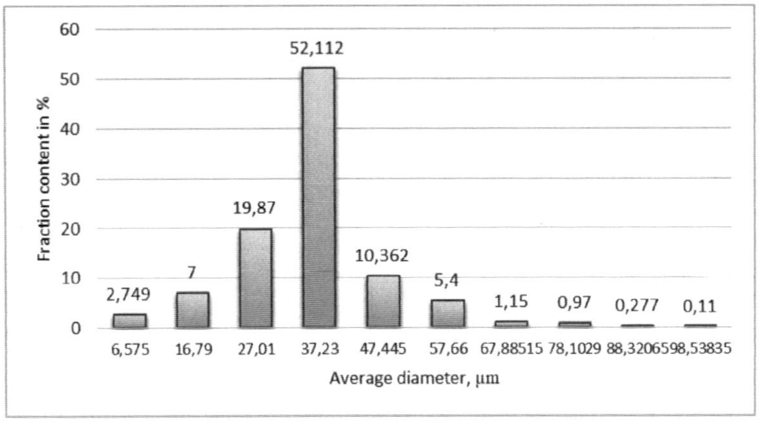

Figure 2. Grain size distribution of 150Cr14 powder particles.

The grain size distribution showed that the amount of 150Cr14 powder particles size up to 12 μm does not exceed 2.749%. The amount of 150Cr14 powder particles size from 10 μm to 40 μm is 79%. The results of the flowability and bulk density investigation of 150Cr14 tool steel powder are presented in table 2.

Table 2. Results of the elemental composition analysis of the 150Cr14 metal powder

Powder grade	Weight, g	Volume, cm^3	Time, s	Flowability, g / s	Apparent density, g/cm^3
150Cr14	137	123.088	52	2.63	1.11

For tensile tests, a batch of 150Cr14 powder samples was produced under various building parameters, taking into account the mathematical experimental design techniques.
Following regime parameters set were used:
- Laser power (P): 125, 175, 228 W.
- Scan speed (V): 400, 550, 700 mm/s.
- Scanning step (t): 50, 100, 150 μm.

The results of uniaxial tensile tests of the SLM manufactured specimens are shown in figures 3 and 4.

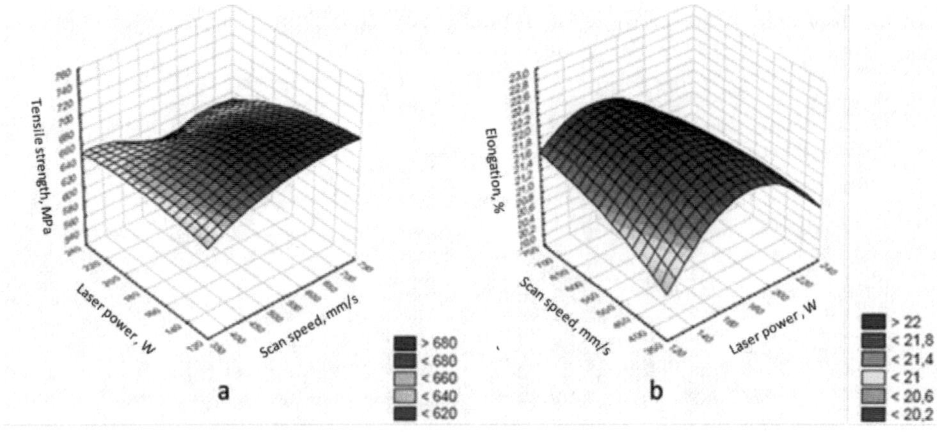

Figure 3. Effect of laser power and scan speed on the tensile strength and elongation: a – the influence of the scanning parameters on the ultimate strength; b – the influence of the scanning parameters on the relative elongation.

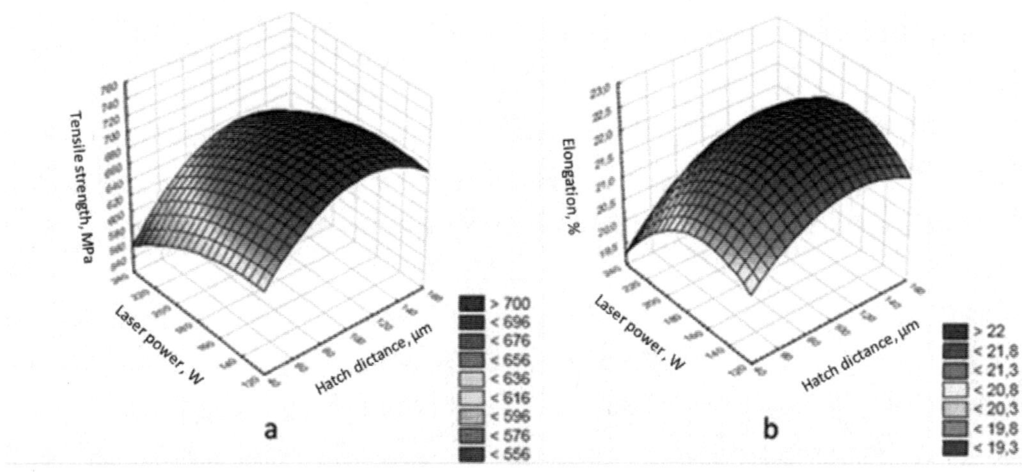

Figure 4. Effect of laser power and scanning step on the tensile strength and elongation: a – the influence of the scanning parameters on the ultimate strength; b – the influence of the scanning parameters on the elongation.

It can be noticed that the best mechanical properties are obtained at a scan speed of 750 mm/s and a laser power of 120 W. The highest value of the investigated parameter is 736.28 MPa is obtained at a scan speed of 700 mm/s and a laser power of 125 W.

The results of hardness measurements of SLM samples are shown in figure 5.

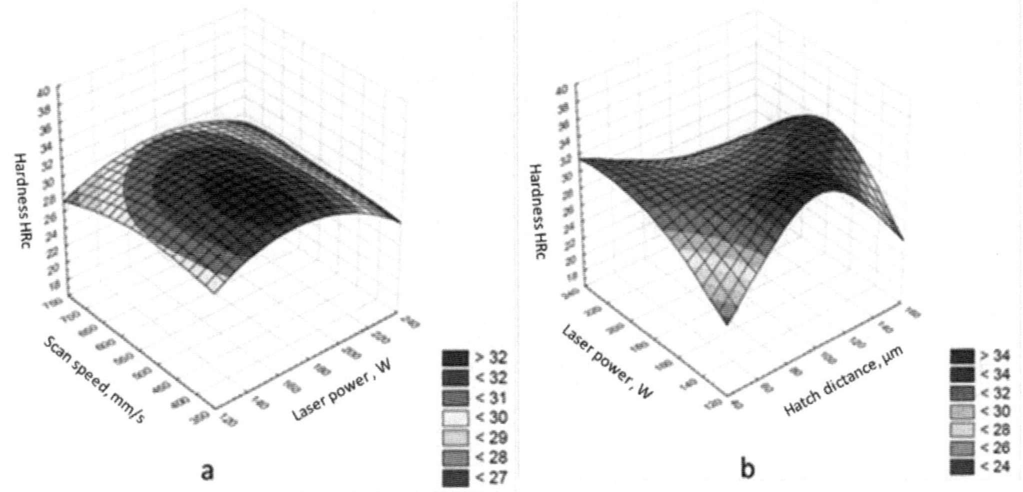

Figure 5. Effect of technological parameters on specimens hardness: a – the influence of laser power and scan speed; b – the influence of laser power and scanning step.

As can be seen from hardness analysis, the specimens hardness reaches 35-38 HRC units. The heat treatment conditions: heating, holding at a temperature of 860°C for an hour, oil quenching, tempering of 180°C for an hour. Based on the measurement results, the specimens hardness after heat treatment reaches 57-61 HRC values at the following SLM process parameters: laser power 125 W, scan speed 550-700 mm/s and scanning step 100 microns.

A study of the roughness of the samples was performed on the samples shown in figure 6.

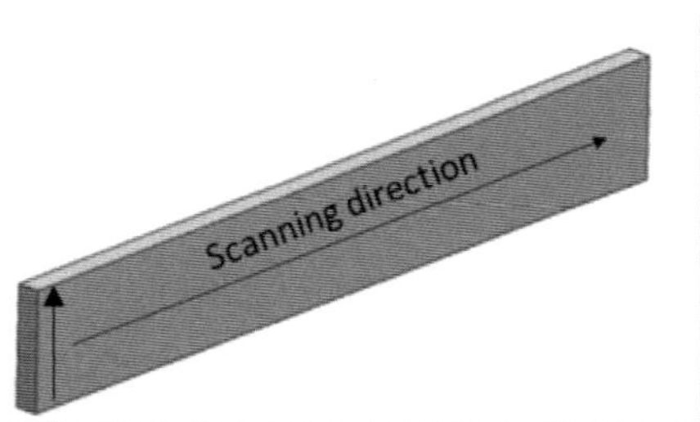

Figure 6. Roughness measurement line.

The results of measuring the SLM samples roughness are shown in figure 7.

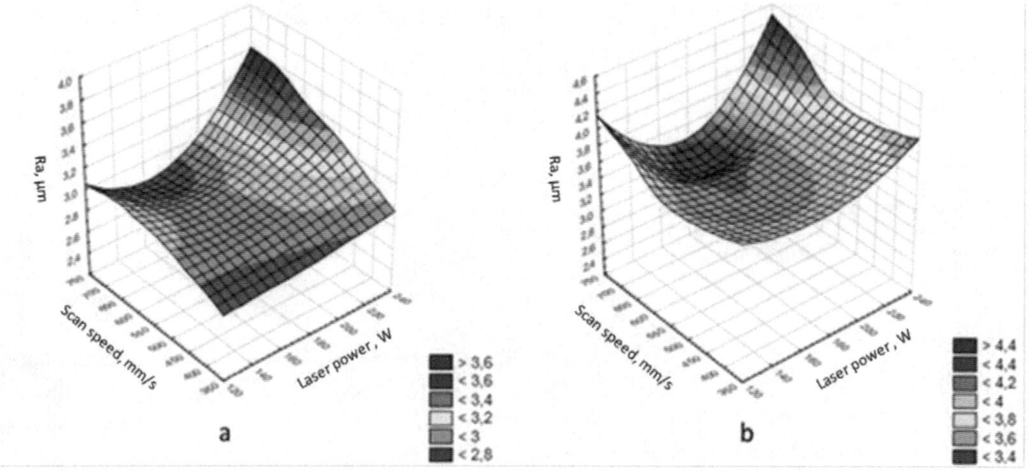

Figure 7. Influence of technological parameters of scanning on roughness of samples: a – roughness of the samples along the scanning direction; b – roughness of samples across the scanning direction.

Based on the conducted researches, it can be concluded that the smallest roughness measured along the scanning direction (figure 7a) is achieved at scan speeds in the range of 400 and 700 mm/s and a laser power range from 125 to 180 W. The lowest roughness is observed at a scanning step of 100 µm. The smallest samples roughness measured across the scanning direction (figure 7b) is achieved at scan speeds in the range from 600 to 7500 mm/s and a power range of 160 to 200 W. The lowest roughness is observed at scanning step of 100 µm.
The results of metallographic researches carried out using optical metallography methods showed that the structure of materials obtained using SLM technology is not uniform in all samples. In some samples there are defects in the form of microcracks. There are also defects in the form of pores. The structure of a sample produced by a laser power of 175 W, a scan speed of 550 mm/s and a scanning step of 150 µm is shown in figure 8. Analysis of the microstructure showed that the particles of the metal powder that retained their spherical shape are not observed. The unmelted particles of metallic powder are absent. In the structure of the material, there are not significant defects of cavities and cracks. Cracks are mainly observed between melting particles.

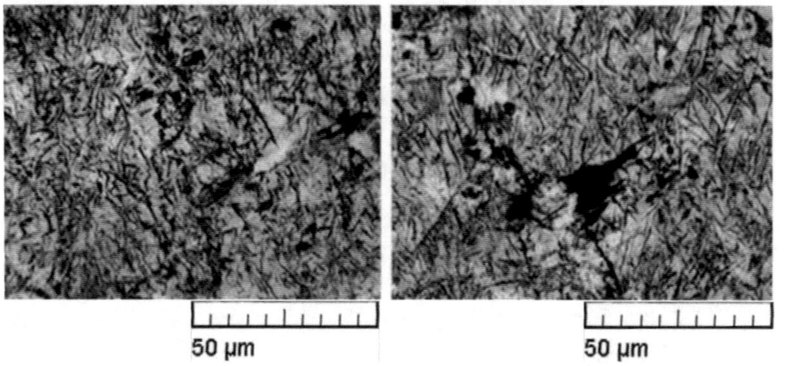

Figure 8. Specimens microstructure.

The microstructure of a sample produced by a laser power of 125 W, a scan speed of 550 mm/s and a scanning step of 150 µm is shown in figure 9. Analysis of the microstructure showed no defects in the form of lack of fusions and cavities. A small number of cracks were found in the microstructure of the material. There is a sandwich structure is characterised for SLM manufactured materials.

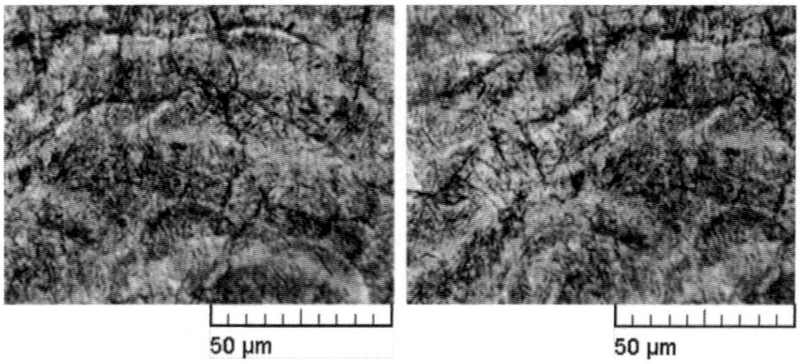

Figure 9. Specimens microstructure.

The parts of the die tooling were manufactured using SLM technology from 150Cr14 tool steel (figure 10). The best values of the material strength and hardness were obtained by following regimes: power 125 W, speed 550 mm/s, scanning step 100 µm. The possibility of manufacturing
die parts from 150Cr14 tool steel powder by SLM technology are shown.

Figure 10. Details of die equipment manufactured by SLM technology.

4. Conclusion

The 150Cr14tool steel powder of the investigated grain size distribution and elemental composition can be used for SLM manufacturing. It is established that the morphology of the initial powder is spherical with small growths-satellites. But after the heat treatment individual cells disappear, and instead of them large grains are formed.
In SLM technology, it is possible to use metal powders with different bulk density. For the engineering production, it is advisable to use metal powders with a greater bulk density.
Strength and ductility of samples obtained by selective laser melting are higher than by traditional manufacturing methods. According to the requirements for die steels, the hardness of the working surfaces and the cutting edges of the die parts must correspond to the values of HRC 54-64, so it is necessary to provide heat treatment after SLM technology.
The effect of building regimes on the roughness showed that, in assigning optimal technological regimes, it is possible to obtain a surface roughness Ra 3.2.

References

[1] Agapovichev AV, Balaykin AV and Smelov V G, 2015 Production technology of the internal combustion engine crankcase using additive technologies. *Modern Applied Science*. **9** 335-343

[2] Vdovin R A and Smelov V G 2014 Elaboration of a casting defects prediction technique via use of computer-aided design systems. *International J. of Engineering and Technology (IJET).* **6 (5)** 2269-2275

[3] Graf B, Gook S, Gumenyuk A, Rethmeier M 2016 Combined Laser Additive Manufacturing for Complex Turbine Blades. *Global Nuclear Safety.* **3 (20)** 34-42

[4] Keicher D M, Miller W D 1998 LENS moves beyond RP to direct fabrication. *Metal powder report.* **53 (12)** 26-28

[5] Sufiiarov V Sh, Popovich A A, Borisov E V, Polozov I A 2015 Selective laser melting of heat-resistant Ni-based alloy. *Non ferrous Metals* **1** 32-35

[6] Grigoriev A, Polozov I A, Sufiiarov V Sh, Popovich A A 2017 In-situ synthesis of Ti2AlNb-based intermetallic alloy by selective laser melting. *Journal of Alloys and Compounds* **704** 434-442

[7] Smirnov G V, Pronichev N D and Nekhoroshev M V 2017 Effect of technological heredity on the fatigue strength in the manufacture of gas turbine engine blades. *IOP Conference Series: Materials Science and Engineering* **177 (1)**

Smelov V G, Sotov A V and Agapovichev A V 2016 Research on the possibility of restoring blades while repairing gas turbine engines parts by selective laser melting. *IOP Conf. Series: Materials Science and Engineering.* **140**

[8] Popovich A A, Sufiiarov V Sh, Polozov I A, Borisov E V, Masaylo D 2016 Producing hip implants of titanium alloys by additive manufacturing. International *Journal of Bioprinting* **2(2)** 78-84

[9] Sato Y, Tsukamoto M, Shobu T, Yamashita Y, Yamagata S, Nishi T, Higashino R, Ohkubo T, Nakano H, Abe N 2018 Preheat effect on titanium plate fabricated by sputter-free selective laser melting in vacuum. *Applied Physics A: Materials Science and Processing.* **124 (4)**

[10] Stepanova E Yu, Burnashev M A, Stepanov Yu S 2017 Additive and hybrid technology in the manufacture of tools and tooling: state of the economy, prospects. *Izvestiya TulGU.* **8 (1)** 141- 146

[11] Safka J, Ackermann M, Volesk L 2016 Structural properties of H13 tool steel parts produced with use of selective laser melting technology. *Journal of Physics: Conference Series.* **709**

[12] Sufiiarov V Sh, Popovich A A, Borisov E V, Polozov I A, Masaylo D V, Orlov A V 2017 The Effect of Layer Thickness at Selective Laser Melting. *Procedia Engineering* **174** 126-134

[13] Shinde M S, Ashtankar K M 2017 Additive manufacturing–assisted conformal cooling channels in mold manufacturing processes. *Advances in Mechanical Engineering.* **9 (5)** 1-14

[14] Mazur M, Brincat P, Leary M, Brandt M 2017 Numerical and experimental evaluation of a conformally cooled H13 steel injection mould manufactured with selective laser melting. *The International Journal of Advanced Manufacturing Technology.* **93** 881-900.

[15] Mazur M, Leary M, McMillan M. Elambasseril J. Brandt M 2016 SLM additive manufacture of H13 tool steel with conformal cooling and structural lattices *Rapid Prototyping Journal.* **22 (3)** 504-518

[16] Grechnikov F V 2014 Manufacturing technology for forging and stamping equipment and die-casting equipment. Samara: Samara University.

[17] Bazyleva O A, Bondarenko Yu A, Morozova G I and Timofeeva O B 2014 Structure, chemical composition, and phase composition of intermetallic alloy VKNA-1V after high-temperature heat treatment and process heating, *J. Metal Science and Heat Treatment.* **56 (5-6)** 299-234

[18] Kurzynowski T, Chlebus E, Kuźnicka B, Reiner *J 2009 Parameters in Selective Laser Melting for processing metallic.* Published in SPIE Proceedings. **8239**

Hardness of products from electroerosive cobalt-chrome powders, obtained by additive technologies

E V Ageev[1], A Yu Altukhov[1], E V Ageeva[1]

[1]FGBOU VO Southwest state University, 94 50 years of October street, Kursk, 305040, Russia.
E-mail: ageev_ev@mail.ru

Abstract. The wide use of the EED method for processing metal waste into powders for the purpose of their reuse and application in additive technologies is hampered by the lack in the scientific and technical literature of full-fledged information on the effect of the initial composition, regimes and media on the properties of powders and technologies of practical application. Therefore, in order to develop technologies for the reuse of electroerosive powders and to evaluate the effectiveness of their use, complex theoretical and experimental studies are required. The purpose of the study is to investigate the hardness of additive products from electroerosive cobalt-chrome powders. Tests on the hardness of the samples on the surface and cross-section were carried out using an automated micro-hardness analysis system DM-8 using the micro-Vickers method with an indentor load of 200 grams per ten prints with a free choice of the site of injection. Based on the results of the conducted studies, it has been experimentally established that the microhardness of additive products obtained from electroerosive powders with an average particle size of 52.5 μm is 5.7 Gpa.

1. Introduction

Additive technologies (AT) for the production of products from materials based on metals and alloys are one of the most promising and actively developing areas of production [1-10].

Proceeding from the peculiarities of the methods for obtaining spherical powders with the aim of obtaining spherical granules of regulated granularity, the electroerosive dispersion technology is proposed, which is characterized by relatively low energy costs and ecological purity of the process[11-12].

The wide use of the EED method for processing metal waste into powders for the purpose of their reuse and application in additive technologies is hampered by the lack in the scientific and technical literature of full-fledged information on the effect of the initial composition, regimes and media on the properties of powders and technologies of practical application. Therefore, in order to develop technologies for the reuse of electroerosive powders and to evaluate the effectiveness of their use, complex theoretical and experimental studies are required.

The purpose of the study is to investigate the hardness of additive products from electroerosive cobalt-chrome powders.

2. Materials and methods

For the implementation of the planned studies, wastes of the cobaltochrome alloy of the brand KHMS "CELLIT" were chosen. As a working fluid, butyl alcohol (butanol-1) was used. For the production of cobalt-chrome powders, a unit for EED of conductive materials was used. Dispersion parameters: voltage 100 V, capacity 48 μF, repetition rate 120 Hz.
The phase composition of the samples was studied by X-ray diffraction on a Rigaku Ultima IV diffractometer in Cu-Kα radiation (wavelength λ = 0.154178 nm) using Soller slits. The diffraction spectrum for the phase analysis is sampled according to the θ-2 схеме scanning scheme with Breguot-Brentano focusing in the angular interval 5 ... 100 deg. 2 Θ. The shooting is carried out in point-by-point mode with scanning step Δ (2θ) = 0.02 deg, speed 0.6 deg / min, working voltage 45 kV, current 200 mA. To refine the profile of the experimental radiographs, the software package PDXL RIGAKU was used.
Subtraction of the background was carried out by the Sonneveld-Visser method, the smoothing of the experimental profile by the Savitsky-Naked method, and the separation of the components kα1 and kα2 by the Racinger method. To describe the diffraction maxima, a superposition of the Gaussian function and the Lorentz function was used. The approximation of each of the reflections in the diffractograms of the samples studied by the pseudo-Voigt function made it possible to accurately determine the position of the reflections, taking into account the displacement caused by the overlap of the reflexes, at half the intensity maximum (FWHM) and intensity. The phase composition of the coatings was determined using an ICCD PDF-2 database (2014).
Dispersing in a liquid with ultrasound. Research Methodology (FR 1.27.2009.06762 "Method for measuring particle size in suspensions, emulsions and aerosols in the nanometer and colloidal ranges using the dynamic light scattering effect"). Sample preparation: dispersion of the sample in a liquid. Background measurement - in order to reduce the influence of the measuring liquid, a background measurement is carried out before each measurement. Any contamination from previous measurements is measured and its effect on the current result is eliminated. Measurement of the particle size distribution: a sample with a volume of about 1-5g was placed in a liquid dispersion module (500 ml volume). The measurement was started automatically, as soon as the absorption value reached the specified value. Measurement parameters: Measurement type - Fraunhofer method; the measurement range is 0.1 [μm] — 1021.87 [μm]; resolution - 102 channels (20/383 mm); the duration of the measurement is 100 (scans); regularization is an average model.

3. Results and discussion

The results of particle size measurement are shown in Figure 1.

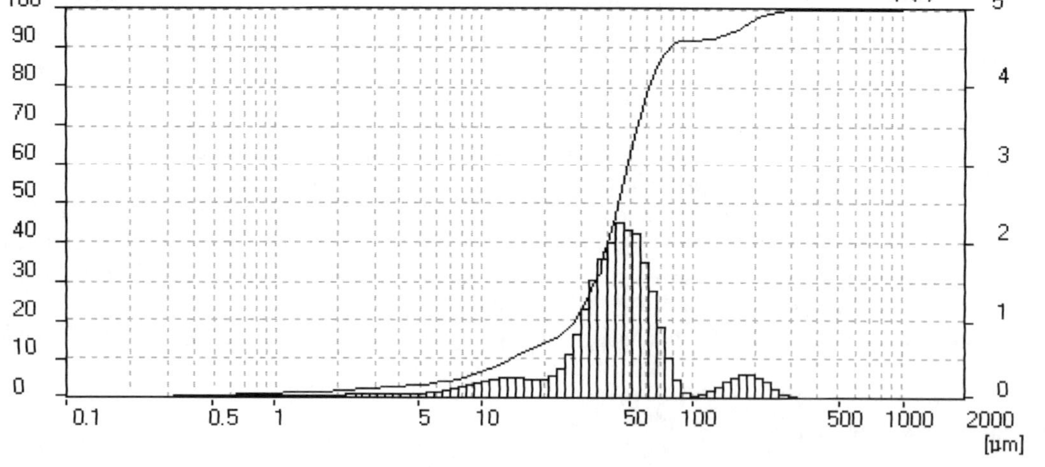

Figure 1. Size distribution of microparticles of "CoCr" powder

Integral curve and histogram: integral curve in coordinates Q3 (x) = f (μm) (left scale) - each point on the curve, how much% of the sample has a particle size less than or equal to a given one. Histogram in the coordinates q3 (×) = f (μm) (right scale) is the amount of a sample with a given particle size

It was found that the average particle size is 52.5 μm, the arithmetic value is 52.496 μm.
Information diffraction patterns can be used not only to determine the particle size, but also to analyze their shape. Particles of a nonspherical shape scatter radiation in their preferred spatial directions. If a small number of particles do not enter the laser beam, an analysis of their shape can be performed on the basis of the information obtained. Due to the unique possibility of moving in a converging laser beam, in analyzing the shape of the particles, the measuring cell is brought closer to the detector, so that the first diffraction maximum of radiation scattered by the medium-sized particles enters the detector regions sensitive to the shape of the particles. Since the diameter of the converging laser beam decreases as it approaches the detector, at this point in the laser beam there is exactly the small number of particles that are necessary for analyzing their shape. Azimuthally (ray-like) located elements of the detector record fluctuations of scattered radiation, on the basis of which the computer program analyzes the shape of particles. The program allows to determine the elongation ratio for the ×50 value of the previously measured distribution, and also to judge the "facetedness" of the particles. Figure 2 shows the shape of the microparticles of the "CoCr" powder.

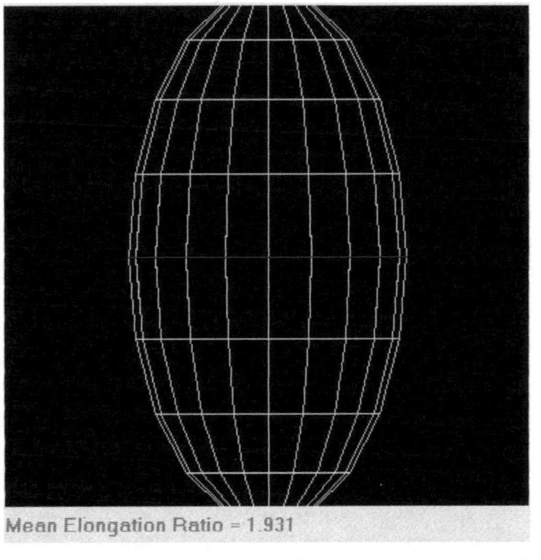

Figure 2. Parameters of the shape of the powder microparticles "CoCr"

It was found that the elongation coefficient (elongation) of particles with a size of 44.128 μm is 1.931.
The AFFRI DM-8 automated microhardness meter (according to Vickers) provides microhardness measurements in automatic mode with the construction of microhardness distribution curves. GOST 9450-76, measurement of microhardness by indenting diamond tips. The use of an automatic image analysis system PRECIDUR® allows manual measurements or fully automates the process of measuring microhardness. For the given device there is a certain set of commands with which help it is possible to set parameters of microindentation taking into account features of each specific research task. For example, you can specify the direction of indentation, specify the number of prints, the distance between them, the load on the indenter, indent on the weld, using the interpolation method, conduct studies of multiphase materials, etc. The built-in digital camera allows not only to visualize

the surface of the sample on the computer monitor, but also to take photographs of residual prints after microindentation. Specifications and additional options: Vickers indiktor, memory is designed for 999 measurements, built-in function of unit conversion, corresponds to ISO, JIS and ASTM standards, common magnifications: 100×, 400×, automatic turret (facilitates the operator by automatically selecting an indentor and lenses), the function of automatic return to the set position at the end of the working cycle.

Tests on the hardness of the samples on the surface and cross-section were carried out using an automatic microhardness analysis system DM-8 using the micro-Vickers method with an indentor load of 200 grams per ten prints with a free choice of the site of the injection in accordance with GOST 9450-76 (Measurement of microhardness by indenting diamond tips). The loading time of the indenter was 15 s. The results of the measurements are given in Table 1.

Table 1. Microhardness according to Vickers

Imprint number	Value
1	523
2	519
3	587
4	624
5	605
6	550
7	601
8	596
9	592
10	601
Average value (units of measure)	580
HV	5684
GPa	5.684

4. Conclusion

Based on the results of the conducted studies, it has been experimentally established that the microhardness of additive products obtained from electroerosive powders with an average particle size of 52.5 μm is 5.7 GPa.

5. Acknowledgments

The work was supported by the Russian Science Foundation. Project number is 17-79-20336.

References

[1] Karlsson J, Snis A, Engqvist H, Lausmaa J 2013 *Journal of Materials Processing Technology* **213** 2109-2118.
[2] Safdar A, He H Z, Wei L Y, Snis A 2012 *Rapid Prototyping Journal* **18** 401–408.
[3] Loeber L, Biamino S, Ackelid U 2011 *Solid freeform fabrication proceedings* **11** 547-556.
[4] Gu D D, Meiners W, Wissenbach K, Poprawe R. 2012 *International Materials Reviews* **57** 133-164.
[5] Wang Z, Guana K, Gaoa M. 2012 *Journal of Alloys and Compounds* **513** 518-523.
[6] Biamino S, Penna A, Ackelid U 2011 *Intermetallics.* **19** 776-781.
[7] Gu D D, Meiners W, Wissenbach K, Poprawe R. 2012 *International Materials Reviews* **57** 133-164.

[8] Song B, Dong S, Zhang B 2012 *Materials & Design* **35** 120-125.
[9] Song B, Dong S, Coddet P 2012 *Surface and Coatings Technology* **206** 4704-4709.
[10] Wang Z, Guana K, Gaoa M 2012 *Journal of Alloys and Compounds* **513** 518–523.
[11] Ageeva E V, Altukhov A Yu, Khardikov S V, Gulidin S S, Novikov A N 2015 *Journal of nano- and electronic physics* **4** 04080(3).
[12] Ageeva E V, Ageev E V, Pikalov S V, Vorobiev E A, Novikov A N 2015 *Journal of nano- and electronic physics* **3** 04058(2)

Substantiation of characteristic bending points of the blade operating body of the geokhod

V V Aksenov[1,4] AB Efremenkov[2] V Yu Sadovets[3,4] D A Pashkov[1]

[1] Institute of Coal FIC UUH SB RAS 650610, Russia, Kemerovo, Leningrad, 10.
[2] Novgorod State University, ul. B. St. Petersburgskaya, 41 173003 Veliky Novgorod, Russia
[3] Kuzbass State Technical University named after T.F. Gorbachev 650000, Russia, Kemerovo, ul. Spring, 28.
[4] Jurginsky Technological Institute, Branch of Tomsk Polytechnic University 652052, Russia, Yurga, ul. Leningradskaya, 26.

e-mail: 55vva42@mail.ru
e-mail: abe@novsu.ru
e-mail: vsadovec@yandex.ru
e-mail: pashkov.d.a@inbox.ru

Abstract. The article substantiates the characteristic bending points of blade operating body of the geokhod. The urgency of the research is considered. The information on the work peculiarities of the operating body of the geokhod is given. The method of calculation of the blade operating body of the geokhod is described, in particular, the entire force of the blocked cutting by a single radial blade of the IO of the geokhod is shown, and this force is also represented as a sum of three constituent forces. To establish the purpose and objectives of the study, it is justified to choose the force to overcome the soil resistance by the front face of the blade P_{cv}, which is proportional to the cross-sectional area of the slot before the front edge of the knife and which depends on the angle and strength of the soil. The limits of integration of the projection of the component of the soil resistance force into the cutting, depending on the width of the cut, on the axis of rotation of the geokhod $P_{0.cv}$ and on the plane perpendicular to the axis of rotation of the $R_{IO.CV}$. Based on the study, the tasks for further research have been determined.

Introduction

Today, new technologies in the field of robotics are being implemented at a rapid pace. However, the less developed area is the creation of devices that can move underground [1, 2]. In this connection, the tasks are to develop new approaches, technology, and machines that allow all processes to be robotically performed to form a cavity in an underground space [3]. One of the directions that allows an underground robot to move underground is the application of geovinchester technology, the basic element of which is the geokhod.

At the current stage of the development of elements of geovinchester technology, there is an urgent need to develop constructive and technical solutions for executive bodies capable of producing cavities in the underground space along rocks with a strength of up to 1 on the scale of M.M.

Protodyakonov [4]. Therefore, the work aimed at justifying the parameters of the executive bodies of geoopds to destroy rocks with a strength of up to 1 is relevant.

Features of the operating body of the geokhod

The peculiar nature of the movement of the geokhod on the face causes the formation of a complex surface shape not only of the bottom face itself, but also of the executive body (IO). The surface of the face, when its IG is destroyed, has the appearance of several helicoidal surfaces with ledges.

Point A (figure 1), located on the periphery of the knife, at the given step of the helical blade h_B of the external propeller, passing for one complete revolution along the circumference of the path $2\pi R_g$, moves to the face at an angle β_1 to the plane perpendicular to the axis of rotation of the geokhod [5, 6, 7], and

$$\beta_1 = arctg\frac{h_B}{2\pi R_g}, \quad (1)$$

where R_g – radius of the head section of the geokhod.

Any point of the knife located at a distance x (Figure 1) from the axis of rotation of the geokhod, moves to the face at an angle [5,6]:

$$\beta_x = arctg\frac{h_B}{2\pi x} \quad (2)$$

The points of the knife, which are closer to the axis of rotation of the geokhod, move to the face at a greater angle than the points on the periphery of the knife. Thus, with the helical movement of the knife IO of the geokhod to the face, the points of the knife form a helicoidal (helical) surface. Consequently, the section of the face in the sector between adjacent knives after their passage takes the form of a screw surface.

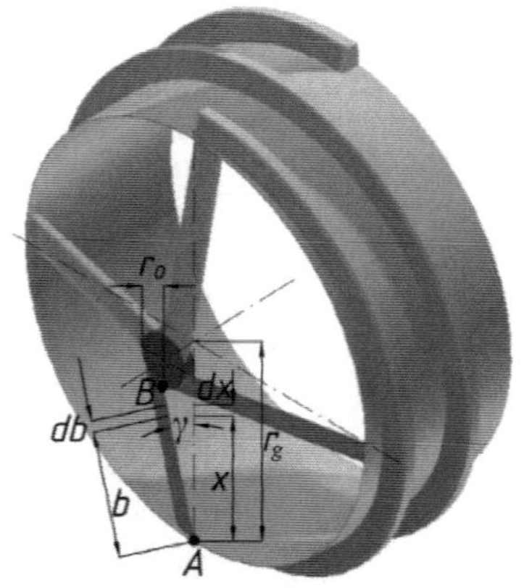

Figure 1. Scheme of the knife executive body.

Each point of the knife in one turnover of the geokhod must be moved to the production face by an amount equal to the step of the external drive of the geokhod generator $h_в$. In this case, the path length for each point in the process of moving will depend on the radius of the location on the knife of the IO of the geokhod. The knife points that are closer to the axis of rotation of the geokhod are moved to the corner at a large angle (point B at an angle β_B) than the points located at the periphery of the knife (point A at an angle β_A) (Figure 2). The value of the cut-off layer for each point of the IO knife of the geokhod will depend on the angle of displacement of the point on the face of the mine.

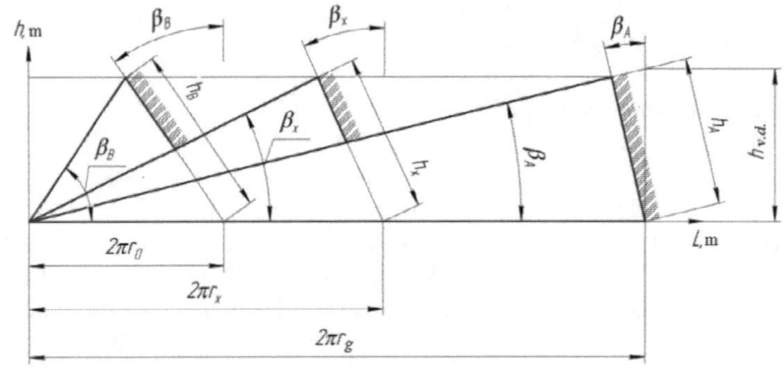

Figure 2. Direction of motion of the knife points IO, depending on the location on the knife along the radius of the geokhod

Thus, the height of the formed shoulder h depends on the pitch of the helical blade, the number of radial blades and the angle of movement of the knife point IO of the geokhod to the workhole [5,6]:

$$h = \frac{h_B}{n}, \tag{3}$$

where n is the number of knives on IO.

Method of calculation of the knife executive body of the geokhod

When cutting a single radial knife for a geokhod engine, the entire force of the blocked cutting can be represented as a sum of three component forces (Figure 3) [8,9,10]:

- force to overcome the resistance of the soil with the front face of the knife P_{cv}, which is proportional to the cross-sectional area of the slot in front of the front edge of the knife and which depends on the angle and strength of the soil;
- forces to overcome the resistance of the ground to the destruction in the lateral extensions of the groove P_{side}, which is proportional to the area of these parts of the slot, which depends on the strength of the soil and which does not depend on the cutting angle and the width of the cut;
- forces to overcome the resistance of the ground to the cut by the lateral edges of the knife at the bottom of the slot of the $P_{side\ mid}$, proportional to the thickness of the cut, depending on the strength of the soil and not depending on the width of the cut and the angle of cutting.

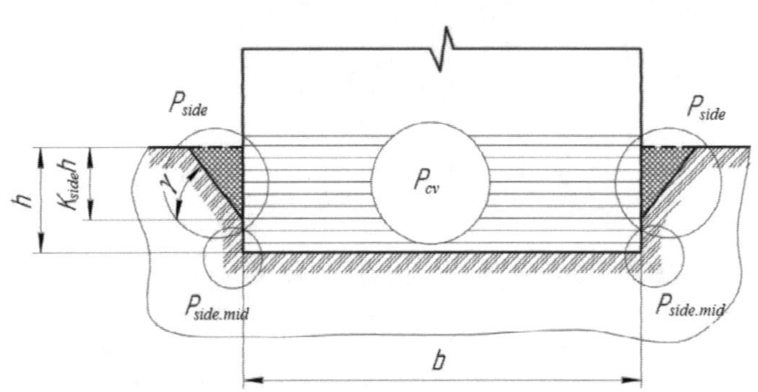

Figure 3. Areas of action of the component sharp knife cutting force.

All force of the blocked cutting by a simple sharp knife [11]:

$$P_{cr} = \varphi m_{cv} bh + 2m_{side} h^2 + 2m_{side.mid} h \quad (4)$$

where $\varphi \ m_{cv} bh$ - the strength of overcoming the frontal resistance of the knife (on Figure 3 $P_{cв}$), H;

$2m_{side} h^2$ - the entire force of soil destruction in the lateral extensions of the slot (P_{side} on Figure 3), H;

$2m_{side.mid} h$ - lateral force ($P_{side.mid}$ on Figure 3), H;

φ - coefficient that takes into account the influence of the cutting angle;

m_{cv} - specific cutting force to overcome the resistance of the ground with the front face at a cutting angle of 45^0, Pa;

b - width of knife, m;

h - depth of cut, m;

m_{side} - coefficient characterizing the strength of soil destruction in the lateral parts of the slot, Pa;

$m_{side.mid}$ - coefficient characterizing the specific shearing force of one of the lateral edges of the knife, H/m.

We divide the force of the blocked cutting into two components, one of which depends on the width of the cut (knife) b, and the other is not. The first component is denoted P_{cv}, the second is P_{side}. Then, on the basis of expression (4)

$$P_{cv} = \varphi m_{cv} hb; \quad (5)$$

$$P_{side} = 2h(m_{side} h + m_{side.mid}). \quad (6)$$

The component of the cutting force, depending on the width of the cut, for the elementary section of the knife IO (db on Figure 1) is determined by the expression

$$dP_{cv} = \varphi m_{cv} h db. \quad (7)$$

An additional cutting force that must be applied to the db knife's elementary section with a wear pad or blunted

$$dP_{p.wear} = P_{wear} db; \quad (8)$$

$$P_{изн} = p_o + p_{condit} \frac{h}{(h + h_{condit})}, \quad (9)$$

where p_o, p_{condit}, h_{condit} - parameters characterizing the resistance to elastoplastic compression. With considering $db = \dfrac{dx}{\cos\gamma}$ we get

$$dP_{cv} = \frac{\varphi m_{cv} h dx}{\cos\gamma}. \quad (10)$$

$$dP_{cv}^{w} = dP_{cv} - dP_{p.wear}; \quad (11)$$

With blunt knives, the normal component of the cutting force

$$dN_{cv}^{w} = ctg(\delta + \varphi_{fr}) \frac{\varphi m_{cv} h}{\cos\gamma} dx - \frac{ctg(\delta + \varphi_{fr})}{\cos\gamma} P_{wear} dx. \quad (12)$$

Figure 4 shows the projections of the total ground resistance force to the cutting on the axis of rotation of the geokhod (P_0) and the plane perpendicular to the axis of rotation of the geokhod (R_{io}).

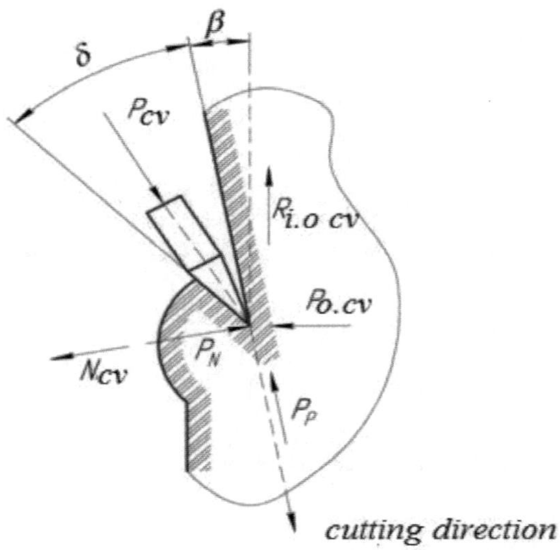

Figure 4. Calculation scheme for determining the resistance force of the soil with the front face of the knife.

Projections of the component of the ground resistance force of cutting, depending on the width of the cut, on the axis of rotation and the plane perpendicular to the axis of rotation of the geodatabase, as well as the moment of resistance to cutting from this component for blades with a wear pad.

$$dP_{o.cv} = \sin\beta_x dP_{cv} - \cos\beta_x dN_{cv}; \tag{13}$$

$$dR_{i.o.sv} = \cos\beta_x dP_{cv} + \sin\beta_x dN_{cv}; \tag{14}$$

After substituting and solving the integrals, we finally have

$$P_{o.cv} = \frac{\varphi m_{cv} h_в^2 + h_в n P_{wear}}{2\pi n \cos\gamma} \cdot \left(\ln \frac{tg\left|\frac{\beta_2}{2}\right|}{tg\left|\frac{\beta_1}{2}\right|} \right) - \frac{h_в}{2\pi}\left(\varphi m_{cv} \frac{h_в}{n} \frac{ctg(\delta+\varphi_{fr})}{\cos\gamma} - \frac{ctg(\delta+\varphi_{fr})}{\cos\gamma} P_{wear}\right) \frac{\sin\beta_2 - \sin\beta_1}{\sin\beta_1 \sin\beta_2} \tag{15}$$

$$R_{i.o.cv} = \frac{\varphi m_{cv} h_в^2 + h_в n P_{wear}}{2\pi n \cos\gamma} \cdot \frac{\sin\beta_2 - \sin\beta_1}{\sin\beta_1 \sin\beta_2} + \frac{h_в}{2\pi}\left(\varphi m_{cv} \frac{h_в}{n} \frac{ctg(\delta+\varphi_{fr})}{\cos\gamma} - \frac{ctg(\delta+\varphi_{fr})}{\cos\gamma} P_{wear}\right) \cdot \left(\ln \frac{tg\left|\frac{\beta_2}{2}\right|}{tg\left|\frac{\beta_1}{2}\right|} \right); \tag{16}$$

Projections of the component of the ground resistance force of cutting, independent of the width of the cut, on the axis of rotation of the geodatabase and the plane perpendicular to the axis of rotation, as well as the moment of resistance to cutting from this component [12]

$$P_{o.side} = \frac{h_B}{n}\left(m_{side}\frac{h_B}{n} + m_{side.mid}\right)\left[-\frac{\cos(\delta+\varphi_{fr}+\beta_1) + \cos(\delta+\varphi_{fr}+\beta_2)}{\sin(\delta+\varphi_{fr})}\right]; \tag{17}$$

$$R_{i.o.side} = \frac{h_B}{n}\left(m_{side}\frac{h_B}{n} + m_{side.mid}\right)\left[\frac{\sin(\delta + \varphi_{fr} + \beta_1) + \sin(\delta + \varphi_{fr} + \beta_2)}{\sin(\delta + \varphi_{fr})}\right]; \tag{18}$$

Methods of research

To find the characteristic points from the bending of the knife IO, we use a force that depends on the width of the cut, i.e. force to overcome the resistance of the ground with the front face of the knife $P_{cv.}$ Figure 4 shows the projections of the component of the ground resistance force of the cutting, depending on the width of the cut, on the axis of rotation of the geokhod ($P_{0.CV}$) and on a plane perpendicular to the axis of rotation of the geodatabase ($R_{IO.CV}$). Based on these projections, four characteristic points can be distinguished:

1) At what x, the projection of the component of the soil resistance force to the cutting, depending on the width of the cut, on the axis of rotation of the geokhod ($P_{0.CV}$) in the range from 0 to x, will be equal to the projection of the component of the soil resistance force of cutting, depending on the width of the cut, on the plane perpendicular to the axis of rotation of the boat ($R_{IO.CV}$) in the range from x to R_g, i.e.

$$\int_0^x P_{o.cv}dx = \int_x^{R_g} R_{i.o.cv}dx\ ; \tag{19}$$

2) At what x, the projection of the component of the soil resistance force to the cutting, depending on the width of the cut, on the axis of rotation of the geokhod ($P_{0.CV}$) in the range from 0 to x, will be equal to the projection of the component of the soil resistance force of cutting, on the axis of rotation of the geodatabase ($P_{0.CV}$) in the range from x to R_g, i.e.

$$\int_0^x P_{o.cv}dx = \int_x^{R_g} P_{o.cv}dx\ ; \tag{20}$$

3) At what x the projection of the component of the soil resistance force of cutting, depending on the width of the cut, on the plane perpendicular to the axis of rotation of the geodatabase ($P_{0.CV}$) in the range from 0 to x, will be equal to the projection of the component of the soil resistance force of cutting, in the plane perpendicular to the axis of rotation of the geodatabase ($R_{IO.CV}$) in the range from x to R_g, i.e.

$$\int_0^x R_{i.o.cv}dx = \int_x^{R_g} R_{i.o.cv}dx\ ; \tag{21}$$

4) At what x, the resultant force of the projection of the component of the co-resistivity of the cutting force, depending on the width of the cut, on the axis of rotation of the geokhod ($P_{0.CV}$) and the projection of the component of the ground resistance force of cutting, depending on the width of the cut, on a plane perpendicular to the axis of rotation ($R_{IO.CV}$) in the range from 0 to x, will be equal to the equal-acting force of the component of the soil resistance force of cutting, depending on the width of the cut, on the axis of rotation of the geodatabase ($P_{0.CV}$) and the projection of the component of the soil resistance force to cutting, depending on breadth s slice, on a plane perpendicular to the axis of rotation of the geokhod ($R_{IO.CV}$) in the range from x to R_g, i.e.

$$\int_0^x R_{cv}dx = \int_x^{R_g} R_{cv}dx\ ; \tag{22}$$

Results

The projection of the component of the soil resistance force to cutting, depending on the width of the cut, on the axis of rotation of the geodatabase ($P_{0.CV}$) in the range from 0 to x

$$P_{o.cv}=\frac{\varphi m_{cv}h_{\text{в}}^2+h_{\text{в}}nP_{wear}}{2\pi n\cos\gamma}\cdot\left(\ln\frac{tg\left|\frac{\beta_x}{2}\right|}{tg\left|\frac{\beta_1}{2}\right|}\right)-$$
$$-\frac{h_{\text{в}}}{2\pi}\left(\varphi m_{cv}\frac{h_{\text{в}}}{n}\frac{ctg(\delta+\varphi_{fr})}{\cos\gamma}-\frac{ctg(\delta_1+\varphi_{fr})}{\cos\gamma}P_{wear}\right)\frac{\sin\beta_2-\sin\beta_x}{\sin\beta_x\sin\beta_2} \tag{24}$$

The projection of the component of the soil resistance force to cutting, depending on the width of the cut, on a plane perpendicular to the axis of rotation of the boat ($R_{IO.CV}$) in the range from 0 to x

$$R_{i.o.cv}=\frac{\varphi m_{cv}h_{\text{в}}^2+h_{\text{в}}nP_{wear}}{2\pi n\cos\gamma}\cdot\frac{1-\sin\beta_x}{\sin\beta_x}+$$
$$+\frac{h_{\text{в}}}{2\pi}\left(\varphi m_{cv}\frac{h_{\text{в}}}{n}\frac{ctg(\delta+\varphi_{fr})}{\cos\gamma}-\frac{ctg(\delta+\varphi_{fr})}{\cos\gamma}P_{wear}\right)\cdot\left(\ln\frac{1}{tg\left|\frac{\beta_x}{2}\right|}\right) \tag{25}$$

The projection of the component of the ground resistance force to cutting, depending on the width of the cut, on a plane perpendicular to the axis of rotation of the boat ($R_{IO.CV}$) in the range from x to R_g

$$R_{i.o.cv}=\frac{\varphi m_{cv}h_{\text{в}}^2+h_{\text{в}}nP_{wear}}{2\pi n\cos\gamma}\cdot\frac{\sin\beta_x-\sin\beta_1}{\sin\beta_1\sin\beta_x}+$$
$$+\frac{h_{\text{в}}}{2\pi}\left(\varphi m_{cv}\frac{h_{\text{в}}}{n}\frac{ctg(\delta+\varphi_{fr})}{\cos\gamma}-\frac{ctg(\delta+\varphi_{fr})}{\cos\gamma}P_{wear}\right)\cdot\left(\ln\frac{tg\left|\frac{\beta_x}{2}\right|}{tg\left|\frac{\beta_1}{2}\right|}\right) \tag{26}$$

Conclusion

For further research it is necessary:

- to determine the meaning of the characteristic bending points of the blade operating body of the geokhod;
- to determine the influence of soil properties on the characteristic points of bending;
- to determine the effect of the parameters of the geokhod on the characteristic points of bending.

References

[1] Nishi S and Seiki T 1997 Planning and design of underground space use *Mem. Sch. Eng. Nagoya Univ* **1**

[2] Maidl B, Schmid L, Ritz W and Herrenknecht M 2008 *Hardrock Tunnel Boring Machines* (Berlin: Ernst& Sohn) 343

[3] Wighman T 1998 *Think deeps – go underground* (ENR: News – Rec.) **4**

[4] Aksenov V, Sadovets V and Pashkov D 2017 The influence of parameters on the generatrix of the helicoid form guide of geokhod bar working body *E3S Web of Conferences. The Second International Innovative Mining Symposium*

[5] Sadovets V, Aksenov V 2011 *Nozhevye ispolnitel'nye organy geokhodov: monografiia* [Blade operating bodies of geokhods: monography] (LAP LAMBERT Academic Publishing GmbH & Co. KG Heinrich-Böcking-Str. 6-8, 66121 Saarbrücken, Germany) 141

[6] Beglyakov V Yu 2012 *Obosnovanie parametrov poverkhnosti vzaimodeistviia ispolnitel'nogo organa geokhoda s porodoi zaboia* [The substantiation of parameters of the surface of interaction of the operating body of the geokhod with the rock of the face] (Thesis of diss.of

cand. tech. sciences, Yurga) 139

[7] Sadovets V Y, Beglyakov V Y and Efremenkov A B 2015 *Applied Mechanics and Materials* **770**, 384

[8] Zelenin A M, Balovnev V I and Keprov I P 1975 Mashiny dlia zemlianykh rabot [Machines for excavation works] (Textbook for universities, Moscow) 424

[9] 1992 *Mashiny dlia zemlianykh rabot* [Machines for excavation works] ed D P Volkov (Moscow) 187

[10] Vetrov Yu A and Baladinskiy V L 1980 *Machines for special excavation works* (Kiev: Kiev University Press) p 308

[11] Vetrov Yu A 1985 *Calculation of forces of cutting and digging of soils* (Kiev: Kiev University Press) p 251

[12] Aksenov V V, Sadovets V Yu and Pashkov D A 2017 Determination of the power parameters of the blade operating body of the geokhod for destruction of rocks of small strength *Vestnik of KuzSTU* **3** 116-126

Temperature influence on the effectiveness of gas-vortex stabilization in plasma torches

S V Anakhov [1,a*], Yu A Pyckin [2,b], A V Matushkin [3,c]
1Russian State Vocational Professional University, Physics and mathematics chamber, 620012 Russia, Yekaterinburg, Mashinostroiteley 11
[2]Ural State Forest Engineering University, Physical and chemical technologies of biosphere protection chamber, 620038, Yekaterinburg, Siberian tract, 37
[3]Ural Federal University, Welding technology chamber, 620078 Russia, Yekaterinburg, Mira str.,19
[a] email: sergej.anahov@rsvpu.ru, [b] email:yappoligon@mail.ru, [c] email:227433@e1.ru

Abstract. The results of the efficiency studies of gas-vortex stabilization systems for metal-cutting plasmatrons are presented. It is noted that the method of efficiency evaluation developed by the authors should be based on the calculation of the uniformity of the gas flow velocity distribution over the section of the gas-heating path of the plasma torch. Various simplified and accurate estimation methods are proposed. The results of calculation of the velocity distribution in the control section for different modifications of plasma torches are presented. Calculations are made on the "cold" model gas flow and its heating by a plasma arc. It is shown that when heated by a plasma arc, the flow rate at the inlet to the nozzle channel of the plasma torch and the degree of irregularity of the velocity distribution in the control section increase. By methods of statistical analysis the main parameter of the effectiveness evaluation of individual gaseous-vortex stabilization was chosen – criterion of the velocity variations. Demonstrated the advantages of the new upgraded torches, including working on technology narrow jet plasma, from the point of view of the effectiveness of gas-vortex stabilization.

1. Introduction

As shown by the authors [1], the development of high-precision plasma cutting technologies should be carried out taking into account the gas-dynamic features of the plasma-forming gas (PFG) flow, which determine the efficiency, quality and safety of the plasma torch. Based on the results of the authors ' analysis of the gas-dynamic features of the gas flow through the gas-air path (GAP) of plasma torches, the authors developed a method for evaluating the efficiency of gas-vortex stabilization systems of the plasma arc [2]. This method bases on the use of automated procedures to determine the criterion of the uniformity of the flow velocity distribution through the cross section of GAP channels. However, the previously presented results of the application of this technique were obtained taking into account the minimum effect of the gas-dynamic flow heating by plasma arc in the nozzle unit of the plasma torch. However, a certain influence of the plasma arc on the PFG velocity in the control section selected at the inlet to the nozzle channel of the plasma torch must be present. In this regard, the authors set the task to assess the degree of this influence and the applicability of the previously proposed method. An additional task was the clarification of criteria of gas-vortex stabilization efficiency, derived using the automated procedures of dynamic analysis.

IOP Conf. Series: Materials Science and Engineering **441** (2018) 012006 doi:10.1088/1757-899X/441/1/012006

2. Technique of researches

As is known, a typical design of a GAP in metal-cutting plasma torch (Fig.1) includes a section of gas supply to the plasma torch, a sedative (expansion) chamber and a gas-vortex stabilization system (GVS), which provides a tangential supply of PFG from the swirler (vortex chamber) to the nozzle unit (electric arc chamber). The method is based on the determination of the uniformity of the flow velocity distribution over the GAP cross section in the electric arc chamber of the plasma torch as one of the main factors affecting the quality and reliability of the plasma torch. This condition, as shown earlier [2], is determined by the design features of GAP plasma torches with a single-flow scheme of the PFG flow and, as a rule, is not fulfilled due to asymmetric and lopsided gas supply to the expansion chamber of the plasma torch. The geometry of individual sections of GAP, as a whole, is the same for most of the known domestic plasmatrons (PVR-402, VPR-410), including the one taken as the basic for the research of the plasma torch PMVR-M.

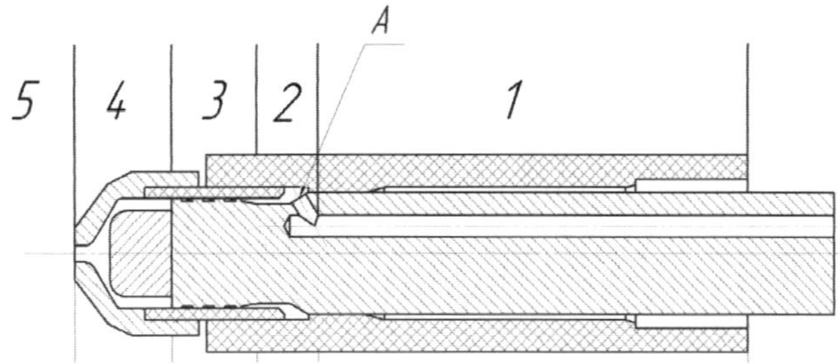

Figure 1. Scheme of air-gas path of metal-cutting plasmatron PMVR-M. Sections of GAP:

1 – PFG supply into the plasmatron, 2- expansion chamber, 3 – swirler, 4 – nozzle unit, 5 – plasma output, A – point of input of PFG into expansion chamber

To improve the uniformity of the flow distribution over the GAP, the authors proposed several design solutions [3]. One of them was implemented in the plasma torch PMVR-2M due to the use of a system of gas-dynamic filters (additional unbroken and perforated walls in the expansion chamber), as well as optimized by gas-dynamic criteria geometry of the expansion chamber and the swirler (Fig.2).

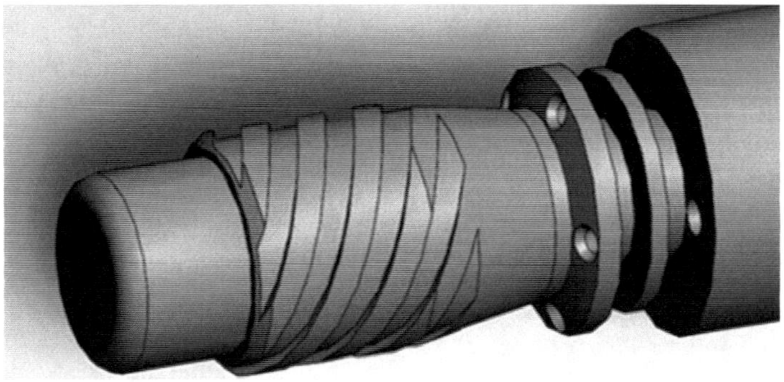

Figure 2. Scheme of the gas path in the plasma torch PMVR-2M with a system of gas-dynamic filters (unbroken and perforated walls in the expansion chamber)

Another effective solution was the development of the GAP design with symmetrical supply of PFG into the expansion chamber, 2 swirlers (forming and stabilizing) and 2 expansion chambers (Fig.3). Application of dynamic analysis methods allows to find the optimal size and location of expansion chambers and swirlers and to show the effectiveness of individual gas-vortex stabilization when using

IOP Conf. Series: Materials Science and Engineering **441** (2018) 012006 doi:10.1088/1757-899X/441/1/012006

this design at the "cold" (without taking into account plasma heating by the arc) flow of PFG. Such method of the velocity equalization can be applied in the plasma torches with both two-and single-flow scheme of the PFG flow organization. In the first case, the considered GVS technology may be used in the forming jet channel of a plasma torch, operating on the principle of a double nozzle and in various embodiments, having the name "narrow-jet, accurate or compressed plasma" [4]. In the second option, it can be used as an additional measure to improve the efficiency of GVS in traditional plasma torches in order to ensure the precision quality of metal cutting.

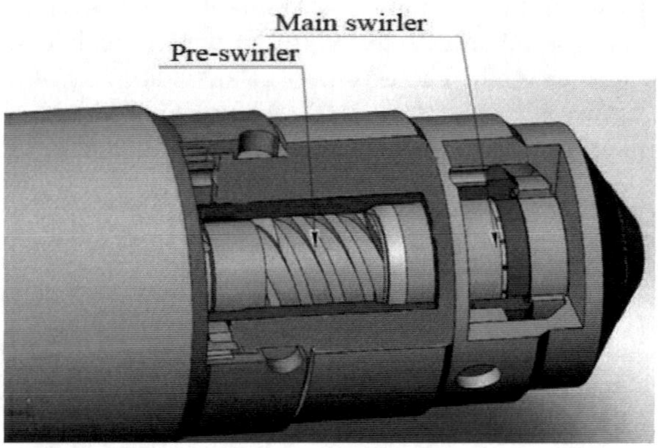

Figure 3. Constructive scheme of the plasma torch with two swirlers in the gas-vortex stabilization system

To assess the uniformity of the gas flow distribution, the authors proposed two methods depending on the hardware capabilities of the software and the required accuracy of calculations. A simplified version of the estimation (at the limited resources) involves the calculation of the PFG flow velocities in 4 symmetrical points in the selected plasma torch cross section, with one of the points selected opposite the point of gas injection into the expansion chamber (Fig.4, a). In this case, it is possible as a criterion for the uniformity degree in the cross section of GAP to use a velocity ratio XI=V1/V2=V1/V3=V1/V4 (Fig.4) [2]. However, there is an error of estimation due to the probability of missing the maximum and minimum deviations from the average values of the velocity in this section with its probable irregular distribution. In this regard, it is more objective to estimate the uniformity by the calculation method over the entire perimeter of the control plane. It is obvious that in this case, for the correctness of the comparison, it is necessary to calculate the velocities along the equidistant ring trajectory from the plasma torch axis (along the middle line of the passage section). At the same time, however, the problem to choose the criterion of the uniformity of the gas flow distribution arises.

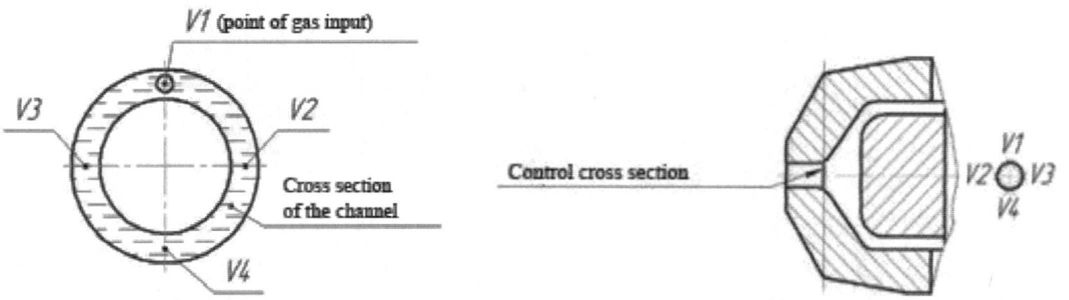

a) b)

Figure 4. The location of points for the PFG velocities calculation: (a) in the cross section of the GAP channels, (b) in the control plane of the nozzle unit.

In order to solve the set tasks there were conducted the investigations of the uniformity of the gas flow distribution in the control section for 3 plasma torches – basic PMVR-M, the upgraded single-flow PMVR-2M and PMVR-5.3 with 2 swirlers in the system of gas-vortex stabilization. The calculations were performed under the same technological conditions (the pressure of PFG at the entrance to the plasma torch P=5 atm, the flow rate of PFG Q=0.011 kg/s) by the SolidWorks software with Flow Simulation application. The velocity calculation was performed along the middle line of the control section, the diameter of which was determined by the size of the nozzle outlet of plasma torch. The number of calculated points along the perimeter depended on the parameters of the computational grid, which were determined automatically by the software (from 40 to 200 points depending on the analysis tasks). To assess the effect of the plasma arc on the uniformity of the gas flow distribution, a comparative calculation of the velocities in the control plane of plasma torches in the absence of arc and in the conditions of the nozzle channel heating by plasma arc was performed. Estimate the temperature for air plasma cutting in the calculations was adopted on the bulk temperature of the arc ≈ 7 000 K for the arc diameter of 2 mm).

3. Results of research and their discussion

The results of velocity calculations in the control sections are presented at Fig.5. The obtained results confirm the conclusions made earlier by the authors about the influence of the GAP design on the gas-dynamic parameters of the PFG flow in the electric arc chamber of the plasma torch. Under the same conditions at the entrance to the plasmatron, the fixed velocities in the control section of the plasmatron PMVR-2M are 20%, and in PMVR-5.3 – 70% lower than of the basic PMVR-M. This result is probably due to two reasons. The first is related to the structural differences of the nozzle used in the considered plasma torches (the output diameter, the length of the cylindrical part of the output channel of the nozzle). The second reason is gas-dynamic losses by GAP of the plasma torch, where the greatest contribution is made by the swirler (up to 40% of the total amount of pressure losses [2]). It should be noted in this regard that the most effective operation of the plasma torches occurs under certain regimes determined by experiment, and, consequently, the presented results should indicate objective differences in the optimal performance ranges for each of the analyzed plasma torches.

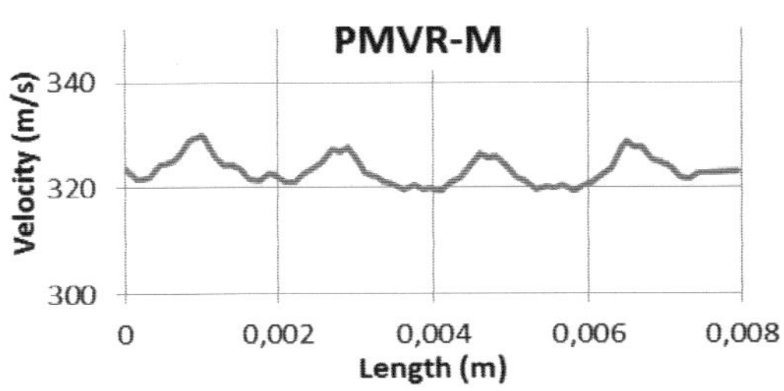

a) no arc (64 points, h = 0.15 mm)

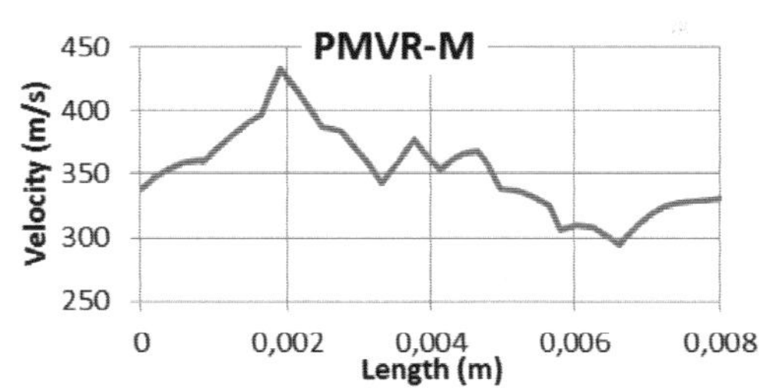

b) heating by plasma arc (41 point, h=0,2 mm)

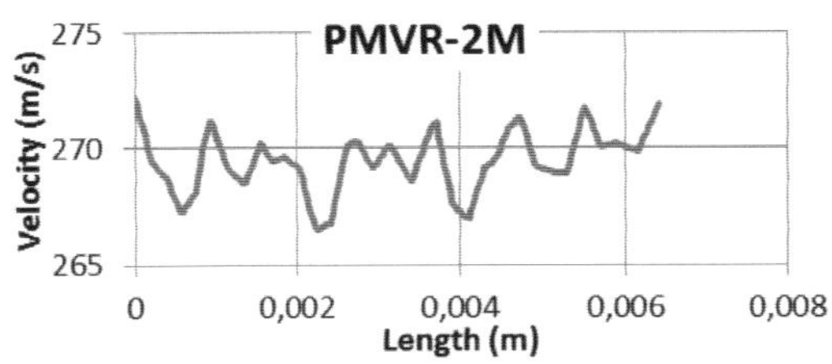

c) no arc (200 points, h = 0.03 mm)

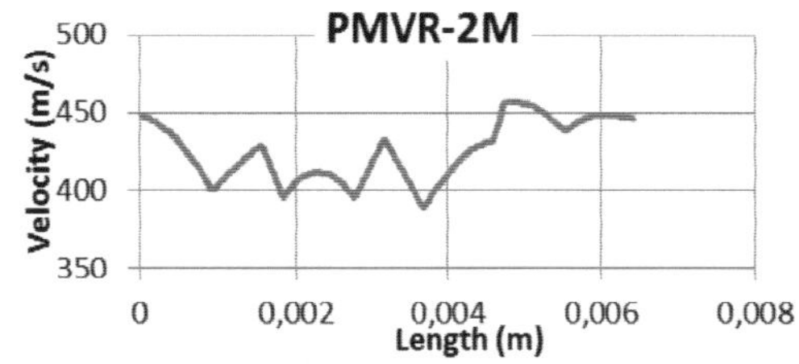

d) heating by plasma arc (200 points, h = 0.03 mm)

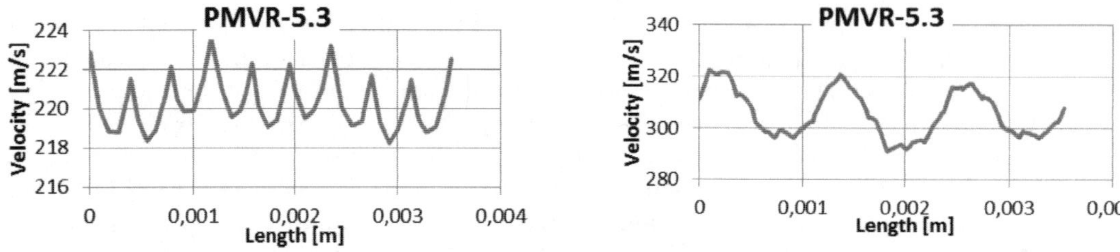

e) no arc (48 points, h = 0.09 mm)

f) heating by plasma arc (124 points, h = 0.03 mm)

Figure 5. Velocity distribution in the control plane of the plasma torches (h - parameter of discretization)

Another important consequence of the obtained results analysis is the revealed effect of plasma arc on the velocity of PFG in the control section. It is obvious that at velocity of 150-300 m/s in the nozzle unit, the residence time of the gas in the arc combustion area from the cathode end to the entrance to the nozzle channel is from 10 to 100 μs, which is clearly not enough to heat the entire flow (this effect is used to protect the channel walls from thermal influence). Nevertheless, the results indicate a small heating of the gas at the inlet to the channel and the velocities increase due to this heating in the points of the control section (in the plasma torch PMVR-M – about 15%, in PMVR-2M –by 55% , in PMVR-5.3 – by 40%). Such differences are obviously also due to the design features of the nozzle units of the analyzed plasma torches, which affect the nature of PFG interaction with the plasma arc. It should be noted that presented in Fig.6 the results of velocity calculations in the nozzle channel and at the outlet of the plasma torch PMVR-5.3 indicate the effective heating of the gas in the nozzle channel and obtaining at the outlet a high-energy flow with velocities of the order of 500-600 m/s for the kinetic effect on the cut metal.

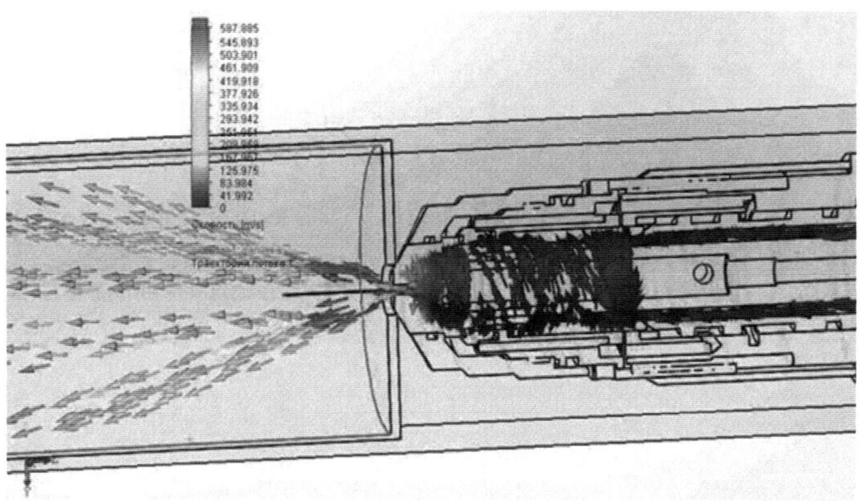

Figure 6. The velocity distribution in the GAP and plasma jet of plasma torch PMVR-5.3

In order to find the criterion for estimating the degree of the flow distribution uniformity in the analysis of a large number of control points, the obtained velocity distributions for different plasmatrons were analyzed by statistical methods. A statistical hypothesis was made that the presented velocity dependences are subject to either the normal or the uniform distribution law of random variables. However, verification of hypotheses by Pearson's χ^2 gave her not to a statistically significant confirmation. It was not confirmed in the analysis of other speed distributions and in the change of frequency calculation intervals. For this reason, the processing presented in Fig.5 results as criteria were considered often used in the processing of large arrays of random variables parameters:

1. Scope of variation $R=V_{max}-V_{min}$,
2. Ratio of variation L= Vmax/Vmin,
3. Average linear deviation (ALD): $a=\sum_{i=1}^{n}|V_i - \overline{V}|/n$,
4. Standard deviation (SD): $S=\sqrt{\sum_{i=1}^{n}(V_i - \overline{V})^2/n}$,
5. Coefficient of variation $F=S/\overline{V}$.

For large sample sizes (in our case n>40) $S\approx\sigma$, were σ - dispersion of the random variable. The results of calculations of these parameters are presented in the table.1.

Analysis of the results, presented in the table.1, indicate a statistically significant increase in the efficiency of gas-vortex stabilization (uniformity of the velocity distribution) achieved in the modernized plasma torch PMVR-2M and PMVR-5.3 compared with the basic plasma torch PMVR-M. This conclusion is confirmed by comparing all the parameters presented for the analysis (R, L, a, S and F). When choosing a single criterion for the effectiveness of GVS, in the opinion of the authors, we should focus on the coefficient of variation F, which gives the expressed percentage of the ratio of the SD (dispersion) to its average value in the analyzed sample. This parameter is easily calculated, for example in Excel, where there is a built-in SD calculation function.

Table1. The results of statistical processing of the velocity distribution of the plasmatrons

Type of plasma torch	Heating conditions	Statistical parameters							
		V_{max}, m/s	V_{min}, m/s	$\overline{V}$ m/s	$R=V_{max}-V_{min}$, m/s	$L=V_{max}/V_{min}$	a (ALD), m/s	S (SD), m/s	F (%)
PMVR-M	no arc	323.4	319.6	330.0	10.4	1.03	2.3	2.8	0.9
	with the arc	353.0	295.0	433.0	138.0	1.47	26	33	9.3
PMVR - 2M	no arc	269.4	266.6	272.2	5.6	1.02	0.9	1.2	0.4
	with the arc	425.0	389.0	457.0	68.0	1.02	17	19	4.5
PMVR - 5.3	no arc	220.5	218.2	223.6	5.4	1.18	1.2	1.4	0.6
	with the arc	306.1	290.9	322.2	21.3	1.11	7.8	8.8	2.9

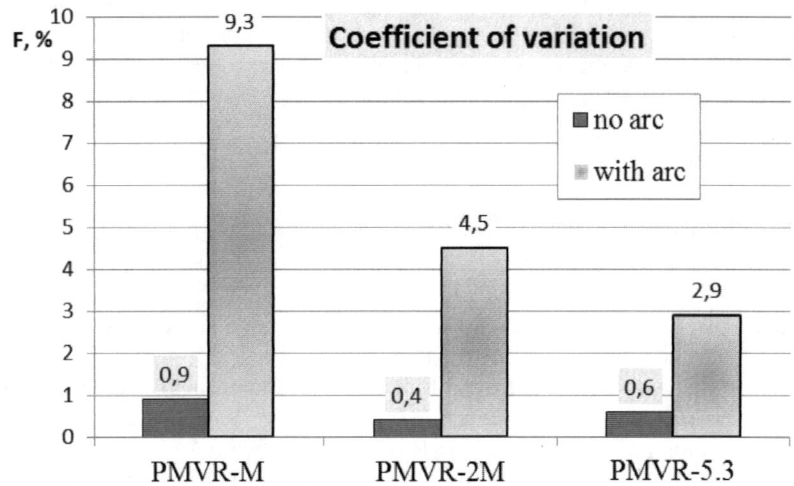

Figure 7. The coefficients of variation of the velocity in the control plane of a plasma torch

4. Conclusions

The analysis of the selected criterion F shows an increase of the GVS efficiency by 1.5-2 times on the "cold" gas (without heating by plasma arc) and 2-3 times at heating by plasma arc in the new plasma torches in comparison with the basic one (Fig.7). The presented results of calculations show, in addition to velocity increase, the effect of the plasma arc temperature on the uniformity of the PFG flow distribution. In plasma torches PMVR-M and PMVR-2M observed increase in the coefficient of variation of about 10 times, and in PMVR-5.3 – about 5 times, which also indicates the constructive advantages of the proposed system of gas-vortex arc stabilization for plasma torches, working by the "narrow-jet plasma" technology.

References

[1] Anakhov S V Principles and methods of designing in electroplasma and welding technologies, *RSVPU, Yekaternburg*. p 144, 2014

[2] Anakhov S V, Pykin YU A, Matushkin A V Investigation of systems for gas vortex stabilization of plasma torches *J Welding International*,Vol 30, **Issue 4**, p. 310-314, 2016

[3] Pyckin Yu A, Anakhov S V, Matushkin A V *Patent for the invention of "Plasmatron"* №. 2584367 RU C1 11.03.2015 (published 20.05.2016)

[4] Anakhov S V, Pykin Yu A, Matushkin A V Narrow Jet Plasma as the Energy Efficient and Safe Technology for Metal Cutting *J Solid State Phenomena (Material Science Forum)*, **Vol 870**, p. 523-527, 2016

Innovative technologies as a means of the development of future engineers' professional mobility abroad

M B Balikaeva[1, a], E L Chizhevskaya[2,b], G Ya Grevtseva[3,c], I O Kotlyarova[4,d], M A Volkova [5,e],

[1,2] Tyumen Industrial University, Lunacharskogo 2, 625002 Tyumen, Russia
[3] Chelyabinsk State Institute of Culture, Orjonikidze 36a, 454091 Chelyabinsk, Russia
[4,5] South Ural State University, Lenin Ave., 76, 454080 Chelyabinsk, Russia
[a]marinabalikaeva@list.ru, [b]chizel76@yandex.ru, [c] yakupovna@rambler.ru,
[d]kotliarovaio@susu.ru, [e] volkovama@susu.ru

Abstract. The article discusses the relevance of innovative technologies in development of the future engineers' professional mobility. In the publication the types of innovative technologies are presented in the universities of Germany, the US, Britain, France, Norway, Poland and other countries. The principles and functions of the innovative technologies are defined in some universities abroad. The forms of innovative technologies contribute to the qualities formation at the basis of the future engineers' professional mobility.

1. Introduction

The innovative technologies become the trend to improve the education of the modern mobile engineer. The leading professors of the best American universities believe that the innovative technologies are a necessary fact of future engineers in their professional training system. Its part is the use of the innovative technologies. It is essential to integrate the innovative technologies in research work, to create conditions for active students' participation. The researchers share this opinion in Western Europe and the United States. Modern integration approaches and innovative technologies in Russia and abroad have some similarities. In both cases, this phenomenon is a relevant, complex planning problem of organization and management to improve the future engineers' training quality using innovative technologies. Their principles are a teaching and researching a combination, an integration, a classroom and out-of-classroom study forms combination in higher school. In this connection the requirements system includes the innovative technologies aimed at the creative development of their personal and professional qualities. They are a part of the professional mobility.

The analysis of philosophers, educators and psychologists views and the own experience of theoretical research allows me to consider the professional mobility definition as an integrative quality in the production conditions on the basic components of professional culture covers (key competencies and qualifications) and professional competency (high intelligence, creative skills, professionalism). It reflects in man's ability to move horizontally and vertically on the social professional structure of society and to change a social, economic and professional (including status) position. It captures the motivational, cognitive, competency-based and operative components, reflecting a certain level of general cultural, general professional skills and competencies [1].

2. Discussion

The innovative technologies in higher education abroad are closely related to the education content and its aim [2]. The aim of future engineers' education is to develop the continuous self-education after graduating from higher school, the formation of future engineers ' productive creative thinking, the development of logical analysis skills and abilities of full information consumption, the educational content orientation on the professional mobility specialist's model. The effectiveness of future engineers' the professional mobility depends on innovative technologies.

Universities use all opportunities for the first year future engineers' involvement in independent research work using the innovative technologies. Some experience in this regard has already accumulated in many higher schools abroad.

For many years, there are seminars at the Dresden Technical University in the form of scientific debate on searching and finding the best argument. Students look for the best arguments to prove the correctness of the thesis. Scientific debate shows how students have mastered the necessary knowledge. As a part of the educational process in many universities in Germany the events are held where "students teach students". These events have a significant impact on didactic training quality and structure. It confirms the experience and the results of educational research [3].

Any full preparation for the seminar is actually the microstudies requiring at least four hours of productive and creative activities with the latest scientific periodicals and Internet sources. There is this form in the universities of the USA and Germany. This preparation is a "weak" students' term paper in some Russian universities. However, the students do once a week on four subjects at least.

There are special courses for the first year students. The aim of this course is to facilitate the transition from school to university. Some interest from this point of view it is a model of the practical training orientation at the Higher Technical School in Darmstadt (Germany). This model includes the future engineers' introduction in the future career occupation, business games, excursions, etc. during the first year of study [3].

In the US the method of small interdisciplinary research groups is widely spread. It has an on-line support. Post-graduate and undergraduate students have the opportunity to exchange views in their groups to find and formulate new, original ways of solving problems. For example, the training process is very individualized in american universities, especially at the undergraduates as the groups are formed for a term to study specific subjects. The students have the right to choose a field of knowledge in accordance with their interests and needs. They can change their specialization and prolong the period of study at the university. No timeframe allows the students to advance in the study according to their individual abilities. Almost all undergraduates study on the individual curricula [3], [4], [5].

Every second of seminars involves students' essay (paper). Weekly each student must complete blogging (diaries) courses at the faculty website, where students have to present their view of the topic passed at the seminar. Every student has the opportunity to prepare reports, presentations at least twice a term.

The special role plays a tutorial, which can also be carried out on-line. A tutorial is a phenomenon adding the training process as one of study means using the innovative technologies. A tutorial is a stimulus for scientific collaboration between a teacher and a student. During a tutorial, they can discuss the research methods, the students' individual opportunities in the scientific work process [5].

In general, in the university practice abroad the most common on-line seminars and discussions are spread. They encourage maximum integration in the training process. At the seminars, the students use various methods of study and research activities, the basics of oratory. The organizational form choice of the US authors explains that the debate [5] helps the students understand better the problem. It facilitates the consideration of the various aspects problem under discussion, allows participants to speak critically and corrects the new information for the group opinion or decision development. The principle "round table" allows all students to see each other and creates a relationship of equality between them and there is an informal environment among them. All these forms develop the ability to predict the actions results aimed at solving specific professional problems and achieve the professional aims, the quick adaption in a professional team, the business and professional communication skills in specific business situations.

The american higher education analysis points out that it is based on the innovative technologies. For example, the US higher education has a group of "open education", "Open University" under the influence of humanistic psychology, where students are free to define their research interests and in the dialogue with tutors to express critically a variety of the scientific positions. In american universities do not have the typical curricula. Each university develops their curriculum. It includes a part of the basic courses, a professional training core with a level indication of competencies development while a significant part of the special and general training courses has a choice form [3].

In US universities, a flexible combination of asynchronous (individual) and synchronous (joint) study is widely used. Many universities in the US and Western Europe use actively the blended learning. The blended learning involves a combination of the different training management techniques and schemes: classroom study, distance study, synchronous distance study [3], [6], [7].

There are some innovative technologies in the universities of the US, Canada, Germany, France and others. The E-learning involves different ways of study using the innovative technologies: distance study, mobile study using mobile devices, virtual study [6]. The E-learning allows students to personalize the training process, develop the skills of knowledge independent acquirement and application, orientation in the flow of information. The most advanced e-learning center is the Massachusetts Institute of Technology (MTI) in the United States. Over 100 million people use the e-learning all over the world [6].

More attention in many universities abroad has been paid to the production process and higher school connection. In Germany the «dual system" and the project method are widely used in the training process [2], [5].

The students simultaneous study is in the walls of the institution (two or three days a week) and in the enterprise (other days). This system is widely used in Germany and the United States. In 2012 over 64,000 students, 1,384 students study programs were invited to the dual system study in German universities [8]. In the Harvard Business School (HBS) this system is called the "field" study (Field Studies) in the leading business companies in the USA. Students write their own business plans for the creation or development of new enterprises [5]. Students have the right to choose a project that will allow them to use the tools and techniques studied in the classroom, to develop the project management skills, to get the real experience of business communication and to improve the ability of collecting and analyzing information, to obtain the industries direct knowledge. The research work on the project is the study in the problem-orientation courses using the innovative technologies. The problem-based study is a combination of students' independent study in the small study groups, which introduce the research and scientific work.

Annually in France, about 35,000 students start distance education, for which the State Center for Distance Learning (CNED), established in 1986 and subordinated to the Ministry of Education, is responsible. Now the Center offers courses and trainings at different levels (licensee, master's degree) and covers the most demanded areas (business, accounting) [9].

One of the programs that the Center actively uses is the "electronic campus". The program provides the opportunity to use various services available in traditional campuses.

Since 1990, Norway has a network of higher education institutions for distance learning. In 2004, this network became a part of the Norwegian Open Universities (LEU), supported by the Ministry of Science and Education. On its website, LEU hosts an extensive database with options for finding educational courses for both distance and inpatient training.

Distance educational programs using ICT are an alternative to traditional courses offered by all educational institutions. There are programs of this type almost in all universities of Norway. As for programs of a certain professional orientation, they have a form of video conferences and a contact with the instructor. In private institutions of distance education, the main aim is to develop offers that meet the requirements of different groups of students, and use flexible methods and forms of education. Most of these courses give an opportunity both to upgrade qualifications and to get a new profession. More than two thirds of such courses cover the following areas: economics and management, technology, communication, social services and health. Private universities of distance education in Norway have been pioneers in the use of television transmissions, videoconferences and the use of computers throughout the education system. Moreover, they have accumulated experience in the organization of distance schools, the development of effective educational modules and a feedback system. The subjects of the offered courses are wide ranging from hobbies to university disciplines [10].

The countries of Eastern Europe have recently begun to learn from the development of effective distance learning. The platform for using e-learning in Polish universities [11] presents possible ways of using e-learning in higher educational institutions. The presentation of the educational material on the network in the form of articles, essays, studies, links to other Internet sources. This form is absolutely passive, because it does not differ from traditional access to educational material except the necessary materials in an electronic database, electronic sources.

Online support: this form of e-learning is used as a more interactive way, as it provides access to discussion forums, chat rooms, newsletters and uses e-mail.

Asynchronous trainings also include mechanisms for monitoring and evaluating the sessions conducted by the instructor-head of the courses. The same tools and methods are used as an on-line support.

There are synchronous trainings through a computer network using the following techniques: teleconference, videoconference, chat or thematic video. Some courses of this kind use a virtual board, or other tools of group work.

In addition to e-learning technologies there is an introduction in higher education institutions as a tool for teaching m-learning (mobile learning) using gadgets as mobile phones, smartphones and tablets. It should be noted that m-learning is effective in carrying out creative, practical works, quickly searching for information, but it can hardly be applied at the stage of knowledge control.

3.Conclusion

In conclusion, it should note that creativity as a component of innovative technologies has a huge impact not only on the specific scientific progress, but also on society by developing its innovative ability.

Thus, the experience of the study practice abroad showed to our problem that the higher education abroad has developed the future engineers' professional mobility process integrated into the training process with using the innovative technologies. Each student must master a research effective style to develop the ability and willingness to the study and the improvement of scientific knowledge and its creative application in the future professional activity. The continuous research study becomes personal needs of each student using the innovative technologies. The innovative technologies impact on solving specific professional problems, achieving professional aims, the group organization of a specific professional project, the quick adaption in a professional team, business skills and professional communication in specific work situations.

References

[1] Balikaeva M B 2017 The Interconnection between Future Engineers' Professional Mobility in Higher School Social and Cultural Environment. *Modern Jour. of Lang. Tea. Meth.*, **7** (11), pp 70–78

[2] Spirin T 2014 Foreign and domestic experience of individualization of training in high school. *World of Sci., Cul., Edu.*, **3**, pp110-113

[3] Martens D Schlüsselqualifikation. Thesen zur Schulung für eine moderne Gesellschaft (Stuttgart, Berlin. Köln. Mainz, 1974) 69 p

IOP Conf. Series: Materials Science and Engineering **441** (2018) 012007 doi:10.1088/1757-899X/441/1/012007

[4] Dmitriev D Anatomy of an American university (Moscow, 2010) 224p
[5] Mizyurova E Key points the educational process efficiency increase in higher school. *Proceed. of the high. ydu.inst. (Volga region. Humanities*, 2011) **2**, pp109-116
[6] Starichenko B Synchronous and asynchronous organization of educational process in higher school on the basis of information technology education model. *Teacher edu. in Rus.*, 2013, **3**, pp 23-31
[7] Novikov A The teaching forms in the modern conditions. Available at: http://www.anovikov.ru/artikle/forms.pdf (accessed 10 February 2018)
[8] Solominsky E Features of higher education in Germany with the Russian professor eyes (an interview with E. Patrick). *Higher edu. in Rus.*, 2014, **2**, pp131- 141
[9] Dobrynin M A, Sukhoi S N 2007 Distance Learning in France. (Minsk: Belarusian Digital Library LIBRARY.BY)
[10] Distance courses in Norway. Available at: https://www.academiccourses.ru/Kursy/Norvegija/ (accessed 6 February 2018)
[11] Distance education in Poland. Available at: http://global-edu.ru/foreign-education/remote-education-in-poland/ (accessed 6 February 2018)

Influence of the supporting surface inclination angle of the external geokhod propulsor on the deflected mode of boundary rock massif

V Yu Beglyakov[1,2], V V Aksenov[1,2], I K Kostinets[3], A B Efremenkov[4] and A A Khoreshok[5]

[1] National Research Tomsk Polytechnic University, Yurga Institute of Technology, Yurga, 652050, Kemerovo Area, Yurga, ulitsa Leningradskaya, 26
[2] Institute of Coal of the Siberian Branch of the Russian Academy of Science 650000, Kemerovo, prospect Sovetski, 18
[3] Branch of Kuzbass State Technical University named after T.F. Gorbachev in Belovo town, 652644, Kemerovo Area, Belovo, urban-type community Inskoy, ulitsa Ilyicha, 32a
[4] Novgorod State University, ul. B. St. Petersburgskaya, 41 173003 Veliky Novgorod, Russia
[5] T.F. Gorbachev Kuzbass State Technical University, 650000, Kemerovo, ulitsa Vesennyaya, 28

E-mail: abe@novsu.ru
E-mail: begljakov@rambler.ru

Abstract. The article describes the interaction model of the external propulsor propeller blade of the geokhod with boundary rock massif. The model makes it possible to investigate the influence of the supporting surface inclination angle of the external geokhod propulsor on the deflected mode of boundary rock massif. The results of modeling and their analysis are given. The ways of optimization of supporting surface geometrical parameters are outlined.

One of the key processes that occur in the underground geodetic mining is the geokhod screwing into rock massif [1-4]. In this case the direction of the interaction force between the propeller blade and the rock is determined by the location of the blade supporting surface (Figure 1), namely, the angle of its inclination in relation to the frontal plane.

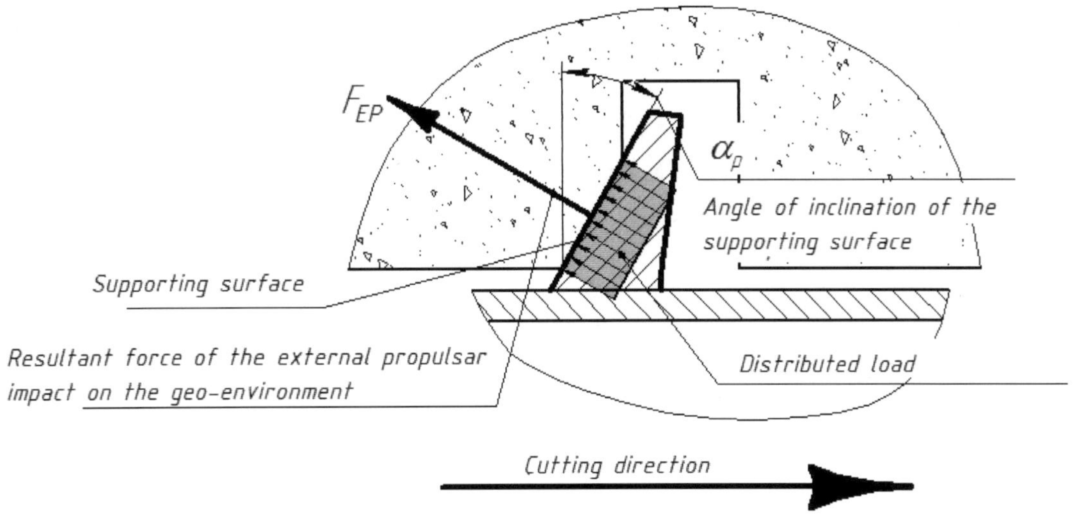

Figure 1. Interaction scheme of the external propulsor propeller blade of the geokhod with geo-environment.

The inclination angle also depends on the value of the interaction force [5, 6]. With $\alpha_p = 0$ the force of interaction is at its minimum, but there is a higher probability of cleavage or shearing of the inter-turn block. As the angle increases, the stresses are shifted towards the compression [7-9], but the forces of interaction increase [5, 6]. That is, the angle $\alpha_p$has an ambiguous effect on the bearing capacity of the boundary rock massif [6].

Therefore, the problem that arises is to investigate the effect of the supporting surface inclination angle α_p of the external propulsor (EP) on the deflected mode of the boundary rock massif.

To assess the influence of the inclination angle $\alpha_p$of the supporting surface, a model of the working area with a two-path system of circumscribed helical channels was created. The scheme of the model is shown in Figure 1.

In modeling, the upper limit of the values of the inclination angle α_p of the supporting surface was assigned to 54°, which corresponded to the theoretical angle of the seizure of the geokhod in the output [6]. The angle α_p varied in the range from 0 to 50° in steps of 5° and in the range from 50 to 54° in steps of 1°.

To exclude their influence, the following parameters remained unchanged: the depth of the boundary channels H = 0.25 m, the pitch of the helical line of the external propulsor $h_в$ = 0.8 m, the length of the supporting surface along the output contour l = 1 m and the position of the free surface.

IOP Conf. Series: Materials Science and Engineering **441** (2018) 012008 doi:10.1088/1757-899X/441/1/012008

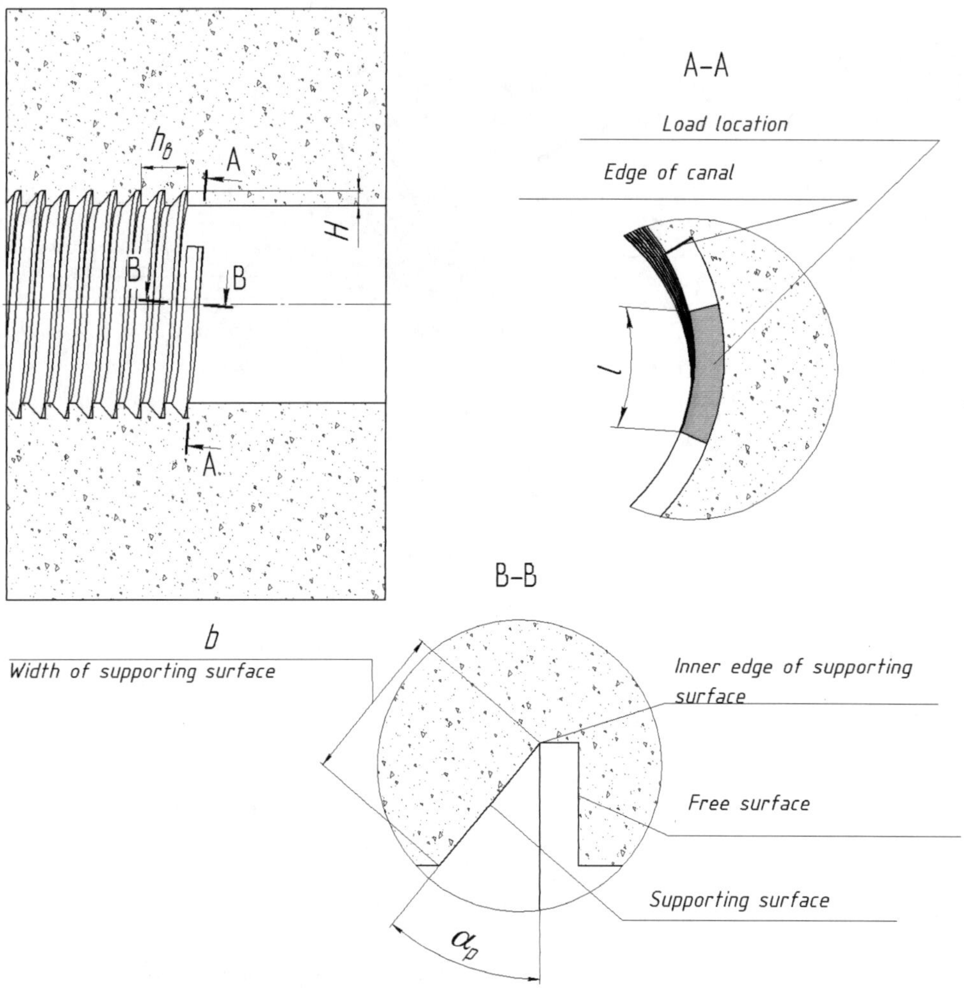

Figure 2. General view of the model for assessing the influence of the supporting surface inclination angle of the external propulsor on the deflected mode of the boundary rock massif.

The model simulated the operating conditions of the two-way external propulsor [10-12], the load was applied to the supporting surfaces of the channels, the leading edge of the supporting surface was located 500 mm from the bottom of the boundary channel.

The models were divided into three 10-node finite elements [13, 15]. The contact surfaces of the boundary rock massif with the external propulsor blades and the adjacent areas were divided into 50 mm elements, the rest of the model was divided into 500 mm elements, the transition of the element sizes from 50 to 500 mm was carried out on four transition layers [16] (Figure 3).

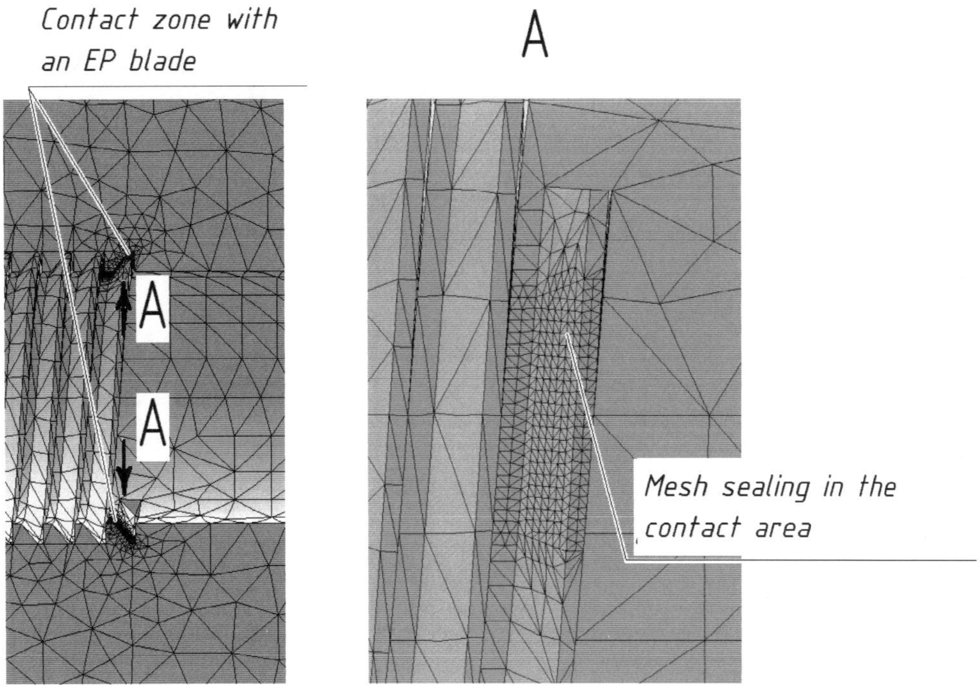

Figure 3. The grid of partitioning the model into finite elements.

A normal and tangential load was applied to the contact surface of the blade with the boundary rock massif (Figure 4) with intensities

$$q_n = \frac{F_{EP}^{N}}{S_{SS}} \text{ and } q_t = \frac{F_{EP}^{FF}}{S_{SS}}$$

Where F_{EP}^{N} is the normal force acting on the side of the blade, F_{EP}^{FF} is the blade's friction force on the rock, S_{SS} the contact area of the blade with the rock. The values of the forces F_{EP}^{N} and F_{EP}^{FF} depended on the inclination angle of the supporting surface and were determined from the mathematical model given in [6], the values of the areas S_{SS} at different angles α_p were measured on the model.

IOP Conf. Series: Materials Science and Engineering **441** (2018) 012008 doi:10.1088/1757-899X/441/1/012008

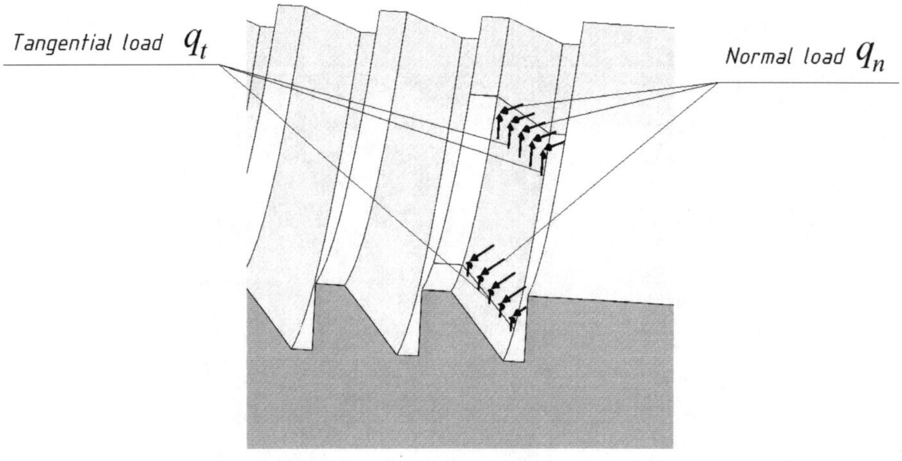

Figure 4. The scheme of applying the load to the model

In the course of preliminary modeling, different types of deflected mode in the area of interaction with the blade were evaluated. Three diametrical cross-sections were considered when assessing the rock deflected mode: at the anterior edge of the blade (III-III), in the middle of the blade (II-I) and at the posterior edge of the blade (I-I) (Figure 5).

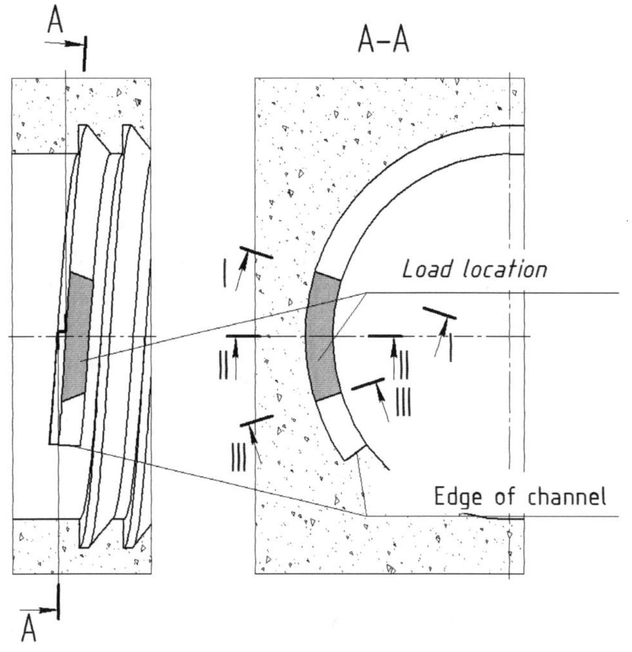

Section I-I - at the posterior edge of the blade, II-II - in the middle of the blade,
III-III - at the anterior edge of the blade

Figure 5. Diagram of diametrical cross sections arrangement on the model

The assessment of the deflected mode was carried out visually according to the diagrams of the main stresses σ_1, σ_1 and tangential stresses τ at angles of inclination of the reference surface from 0° to 40° in steps of 10°. Stress diagrams in section II-II for some angles α_p are shown in Figure 6.

From stress diagrams in diametrical cross sections, it can be seen that the maximum stresses occur on the contact surface and decrease with removal in depth. At values of the angle $\alpha_p > 20°$ the stress moduli increase with increasing angle.

Also, during preliminary modeling, the stress distribution along the reference surface was visually assessed.

Figure 7 shows the diagrams of the distribution of principal and tangential stresses over the contact surface.

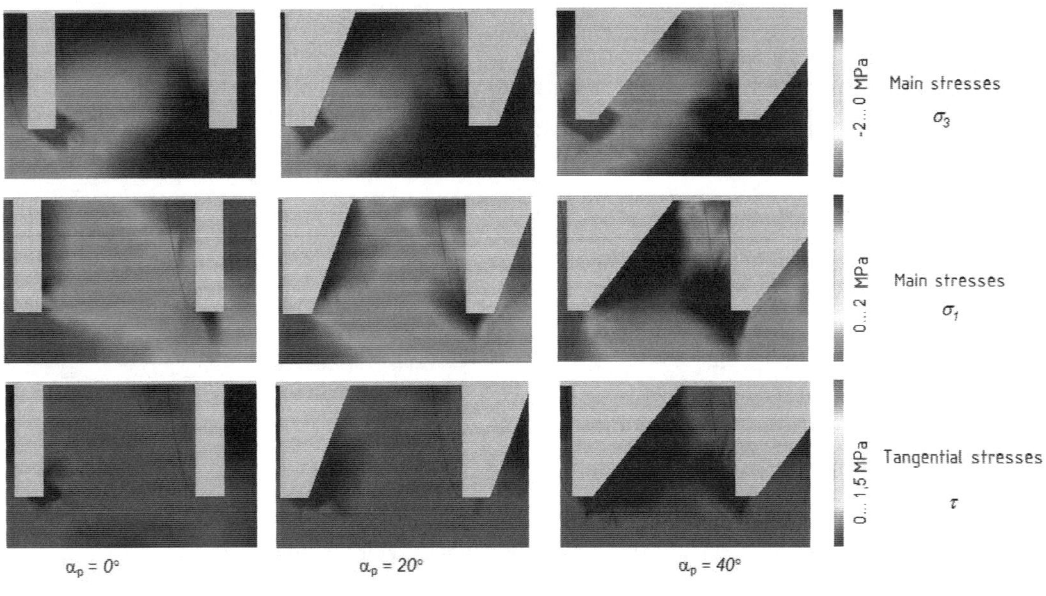

Figure 6. Stress diagrams in the middle section of the reference channel surface

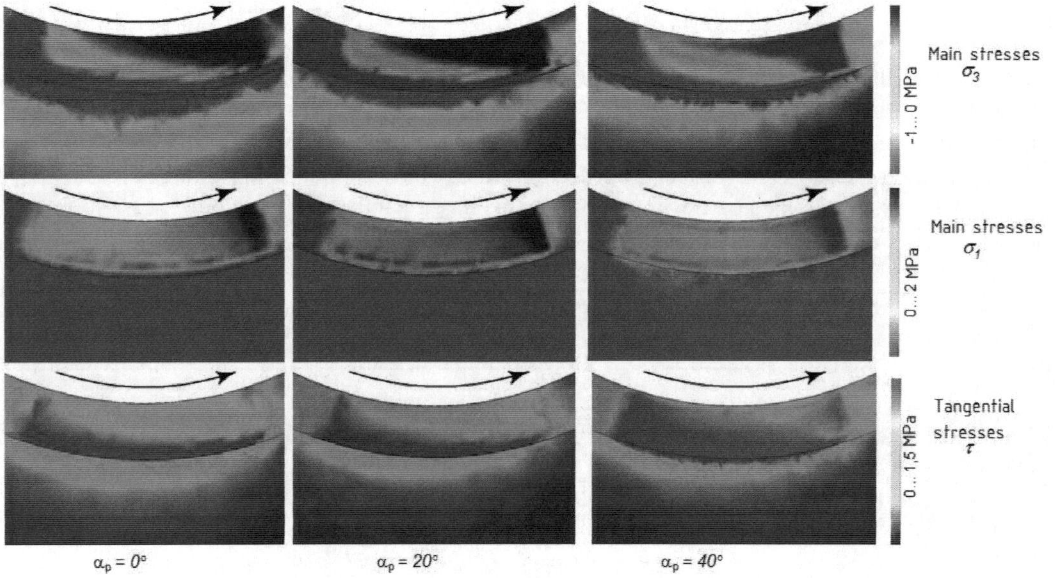

Figure 7. Distribution of stresses along the supporting surface of the boundary channel

From the stress distribution diagrams along the supporting surface of the channel it is seen that the compressive stress prevails in the area of the anterior edge of the blade, and the blades in the area of the posterior edge of the blade are stretching. Tangential stresses grow from the anterior to posterior margin of the blade. The change in stress from the anterior to posterior edges of the blade is monotonous. At values of the angle $\alpha_p > 20°$ the stress modules increase with increasing angle.

With further modeling, the influence of the angle $\alpha_p$on the stresses in the rock at various points of the reference surface was investigated.

Figure 8 shows the layout of the control points on the reference surface.

During the simulation, the main stresses σ_1 and σ_3, tangential stresses τ and equivalent stresses σ_{equV} were determined on the interaction surface of the blade with the boundary rock massif, and the stress distributions over the surface were compared for different angles of inclination α_p.

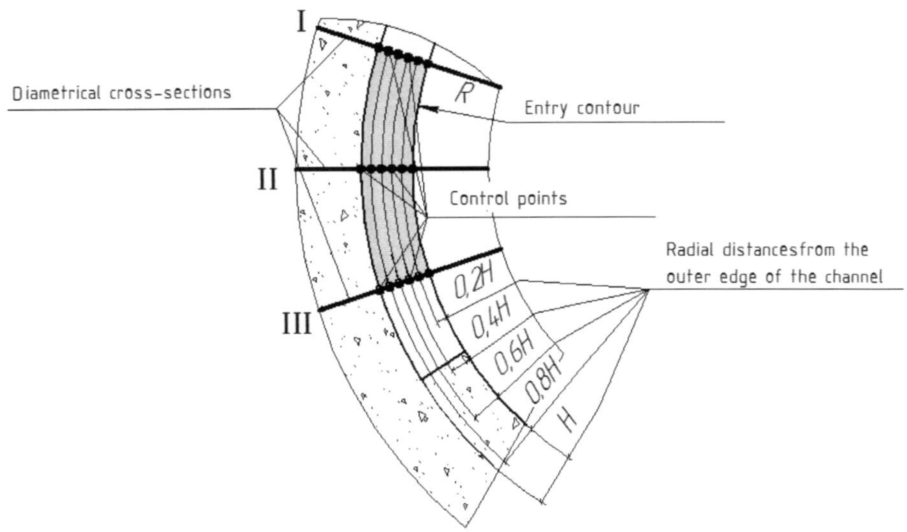

Figure 8. Scheme for placing control points on the supporting surface of the boundary screw channel.

Equivalent stresses were determined by the theory of Mohr strength according to the formulas [17]:
- for areas with positive stresses σ_1 and negative ones σ_3:

$$\sigma_{equV} = \sigma_1 - \frac{[\sigma]_C}{[\sigma]_P}\sigma_3$$

- for triaxial compression areas: $\sigma_{equV} = \sigma_1 - \sigma_3$

- for triaxial stretching areas: $\sigma_{equV} = -\frac{[\sigma]_C}{[\sigma]_P}\sigma_3$

where $[\sigma]_C$ and $[\sigma]_P$ are the ultimate strength of the rock massif to uniaxial compression and stretching, respectively.

Numerical results of the simulation are presented in the form of graphical dependencies in Figures 9 ... 12, which reflect the influence of the angle α_P of the supporting surface inclination of the external propulsor propeller blade on the distribution of stresses along the supporting surface of the screwed channel.

IOP Conf. Series: Materials Science and Engineering **441** (2018) 012008 doi:10.1088/1757-899X/441/1/012008

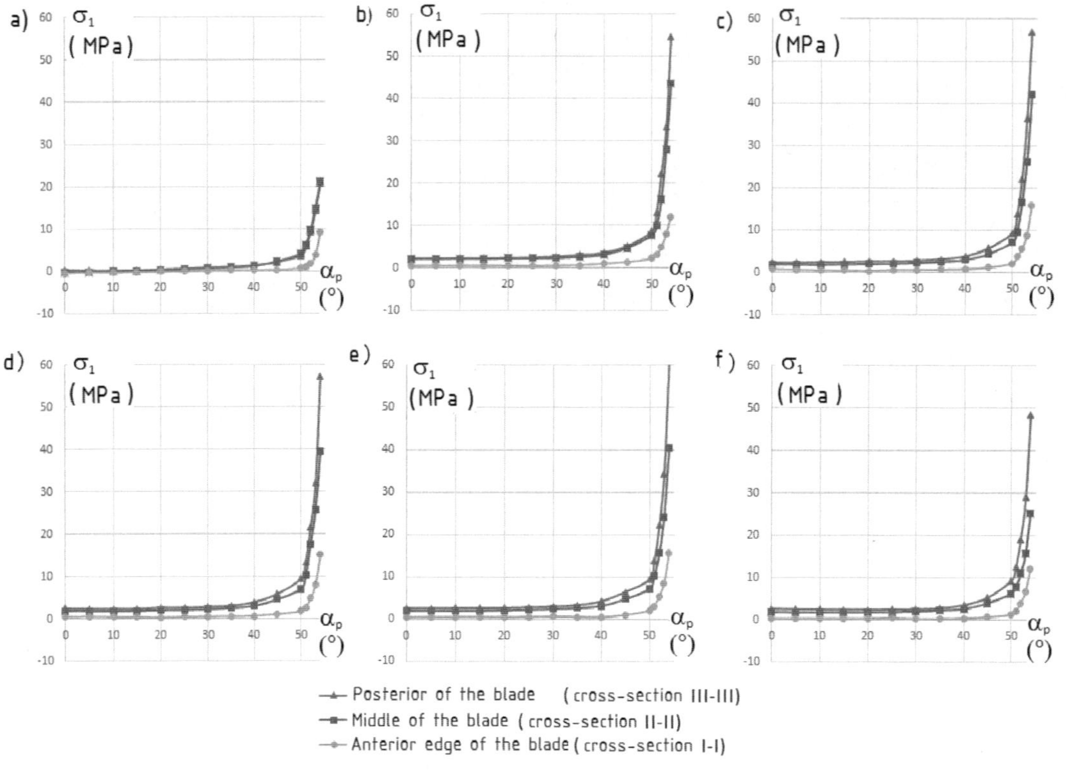

a) at the inner edge, b) at a distance of 0.8 H from the outer edge, c) at a distance of 0.6 H from the outer edge, d) at a distance of 0.4 H from the outer edge, e) at a distance of 0.2 H from the outer edge, f) in the area of the outer edge

Figure 9. Dependence of the main stresses σ_1 on the inclination angle of the supporting surface α_P .

The family of curves in Figure 9 reflects the dependence of the principal stresses σ_1 on the inclination angle α_P of the supporting surface of the external propulsor propeller blade. It can be seen from the graphs in Figure 6 that with increasing angle α_P in the range from 0 to 25 degrees, the values of the principal stresses σ_1 decrease slightly, with further increase in the angle α_P, the stress values σ_1 increase. In the range of angles α_P from 0 to 30 degrees, the main stresses σ_1 are slightly dependent on the angle α_P, at angles α_P greater than 50 degrees there is a sharp increase in stress. In the area of the inner edge at angles α_P less than 20 degrees there are negative stress values σ_1, which indicates the presence of areas with triaxial stretching in this area.

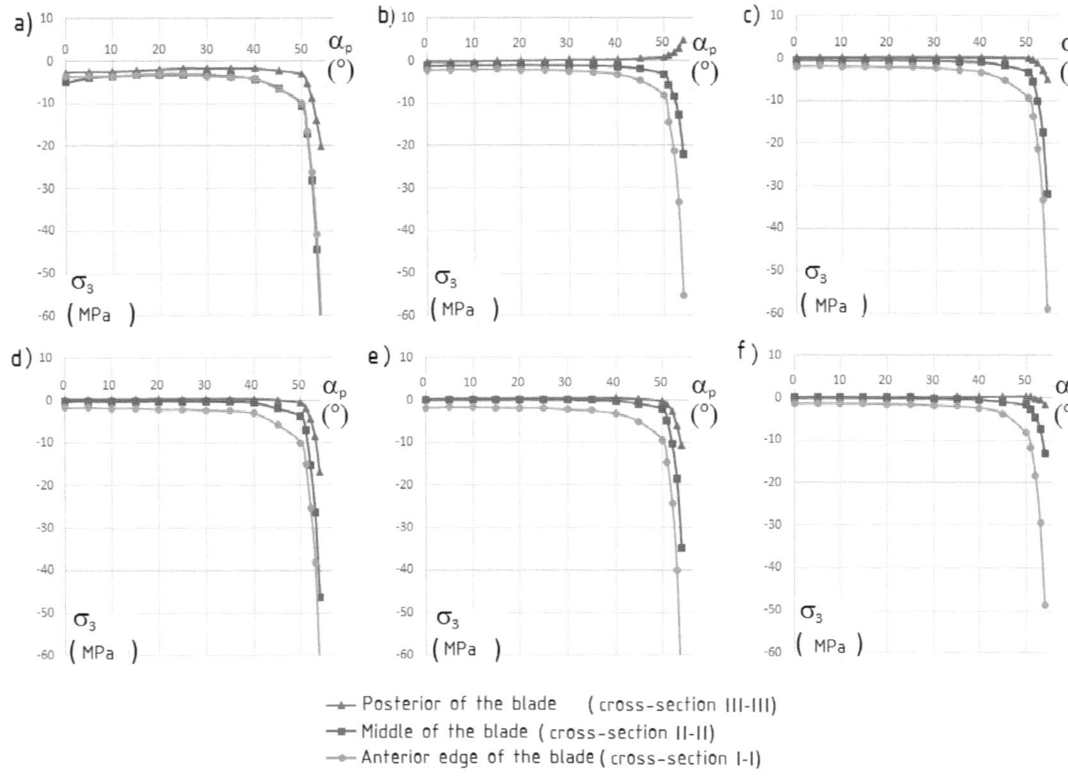

a) at the inner edge, b) at a distance of 0.8H from the outer edge, c) at a distance of 0.6H from the outer edge, d) at a distance of 0.4H from the outer edge, e) at a distance of 0.2 H from the outer edge, f) in the area of the outer edge

Figure 10. Dependence of the main stresses σ_3 on the inclination angle of the supporting surface α_P

The family of curves in Figure 10 reflects the dependence of the main stresses σ_3 on the inclination angle α_P of the supporting surface of the external propulsor blades. From the graphs in Figure 7, we conclude that with increasing angle α_P in the range from 0 to 25 degrees, the values of the principal stresses σ_3 shift slightly in the direction of compression, with further increase in the angle α_P, the stress values σ_3 shift towards stretching. In the range of angles α_P from 0 to 30 degrees, the main stresses σ_3 are weakly dependent on the angle α_P, at angles α_P greater than 50 degrees there is a sharp increase in the stress modules, the stresses are shifted towards the stretching. In the area of the leading edge of the external propulsor blades, positive stress values σ_3 occur, indicating that there are areas with triaxial compression in this area.

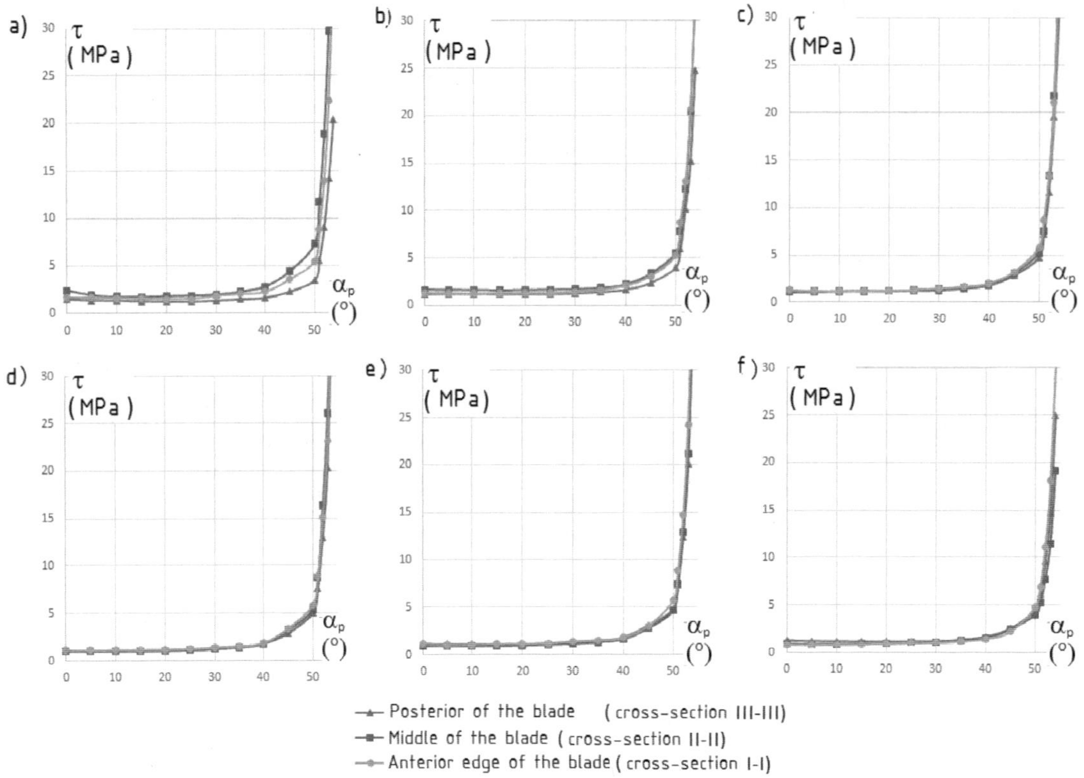

a) at the inner edge, b) at a distance of 0.8 H from the outer edge, c) at a distance of 0.6 H from the outer edge, d) at a distance of 0.4 H from the outer edge, e) at a distance of 0.2 H from the outer edge, f) in the area of the outer edge

Figure 11. Dependence of tangential stresses τ on the inclination angle of the supporting surface α_P

The family of curves in Figure 11 reflects the dependence of tangential stresses on the inclination angle of the supporting surface of the external propulsor blade. From the graphs in Figure 8, we conclude that with increasing angle in the range from 0 to 15 degrees, the tangential stress values decrease slightly, with further increase in the angle, the stress values increase. In the range of angles from 0 to 30 degrees, the tangential stresses are slightly dependent on the angle, at angles greater than 50 degrees there is a sharp increase in stresses. When moving from the leading edge of the blade to the rear, there is an insignificant increase in tangential stresses.

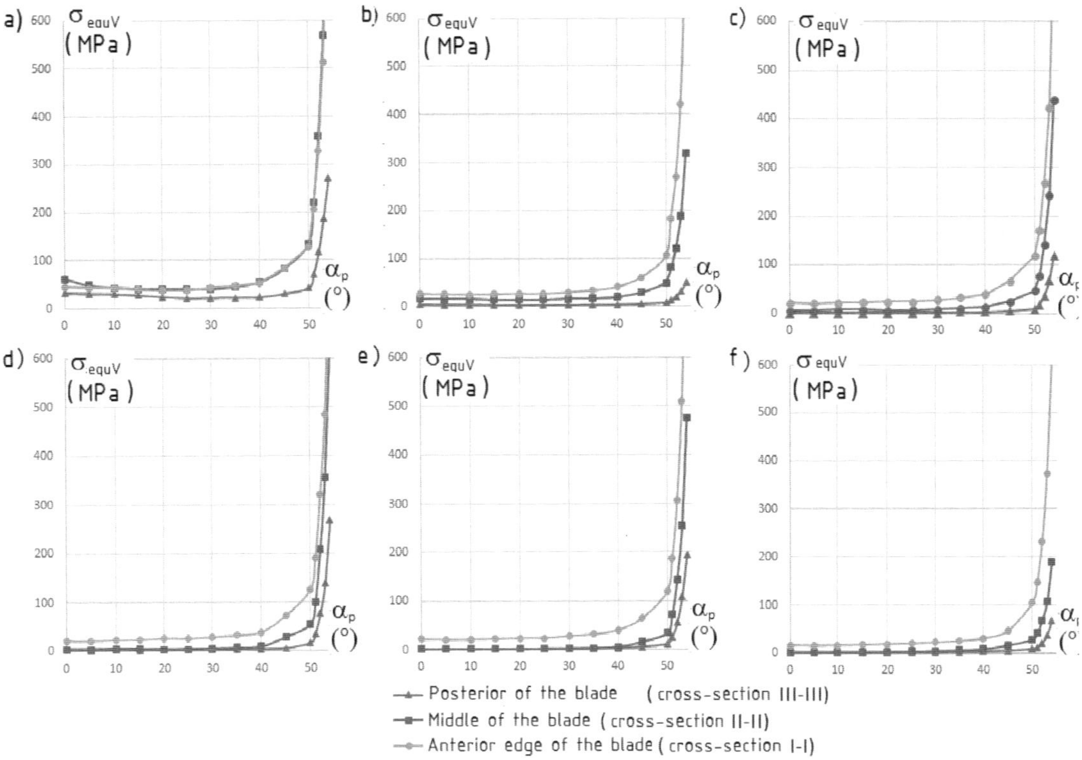

a) at the inner edge, b) at a distance of 0.8 H from the outer edge, c) at a distance of 0.6 H from the outer edge, d) at a distance of 0.4 H from the outer edge, e) at a distance of 0.2 H from the outer edge, f) in the area of the outer edge

Figure 12. Dependence of equivalent stresses σ_{equV} on the inclination angle of the supporting surface α_P .

The family of curves in Figure 12 reflects the dependence of the equivalent stresses on the inclination angle of the supporting surface of the external propulsor blades. From the graphs in Figure 12 we conclude that with increasing angle in the range from 0 to 20 degrees, the values of equivalent stresses decrease slightly, with further increase in the angle, the stress values increase. In the range of angles from 0 to 30 degrees, equivalent stresses are weakly dependent on the angle, at angles greater than 50 degrees there is a sharp increase in stresses. When moving from the leading edge of the blade to the rear, equivalent stresses grow.

The following conclusions can be drawn from the studies:

- when designing external geostatic propulsors it is advisable to take measures to shift the maximum loads to the front edge of the blade and partially unload the rear edge of the blade.
- the dependence of stresses on the inclination angle of the supporting surface is nonmonotonic in nature, i.e. there is an angle at which the crushing probability of the reference surface of the channel is minimal, this angle depends on the point position on the support surface;
- the use of inclination angles of more than 30 degrees is inappropriate, since it leads to an unconditional increase in stresses in the rock and interaction forces;
- with further research it is advisable to consider the range of angles from 0 to 30 degrees.

References

[1] Aksenov V V, Khoreshok A A , Efremenkov A B , Kazantsev A A , Beglyakov V Yu and Walter A V 2015 Sozdanie novogo instrumentariya dlya formirovaniya podzemnogo prostranstva [Creation of a new instrument for the underground space formation] *Gornaja tekhnika [Mining machinery (Catalog-reference book)]* (St. Petersburg) **1**(15) 24-26

[2] Begljakov V Yu and Aksenov V V 2012 *Poverhnost' zaboja pri prohodke gornoj vyrabotki geohodom [A surface of the face during the mining of the mine by the geokhod]* (Saarbrucken: LAP LAMBERT Academic Publishing GmbH & Co. KG) 139

[3] Aksenov V V, Sadovets V Yu and Beglyakov V Yu 2009 Influence of the dynamic processes formed in operating conditions on the power parameters of the knife executive body *Mining information-analytical bulletin* (scientific and technical journal). Vol. 10, **12**, 91-106.

[4] Aksenov V V, Beglyakov V Yu, Kazantsev A A, Walter A V and Efremenkov A B 2016 Opyt uchastija v proekte po organizatsii vysokotehnologichnogo proizvodstva [Experience in the project on the organization of high-tech production] *Mining equipment and electromechanics*. **8** (126) 8-15

[5] Beglyakov V Yu, Aksenov V V, Kostinets I K and Khoreshok A A 2017 Shemy nagruzhenija pri modelirovanii protsessa vzaimodejstvija vneshnego dvizhitelja geohoda prikonturnym massivom porod [Loading schemes for modeling the process of interaction of an external propeller of the geodatabase by an boundary rock massif of rocks] *Mining sciences and technologies* **3** 3-10

[6] Beglyakov V Yu, Aksenov V V, Kostinets I K and Khoreshok A A 2017 Determination of the interaction forces of the main systems of the geokhod with the geoenvironment and with each other *Mining sciences and technologies* **4** 23-30

[7] Aksenov V V, Kostinets I K, Beglyakov V Yu 2012 Influence of the inclination angle of the interaction surface between the executive part of the geokhod and the rock of the face on its stress-strain state *Mining information analytical bulletin* (scientific and technical journal) **2** 30-36.

[8] Aksenov V, Sadovets V Yu, Buyalich G D and Beglyakov V Yu 2011 Influence of the ledge on the deflected mode of the bottomhole part of the mine workings *Mountain Information and Analytical Bulletin* (scientific and technical journal) Ed. **2**: Mining machinery 55-67

[9] Aksenov V V, Efremenkov A B and Beglakov V Yu 2011 Modeling of the stressed-deformed state of the rock created by the action of the executive body of the mining machine on it *Mining information-analytical bulletin* (scientific and technical journal) S**5** 9-14.

[10] Aksenov V V, Beglyakov V Yu, Kazantsev A A, Kostinets I K and Koperchuk A V 2016 Classification of geometric parameters of an external propeller of a geokhod *Mining equipment and electromechanics* **8** (126) 33-39

[11] Aksenov V V, Kostinets I K, Beglyakov V Yu 2012 Substantiation of the need to create an external propulsion device for emergency rescue workings *Mining Information and Analytical Bulletin* (scientific and technical journal) S**6** 110-114.

[12] Aksenov V V, Kostinets I K, Beglyakov V Yu 2013 Work Features of the geokhod external propeller *Mining information-analytical bulletin* (scientific and technical journal) S**6** 419-425

[13] Gallagher R 1984 *Metod konechnyh `elementov. Osnovy* [*The finite element method. Fundamentals]* (Moscow, Publ Mir [The World]) 428

[14] Alyamovskiy A A 2004 *SolidWorks. Engineering analysis by the finite element method* - Moscow: DMK Press] 432

[15] Fadeev A B 1987 *Metod konechnyh `elementov v geomehanike* [*Finite Element Method in Geomechanics]* (Moscow, Publ Nedra) 221

[16] Beglyakov V Yu, Aksenov V V, Kazantsev A A, Kostinets I K 2017 Development of a perimeter support-propulsion system of geokhods *Bulletin of the Kuzbass State Technical University* **6** (124)

[17] Aleksandrov A V, Potapov V D and Derzhavin B P 2009 *Soprotivlenie materialov* [Strength

of materials] *(*Moscow: Higher School) 560

Use of the superplasticity effect of materials in engineering technology

E G Berdichevskij

Yaroslav-the-Wise Novgorod State University, ul. B. St. Peterburgskaya, 41, 173003 Veliky Novgorod, Russia
E-mail: bersev@mail.ru

Annotation. Technological aspects of using the effect of superplasticity in the manufacture of many mechanical engineering products by pressure treatment methods are considered. Superplasticity, which manifests itself under special conditions, allows the formation of complex forms of specific materials with heat and deformation resistance, hardness and brittleness, and atypical physical and technical properties at low costs.
The conditions and factors of the manifestation of superplasticity effect for traditional and innovative materials, such as metallic glass and nanoceramics, are revealed.
New mold forming technologies that do not require press equipment are considered, and the areas of optimal use of the capabilities of the superplastic state of materials are specified.

The leading processes in the manufacture of many parts are punching materials by pressure (PMP). However, traditional PMP is often resource-intensive; it requires special measures to combat hardening to reduce friction in the deformation zone [1]. In addition, critical parts are made of materials and alloys with special properties. These materials, as a rule, are low-ductile and deformation resistant, which forces technologists to use alternative, but high-cost foundry technologies.

The effect of superlasticity of materials is the resource of increasing the applicability of PMP technology in engineering. The present work is devoted to the analysis of main aspects of this problem.

By superplasticity we mean nonlinear viscous behavior of a number of materials possessing a hyperfine grain microstructure [3].

The state of superplasticity can be determined by a combination of characteristics:

1) Increased sensitivity of the yield stress of materials to the change in the rate of deformation;

2) Extremely low value of strain hardening;

3) Very high resource of deformation ability;

4) The yield stress of material in superplasticity state is several times less than the yield point, which characterizes the plastic state of a particular material.

The effect of superplasticity in the case of PMP manifests itself under certain conditions, such as the structural state of the deformable metal, metal temperature, and the rate of deformation. Most often, superplasticity manifests itself at the temperature (0.4-0.8) T melting. The rate of deformation is between $10^{-5} - 10^{-3}$ sec^{-1}

The effect of superplasticity has two varieties: structural and phase. The use of phase superplasticity is under study. Structural superplasticity has been studied quite well for practical use.

To realize the structural superplasticity, the treated metal must be given a special equiaxial fine-grained structure. The grain size does not exceed 10 microns in diameter. In the process of heating to the PMP temperature, the grains must be resistant to growth.

Alloys can be converted into superplastic state by preliminary thermal or thermo-mechanical treatment.

The temperature of the manifestation of superplasticity effect must be maintained constant over the volume of the deformable metal during the entire processing period to ensure even flow of the material.

The advantages of using the effect of superplasticity in engineering technology are the following:

1) PMP of low-plastic and deformation resistant alloys based on Ni, Ti, Al, Fe, Mg becomes possible. It is possible to process the part in one technological transition.

2) It is possible to forge details of particularly complex shape (thin-walled parts, details with ribs, details with relief texture, etc.).

3) The forging force and the power of the equipment used are reduced.

4) The quality of the finished product is improved due to the better filling of the stamping die, the dimension accuracy is improved, the surface defects are reduced, and the microgeometry of the surfaces of the proccessed parts is improved.

5) The cost of manufacturing products with the use of superplasticity effect is reduced by 30-40% compared to punching metal in ordinary state.

Figure 1 shows a comparative analysis of the effectiveness of various types of processing in the manufacture of parts of complex shapes [4]. In the example considered, the stamping slopes are 1^0. The coefficient of metal use is 0.8, whereas for stamping under normal conditions it is 0.2-0.5.

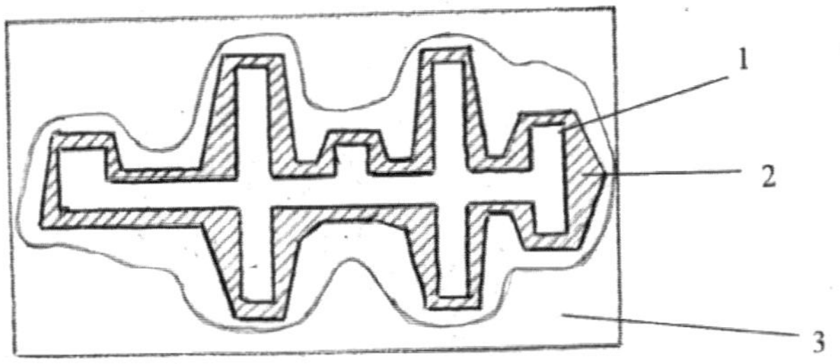

Figure 1. Comparative analysis of the efficiency of the superplasticity process: 1-component; 2 – PMP under superplasticity conditions; 3 - casting.

Along with the above mentioned advantages, innovative technologies that do not require press equipment are implemented together with the effect of superplasticity. Such technologies include free-flow drawing, gas-static molding from sheet material, thermoelastic stamping.

Free-flow drawing based on superplastic behavior of the material combines loading only with tensile forces with localized heating. The tubular blank (Figure 2) is clamped from one end, and the other end is pulled at a controlled speed V2.

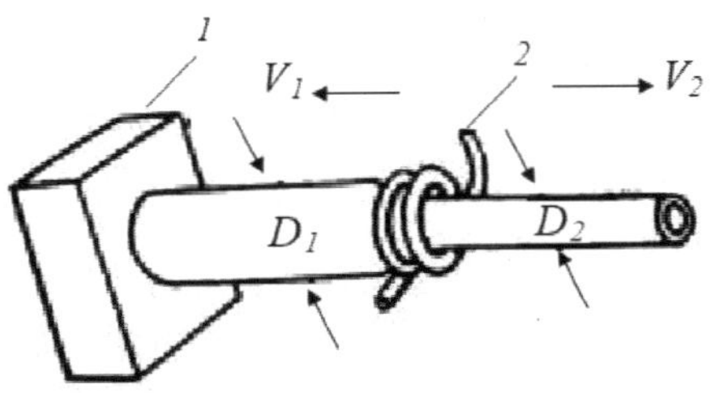

Figure 2. Scheme of the free-flow drawing: 1 - fixture; 2 - inductor

The inductor heats the localized section of the pipe to the superplastic state and moves with velocity V_1. The resulting reduction is determined according to the formula:

$$(D_1 / D_2)^2 = (V_2 / V_1) - 1$$

In one pass, a reduction of up to 50% was achieved. Multiple passes allow to obtain large total reductions. This scheme is promising for cases when it is necessary to preserve the axial symmetry in the reduction of blanks [4]. Similarly, sheets or plates of any given size can be rolled with high precision, while for local heating during deformation the contact resistance in the deformation zone can be used.

The scheme of gasostatic molding of hollow products from sheet material is shown in Figure 3.

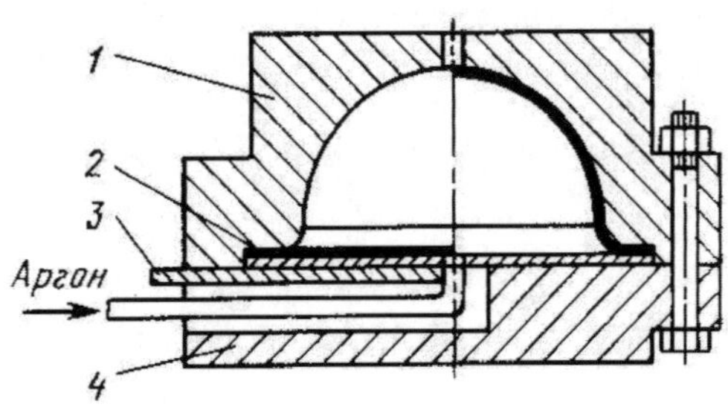

Figure 3. Scheme of superplastic pneumatic shaping of hemispheres: 1 - matrix; 2 - blank; 3 - valve; 4- Clamping cover

Forming of the sheet blank is carried out under the influence of the difference in pressure created on both sides of the workpiece. The sheet blank 2 of the superplastic material is rigidly clamped between the clamping cover 4 and the matrix1 and heated together with the die to the desired temperature. Then, gas (argon, carbon dioxide, compressed air) is fed into the cavity of the lid 4 at a pressure of 0.1-0.5 MPa. Under this pressure, the blank is expelled into the cavity of matrix 1 and then fills the whole space.

The innovative complex technology of isothermal punching of high-precision products from super heat-resistant alloys is described in [6]. Unorthodox technique for forming the alloy structure by high-gradient directional crystallization is proposed. When developing such complex technologies, it is necessary to use the construction of analytical models [7].

An example of complex optimization of the modes of superplastic forming of shells from titanium alloy VT6 is given in [8].

A significant number of parts and elements of precision instrumentation is made of ceramic materials, amorphous alloys, composites on a ceramic matrix. Superplastic ceramic materials with a stable submicron grain are of particular interest. The material exhibits superplasticity under certain temperature-velocity strain conditions during technically acceptable time.

There are two types of superplastic ceramics: single-phase ceramics and composites. The principal difference between the superplastic deformation of metals and ceramics is the required size of the structural constituents of the material. For ceramics, the grain size should be 0.1 - 1.0 μm, i.e. an order of magnitude less than for most superplastic metals. The preparation of ceramic blanks for subsequent superplastic deformation is carried out from nanocrystalline powders by high-temperature consolidation.

The physical-and-technical properties of products made from superplastic ceramics are greatly influenced by the process of evaporation which occurs during plastic deformation.

In superplastic deformation in ceramics, powder formation is most pronounced at higher strain rates, and in metals, on the contrary, at the lowest rates.

Amorphous alloys are among modern and very promising materials. A large group of amorphous alloys characterized by a low rate of amorphization form a special kind of glass-forming systems called metal glass. These materials have excellent mechanical, magnetic and anti-corrosion properties, but are extremely fragile. At the same time, they behave like superplastic materials in a super cooled liquid state.

Some of these materials can be obtained in the form of dimentional amorphous blanks. In supercooled liquid state they are characterized by very low viscosity and high deformability which can be used for punching products of complex shape using conventional PMP technologies.

When evaluating the ability of amorphous alloys for superplastic deformation, two features must be taken into account, namely: glass formation ability and the stability of the amorphous state. Glass-forming ability is characterized, mainly, by the critical cooling rate of the melt and the relative glass-transition temperature.

Metallic glass is in a supercooled liquid state at a temperature above the glass transition temperature. This state is internally equilibrium and metastable regarding the crystallization process. In this state, they exhibit a linearly viscous flow over a wide range of deformation rates. This leads to a very high deformation capacity of the materials. Experimental samples achieved elongation of about 20,000% at 200^0 C and deformation rate of 5×10^5 sec $^{-1}$ [8].

The field of application of metallic glass is still limited by the fact that they can be obtained by rapid cooling from the liquid state only in the form of thin ribbons below 60 μm thick and not more than 200 mm wide or a wire with a diameter of 0.5-20 μm. However, there are broad prospects for the development of materials of this group. Soft magnetic amorphous alloys are already widely used in various industries.

Conclusions

The effect of superplasticity of many materials widely used in engineering and instrument making can be fundamental for resource-saving technologies based on plastic deformation.

References

[1] Berdichevskij E 2011 Trenie i smazka pri plasticheskoj obrabotke metallov *(Friction and lubrication in the plastic processing of metals: monograph)* (Velikij Novgorod)

[2] Saprykina N A, Saprykin A A and Matrunchik M S 2014 Formation of Surface Layer of Cobalt Chrome Molybdenum Powder Products with Differentiation of Laser Sintering Modes *Applied Mechanics and Materials* **682** pp 294-298

[3] Tihonov A S 1978 Effekt sverhplastichnosti metallov i splavov *(Superplasticity of metals and alloys)* (Moskva: Nauka)

[4] Kokorin V, Titov Y, Tamoverov V and Fedorov L V 2006 Special'nye sposoby obrabotki metallov davleniem *(Special methods of metal forming by pressure: textbook)* (Ul'yanovsk: ULGTU)

[5] Kablov E, Osvennikova O and Lombert B 2012 Kompleksnaya innovacionnaya tekhnologiya izotermicheskoj shtampovki na vozduhe v rezhime sverhplastichnosti diskov iz superzharoprochnyh splavov (Integrated innovative technology of isothermal stamping of disks made from super-high-temperature alloys in superplasticity mode) *80 let. Aviacionnye materialy i tekhnologii* (Moskva: VIAM) pp 3-23

[6] Chumachenko E, Portnoj V K and Logashina I 2014 Mekhanicheskie ispytaniya i postroenie analiticheskih modelej povedeniya materialov v usloviyah sverhplastichnosti *(Mechanical tests and construction of analytical models of materials behavior under superplasticity conditions)* **1** (Moskva: Metallurg) pp 68-71

[7] Zung N and Pol'kin V 2014 Optimizaciya rezhimov sverhplasticheskoj formovki obolochek iz titanovogo splava (Optimization of regimes of superplastic forming of shells made of titanium alloy) VT6 *Tekhnologiya legkih splavov* 1 pp 91-96

[8] *Metallicheskie stekla* (Metallic glass) Perevod s angl 1984 (Moskva: Mir)

IOP Conf. Series: Materials Science and Engineering **441** (2018) 012010 doi:10.1088/1757-899X/441/1/012010

Friction Process Control in Forging and Hot Stamping Technologies

E. G. Berdichevsky

Yaroslav the Wise Novgorod State University, ul. B. St. Petersburgskaya, 41, 173003 Veliky Novgorod, Russia

E-mail: bersev@mail.ru

Abstract. Theoretical and applied aspects of friction and lubrication process control in leading engineering technologies of metal forming have been considered. Innovative modifiers of friction for process lubricants in forging and hot stamping have been suggested. The efficacy of pyro- and nanomodifiers has been confirmed experimentally. Compositions of lubricants as well as the ways of studying tribological phenomena in hot metal deformation have been suggested.

1. Formulation of the problem

The purpose of the given paper is to determine scientifically the friction and lubrication process control in hot metal finishing. Our arguments are founded on modern tribological concept of multi-level area of frictional contact between an instrument and a product being deformed [1] as well as structural and energetic model of friction and wear-out processes [2].

There are three large-scale levels in an extremely heterogeneous contact friction area: macrolevel (10^{-3} m), mesolevel (10^{-6} m), nanolevel (10^{-9} m). Despite the fact that a distinct border between the levels is difficult to draw, each level possesses its own specific pattern. The interaction of processes taking place at different large-scale levels should not be excluded.

The influence of friction and technological lubricants (TL) is specific at each large-scale level. Table 1 gives the results of isolation of main TL properties, which determine the tribological effect at each large-scale level. Main functional TL properties are divided into two groups: individual and contact. Individual features characterize the physical and chemical properties of TL, which can be determined beyond the friction act and plastic deformation. These properties depend only on composition, structure and chemical nature of TL. The contact properties of TL manifest themselves interacting with the surfaces of participants of frictional and plastic contact.

We may suggest that an optimal TL is a composition demonstrating a positive effect at each large-scale level. The majority of TL used are intended to achieve a tribological effect at the macrolevel. However, the dispersity of TL components has often been ignored. For example, molybdenum disulfide or graphite preparations widely used in the processes of forging and hot stamping can demonstrate high antifrictional and anti-wear activities only in the colloidal state [3].

So, when choosing innovative TL, you should consider their tribological efficacy at three large-scale contact levels, in the first place - at meso- and nanolevels. Also, you should bear in mind that components producing necessary properties at any large-scale level may interact both antagonistically and synergistically [2].

Table 1. Main tribological TL properties demonstrated at different large-scale levels

Large-scale level	Individual properties	Contact properties
macro	density, viscosity, surface tension, pH and	contact angle of wetting, spreading,

	electrical conductivity (for water solutions), dielectric constant, thermal conductivity, thermal capacity, thermal diffusivity	heat transfer coefficient, thickness of original lubricant film
meso	volatility, solubility of gases, viscosity piezoelectric coefficient, thermal stability, chemical potential	microcapillary properties, adhesion work, adsorptive and chemisorptive activity, activation energy for formation of chemically modified layers, nanostructure of subtle layers at interfaces;
nano	dispersity of initial nanocomponents	activation energy of nanostructures

2. Experimental research

The variation of TL functional properties is easy to perform by means of so-called friction modifiers presented with additives for TL of various chemical and phase composition.

In Novgorod State University, we have isolated two kinds of friction modifiers for standard main bases of TL: pyromodifiers and nanomodifiers. In general, an optimal TL contains simultaneously both modifiers, provided that they are not antagonistic but, on the contrary, can interact synergistically.

Pyromodifiers of friction influence the process of thermal degradation (pyrolysis) of TL components in the area of plastic deformation during forging or hot stamping. As the result of pyrolysis, numerous volatile components are formed, producing high gas-static back pressure in microcavities of stamp (hammer) surfaces and billet. The pressure in the gas interlayer composed by volatile components can reach up to 100 mPA (Figure 1). In this case, within the processing area, normal and tangent tensions decrease significantly and so does the coefficient of plastic friction.

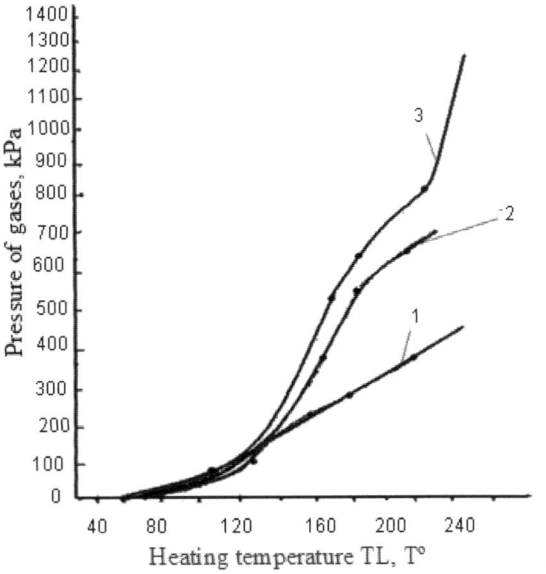

Figure 1. Investigation of the gaseovs ability of lubricants:
1 – Aluminol, 2 – Pressol KF, 3 – Grafitol E21

Volatile components of pyrolysis possess high chemical activity as they result from thermodestruction of high-molecular chains. They are instantly adsorbed by resultant fresh-exposed microareas of contact surfaces produced by friction, the formation of a protective layer following.

Solid coke-like sediment produced by pyrolysis can play the role of a solid lubricating film shielding the working surface of a stamp (hammer). The forming of a solid coke-like film during the TL pyrolysis is influenced by the nature of metallic contacting surfaces. By their ability to form a film, metals can be ranked as follows: Fe>Ni>Ti>Zr>Cu>W. Now it is clear why stamping of stainless steels containing significant amount of chromium is followed by high wear-out of the stamp and low efficacy of standard TL.

We suggest using MCC-1 microcellulose and technical cellulose of various brands as a friction pyromodifier.

The laboratory testing of pyrolytic activity of MCC-1 microcellulose has shown that it can significantly soften severe contact friction regimes, if it is added into synthetic solutions and emulsions (through emulsol) in the form of powder. Bearing capacity of the lubricating layer increases by 40%, its ultimate strength – by 80%. Load wear index increases by 14%. Combined with 2% MCC-1 microcellulose, antifrictional properties of compositions change insignificantly. An optimal result can be obtained when a standard antifrictional additive – colloidal graphite (2%) with microcellulose (2%) – is used in a standard emulsion. In this case, a synergistic effect between the components is evident (Table 2).

Table 2. Tribological characteristics TL with pyrolitic active components

TL	$P_к$, H (critical load)	P_c, H (welding load)	I_z (index tease)	Coefficient of friction
Aqueovs dispersion of graphite – 3%	63	126	36	0,177
Aqueovs dispersion of graphite (3%) and MCC-1 (2%)	92	240	68	0,163
Aqueovs dispersion of MCC-1 – 2%	85	200	52	0,171
Aqueovs dispersion of emulsol – 3%	56	105	44	0,155

The next group of friction modifiers is that of nanomodifiers sprung to life by achievements of nanotechnologies. Nanomodifiers are successfully used as the components of motor and transmission oils. We can also remember the positive influence of TL enriched by nanoproducts in the processes of hot stamping.

Specific features of nanoproducts include small dispersion determining their high chemical and responsive activities. Nanoparticles can play the role of catalysts of unique chemical reactions proceeding with superhigh speed. This characteristic gives unpredictable properties to nanomaterials. For example, under certain conditions nanomaterials can organize themselves, i.e. atoms act together to form regulated structures [4-8].

Nanoclustered carbon materials, such as fullerenes, fullerene soot and fullerene black, carbon nanotubes, nanofibers, etc., may be applied practically.

Fullerenes, a new form of carbon, underlie the majority of technical nanomaterials. The most common is fullerene C_{60}, a molecule composed of 60 carbon atoms forming a closed spherical surface made up of regular hexa- and pentagons. The main characteristic of fullerenes is their increased responsive activity. They catch atoms of other substances easily and form materials with completely new attributes.

Considering the high cost of fullerene C_{60}, manufactured in Russia, fullerene soot (FS) and fullerene black (FB) have been tested as TL nanomodifiers for hot stamping. There is a concept that suggests inhibiting friction and wear-out at all three large-scale levels without fail. In order to prove this, a complex of oil additives IS-12 was tested. The oil was alloyed with nanomodifiers as well as standard additives: graphite, molybdenum disulfide, chloroparaffin and zinc dialkyldithiophosphate (DF-11).

The results of tribological and bench tests are given in table 3.

3. The discussion of the results

The table demonstrates that the tribological efficacy of fullerene black gives little to that of fullerene soot. $P_к$ (critical load) and P_c (welding load) of fullerene black and molybdenum disulfide combined exceed largely the indices of the oil with fullerene soot only. Welding load is impacted by nanocarbons most of all. FS and FCh additives literally break off the catching of balls. Frictional characteristic of oils alloyed with nanocarbons almost doubles that of oils with graphite, and is 80 % higher than that of oils with molybdenum disulfide.

Table 3. Tests of nano-containing lubricating components

Lubricating composition	Tribological characteristics			
	$P_к$, H (critical load)	P_c, H (welding load)	Wear Scar Diameter, mm	Coefficient of friction
Aqueous dispersion of graphite – 3% Aerosol spray	220	310	5,2	0,38
Aqueous dispersion FS – 3% Aerosol spray	280	450	4,6	0,32
Aqueous dispersion FCh – 3% Aerosol spray	270	420	4,7	0,32
Aqueous dispersion of FCh (2%) and graphite (3%) Aerosol spray	310	480	4,1	0,28
Oil IS–20 +5% MoS_2	450	730	3,7	0,28
Oil IS–20 +5% graphite	410	640	3,9	0,32
Oil IS–20 +3% FS	630	920	2,8	0,22
Oil IS–20 +3% FCh	600	990	2,9	0,24
Oil IS–20 +3% FCh + 3% MoS_2	710	1100	1,7	0,18
Oil IS–20 +3% FCh + 3% MoS_2 +2% chlorparaffin + 2% DF–11	780	1220	1,5	0,12

In the aqueous dispersions, nanocarbons do not influence significantly the coefficient of friction (an exception is the composition of FCh + graphite), but they increase largely the welding load. Frictional characteristics of the compositions are improved markedly, when standard lamellar additives are replaced with nanocarbons. FS and FCh in aqueous dispersions are practically interchangeable.

Since FCh is much cheaper than FS, it can be considered to be more promising and up-to-date friction modifier. The composition of molybdenum disulfide, fullerene soot, chlorparaffin and complex phosphate has been proved to be extremely effective in severe regimes of boundary friction. This composition can serve as a base for developing a series of innovative technological lubricants.

Since FS and FCh powders are unstable and tend to agglutinate, they have been supposed to be used in a microcapsular form. When being tested in industry, microcapsules were manufactured by means of phase separation in a gelatinous shell with 300-500 mcm diameter. Dispersion from microcapsules can be easily performed with a standard aerosol spray. Microencapsulation allows saving up to 30 % of nanocarbon materials.

References

[1] Berdichevskij E 2011 *Trenie i smazka pri plasticheskoj obrabotke metallov* (Velikij Novgorod) (Friction and lubrication in metal plastic finishing).

[2] Pogodaev L and Kuz'min V 2006 *Strukturno-ehnergeticheskie modeli nadezhnosti materialov i detali mashin* (Sankt-Peterburg: Akademiya transporta Rossijskoj Federacii) (Structural and energetic models of material reliability and machine details).

[3] Berdichevskij E 2010 Modifikatory treniya kak osnova smazochnyh kompozicij dlya metalloobrabotki *Fundamental'nye i prikladnye problemy tekhniki i tekhnologii* **5** pp 43-46 (Friction modifiers as base of lubricating compositions in metal finishing *Theoretical and applied problems of technics and technology).*

[4] Bokarev D, BakuninV, Kuz'min G and Parenago O 2004 Vysokooehffektivnye modifikatory treniya na osnove nanorazmernyh materialov *Himiya i tekhnologiya topliv i masel* **4** pp 28-30 (Highly effective friction modifiers based on nanoscale materials *Chemistry and technology of fuels and oils.*

[5] Malushin1 N N, Valuev D V, Il'yaschenko D P, Gizatulin R A, Nekrasov R Yu 2018 Improvement of Research Procedures into the Highly Rigid Welded Metal Using Thermal Microscopy Facilities *J. Materials Science Forum* **927** 148-152

[6] Valuev D V, Danilov V I, Ivanov V I 2007 On the causes of formation of non-optimal structures in a pressure-treated low-carbon steel *J. Russian Physics Journal* Issue **10 (50)** 964-968.

[7] Kryukov R E, Kozyrev N A, Galevsky G V, Bendre Y V, Goryushkin V F, Valuev D V 2015 Some aspects of oxidation-reduction reactions under carbon-bearing flux welding *IOP Conference Series: Materials Science and Engineering*. **91**

[8] Malushin N N, Valuev D V, Valueva A V, Serikbol A 2014 Kinetic study of the effect of plasticity and its role in stress relaxation in the weld speed steel during the martensitic transformation *Applied Mechanics and Materials* **682** 58-63

Optimal selection of the parameters of the kinematic chain of the main drive of the cold rolling mill

Yu B Chechulin[1], S K Buynachev[1], A V Bakin[1]
[1]620002, 19 Mira street, Ekaterinburg, Russia Ural Federal University, Russia
email: uchechulin@mail.ru

Abstract. The article is devoted to the choice of optimal parameters of the system of dynamic balance of loads in the mechanism of the cyclic action of the drive of the working stand of the cold rolling mill.

Introduction

The balancing of dynamic loads in the main drive line of cold pilgrim rolling mills that arise as a result of large accelerations in the extreme positions of the work stand during its reciprocating motion is the main reason for limiting their speed [1-4]. The share of the main drive mechanism accounts for up to 65% of the total number of failures of cold rolling mills (CPT) [5]. The solution of the problem of increasing the structural and operational reliability requires the systematization of general patterns of changes in internal loads, the establishment of functional dependences of the conversion of power flows, and the parametric control of their values. The resulting loading of the main drive system of the mill is determined by the technological and inertial loads arising during the high-frequency reciprocating motion of the work stand [1]. When using the mechanisms of dynamic balancing, the solution of this problem is to determine their rational parameters, taking into account the phase shift of the technological and inertial components.

Depending on the size of the mills, various variants of balancing of inertial loads are used [1]: the type of the main moment balancing by placing the rotating counterweight on the drive shaft of the crank-slider mechanism; the balancing system of the stand with a mass m1, complemented by a synchronous rotating mass m_3, which allows to reduce the values of the rotating counterweights; the balancing system of the stand drive by an additional load, whose crank is offset relative to the crank of the stand by 90^0.

Let us consider the solution of the optimal choice of balancing parameters for the lever-type mechanisms on the drive crank in the vertical plane used to convert the rotational motion of the main drive shaft into reciprocating motion of the work stand (Figure 1).

The lengths of connecting rods are the same 1.75 meters. The deaxial of the balancing chain is absent, and in the chain of the stand it is 0.15 meters. The radius of the drive of the OA stand is 0.25 meters. The radius of the balancing circuit is more than 0.3 meters for design reasons.

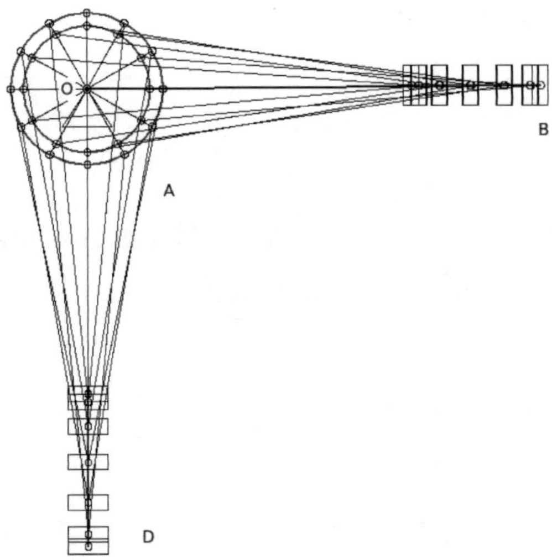

Figure 1. Kinematic chain of the main drive of the cold rolling mill

The equation of motion of the machine aggregate is represented by the equation

$$J\ddot{\varphi} + \frac{1}{2} J'_{\varphi}\dot{\varphi}^2 = M_d - M_C - M_S,$$

where

J и J'_{φ} - The reduced moment of inertia and its derivative with respect to the angle of rotation;

$\ddot{\varphi}, \dot{\varphi}, \varphi$ - angular acceleration, angular velocity and crank angle;

M_d - motor torque;

M_C - The given moment of technological loading;

M_S - reduced moment of the weight of the balance weight.

Equalizing the torque on the crank will reduce load and speed fluctuations. Ideally, the acceleration of the crank will be assumed to be zero, and then

$$\Im = \int_0^{2\pi} (M_d)^2 d\varphi,$$

$$M_d = M_I + M_C + M_S,$$

It means $$\Im \to \min_p ,$$

where $p = (m_g, beta, e, l)^T$.

Here $J'_{\varphi} = \vartheta^T \Theta \vartheta_{\varphi}$ is a quadratic form in the coordinates

$$\vartheta, \vartheta^T = (\vartheta_{Bx}, \vartheta_{Dy})^T.$$

Coordinate of the stand $\vartheta_{Bx} = R\cos(\varphi) + \{L^2 - [R\sin(\varphi) - E]^2\}^{0.5}$,

Coordinate of cargo $\vartheta_{Dy} = r\sin(\varphi + \beta) - \{l^2 - [r\cos(\varphi + \beta) - e]^2\}^{0.5}$,

m_g– balancing weight,

beta– the difference of the turns of the cranks,

e–dexial of the movement of the counterbalancing load,

l–length of connecting rod of counterbalancing load.

Vector $\vartheta_{\varphi} = \dfrac{d\vartheta}{d\varphi}$ - is the derivative ϑ of the coordinate of the crank angle

The matrix of a quadratic form is diagonal $\Theta = \begin{bmatrix} m_k & 0 \\ 0 & m_g \end{bmatrix}$.

Let us consider the limiting case for cold rolling. For such a stroke length, the frequency ω=25 radians per second. The weight of the stand is 1450 kg. The given moment of the technological load will be considered as the maximum.

The interval boundaries for all parameters are chosen from the design features of the equipment.

The moment of inertia forces of the second kind M_I= has a constant phase and amplitude at a given speed and mass of the counterbalancing load, but the moment of inertia changes the phase due to the angle β and the difference between the stand mass and the counterbalance load. The moments M_C and M_S on the interval [0:2π] remain at fixed positions. Parameters are selected taking into account the previously assigned values that ensure maximum loads.

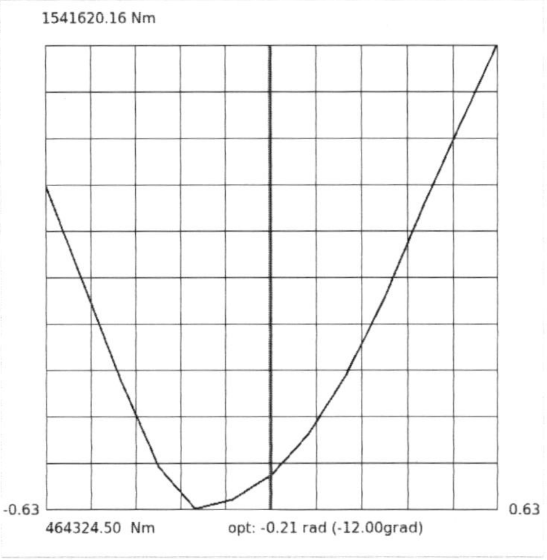

Figure 2. The dependence of the moment on the difference in the angles of rotation of the cranks of the crane drive and the balancing weight

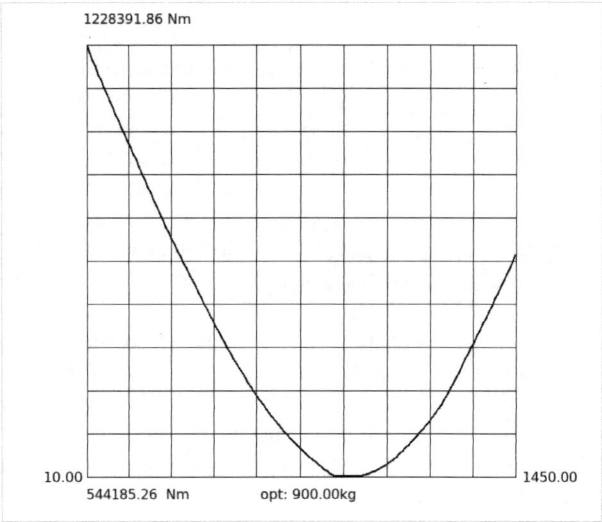

Figure 3. The dependence of the moment on the mass of the balanced cargo All the parameters of the balanced circuit are related to each other. For the values obtained, the best connecting rod length of the counterbalancing load is 1.1 meters.

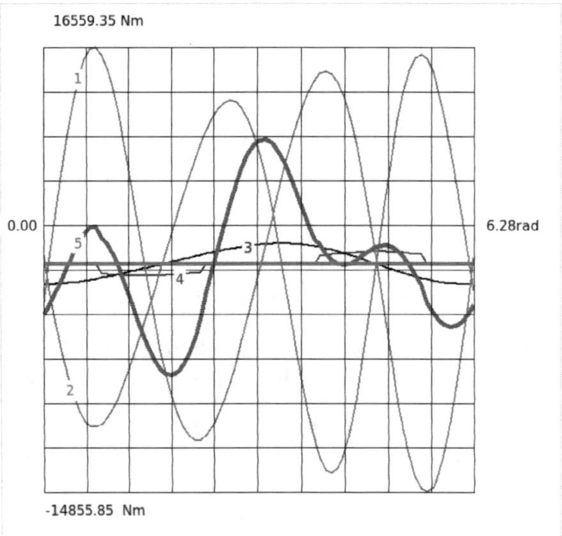

Figure 4. Kinematic diagram with the selected parameters and graphs of the moment dependencies on the selected parameters (1 is the moment of inertia of the second kind of stand, 2 is the moment of inertia of the second kind of the counterbalancing load, 3 is the static moment of the counterbalancing load, 4 is the technological moment, and 5 is the total moment).

Conclusion

All points are brought to the OA crank. Depending on the speed of the mill and the magnitude of the technological load moment, the values of the parameters change. Obviously, the weight of the counterbalancing load and the angle β had the greatest influence on the change of the engine torque. Less other parameters are affected by the length of the connecting rod and the de-axial chain of the counterbalancing load.

References

[1] Y Chechulin, L Kondratov, G Orlov 2017 *Cold rolling of pipes*, (M. Metallurgizdat)
[2] R Adamia 1978 *Optimization of dynamic loads of rolling mills*., (M. Metallurgy)
[3] V Frolov, V Danchenko, Y Frolov2005 *Cold pilger rolling of pipes*, (Dnepropetrovsk: Thresholds)
[4] A Nazarenko 1961 *Mathematical modeling and investigation of the dynamics of the crank-connecting rod drive of cold rolling mills* **3** (Tr. VNIIMETMASH)
[5] V Grebennik, V Tsapko 1980 *Reliability of metallurgical equipment*, (Moscow: Metallurgy)

Increasing the working efficiency of the forholler of the pilger rolling mill

Yu B Chechulin[1], **Yu A Popov** [1]
[1]620002, 19 Mira street, Ekaterinburg, Russia Ural Federal University, Russia
email: uchechulin@mail.ru

Abstract. The article examines the issues of increasing the reliability and efficiency of the feeding mechanism (forholler) of Piler rolling mill. A new design of the trolley and the piston system of the brake chamber is proposed.

Introduction

Pilger mills of periodic rolling [1] are used for the production of hot-rolled seamless pipes of large diameter. The process of deformation is carried out on the mandrel in rolls with a variable radius of the caliber (Figure 1) rotating opposite to the direction of supply of the sleeve (workpiece).

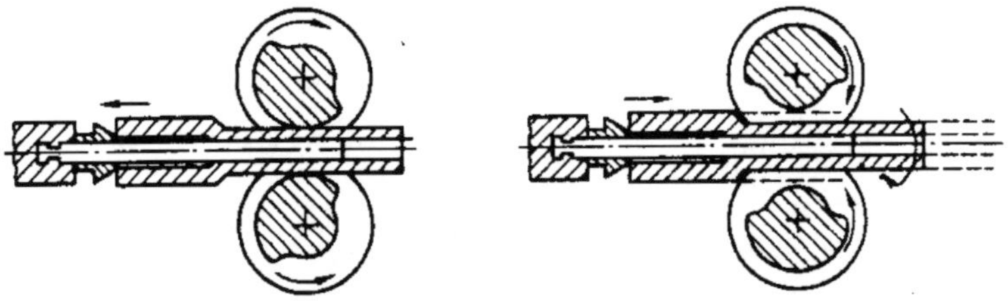

Figure1. The scheme of hot pilger rolling

During each revolution of the working rolls, the deformable sleeve makes a return movement to the length of the working cone (up to 1500 mm). The end of the reduction cycle is accompanied by an increase in the radius of the calibers, which ensure that the remaining angle of rotation of the rolls allows the workpiece to return to the deformation zone by sliding it in the direction of the outlet of the finished pipe by an additional one-time feed (30-60 mm). The procedure for the return and delivery of the workpiece, as well as its canting around its own axis at an angle of up to 900, is performed by a feeding device, the so-called forholler. Forholler is the most responsible mechanism of the pilgrim mill, the reliability of which mainly determines the uninterrupted operation of the equipment and the quality of the pipes produced. The return movement of the billet together with a mandrel weight of up to 10 tons for a distance of up to 1500 mm in a short time interval not exceeding 0.5 s is accompanied by large dynamic loads. This explains the continuous attempts to further improve the design and search for optimal parameters of the law of motion of feeders [2]. The problematic issues of operation of the gun are:

Large dynamic loads during the half-cycle of the cyclic return of the workpiece to the rolling zone, the duration of which does not exceed half the half-cycle of the single-reduction cycle

IOP Conf. Series: Materials Science and Engineering **441** (2018) 012012 doi:10.1088/1757-899X/441/1/012012

- the need for rapid acceleration and soft braking of the piston system at the end of the return movement of the workpiece, i.e. achieving a minimum, close to zero final speed and acceleration, excluding the removal of the liner from the mandrel in front of the compression bar and the cast of the workpiece relative to it;
- ensuring low-cycle strength at the final extraction of the mandrel from the liner and fatigue strength at high-frequency cyclic loading of the screw pusher, its movable traverse and the details of their connection.
- achievement of strict synchronization of the end of the workpiece movement with the moment of its gripping by working rolls, excluding both premature stopping and moving it during rolling;
- problems of sealing large-sized moving parts of the gun and their fatigue life are among the hard-to-solve problems.

At the present time, the most widely used are feeders with an air-hydraulic propulsor. When rolling in each single cycle of reduction, the workpiece moves opposite to the direction of exit of the finished pipe, ensuring a multiple increase in the air pressure in the chamber of the pneumatic cylinder *11* (Figure 2), designed to accelerate it during the return movement during the opening of the curved creepers of constantly rotating working rolls *1*.

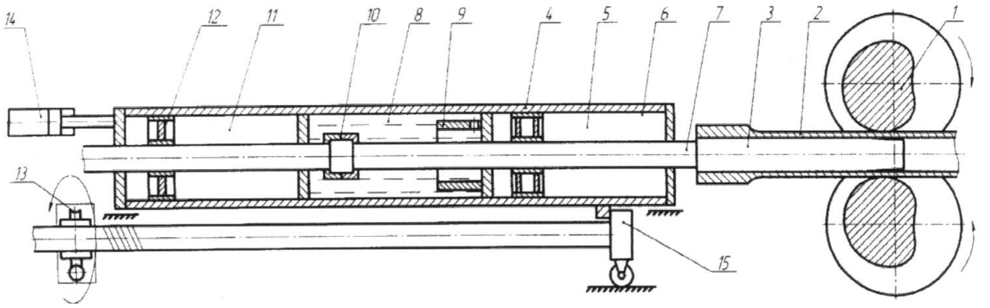

Figure2. Scheme of movement of the pistons of the forholler during crimping of the workpiece

(1 – stand of pilger mill; 2 - rolled pipe; 3 – mandrel; 4 – body of forholler; 5 - guide chamber; 6 - guide piston; 7 – stock; 8 – hydrocamera; 9 - brake-axle; 10 - hydraulic piston; 11 - air chamber; 12 - air piston; 13 – feeder; 14 - clamping cylinder; 15 – traverse)

To guarantee the stopping of the return movement of the workpiece, before using the working part of the crochets, it is used brake devices of hydraulic type. They include a brake cylinder and a periodically inserted piston *10* with a constant or alternating diametral clearance with respect to the mirror of the brake sleeve (chamber), as well as a system of adjustable throttling outlets [3].

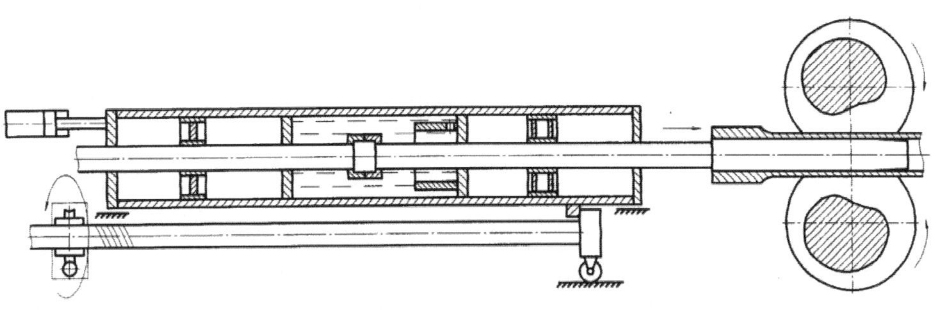

Figure3. Scheme of the return movement of the forholler

During the period of the workpiece return to the deformation zone, its movement represents the sum of the movements of the trolley with the stop *5* by the amount of the single feed and the rod *7* together with the mandrel 3 and the workpiece 2 by the total amount of rollback during rolling. Determination of the rational law of the return motion of the billet (Fig. 3) requires a reasonable choice of initial settings, including the initial pressure in the pneumatic cylinder and the parameters of the brake chamber chambers agreed with it, taking into account the change in the temperature regime of the working fluid and changing the rolling conditions of one billet, and the material of the original sleeve. The movement of the trolley by the supply amount is provided by a central screw mechanism, the nut of which is fixed in the worm drive wheel *13*. A disadvantage of the known feeding mechanism is insufficient longitudinal stability and transverse deformation of the rod and screw, which worsens the work of the screw-nut pair, causing wear of the latter and reducing the service life of the feeder mechanism as a whole. In addition, in the process of operation of the forholler and previous measurements, its operation was found to be flawed, including the unstable braking of the moving parts before the cycle of reduction of the workpiece and breakdown in the fastening unit of the traverse-stop *5* of the feed mechanism. The connection of the feed screw with the traversing stop is shown in Figure 4 [5].

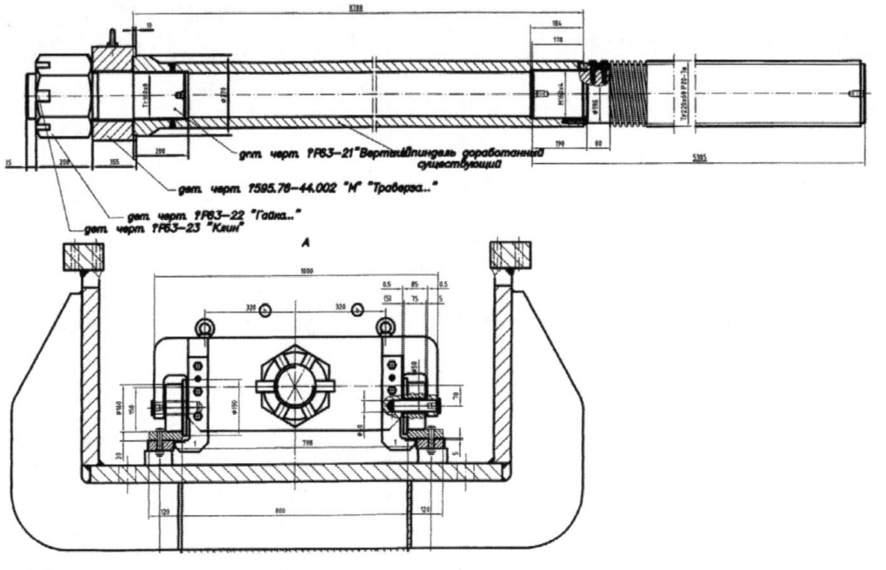

Figure 4. Trolley and feed screw feeder of forholler's the initial design

A complex of full-scale measurements, including pressures in the pneumatic and brake chambers, strain gauging measurements on the rod of the forholler and the feed screw was carried out. It is established that the feed nut forms an unstressed connection with the supply screw, loaded during the rolling period by cyclic loads varying according to a law close to a symmetrical one with a duration of 1.5 s. The source of maximum values of cyclic loads is the peak pressure in the brake chamber of the forholler, which determines the maximum value of the longitudinal force transmitted through the carriage of the forholler and the traverse-stop on the screw and nut of the feeder. Recorded in the process of full-scale measurements, the maximum of these forces reaches 600 kN per cycle [6-8].
In the process of extracting the mandrel from the rolled tube, the load on the transmission of the screw-nut of the feed mechanism is determined by the forces developed by the additional hydraulic cylinders and by the feed screw drive with the fixed carriage of the forholler. The period of recurrence of these loads is 250-300 times higher than the duration of the rolling cycle. Under these conditions it is permissible to consider the loading regime as static. The measured values of the maximum forces during the extraction of the mandrel were 1000 ... 1100 kN, of which 80% are the forces developed by additional hydraulic cylinders.

In order to improve the reliability of the feeding mechanism, a new design of the trolley and the piston system of the brake chamber is proposed.
According to the proposed variant shown in Figure 5, the travel stop of the bogie is made in the form of two parallel screws connected to a rigid restraint frame located in the guides of the base under the carriage, the screws being rotatably disposed from the nuts mounted in the drive gears cooperating with the central driving gear of their driving device.

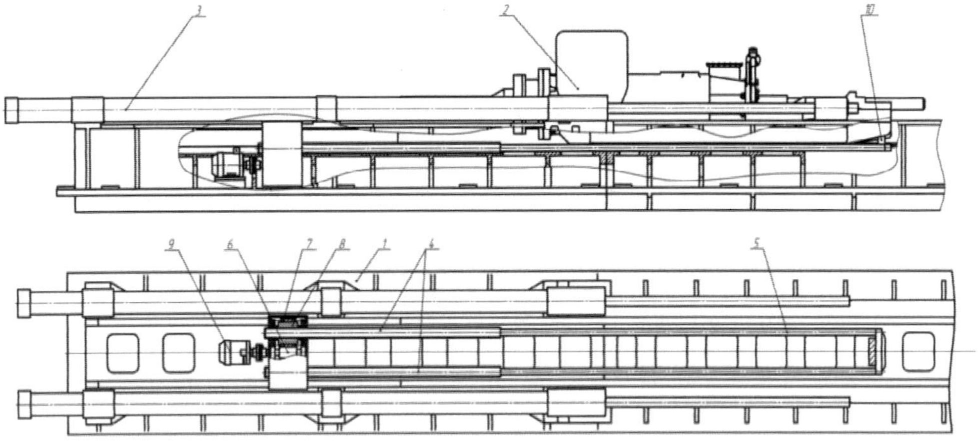

Figure 5. Feed restriction mechanism

The feeding mechanism of the pilger mill operates as follows (Figure 5). Before the start of the rolling of the next workpiece, the rigid bounding frame *5* is set to its initial position by moving it by the return movement of the screws *4* to the stop *10* of the carriage. To carry out the feed of the workpiece, the electric motor *9* is moved to the operating position, communicating the translational motion at a predetermined feed rate to the screws *4* through the drive gear *6*, the driven gears *7* and the nuts *8* fixed therein. In this case, the rigid limiting frame *5* connected to the screws *4* moves in the guides of the base *1* under the carriage *2*, limiting the movement of the stop *10*. In this case, the rods of the hydraulic cylinders *3* fixed to the carriage *2* are made, by the fluid pressure, by the constant contact of the abutment *10* with the frame *5*. After the end of the rolling of the next workpiece, the carriage *2* is retracted to the rear starting position by the rods of the hydraulic cylinders *3*, and the frame *5* is returned by the screws *4* due to the reverse with the increased speed of the engine *9* to the touch with the stop *10* of the frame *5*. The movement of the rigid frame *5* in the guides of the frame *1* eliminates the loss of its stability and transverse vibrations, providing an increase in the efficiency of the feeder.
For the design justification of the structural dimensions of the details of the proposed mechanism, let us consider the loads acting on the system during the operation of the feeder. The carriage *2* (Figure1) moves under the action of the force ΔF of the hydraulic cylinders *3*, which results from a difference in the forces of the piston and rod cavities. When the carriage moves, a friction force F_{TP}, directed against the movement, occurs. During rolling and during the acceleration of the moving masses, the force from the pneumatic chamber $F_{ВОЗД}$ acts on the carriage, which in turn is transferred to the rods of the hydraulic cylinders, causing additional deformation of the compression. During the period of active hydro-braking, a force F_{TOPM} arises, causing a deformation of the extension of the rod of hydraulic cylinders. Frame *5* has the algebraic sum of the F_{FRAME} of all external forces.
During the rolling and acceleration period, the equilibrium equation takes the following form:

$$\Delta F - F_{TP} + F_{ВОЗД} - F_{FRAME} = 0 \quad (1)$$

During the period of active braking the equation of equilibrium takes the form:

$$\Delta F - F_{TP} - F_{TOPM} - F_{FRAME} = 0 \quad (2)$$

Here, $F_{ВОЗД}$ and F_{TOPM} are determined from the solution of equations (1) and (2).

Carrying out the limiter of the trolley in the form of parallel screws connected to the rigid bounding frame makes it possible to increase the rigidity of the screws in the transverse direction, which increases the reliability of the screw-nut pair and, as a result, the reliability of the entire feeding mechanism. The design of the elongated U-shaped frame allows manipulation to remove the mandrel and is independent of the drive unit for returning the cart of the forholler to its initial position when the next workpiece is rebooted.
Figure 6 is a diagram of the forces acting on the frame of the feed limitation mechanism for the entire rolling cycle. As can be seen from the diagram, the maximum load on the frame, equal to the sum of the forces from the feed cylinders and braking forces, occurs in the active braking phase of the moving parts of the forholler when the workpiece is returned to the stand.

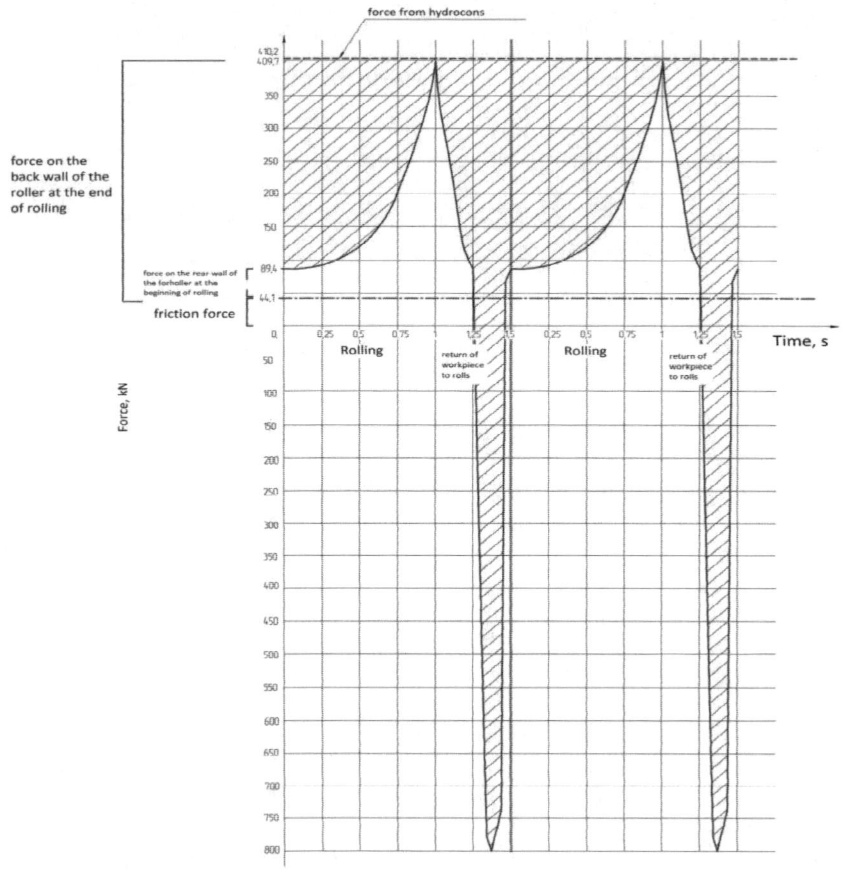

Figure 6. The forces perceived by the traverse of the feed restriction mechanism

The complex of design changes in the feeding device is designed to improve the reliability of the Pilger Rolling Mill as a whole, in order to avoid spontaneous fluctuations in the values of single feeds and, thus, to increase the geometric accuracy of the dimensions of the rolled pipes.

Conclusion
So, the main directions of increasing the efficiency of the newly proposed design of the forholler of the pilgrim rolling mill are [9]:

- Increase the load capacity of the mechanisms of moving the frame structure due to the bifurcated two-screw reinforced line of the drive device;

- Increase the maneuverability of the mandrel-mandrel extraction devices in the final rolling stage by releasing the inner space for the carriage withdrawal, removing the load from the beam of the frame structure and completely transferring the mandrel extraction force to an additional hydraulic drive;
- Increase of stability of movement and stability of values of cyclic single feeds of a preparation in a working cage.

References

[1] Emelyanenko P T 1949 *The theory of oblique and pilger rolling* (M. Metallurgizdat)
[2] Teterin P K 1978 *The theory of periodic rolling* (M: Metallurgy)
[3] Kolikov A P 1998 *Machines and units of pipe production: Educational manual for universities*, (Moscow: MISiS)
[4] Danchenko V N, Kolikov A P, Romantcev B A, Samusev S V 2002 *The technology of pipe production* (M-Intermet Engineering)
[5] Filatov S N, Sokolinsky B Y, Chechulin Y B, Popov Y A 2015 *Forholler of the pilger mil* (Patent of the Russian Federation)
[6] Kozhevnikov S N, Holidays A V, Ioffe A M 1974 *The equipment of shops with pilgrim pipe-rolling plants* (M.: Metallurgy)
[7] Chechulin Y B, Popov Y A, Filimonov I A 2011 Research of the forholler Pilger Mill work *Steel* **12**. 32 – 34
[8] Yury B. Chechulin, Yury A. Popov, Sergey N. Filatov, Boris Y. Sokolinsky 2013 The choice of rational parameters of the feeding apparatus Pilger Mill *Steel* **12** 36 – 39
[9] Chechulin Yu B, Popov Yu A 2015 Dynamic analysis of hydraulic pneumatic propulsion on the example of the forholler of the pilger rolling mill *Modern problems of science and education* **1**

Modelling of technological machines motion

Yu B Chechulin[1], S K Buynachev[1]
[1]620002, 19 Mira street, Ekaterinburg, Russia Ural Federal University, Russia
email: uchechulin@mail.ru

Abstract: The software for modelling dynamic processes of technological machines drive mechanisms using algorithmic language PYTHON.

Introduction

Dynamic analysis of transition and stable power processes in drive mechanisms of technological machines is based on solving motion equations. The final task of defining optimal characteristics of variable factors delivering minimal of power impacts at the design stage suggests mathematic modeling. Motion modeling for a large complex system is always related to speed of action, memory amounts and universal code. The use of ready packages in mathematic modeling is limited by many factors.

We believe Python algorithmic language and the following approach offer a solution to that. For kinematic scheme figure 1 see graph figure 2.

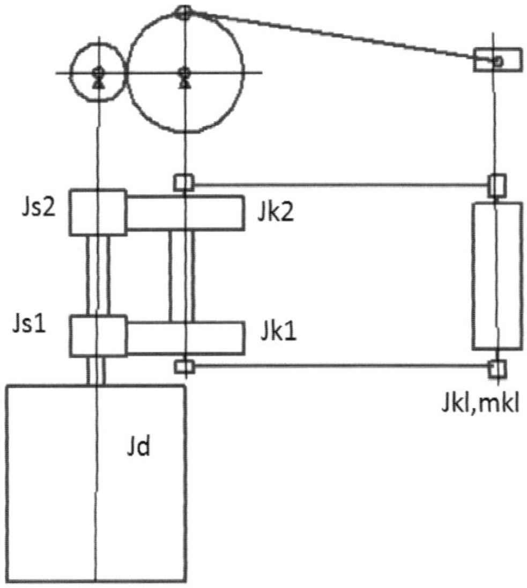

Figure 1. Kinematic scheme of CPR mill.

The graph describing the structure of this machine at figure 2. Peaks are masses and connections are elastic connections. We define

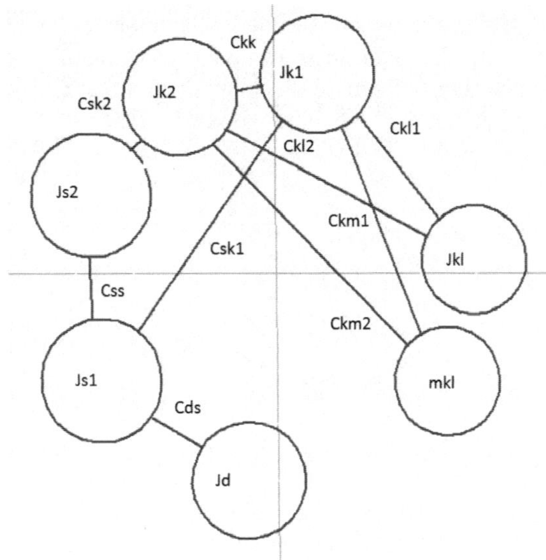

Figure 2. Graph of machine structure (ϑ_I-mass coordinates (peaks), η_j-mass coordinates difference.

We use system of coordinates one of which describes position of masses,
$\vartheta=(\vartheta_1,..,\vartheta_n)^T$,
And the other
$\eta=(\eta_1,..,\eta_n)^T$,
– connections between them. These coordinates are expressed via functions with generalized coordinates
$\tau(\vartheta,q)=0$
$\pi(\eta,q)=0$
Free oscillation equation is

$$.\tau_q^T[\tau_\vartheta^T]^{-1}\Theta\vartheta''+\pi_q^T[\pi_\eta^T]^{-1}F=0.$$

This equation requires Jacobi matrixes, reverse matrixes, then we formulate second derivatives of quasi-inertial coordinates and calculate the strength of elastic connections. To simplify the system we use graph theory [1]. However as a result of linearization and simplification system adequacy is lost. It must be mentioned that nonlinearity of lever systems and gaps. This equation provides direction to several simplifications. First of all each connection equation can be transformed as an expression of one coordinate
$\vartheta_i+\tau_i(q)=0,\ i=1,n$
$\eta_i+\pi_i(q)=0,\ i=1,m.$
Secondly we take quasi-inertial coordinates as generalized coordinates
$\vartheta+q=0.$
As a result the equation of free oscillation takes a simpler form,
Expression F we will analyze later
$\Theta\vartheta''+\eta_\vartheta^TF=0$
In which matrix Θ is diagonal and elastic forces result in coordinates ϑ. Lagrange equations of the Ist type also simplify and look as
$\Theta\vartheta''-\Lambda_T=0,$
$F-\Lambda_I=0,$
$\Lambda_T+\eta_\vartheta^T\Lambda_I=0.$
Value of undetermined multipliers Λ_I depends on coordinates η, which are expressed from masses via generalized coordinates. To simplify programming connections between peaks of the graph and

coordinates η we can use splitting procedure [3]. Into each connection we put the peak of new type and bind with the peaks of coordinates ϑ. Number of new peaks equals the number of old connection equations. We get a two-segment graph figure 2. Each segment of the graph has one class that can be developed into the necessary direction.

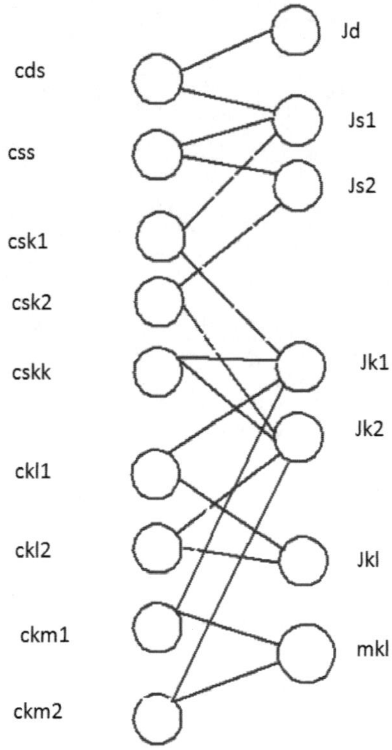

Figure 3. Two-segment graph

Right segment class can be described as [4].

```
class M:
def __init__(self,m):
self.m,self.f,self.x,=m,0.0,[0.0,0.0,0.0]
def step(self,dt):
if self.m==0: return
self.x[2]=self.f/self.m
self.x[0]+=0.5*dt*self.x[1]
self.x[1]+=dt*self.x[2]
self.x[0]+=0.5*dt*self.x[1]
```

Left segment class looks as follows.

```
class C:
def __init__(self,a,b,c,r=0.0):
self.a,self.b=a,b
self.c,self.r=c,r
self.f,self.ds,self.dv=0.0,0.0,0.0
def F(self):
ds,dv=self.a.x[0]-self.b.x[0],self.a.x[1]-self.b.x[1]
self.f = ds*self.c+dv*self.r
```

IOP Conf. Series: Materials Science and Engineering **441** (2018) 012013 doi:10.1088/1757-899X/441/1/012013

```
self.a.f-=self.f
self.b.f+=self.f
```
Creation of right segment objects.
```
Jd,Js1,Js2,Jk1,Jk2 =M(150),M(0.1),M(0.1),M(0.5),M(0.5)
Jkl,mkl =M(10.0),M(1450.0)
V=[ Jd,Js1,Js2,Jk1,Jk2,Jkl,mkl]
```
Creation of left segment objects.
```
cds,css=C(Jd,Js1,3540.0,12.5), C(Js1,Js2,3140.0,12.5)
csk1,csk2,ckk=C(Js1,Jk1,3540.0,12.5), C(Js2,Jk2,3140.0,12.5)
ckkl1,ckkl2= C(Jk1,Jkl,3540.0,12.5), C(Jk2,Jkl,3140.0,12.5)
ckm1,ckm2= C(Jk1,mkl,3540.0,12.5), C(Jk2,mkl,3140.0,12.5)
E=[cds,css,csk1,csk2,ckk,ckkl1,ckkl2,ckm1,ckm2]
```
Here programming stops and two lists V and E should be scanned. As a result we present graphs at figure 4.

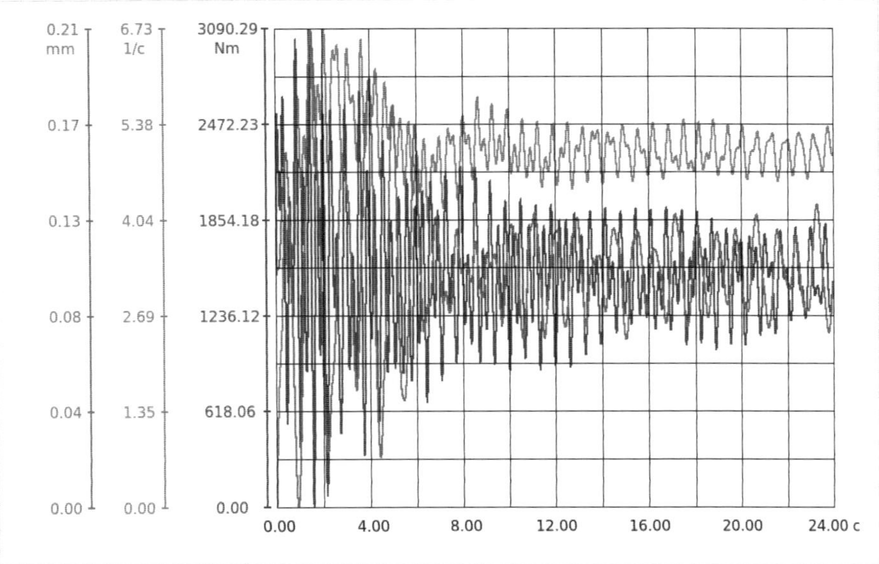

Figure 4. Motion and load at acceleration and stable motion (Red - speed of engine rotor, blue - engine rotor momentum, green - oscillation of mill crate position in a horizontal plane).

Results and Discussion

In this case connections in the graph have a meaning of information flows. Matrix transformations are not needed. As a result we get a solution reflecting mill motion. For the first time we get a description of crate in a horizontal plane so easily and precisely.

Conclusion

Suggested modeling solution has the following advantages:
1. Number of masses and hardnesses is not limited.
2. No limitation in nonlinearity description.
3. Simplicity of programming.

References

[1] Veitz V L 1969 *Dinamika mashinnykh agregatov* (Mashinostroenie)
[2] Evstigneev V A 1985 *Primenenie teorii grafov v programmirovanii* (M. Nauka)
[3] Thulasiraman M, Grafy K 1984 *Seti i algoritmy: translated from Engl* (M. Mir)

[4] Bizli D 2000 *Yazik programmirovanija Python* (K. DiaSoft)

Research of crankshaft with three crankpins warping due to fillets stamping

V N Emel'yanov[1] and V I Berezhnov[1]

[1] Novgorod State University, ul. B. St. Petersburgskaya, 41, 173003 Veliky Novgorod, Russia

E-mail: valem1219@mail.ru

Abstract. The paper presents the results of the experimental study of the process of warping of dimensioned crankshafts with three crankpins during hardening their fillets by stamping

Introduction

Crankshafts (CS) are widely used in modern engineering. However, despite the extensive experience gained by specialists in the design, manufacture and operation of CS, they are often destroyed in operation due to insufficient supply of fatigue resistance. According to Prof. V. P. Uskov, such destruction is observed in 2-8% of engines coming for repair [1]. Destruction occurs mainly along fillets. The most effective and simply implemented in terms of increasing the resistance to fatigue of CS is the operation of hardening their fillets by surface plastic deformation (SPD): roller burnishing, hammering, etc. [2]. However, as practice shows and numerous experiments, processing fillets of CS by SPD leads to warping of CS [3]. Moreover, the more effective way in terms of increasing the resistance to fatigue, the more it leads to CS warping. With intensification of hardening mode resistance to CS fatigue, increases but at the same time increases the size of its warping.

At the department of engineering technology of Yaroslav-the-Wise Novgorod State University the basic provisions of the theory of CS warping have been developed when hardening their fillets by SPD [4]. In particular, the experimental and computer simulation was carried out of the process of warping of spatial crankshafts with six crankpins (CP), which are in pairs located at angles of 120 ° [5] and with five CPs, located at angle of 72 ° [6, 7]. The purpose of this work is to study the regularities of the spatial CS warping process with three CPs, located at angle of 120 ° relative to each other, which has not been previously studied.

Results of the experiment

As the object of the research the CS of SMD-60 engine was selected, the sketch of which is presented in Figure 1. This CS has three CPs, located at an angle of 120 ° to each other. The diameter of the crankshaft necks (CN) is 90.5 mm, the diameter of the CP equals to 84.5 mm. The radius of the crank is 57.9 mm. One cheek-piece (to the left in Figure 1) of CS has the thickness of 34 mm, other cheeks of the shaft have the same thickness of 26.5 mm. The radius of fillets of the CN, and crankpin equal to 4.5 mm. The distance between supports is L = 551 mm.

IOP Conf. Series: Materials Science and Engineering **441** (2018) 012014 doi:10.1088/1757-899X/441/1/012014

Experiments were conducted at the consistent hammering of the fillets. Fillets were processed manually with the help of striker with spherical head and hammer in such sequence: fillet No 2, 3, 4, 5, 6, 7, 8, 9, 10, 11 (fillet 1 and 12 were not processed). The energy of impact E_I was approximately 38 J. The magnitude and direction of the CS wobbles vector was measured after the processing of each fillet with the help of indicator, the measuring rod of which was entered in contact with the third on the left (in Figure 1) CN, between fillet 8 and 9. The direction of the wobbles vector was characterized by an angle of α_i that is the angle between the y axis and the wobbles vector. The size and direction of the wobbles vector of CS from the processing of each fillet counted as the difference of the wobbles vectors after and before the processing of given fillet. The direction of the wobbles vector was characterized by an angle of α_i that is the angle between y axis and the wobbles vector.

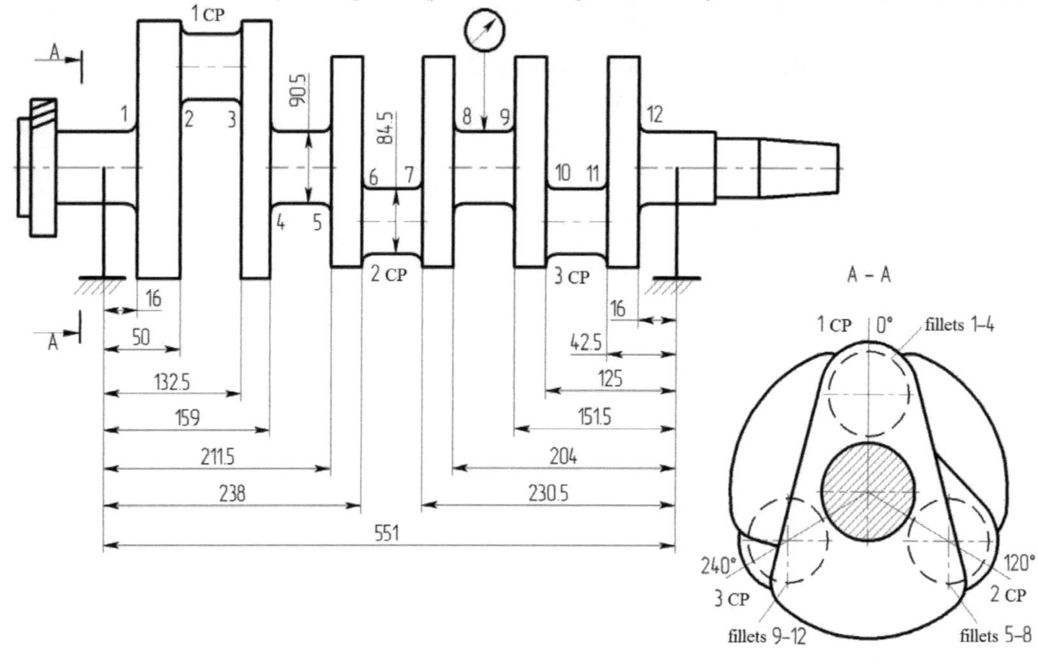

Figure 1. Sketch of SMD-60 engine CS.

Before the beginning of the experiment the initial wobble of CS δ_i was 0.21 mm and the vector of wobbles was directed at an angle α_i = 190 °. After hammering fillets 2-11 the final wobble of CS α_e was 0.18 mm, and the vector of wobbles is directed at angle α_e = 220 °. The total value of warping δ_{sum} after hammering fillets 2-11 was 0.115 mm. By results of experiment the epure of CS warping vector was ploted which is shown in Figure 2.

The analysis of the given plot has shown, that the majority of CS warping vectors from hardening of separate fillets lie in the plane of cranks with an error not exceeding 10 – 15% (except for vectors from hammering fillet 2, 5, 6).

On the basis of the received data the curve of dependence of CS warping value from distance from fillets to supports, i.e. from fillet number (Figure 3) has been constructed.

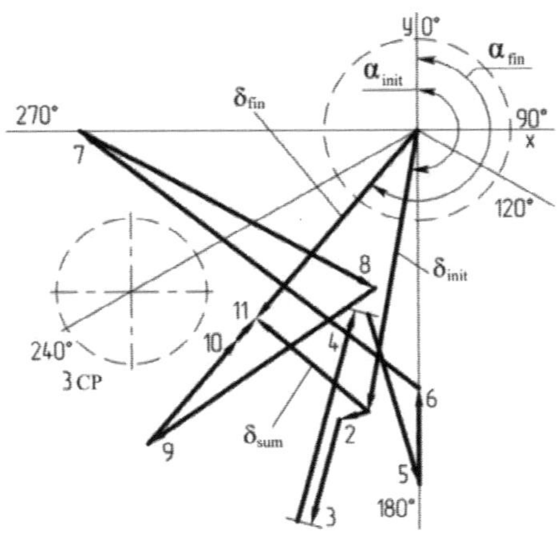

Figure 2. Epure of CS warping vectors for engine SMD-60.

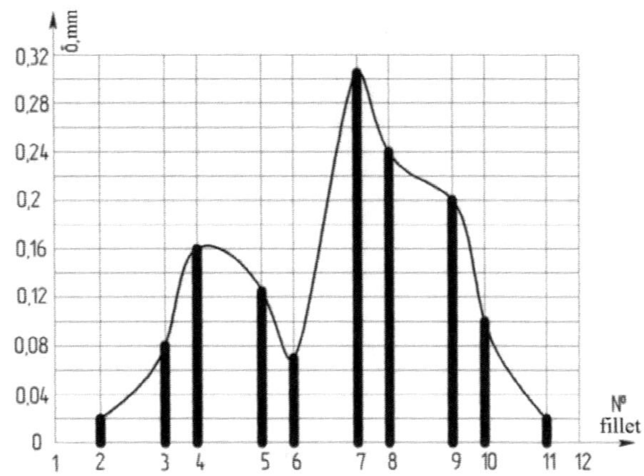

Figure 3. The graph of the dependence of CS warping for engine SMD-60 from the distance from fillets to the supports (from the fillet number).

During the experimental study of the CS warping process with five CPs also was obtained undulating graph of the dependence of CS warping from the distance from fillets to the supports (from the fillet number) (Figure 4) [6, 7].

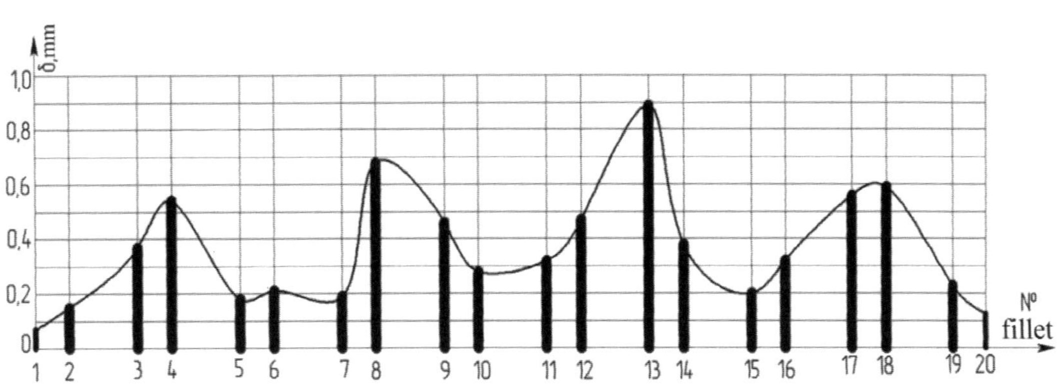

Figure 4. The graph of the dependence of CS warping for engine SMD-60 from the distance of fillets to the supports (from the fillet number).

The authors explain the undulating nature of the curves in the following way. In publication [8] formula (1) for calculation of CS warping value at hammering the fillets is given:

$$\delta = K\frac{HB^{0,75}(1540-HB)(1-\upsilon)}{E}\frac{R}{h^2}\frac{c}{2}\sqrt[4]{E_I D_1}\Psi , \qquad (1)$$

where K is the coefficient determined by experiment;

HB is Brinell hardness;

ν is the Poisson coefficient;

E is the modulus of elasticity of I-type, MPA;

R is the radius of the CS crank, mm;

h is thickness of the cheek, mm;

C is the distance from the left support (at processing fillets 2 – 6) or from the right support (at processing fillets 7 – 11) up to the processed fillet;
E_I is the energy of striker impact, J;
D_1 is diameter of the spherical head of the striker, mm;
Ψ is the coefficient that takes into account the incomplete processing of the cheek by hammering; It is counted according to formula (2) [8-11]:

$$\Psi = \frac{F_i}{F_R} \leq 1, \quad (2)$$

where F_i is the area of the cheek of the crankshaft, where the cheek bend occurs (this changes the macrogeometry of the crankshaft);
F_R is the total area of spherical prints of striker on the surface of fillet (and partially cheeks), which in turn depends on the diameter of the neck and the radius of the fillet.

Thus, "Ψ" value of each fillet also depends on the diameter of the neck and the radius of the fillet. At transition from processing one fillet to processing the next fillet at the same time value "C" and value "Ψ" are changing simultaneously. With the increase of "C", value "δ" gradually increases up to the middle of the shaft. And value "Ψ" increases, then decreases, because CN and CP for the given CS alternate. This circumstance leads to the appearance of wave graph of CS warping dependence on the fillet number .

References

[1] Uskov V P 1998 *Spravochnik po Remontu Bazovykh Detalei Dvigatelei* (Briansk: Klincovskaya Gorodskaya tipografiya) p 589
[2] Emel'ianov V N and Efremov S A 2003 Ekonomicheskaia effektivnost' uprochneniia galtelei kolenchatykh valov PPD *Avtomobil'naia promyshlennost'* **4** pp 28-31
[3] Emel'ianov V N 1998 Koroblenie kolenchatykh valov pri uprochnenii galtelei i sposoby ikh pravki *Avtomobil'naia promyshlennost'* **1** pp 25-27
[4] Emel'ianov V N 1997 Osnovnye polozheniia teorii korobleniia kolenchatykh valov pri uprochnenii ikh galtelei i pravke poverkhnostnym plasticheskim deformirovaniem *Vestnik NovGU Ser. Estestv. i tekhn. nauki* **5** pp 29-34
[5] Emel'ianov V N and Romanov I V *Koroblenie Prostranstvennykh Kolenchatykh Valov pri Chekanke Galtelei* 2011 Saint-Petersburg **2 (2)** (Saint-Petersburg: Politekhn. university) pp 118-121
[6] Emel'ianov V N and Kostin P V 2013 Komp'yuternoe modelirovanie protsessa korobleniya prostranstvennykh kolenchatykh valov s pyat'yu shatynnymi sheikami *Vestnik NovGU. Ser. Estestv. i tekhn. nauki* **75 (1)** pp 67-74
[7] Emelyanov V 2014 Research on hogging process of crankshafts with five rod journals because of fillets stamping *Journal of Engineering And Technology Research* **2 (2)** pp 65-
[8] Zajdes S A and Emel'yanov V N 2017 *Vliyanie Poverhnostnogo Plasticheskogo Deformirovaniya na Kachestvo Valov* (Irkutsk: IRNITU) p 380
[9] Malushin1 N N, Valuev D V, Il'yaschenko D P, Gizatulin R A, Nekrasov R Yu Improvement of Research Procedures into the Highly Rigid Welded Metal Using Thermal Microscopy Facilities *J. Materials Science Forum* **927** 148-152

[10] Gusev A I., Kozyrev N A, Osetkovskiy I V, Kozyreva O A, Valuev D V 2017 Quality of metal deposited flux cored wire with the system Fe-C-Si-Mn-Cr-Mo-Ni-V-Co *J. Key Engineering Materials* **736** 23-28

[11] Malushin N N, Osetkovsky V L, Valuev D V, Valueva A V, Serikbol A 2014 Selection criteria parameter thermal cycle during surfacing heat resistance high hardness parts of metallurgical equipment *J. Applied Mechanics and Materials* **682** 69-74

IOP Conf. Series: Materials Science and Engineering **441** (2018) 012015 doi:10.1088/1757-899X/441/1/012015

Invalid vehicles: mechanisms for changing the overall size across the width

V. G. Karpushkin[1], V. V. Timofeev[2], V. N. Emelyanov[1]

[1] Novgorod State University, ul. B. St. Petersburgskaya, 41 173003 Veliky Novgorod, Russia
[2] Duma of Velikiy Novgorod, ul. Bolshaya Vlasyevskaya, Velikiy Novgorod, 4, 173007, Russia

E-mail: karpushkin.viktor@yandex.ru

Abstract. The paper presents the research results in the field of transformable invalid vehicles with the possibility of changing the overall size according to the width. The estimation of the general situation on this issue in the printed media and on the Russian market of invalid equipment is described in this document. The variants of using various mechanisms for changing the overall size according to the width are considered to be taken into account in the design of the wheelchair and the location of the drive wheels. According to the test results, the most promising model of a wheelchair and the mechanism used for changing the overall size across the width is described. The presented comparative analysis confirms the market demand and relevance of the presented research subject.

1. Introduction

The world and Russian market of wheelchair technology have shown steady growth in recent years. About one-third of the rehabilitation equipment market is wheelchairs. The relevance of new developments for this market, its quantitative characteristics, and features of the Russian market were mentioned earlier in [1-2].

Studies in the prospective area of developing invalid vehicles with various characteristics are being conducted at the Department of Engineering Technology of Novgorod State University named after Yaroslav the Wise for about 30 years. The main contribution to the development of this topic was made by Professor Shchegolev V. A., Professor Emelyanov V. N., Professor Timofeev V. V., Associate Professor Bordashev K. A., and others.

The accumulated experience and information made it possible, by 2013, to formulate a specific technical task for a transformable invalid electric vehicle (TIEV), which, taking into account the conditions of Russian operation and the situation on the Russian market, should allow to solve the tasks of social and professional rehabilitation of disabled people, mainly with disabilities of the musculoskeletal system, as well as central and peripheral nervous systems.

2. Solutions to the task

To solve the abovementioned tasks, the TIEV should meet the following requirements:

1. TIEV should be equipped with a mechanism to change the overall size to the width to overcome doorways with a width of 600 mm, which is especially important for Russian operating conditions.

2. The TIEV should be equipped with a backrest and armrests that fold into one level with the seat, which will allow moving from the seat plane in any direction, increasing the mobility of the TIEV user.

3. TIEV should have a reduced weight (less than 40 kg) and the cost of the final product up to 100 thousand rubles. That will allow it to compete in the low price segment with analogues of Chinese production that do not possess the abovementioned properties.

4. Transformation processes must be carried out by the user of TIEV independently without assistance and additional tools.

We carried out an analysis of the design of electrically powered wheeled vehicles (IV) [3 - 9]. It can be argued that the main problem considered by modern specialists relates to the electric drive of the wheelchair, its subsystems, and operational parameters. In this case, we are interested in the mechanism of changing the overall size in width and its integration into the overall design of a wheelchair, taking into account the location of the drive wheels.

The problem of passing through narrow doorways is considered mainly in the context of changing the environment for the needs of disabled people [10 - 13], rather than the design of the IV itself, and does not raise the issue of the location of the mechanism for changing the overall dimensions along the width and arrangement of the drive wheels on the IV.

As mentioned earlier [14], within the framework of the present research, the variants of the formulas of three-wheeled wheelchairs are not considered, because such an arrangement is not capable of providing the necessary static and dynamic stability [15 - 17] when fulfilling a number of dimensional requirements given in the terms of reference. Proceeding from this, the developments were based on the four-wheel formula of TIEV.

Note that most of the models on the market have a drive to the rear wheels, but this design feature is not determinative. The leading models of the market-leading company Otto Bock can have a drive, both on the rear and the front axle of the IV. Each of the drive circuits has its advantages and disadvantages.

3. Design options and design of the mechanism for changing the overall size across the width

The design with four driving wheels is the most effective in overcoming obstacles, however, the high cost of its production, the complexity of the design, the increase in the total mass of the IV and the reduction in the power reserve do not allow considering this scheme in the framework of this work.

The arrangement of the driving wheels on the front axle of the wheelchair makes it possible to simplify the structure of the carcass and the mechanism for changing the overall dimensions along the width. At the same time, the front-wheel-drive TIEV can have an insufficient grip of the front wheels with the surface, especially when lifting to the ramp (with a lifting angle exceeding the normative - more than 10 °), which leads to slippage. The occurrence of such a problem depends on the location of the user on the seat, the coefficient of adhesion of the driving wheels to the road surface and a number of other conditions.

The location of the drive wheels on the rear axle avoids slipping when climbing the ramp, but in a rear-wheel drive design, it is more difficult to implement a mechanism for changing the overall width dimension while maintaining high reliability and low cost. In addition, when the width is changed, inversion of the TIEV control via the joystick occurs, which causes inconvenience in the control and creates a number of additional problems.

There are known structures of IV on muscular strength with a mechanism for changing the width (for example, model KY-901), in which changes in the overall size of the width are made by reducing the width of the seat, which imposes severe restrictions on the operation of such IV wheelchair users with excess weight or with a wide the hip joint. Such mechanisms do not allow changing the overall width during driving, and sometimes without the assistance of an accompanying person and/or a special tool. This option does not allow talking about the independence and mobility of disabled wheelchair users in the Russian conditions.

The option of creating a wheelchair with a constant width of less than 600 mm is also ineffective. This design can only be used in the models of children's IV. Otherwise, the design will not meet the above requirements for static and dynamic stability and will not be able to provide comfortable operation for the disabled in terms of placement on the seat. In addition, passing through a narrow doorway requires the availability of a margin in width between the IV clearance points and the door frame.

Based on the analysis of models of electric IV present on the Russian market [18-23] and based on the experience of earlier domestic developments [24,25], it was decided to consider both versions of the TIEV drive schemes and two options of the mechanism for changing the overall width. The final choice between them was to be made at the prototype testing stage.

In the process of searching for the most effective and relevant constructive solutions to this issue, various options of the designs and mechanisms for changing the overall size in width, presented in the figures, were considered.

For TIEV model 3.2 with a drive to the rear wheels, a previously developed and patented [26-28] mechanism was used (Figure 1). Due to the swivel supports, on which the asymmetrical forks for attaching the wheels are located, and the overall size is changed in width. The axis of symmetry of the fork is remote from the axis of symmetry of the wheel by a distance e, which allows the stable construction of the stroller in the decomposed position A (Lmax ~ 700 mm). In position B (Lmin ~ 500 mm), the wheelchair is able to overcome the narrowest door openings, from those encountered in real operation.

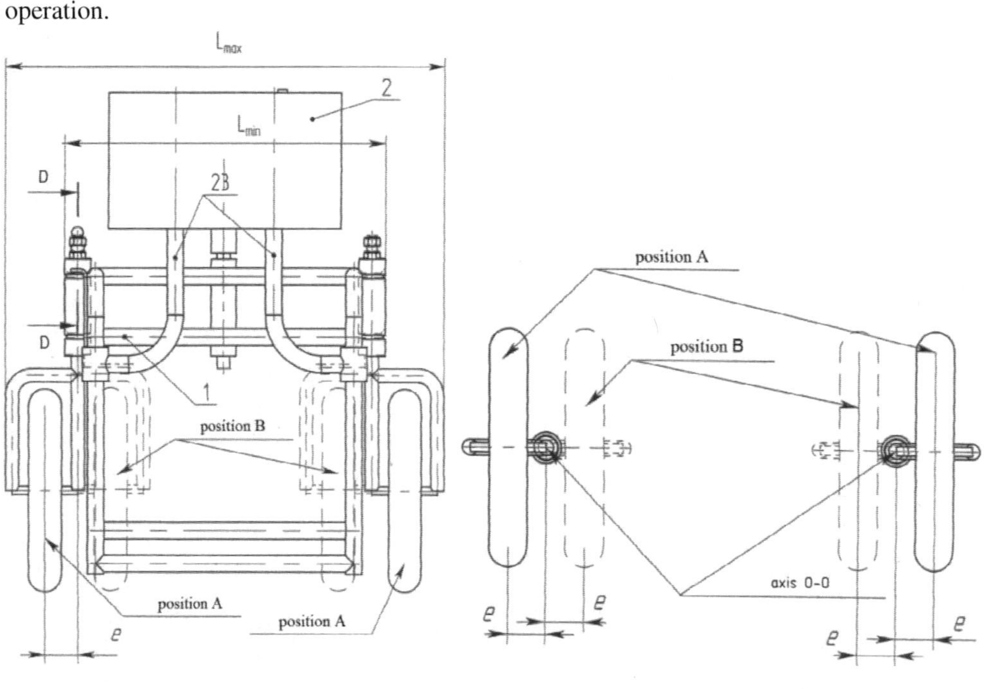

Figure 1. Mechanism of changing the overall width due to the console fork of the rear drive wheels

When implementing the IV scheme with front-wheel drive, the rear wheels are self-aligning, and as a consequence, various mechanisms for displacing the vertical axes of the self-aligning wheels can be used to vary the width of the track of the rear wheels. The application of a similar circuit, as well as for the model with a drive to the rear wheels, is possible and is represented in Figures 2 and 3.

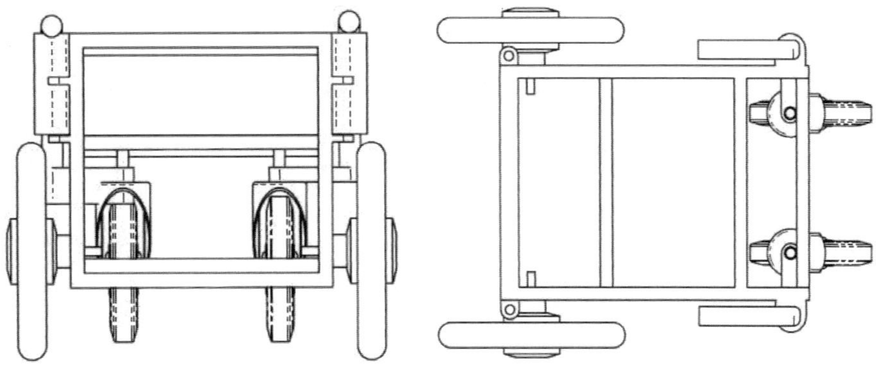

Figure 2. Mechanism for changing the overall dimensions in width due to the console fork of self-aligning (rear) wheels (minimum overall dimension)

The main drawback of this mechanism is an insufficient distance between the tilting wheels in a folded state, which leads to loss of stability of the whole structure, and there is a risk of lateral tilting of the IV. Further, excessive displacement vane wheel under the seat and backrest of the wheelchair reduces the useful space under it, causing problems with folding backrest in a level with **the seat.**

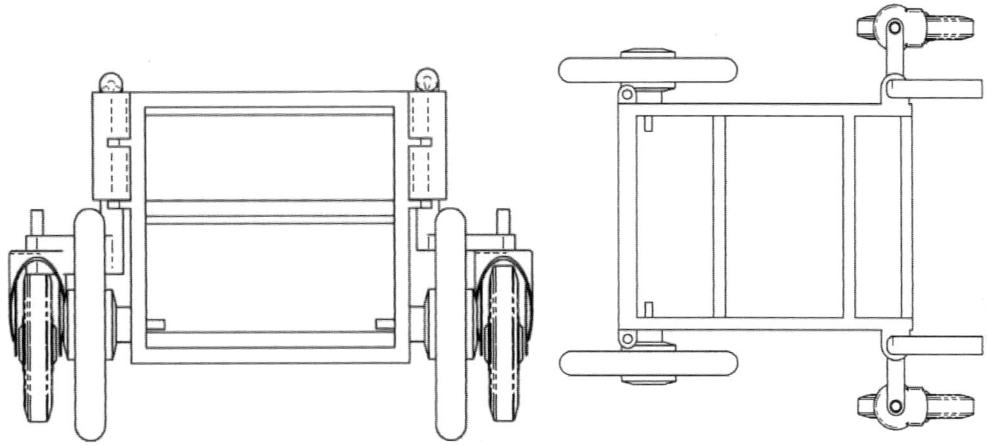

Figure 3. The mechanism for changing the overall width due to the console fork of self-aligning (rear) wheels (maximum overall size)

As the size of the console on which the vane wheel is fixed decreases in order to solve the problems described above, the difference between the maximum and minimum overall dimensions across the width decreases significantly, which makes the mechanism extremely ineffective in terms of the stated problem of maintaining stability in the unfolded (maximum) position of the weathervane wheels. Due to the listed shortcomings of such a mechanism, its use had to be abandoned.

For the scheme with the front wheel drive, another version of the mechanism for changing the overall width, shown in Figure 4, was developed.

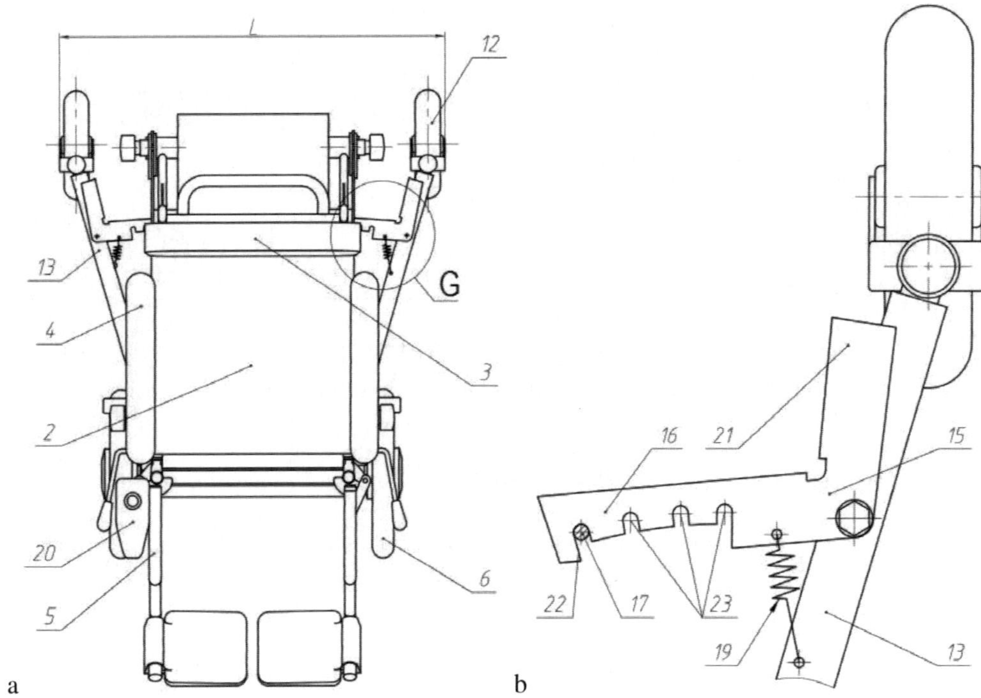

Figure 4. Mechanism of changing the width at the TIEV with front wheel drive. a - top view; b - view of G by b (increased). 2 - the seat; 3 - backrest, 4 - armrests; 5 - feet support; 6 - front driving wheel; 12 - rear wheel; 13 - the upper rod of the side frame 8; 15 - two-arm lever; 16 - shoulder of double-armed lever 15 with grooves 22-23; 17 - retainer; 19 - tension spring; 20 - joystick; 21 - the shoulder of the two-arm lever 15; 22, 23 - grooves of the shoulder 16 of the two-arm lever 15.

The control of each mechanism for changing the distance between the rear wheels of the wheelchair is as follows. When moving on a free platform or over wide aisles, the arm 16 of the two-arm lever 15 is in the position shown in Figure 4b. The distance L (figure 4a) between the rear wheels 12, in this case, will be maximum. Before the narrow passage (corridors, doorways in the bathroom, toilet, elevator, etc.), the person driving the wheelchair grips the upper shaft 13 (figure 4b) of the side frame 8 and at the same time the shoulder 21 of the two-arm lever 15. Then it stretches the tension spring 19, thereby removing the groove 22 of the arm 16 of the two-arm lever 15 from engagement with the latch 17. Further, the side frame 8 rotates about the main vertical axis 7 by the upper shaft 13 at the desired angle. Then one of the slots 23 is brought into contact with the latch 17. At the same time, the distance L (figure 4a) between the rear wheels 12 is fixed at a lower value, in which the previous case. The distance L (respectively one or the other of the slots 23) is selected in accordance with the width of the passage.

4. Results and conclusions

Thus, in the framework of this study, prototypes of the TIEV model 3.1 (Figure 5) and TIEV model 3.2 (Figure 6) were developed and manufactured with the above-mentioned mechanisms for changing the overall size in width for subsequent field tests in accordance with GOST R ISO 7176.

IOP Conf. Series: Materials Science and Engineering **441** (2018) 012015 doi:10.1088/1757-899X/441/1/012015

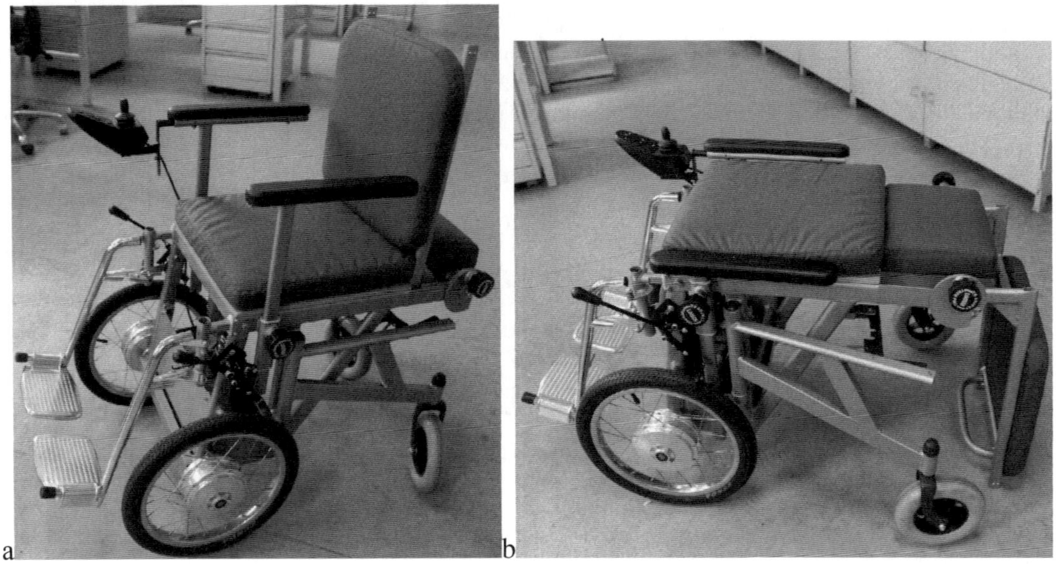

Figure 5. TIEV model 3.1 with front drive before transformation (a) and in the transformed state (b)

Analysis of the results of the tests allows us to say that in terms of the static and dynamic stability indicators of the TIEV model 3.1 with the limits of the ramp up to 18°, it exceeds the TIEV model 3.2 (the limiting angle is 13°). In addition, the TIEV model 3.1 has a more ergonomic design, more understandable control of the transformation mechanism, and there is no inversion of the control signal when the overall size changes in width. A very important advantage of this mechanism in comparison with TIEV model 3.2 is the possibility of fixing the mechanism in different positions depending on the operating conditions, whereas the mechanism with an asymmetric rotary fork has only two extremes (maximum and minimum).

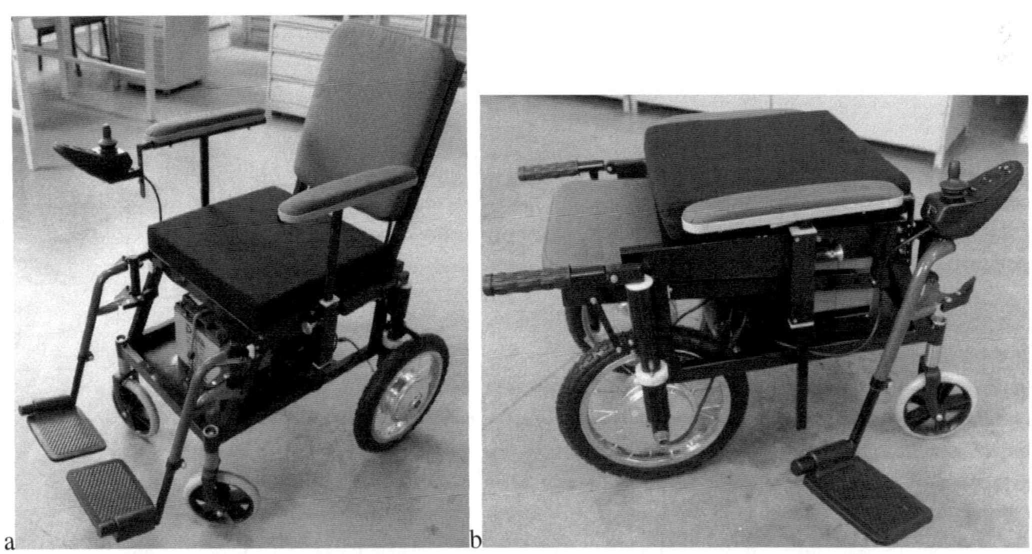

Figure 6. TIEV model 3.2 with rear drive before transformation (a) and in the transformed state (b)

Within the framework of the trilateral cooperation of Novgorod State University named after Yaroslav the Wise, the Innovation Support Fund and the Kvant Scientific and Production Association, it is planned to mass-produce a transformable invalid electric vehicle that is structurally close to the prototype TIEV model 3.1. Currently, preparations are carried out for the production and certification of this TIEV model.

The design of the mechanism for changing the overall dimensions along the width is protected by patent for utility model No. 175141 "Wheelchair with variable distance between the rear wheels" [27-28]. The design of this mechanism allows you to talk about the achievement of the planned properties and characteristics of the wheelchair being developed, namely:

1. The patented mechanism for changing the overall clearance along the width allowing to overcome doorways with a width of 600 mm, which is especially important for Russian operating conditions.
2. Convertible to the same level with the seat backrest and armrests allowing to move from the seat plane in any direction, which increases the mobility of the TIEV user .
3. TIEV has a weight of less than 40 kg and an estimated final cost of up to 100 thousand rubles.
4. All the transformation processes are carried out by the user of the IV independently without assistance and an additional tool.

In support of the relevance of the development in Table 1, the comparative characteristics of the TIEV model 3.1 with the main competitors in the market of wheelchair electric wheelchairs of Russian production are given, which is especially important in the context of the import substitution policy.

Table 1 - Comparative characteristics of TIEV 3.1-A with models of "Observer" and "CaterWil" companies

IV Name	Width in operation, mm	Weight, equipped, kg	Can fold the back and arm support to the seat level	Capable to pass the stairs	Price, thousand rub.	Production
Observer Standart	645	61,5	no	no	97	Russia, Kaliningrad
Observer Optimus OB-EW-012	740	186	no	yes	530	Russia, Kaliningrad
CaterWil GTS3	645	115	no	yes	504	Russia, Novosibirsk
TIEV model 3.1	575	38	yes	no	from 80	Russia, Velikiy Novgorod

5. Aknowledgements
This research was carried out within the framework of the program "Cooperation" of the Fund for the Promotion of Innovation (contract No. 12622AR / 10453).

References
[1] Karpushkin V.G., Kubiak P., Timofeev V.V. Invalid vehicles: features of operation and the state of the Russian market. Proceedings of SWorld. Issue 4 (37). Volume 2. Odessa, 2014. pp. 48-54. (In Russian).
[2] Karpushkin V.G. Development of new types of IV: relevance for the Russian market. Actual problems of technical sciences in Russia and abroad. Collection of scientific papers on the results of the international scientific and practical conference. №2. Novosibirsk, 2015. pp.

71-74. (In Russian).
[3] Al Masoud T. Individual vehicle with electric drive and capacitive energy storage: dis. Cand. those. Science. Moscow, 1995. 16 p. (In Russian).
[4] Vasiliev N.F., Loginov A.L., Batayev A.V. Microprocessor control system for wheel-drive wheelchair. Electrical engineering. 1994. no. 11. pp. 32-34. (In Russian).
[5] Voronin S.G., Koshcheev E.I., Posokhov G.N., Tymanov A.V. Problems of creating reliable and safe vehicles for disabled people. Electrotechnical complexes of autonomous objects: theses dokl. sci-tech. conf. Moscow, 1997. pp. 34-36. (In Russian).
[6] Voronin S.G., Korobatov D.V., Sogrin A.I. The electric drive of a wheelchair: principles of construction and implementation problems. Herald of SUSU. Series "Energy". 2001. issue 1. no. 4 (04). pp. 84-88. (In Russian).
[7] Korobatov D. V. The motion control system for a wheelchair with an electric drive . Dis. PhD. tech. Sciences, Chelyabinsk, 2007. 169 p. (In Russian).
[8] Petlenko A.B. Wheelchair with separate electric wheel drive and combined power plant . Dis. PhD. tech. Sciences. Moscow, 1997. 263 p. (In Russian).
[9] Sogrin A.I. Emergency braking system for a wheelchair with an electric drive. Dis. PhD. tech. Sciences. Chelyabinsk, 2007. 130 p. (In Russian).
[10] Typical instruction for providing wheelchair users with wheelchairs in public buildings projects: planning and building of populated areas, 1988 (In Russian).
[11] Fedutinov Y.A., Shklyaev N.A. Ensuring the mobility of disabled people and the elderly in large cities. Problems of large cities. Sat. articles. Moscow, 1989. pp. 3-5. (In Russian).
[12] Kuzmenko S.N., Timofeev V.V. Methodological recommendations on creating an accessible environment for disabled people. Velikiy Novgorod. 2012. 95 p. (In Russian).
[13] Passport of the Russian Federation state program "Accessible Environment" for 2011-2015. (In Russian). Available at: http://www.rosmintrud.ru/ministry/programms/3/0 (accessed 19.07.2018)
[14] Karpushkin V.G. Invalid vehicles: a prototype TIEV-1. Topical issues of science and technology. Collection of scientific papers on the results of an international scientific and practical conference. vol. 3. Samara, 2016. pp. 97-100. (In Russian).
[15] GOST R 51083-97. Wheelchairs. General specifications. Moscow. Standsrtinform Publ., 1997. 24 p. (In Russian).
[16] GOST R ISO 7176-1-2005 Wheelchairs. Part 1. Determination of static stability. Moscow. Standsrtinform Publ., 2005. 16 p. (In Russian).
[17] GOST R ISO 7176-2-2005 Wheelchairs. Part 2. Determination of dynamic stability of wheelchairs with electric drive. Moscow. Standsrtinform Publ., 2005. 20 p. (In Russian).
[18] Armed. Available at: https://www.armed.ru/product-category/texnicheskie-sredstva-reabilitacii/ (accessed 19.07.2018)
[19] OrtoOttoShop. Available at: https://ortoottoshopru.tiu.ru/product_list (accessed 10.07.2017)
[20] AirMedic. Available at: http://air-medic.ru/catalog/invalidnye-kolyaski/invalidnye-kolyaski-s-elektroprivodom/ (accessed 10.07.2017)
[21] Medspros. Available at: http://www.medspros.ru/shop/kreslo_kolyaska_s_elektroprivodom_i_ruchnym_upravleniem_dlya_invalidov_kar_4_1.html (accessed 19.07.2018)
[22] Orthobox. Available at: https://ortobox.ru/catalog/section/invalidnie-kolyaski/ (accessed 19.07.2018)
[23] MET. Available at: http://www.met.ru/electric/ (accessed 19.07.2018)
[24] Ilyinov V.S. Wheelchair for the disabled. Patent USSR, no. 862947, 1981.
[25] Timofeev V.V., Boldyshev P.P. Multidisciplinary vehicle for the disabled. Patent RF. no. 2445057, 2012.

[26] Emelyanov V.N., Karpushkin V.G., Timofeev V.V. and others. Wheelchair with a variable base. Patent RF, no. 2542557, 2015.
[27] Emelyanov V.N., Karpushkin V.G. Wheelchair with variable distance between the rear wheels Patent RF, no. 175141, 2017.
[28] Malushin1 N N, Valuev D V, Il'yaschenko D P, Trifonov V A, Nekrasov R Yu Technological Improvement of Surfacing of Parts of Hammer Crushers Used in Coke-Chemical Industry *J. Materials Science Forum* **927** 168-175

IOP Conf. Series: Materials Science and Engineering **441** (2018) 012016 doi:10.1088/1757-899X/441/1/012016

Modeling the distribution of temperature field within the micro heat-loss anemometer based on silicon carbide

D A Evstigneev [1, 2], V A Karachinov [1], A S Varshavskiy [1] and V A Manykhin[1]

[1] Division of Microelectronics, Novgorod State University, ul. B. St. Petersburgskaya, 41 173003 Veliky Novgorod, Russia

[2] Email: Danya2allstars@yandex.ru

Abstract. The results of modeling the thermal characteristics of microsystems are presented and mutual thermal influence of the elements depending on the type of construction has been evaluated.

1. Introduction

The hot-wire method, based on the dependence of the electrical resistance of a temperature-sensitive element on the flow velocity of a liquid or gas, is widely used in various fields [1]. There are thermal microsystems that allow recording the velocity of high-temperature gas flows [2]. The basis of the construction of such microsystems is basically a beam structure, which from the point of view of the propagation of heat flows is not a losing option, and since the greatest interest at the present time is the speed and accuracy of measuring physical quantities, the need to create more advanced microsystems that meet these requirements will be at a high level. Reducing the heat flow, spreading from one element to another, increasing the performance indicators, these problems constantly require improvement of the design base. At the same time, continuous development of technology and a variety of operating conditions leads to the creation of designs of thermo anemometers of complex shape. An experiment is needed to carry out the design optimization study. But in view of the fact that it is not possible to carry out natural experiments due to the high cost of the material, the necessary stage of the study is the simulation of the thermal characteristics of the microsystem [3, 4]. The results of this simulation are presented in this article.

2. Research methodology

Heat – loss anemometer, which is a microsystem incorporating a thermometer, is made of silicon carbide 6H polytype with a concentration of uncompensated donors $N_D - N_A \cong 3 \cdot 10^{18}\ cm^{-3}$. The simulating process in the ANSYS environment was performed using the finite element method.

3. The results of the study and their discussions

The study of the thermal regime of the microsystem, in which the thermal bonds between the elements are determined mainly by the conduction process, the properties of the material and the design features, is a model analysis of the distribution of the temperature field from a strongly heated element to a slightly heated one.. For comparison, 2 variants of the design and the thermal model of the microsystem, presented in Figure. 1, are proposed.

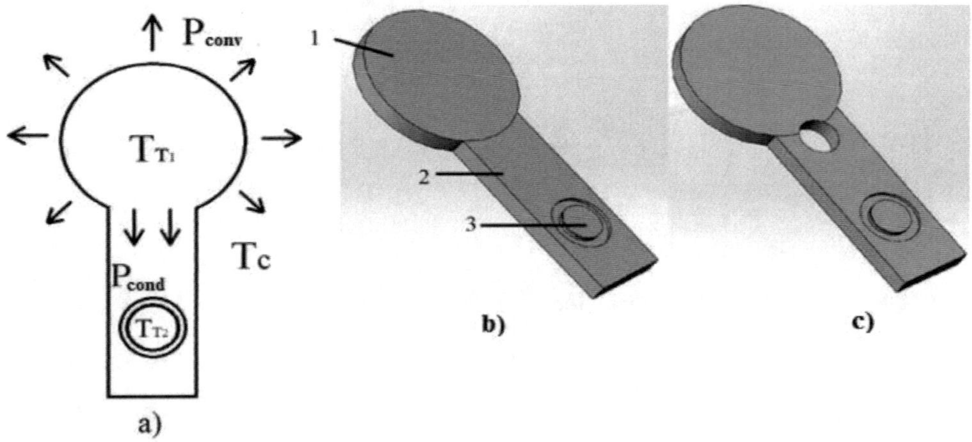

Figure 1. Thermal model of microsystem (a), microsystem design without hole (b), microsystem design with hole (c). 1 - the site of the heat-loss anemometer; 2 - the leg; 3 - the site of the thermometer.

We write down the original Navier-Stokes equations [6]:

$$\rho\frac{Du}{Dt}=X-\frac{\partial\rho}{\partial x}+\frac{\partial}{\partial x}\left[\mu\left(2\frac{\partial u}{\partial x}-\frac{2}{3}divm\right)\right]+\frac{\partial}{\partial y}\left[\mu\left(\frac{\partial u}{\partial y}+\frac{\partial v}{\partial x}\right)\right]+\frac{\partial}{\partial z}\left[\mu\left(\frac{\partial\omega}{\partial x}+\frac{\partial u}{\partial z}\right)\right],$$
$$\rho\frac{Dv}{Dt}=Y-\frac{\partial\rho}{\partial y}+\frac{\partial}{\partial y}\left[\mu\left(2\frac{\partial v}{\partial y}-\frac{2}{3}divm\right)\right]+\frac{\partial}{\partial z}\left[\mu\left(\frac{\partial v}{\partial я}+\frac{\partial\omega}{\partial y}\right)\right]+\frac{\partial}{\partial x}\left[\mu\left(\frac{\partial u}{\partial y}+\frac{\partial v}{\partial x}\right)\right], \quad (1)$$
$$\rho\frac{D\omega}{Dt}=Z-\frac{\partial\rho}{\partial z}+\frac{\partial}{\partial z}\left[\mu\left(2\frac{\partial\omega}{\partial z}-\frac{2}{3}divm\right)\right]+\frac{\partial}{\partial x}\left[\mu\left(\frac{\partial\omega}{\partial x}+\frac{\partial u}{\partial z}\right)\right]+\frac{\partial}{\partial y}\left[\mu\left(\frac{\partial v}{\partial z}+\frac{\partial\omega}{\partial y}\right)\right].$$

Here are the components of the velocity field in the x, y, z coordinate system; ρ - is the density; t - is the time constant.

To these equations, we must add the continuity equation, which in the expanded form has the following form:

$$\frac{\partial\rho}{\partial t}+\frac{\partial(\rho u)}{\partial x}+\frac{\partial(\rho v)}{\partial y}+\frac{\partial(\rho\omega)}{\partial z}=0, \quad (2)$$

In addition to modeling with the help of equations (1) - (3) of heat diffusion in a fluid, heat transfer in solids is also modeled using the equation:

$$\frac{\partial\rho\cdot e}{\partial t}=\frac{\partial}{\partial x_i}\cdot\left(\lambda\cdot\frac{\partial T}{\partial x_i}\right)+Q_n \quad (3)$$

Where $e=c\cdot T$(c is the specific heat; T-temperature); λ - thermal conductivity; Q_n - specific (per unit volume) heat release of the heat source.

Convective heat exchange between the surface of solids and the fluid is simulated when modeling the boundary layer of the fluid flow.

The radiating surfaces are defined by absolutely black, absolutely white, or ideally gray, so that, in accordance with Lambert's law, their radiation is assumed to be diffuse, i.e, with a brightness that does not depend on the direction of radiation. As a result, in accordance with the Stefan-Boltzmann law, the heat emitted from a surface unit is defined as:

$$Q_n = \varepsilon \cdot \delta_0 \cdot T_w^4, \tag{4}$$

where ε is the degree of blackness of the surface; - the Stefan-Boltzmann constant; Surface temperature. Correspondingly, the absorption and / or reflection of radiative heat by the surfaces participating in radiative heat exchange are modeled [5]. These equations were solved by the finite volume method.

For the process of distribution of the temperature field studied, the following boundary conditions were set: on the surface of the model, a boundary condition of the third kind is given.

$$\lambda_n \frac{\partial T}{\partial n} = -\alpha_k (T - T_c) - \beta (T^4 - T_c^4) \tag{5}$$

Considering that the edge of the leg is a drain of thermal energy due to conductive connections with the "cold" environment, its temperature was set as a constant. (A boundary condition of the first kind $(T = T_c)$).

The necessary boundary conditions for modeling were set as follows:

1. The temperature of the "cold" edge of the investigated model $T = 20$°C
2. The value of the electric current passing through the circular area of the heat – loss anemometer, with the following values: 1; 5; 10; 15; 20; 30; 40; 60; 80; 100; 120 mA.
3. Ambient temperature (air) t = const = 22°C,
4. The coefficient of thermal conductivity (at t = 300 °C) λ = 490 Wt/m · K;
5. Seebeck coefficient (at t = 300°C) $\alpha = 0{,}009\ W/K$;
6. Electrical resistance of silicon carbide R = 0,01 Om · mm.

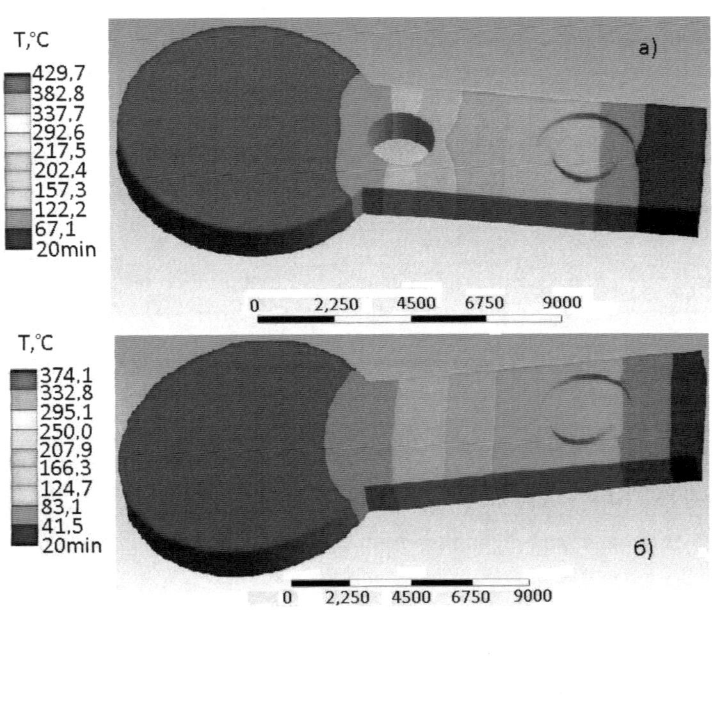

Figure 2. Temperature field in the microsystem.

Qualitative analysis of the temperature field on the color images, the shape and nature of the distribution of isotherms in the microsystem allowed us to identify the following features (see Fig.2). In the limit of the radiating site, the temperature field can be characterized as uniform. While in the area of the base of the structure there is clearly seen a strong unevenness in the distribution of temperatures. At the same time, the maximum temperatures in both cases differ significantly (374 ^{0}C in the first case and 427 ^{0}C in the second case). This is due to the presence in one of the models of a through hole, which prevents the free flow of heat flow, which, when encountering an obstacle, lingers on a circular platform. It should be emphasized that the temperature dependence of the SiC thermal conductivity, in particular its decrease with increasing temperature, plays an essential role in the processes considered [7].

According to the obtained data, curves for the dependence of the electric current on the temperature of the model under study at two points, the hottest point (1), and the coldest point (2) itself, shown in Fig. 3, are plotted. Figure 3 shows that the thermal model with a through hole is preferable, in consequence of the fact that the temperature of the thermo-anemometer in working condition has less effect on the thermometer located on the foot of the microsystem. In this case, the inverse thermal effect is excluded from the calculation, since the working temperature of the thermometer is assumed to be equal to the temperature of the medium being measured.

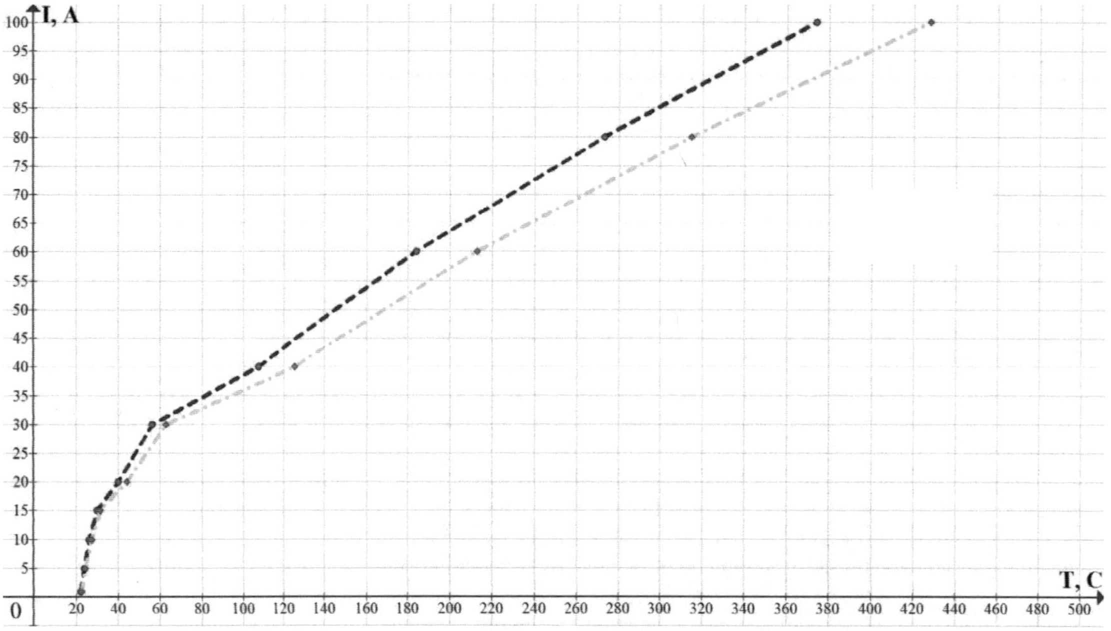

Figure 3. Graph of the temperature dependence of the microsystem at the "hot spot" itself on the magnitude of the electric current.

The obtained results can be regarded as a prerequisite for experiment on real prototypes.

Conclusions

In the course of the study, a method for calculating the micro-system was proposed, showing the mutual thermal effect of the elements.

According to the data of the study, curves for the dependence of the electric current on temperature are plotted, as shown in Fig. 3.

Based on the results of the simulation, it can be judged that the tested versions of the microsystem designs have significant temperature differences >50 °C in the presence of small design solutions.

References

[1] Baicar R I, Varshava S S, Potapchyk G N and Chekyrin V F 1994 Pribori I tehnika exsperimenta *Devices and experiment technique* **3** pp 158-163

[2] Karachinov V A 1995 Termoanemometer na osnove karbida kremnija *Silicon carbide thermoanemometer* Works of the international seminar «Semiconductor silicon carbide and devices based on it» Novgorod pp 72-73

[3] Yanyushkin A S, Medvedeva O I, Saprykina N A 2014 Mechanism of Protective Membrane Formation on the Surface of Metal-Bonded Diamond Disks Applied *Mechanics and Materials* **682** pp 327-331

[4] Saprykin A A, Ibragimov E A, Babakova E V 2016 Modeling the Temperature Fields of Copper Powder Melting in the Process of Selective Laser Melting *IOP Conference Series: Materials Science and Engineering* **142(1)** 012061

[5] Aljamovskijj A A, Sobachkin A A and Odincov E V 2005 Solid Works *Komp'juternoe modelirovanie v inzhenernojj praktike*[Computer simulation in engineering practice] SPb pp 799

[6] Shlikhting G 1974 Teoria pogranichnogo sloya *Boundary layer theory* Moscow Science pp 70-72

[7] Lykov A V 1978 Teplomassoobmen: Spravochnik *Heat and mass transfer: a directory* Moscow Engineering pp 560

IOP Conf. Series: Materials Science and Engineering **441** (2018) 012017 doi:10.1088/1757-899X/441/1/012017

Technology of production and magnetoelectric characteristics of multilayer structures nickel-tin on the gallium arsenide substrate

D A Filippov[1], I N Manicheva[1], K A Bordashev[1], V M Laletsin[2], T A Galichyan[3]

[1]Novgorod State University 41, B. St. Peterburgskaya, Velikiy Novgorod, 173003, Russia
[2]Institute of Technical Acoustics National Academy of Sciences of Belarus 13, Lyudnikava Praspekt, Vitebsk, BY-210023 Belarus
[3] Institute of Mechanics, National Academy of Sciences of Armenia, 24B Baghramyan Ave., Yerevan, 0019 Armenia

E-mail: Dmitry.Filippov@novsu.ru

Abstract The technology of fabrication and the results of the magnetoelectric effect investigation in sandwich structure manufactured by galvanic deposition tin and nickel on the gallium arsenide substrate are presented. It is shown that the use of tin as an intermediate layer lead to reduces the mechanical stresses resulting on the interface nickel and gallium arsenide. It is possible to obtain qualitative structures with nickel layer thickness on the order of 100 microns. Experimental results of the frequency dependence of the magnetoelectric voltage coefficient in the region of electromechanical resonance are presented. The resonance value of the magnetoelectric voltage coefficient reached 40V/(cm·Oe) with the Q-factor ≅ 700, which significantly exceeds the characteristics of similar structures obtained by bonding.

1. Introduction

Composite magnetostrictive-piezoelectric materials attract attention by the fact that the magnitude of the magnetoelectric (ME) effect in them is several orders of magnitude greater than in single crystals. This is explained by the fact that the mechanism of the ME effect in single crystals is the joint action of spin-orbital interaction, the interaction of the electron with the external and intracrystalline electric field [1]. Despite the fact that currently there are dozens of single crystals in which the ME effect is found [2], however, due to its smallness, the ME single crystals have not found wide application in technology. The mechanism of the ME effect in composites is the mechanical interaction of magnetostrictive (MS) and piezoelectric (PE) phases. An alternating magnetic field causes mechanical deformations in the MS component, which are transmitted across the interface to the PE component, which leads to a change in polarization and the appearance of electrical voltage. Composite materials can be divided into groups: bulk and layered composites. Bulk composites are mechanically linked mixtures of MS and PE phase powders [3]. Layered composites are structures consisting of alternating layers of magnetic and piezoelectric [4]. Layered structures have several advantages compared to bulk composites: easily polarized, are small in leakage currents [5], as the MS phase, it's possible to use metals having a large coefficient of magnetostriction [6-8]. At the same time, they have a number of disadvantages due to the boundary between the MS and the PE phases. Most layered me structures are obtained by gluing, which leads to a decrease in the quality of the structure, the weakening of the ME effect, undesirable high temperature dependence. In [9.10] the results of an investigation of the ME effect in structures were reported, where the magnetostrictive phase was applied to the piezoelectric

substrate by the sputtering method.
The resulting structures had good adhesion, but had a small value of the ME effect. As shown in [11] the maximum of the ME effect is observed under the condition of equality ${}^{p}t\sqrt{{}^{p}Y} = {}^{m}t\sqrt{{}^{m}Y}$, where ${}^{p}Y, {}^{m}Y$ are the Young's moduli of the piezoelectric and magnetic phases, and ${}^{p}t, {}^{m}t$ are their thickness. Since the Young's moduli of a magnetic and a piezoelectric phase do not differ by more than a factor of two, the maximum effect is observed when the thickness of the magnetic phase is commensurable with the thickness of the piezoelectric layer. It is impossible to fabricate such structures by the sputtering method. Using the method of electrolytic deposition of metal on the piezoelectric substrate allows to obtain layers of a magnetic phase whose thickness is commensurable with the thickness of the piezoelectric phase. In order to improve the adhesion between phases during fabrication, it is advisable to use structures preliminarily deposited on a GaAs substrate by Au-Ge-Ni sublayers [12]. However, due to the incommensurability of the lattice parameters of Ni and GaAs at large layer thicknesses, mechanical stresses arise, leading to warping of the structure and its destruction. One of the methods for eliminating these stresses, proposed in this paper, is the method of creating a sandwich structure in which the nickel layer alternates with a buffer layer of tin.

2. Technology of manufacturing structures

Samples in the form of a parallelepiped with dimensions $11 \times 5 \times 0.4$ mm were cut from GaAs plates with a surface orientation (100), the long side of which coincided with the direction of the <011> crystal (Figure 1). To improve the adhesion to the samples, Au-Ge-Ni sublayers were previously deposited. In the production of the multilayer structure, electrolytic deposition was used alternately in the sulfuric acid electrolyte of nickel, with a cathode current density of 1 A/dm^2 and an electrolyte temperature of 55-65 °C, and then electrodeposition in the tin electrolyte at room temperature and a cathode current density of 2 A/dm^2. As a result, the resulting multilayer structure consisted of sixteen layers of nickel on each side of the sample, with a total thickness of 100 µm and seventeen layers of tin, with a total thickness of 200 µm. The total thickness of the sandwich structure, taking into account the substrate, was 1 mm. The layers on gallium arsenide had an even, matte surface, without visible defects.

3. Magnetoelectric effect

The magnetoelectric (ME) effect in the structure was studied by measuring the voltage on the sample when placed in a constant (magnetizing) and alternating magnetic fields directed along the long side of the sample (Figure 1).

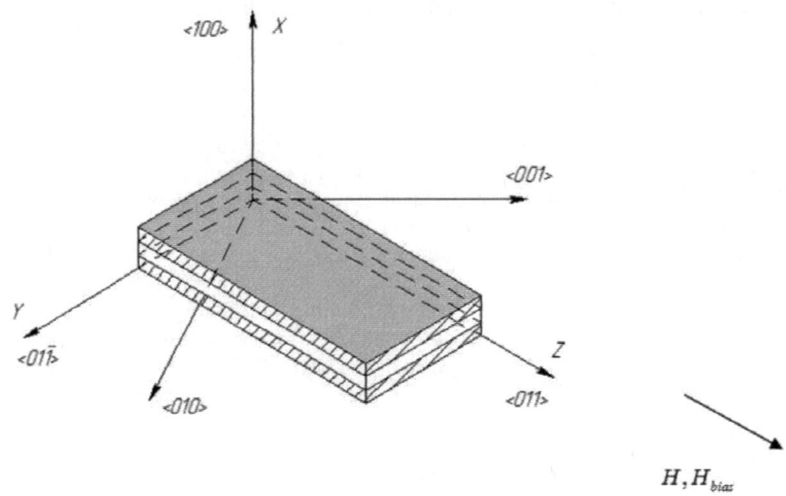

Figure1. The schematic drawing of the sandwich structure

At first, the field dependence of the low-frequency ME signal was investigated. At a constant value of the strength of the alternating magnetic field *H=1 Oe*, the dependence of the ME coefficient on the strength of the bias field H_{bias} was measured. Then, with the magnetization field strength corresponding to the maximum of the effect, the frequency dependence of the ME coefficient in the electromechanical resonance region was investigated. In gallium arsenide have nonzero components $d_{14}= d_{25}= d_{36}$ of the piezoelectric tensor, therefore, the electric voltage induced on the sample plates arises from shear deformations rather than deformations tension-compression as in PZT [11]. The alternating magnetic field directed along the long side of the sample (the *Z* axis) induces deformations of tension-compression in the magnetic component whose tensor, in the coordinate system associated with the sample, will be denoted by S_{zz}. These deformations are transferred to the piezoelectric, which leads to the appearance of an electrical voltage on the plates. To find the ME characteristics of the sandwich structure, we use the method developed earlier for a two-layer structure [13]. We use the fact that the thickness of the piezoelectric, magnetic and buffer layer is much smaller than the length of the sample, so in the first approximation it can be assumed that the displacements of the layers are the same and do not vary in the thickness of the sample. In this approximation, the equation of motion for the z-projection of the displacement vector of the medium is written in the form:

$$\bar{\rho}\frac{\partial^2 u}{\partial t^2} = \frac{\partial \bar{T}_{zz}}{\partial z}, \qquad (1)$$

where $\bar{\rho}$ is the mean value of the sample density, and $\bar{T}_{zz}$ is the mean value of the stress tensor in the sample. The equations for components of the strain tensor of piezoelectric ${}^{p}S_{zz}$ and magnetic ${}^{m}S_{zz}$, and electric induction ${}^{p}D_{x}$ are as follows:

$$ {}^{p}S_{zz} = \frac{1}{{}^{p}Y}\,{}^{p}T_{zz} + {}^{p}d_{x,zz}\,{}^{p}E_{x}, \tag{2}$$

$$ {}^{m}S_{zz} = \frac{1}{{}^{m}Y}\,{}^{m}T_{zz} + {}^{m}q_{z,zz}\,{}^{m}H_{x}, \tag{3}$$

$$ {}^{p}D_{x} = {}^{p}\varepsilon_{xx}\,{}^{p}E_{x} + {}^{p}d_{x,zz}\,{}^{p}T_{zz}, \tag{4}$$

where ${}^{p}T_{zz}$, ${}^{m}T_{zz}$ are components of the stress tensor in the piezoelectric and magnetostrictive phase; ${}^{p}Y$, ${}^{m}Y$ are the Young`s modulus in the piezoelectric phase along the <011> (Z-axis) direction, and the magnetic phase, respectively, ${}^{p}d_{x,zz}$ is the piezoelectric tensor in the XYZ coordinate system (Figure 1), ${}^{p}\varepsilon_{xx}$ is the dielectric tensor and ${}^{p}E_{x}$ is the x component of the electric field strength vector, ${}^{m}q_{z,zz}$ is the piezomagnetic coefficient, ${}^{m}H_{x}$ is the magnetic field strength.

The components of the piezoelectric tensor ${}^{p}d_{x,zz}$ in the XYZ coordinate system are related to the components of the piezoelectric tensor ${}^{p}d_{\alpha,\beta}$ in the crystallographic coordinate system by the relation:

$$ d_{x,zz} = d_{14}\beta_{z2}\beta_{z3}, \tag{5}$$

where β_{z2}, β_{z3} is the matrix of cosines between the Z axis and the axes 2 and 3 (<010> and the direction <001>).

The solution of equation (1) for the displacement vector of the medium is represented in the form of plane waves propagating along the length of the sample:

$$ u(z) = A\cos(kz) + B\sin(kz), \tag{6}$$

where A and B are the integration constants.

Substituting expression (6) into equation (1), we obtain for them the dispersion relation in the form:

$$ \omega = \sqrt{\frac{\bar{Y}}{\bar{\rho}}}\,k = \sqrt{\frac{{}^{m}Y\,{}^{m}t + {}^{l}Y\,{}^{l}t + {}^{p}Y\,{}^{p}t}{{}^{m}\rho\,{}^{m}t + {}^{l}\rho\,{}^{l}t + {}^{p}\rho\,{}^{p}t}}\,k, \tag{7}$$

where $\bar{Y} = \dfrac{{}^{m}Y\,{}^{m}t + {}^{l}Y\,{}^{l}t + {}^{p}Y\,{}^{p}t}{{}^{m}t + {}^{l}t + {}^{p}t}$ the average Young's modulus of the sandwich structure.

4. Magnetoelectric voltage coefficient

The magnetoelectric voltage coefficient is defined as the ratio of the average intensity of the electric field <E> to the intensity of the alternating magnetic field H, t. it is.

$$ \alpha_E = \langle E\rangle / H, \tag{8}$$

where $\langle E\rangle = U/({}^{m}t + {}^{l}t + {}^{p}t)$ is the average value of the electric field in the structure and U is the voltage induced between the electrodes.

To obtain the expression for the ME coefficient, we use the method developed earlier in (Filippov, 2013, 2014). From the condition of mechanical equilibrium at the free sidewalls of the sample, i.e. at the points $z = \pm L/2$ we have the following boundary conditions:

$$ \int_{0}^{{}^{p}t} {}^{p}T_{zz}(\pm L/2, x)\,dx + \int_{{}^{p}t}^{{}^{l}t} {}^{l}T_{zz}(\pm L/2, x)\,dx + \int_{{}^{l}t}^{{}^{m}t} {}^{m}T_{zz}(\pm L/2, x)\,dx = 0 \tag{9}$$

Using these boundary conditions, for integration constants we obtain:

$$ A = 0, \qquad B = \frac{{}^{m}Y\,{}^{m}t\,{}^{m}q_{z,zz}\,{}^{m}H_{x} + {}^{p}Y\,{}^{p}t\,{}^{p}d_{x,zz}\,{}^{p}E_{x}}{k\cos(\kappa)}\,\frac{1}{({}^{m}t\,{}^{m}Y + {}^{l}t\,{}^{l}Y + {}^{p}t\,{}^{p}Y)}, \tag{10}$$

where a dimensionless parameter $\kappa = kL/2$ is introduced. Having expressed the component of the stress tensor in terms of the components of the strain tensor from Eq. (2) and substituting the resulting expression into the equation for the normal component of the electric induction vector, we obtain to the expression in the form:

$$^{p}D_{z} = {}^{p}\varepsilon_{xx}\,{}^{p}E_{x} + {}^{p}Y\,{}^{p}d_{x,zz}\frac{\partial\,{}^{p}u_{z}}{\partial z} - {}^{p}Y({}^{p}d_{x,zz})^{2}\,{}^{p}E_{x}. \tag{11}$$

To determine the electric field $^{p}E_{x}$ induced in a piezoelectric, we use of the open-circuit condition:

$$I = \iint \frac{\partial D_{x}}{\partial t} dz dy = 0. \tag{12}$$

Substituting expression (11) into equation (12) and carrying out the integration, we obtain:

$$^{p}E_{x} = \frac{^{p}Y\,{}^{p}d_{x,zz}\,{}^{m}q_{x,zz}}{^{p}\varepsilon_{xx}}\frac{^{m}H_{x}}{\Delta}\frac{^{m}Y\,{}^{m}t}{(^{m}t\,{}^{m}Y + {}^{l}t\,{}^{l}Y + {}^{p}t\,{}^{p}Y)}\frac{\tan(\kappa)}{\kappa}, \tag{13}$$

where the notation

$$\Delta = 1 - k_{p}^{2}\frac{^{p}t\,{}^{p}Y}{(^{m}t\,{}^{m}Y + {}^{l}t\,{}^{l}Y + {}^{p}t\,{}^{p}Y)}\frac{\tan(\kappa)}{\kappa}. \tag{14}$$

Using the definition of the magnetoelectric coefficient (8), taking into account the fact that the voltage induced between the electrodes is $U = {}^{p}E_{x}\,{}^{p}t$. Hence we obtain the expression for the ME voltage coefficient

$$\alpha_{E} = \frac{^{p}Y\,{}^{p}d_{x,zz}\,{}^{m}q_{z,zz}}{^{p}\varepsilon_{xx}(^{m}t\,{}^{m}Y + {}^{l}t\,{}^{l}Y + {}^{p}t\,{}^{p}Y)}\frac{^{m}Y\,{}^{m}t}{\Delta}\frac{\tan(\kappa)}{\kappa} \tag{15}$$

Equation (15) determines the frequency dependence of the ME voltage coefficient. This dependence is determined by the parameter $\kappa = kL/2$, and the wave number k is determined by the dispersion relation (6). Therefore, changing the composition it is possible to obtain different frequency characteristics. In Figure.2 the results of experimental measurements of the frequency dependence of the ME voltage coefficient for a structure consisting of sixteen layers of nickel on each side of the sample, a total thickness of 100 microns and seventeen layers of tin, a total thickness of 200 microns. The total thickness of the sandwich structure, taking into account the substrate was 1 mm. Layers on gallium arsenide had a smooth, matte surface, without visible defects.

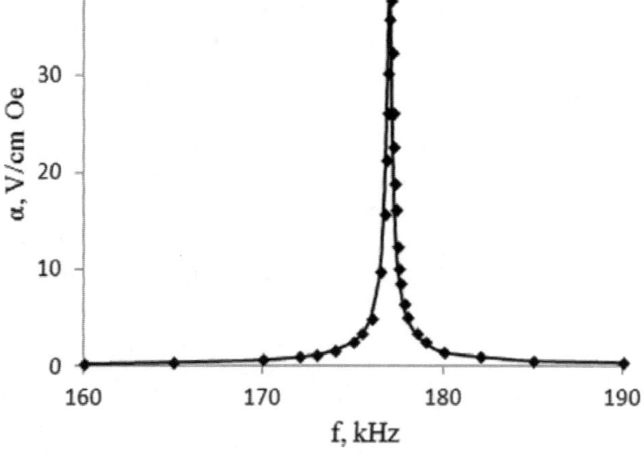

Figure2. Frequency dependence of the sandwich structure gallium arsenide-nickel-tin-nickel.

The bias field H_{bias}=360 Oe

As follows from the figure, the frequency dependence has a resonant character. The value of the ME voltage coefficient is somewhat lower than in the pure structure of nickel-gallium arsenide [12]. This structure has high the quality factor Q≅700, which is much better than the quality factor of the samples obtained by the gluing method and is comparable to the quality factor of bulk composites.

5. Conclusion

The use of an intermediate tin layer in the electrolytic deposition of nickel on an arsenide-gallium substrate makes it possible to obtain structures with a nickel layer thickness of up to 100 μm. These structures have good adhesion between layers, have good mechanical strength. Such multilayer structures are promising for designing devices based on the ME effect.

Reference

[1] Bichurin M I, Filippov D A 1997 *Ferroelectrics* **204**, 225-232
[2] Zvezdin A K, Pyatakov A P 2012 *Phys. Usp.* **55**, 557-581
[3] Laletin V M, Srinivasan G 2002 *Ferroelectrics* **280**,177-185
[4] Srinivasan G, Rasmussen E T, Gallegos J, Srinivasan R, Yu I, Bokhan I, Laletin V M 2001 *Physical Review B*. **64**, 214408 (1-6)
[5] Filippov D A, Laletin V M, Srinivasan G 2012 *Technical Physics* **57**, 44-47
[6] Fetisov Y K, Petrov V M, Srinivasan G, *J. Mater.Res.***22**, 2074-2081
[7] Fetisov Y K, Kamentsev K E, Chashin D V, Fetisov L Y, Srinivasan G 2009 *J.Appl.Phys.***105**, 123918(1-4)
[8] Filippov D A, Radchenko G S, Laletin V M 2016 *Physics of the Solid State* **58**, 508-514
[9] Gridnev S A, Kalinin Y E , Kalgin A V, Grigor'ev E S 2015 *Physics of Solid State* **57**, 1372-1376
[10] Laletin V M , Stognij A I , Novitskij N N, Poddubnaya N N 2014 *Technical Physics Letters*.**40**, 969-971
[11] Filippov D A, Laletin V M, Galichyan T A 2013 *Physics of Solid State* **55**, 1840-1845
[12] Filippov D A, Firsova T O, Laletin V M, Poddubnaya N N 2017 *Technical Physics Letters* **43**, 313-315
[13] Filippov D A, Laletin V M, Galichyan T A 2014 *Applied Physics A*. **115**, 1087-1091

Using AR(1) model to simulate strictly stationary random sequences

Sergey Garbar

KPMI, Novgorod State University, ul. B. St. Petersburgskaya, 41 173003 Veliky Novgorod, Russia
Sergey.Garbar@novsu.ru

Abstract. We consider using AR(1) model for simulating strictly stationary random sequences and propose a method finding corresponding PDF of independent random variable in the model. Random sequences with normal, uniform, exponential and Erlang distributions are the considered examples of using that algorithm.

1. Introduction

Simulated random processes can be used to model objects and phenomena of different nature. Stationary random processes have an important role for describing time series with parameters such as the mean, variance and autocorrelation structure that do not change over time and do not follow any trends.

Autoregression model of n-th order $AR(n)$

$$Y(t) = a_1Y(t-1) + a_2Y(t-2) + \cdots + a_nY(t-n) + bX(t)$$

can be used to describe wide-sense stationary random processes if the complex roots of the characteristic polynomial $\lambda^n - \sum_{i=1}^{n} a_i\lambda^{n-i} = 0$ lie inside the unit circle. This model can be considered to be an algorithm that transforms a sequence $\{X(t)\}$ of independent identically distributed random variables $X(t)$ into a sequence $\{Y(t)\}$ with some non-zero correlation function $R(\tau)$.

If the distribution of the added random variable $X(t)$ is normal, then the process will also be normal and strictly stationary. That happens because next member of the sequence is determined as a linear combination of previous and an added random variable, and it is known that the linear combination of normally distributed random variables is also normally distributed. In [1] authors propose to transform strictly stationary normal sequences using probability integral transform. The necessity to inverse non-elementary cumulative distribution function of Gaussian is a downside of that approach.

In [2] author proposes to simulate strictly stationary random sequences using autoregression coefficients with stochastic binary orthogonal coefficients. The downside of resulting sequences is that the next value is either independent of all the previous or equals to one of them.

Below we consider autoregression model of first order $AR(1)$:

$$Y(t) = aY(t-1) + bX(t),$$

where $a, b \geq 0$, $a + b = 1$. With these limitations the simulated sequence will be stationary in wide sense. Correlation coefficient between the adjacent sequence members is $R(1) = a$. Correlation function is exponential

$$R(i) = a^i.$$

The task of this article is finding the probability density function (PDF) $f(x)$ of added random variables $X(t)$ such that the PDF of next sequence member $Y(t)$ will be a preassigned $g(y)$. It continues the work described in [3]. To examine the properties of generated sequences the joint probability density functions of adjacent sequence members are found.

2. Using AR(1) model to simulate a strictly stationary random sequence

Note that random variables $Y(t-1)$ and $X(t)$ are independent, therefore PDF for their linear combination used in AR(1) model $Y(t) = aY(t-1) + bX(t)$can be found fairly straightforward. As we require strict stationarity of resulting sequence, PDFs of $Y(t-1)$ and $Y(t)$ are the same, and we can write a relation that can be solved as an integral equation on $f(x)$

$$g(y) = \int_{-\infty}^{+\infty} g\left(\frac{y-bx}{a}\right) f(x) \frac{1}{a} dx.$$

That equation can be solved directly for some PDFs $g(y)$, but the use of characteristic function properties gives a convenient solution for some PDFs.

Let $\psi(t)$ be a characteristic function of $Y(t-1)$ and $Y(t)$ and $\varphi(t)$ — a characteristic function of $X(t)$. As random variables $aY(t-1)$ and $bX(t)$ are independent, their product's characteristic function is a product of corresponding characteristic functions

$$\psi(t) = \psi(at) \cdot \varphi(bt).$$

Therefore a characteristic function $\varphi(t)$ of $X(t)$ can be expressed as

$$\varphi(bt) = \frac{\psi(t)}{\psi(at)}.$$

Having this expression we can write out a relation for $f(x)$ in a form

$$\frac{1}{b} \cdot f\left(\frac{x}{b}\right) = \frac{1}{2\pi} \int_{-\infty}^{\infty} e^{-itx} \frac{\psi(t)}{\psi(at)} dt.$$

If $g(y)$ is symmetrical relatively to some x_0, taking $(2y_0 - Y(t))$ instead of $Y(t)$ won't change the PDF, but the sign of correlation coefficient $\mathrm{corr}(Y(t-1), 2y_0 - Y(t))$ will be changed. That way we can simulate random sequences with negative $R(1)$ in certain cases. Important example of such distribution is a uniform distribution.

2.1. Examples

Shown below are some examples of finding PDF for added random variable in AR(1) model-based algorithm for simulating strictly stationary random sequences.

2.1.1. *Normal distribution* is one of the most important probability distributions due to central limit theorem.

PDF of normal distribution is

$$f(x) = \frac{1}{\sigma\sqrt{2\pi}} e^{-\frac{(x-m)^2}{2\sigma^2}}.$$

Below we consider standard normal distribution with parameters $m = 0$ and $\sigma = 1$ as other their values can be attained with a linear transformation that does not alter the correlation properties.

When using AR(1) model to simulate random sequence with standard normal distribution, added random variable $X(t)$ should have PDF

$$f(x) = \frac{\sqrt{1-a}}{\sqrt{2\pi}\sqrt{1+\mathrm{a}}} e^{-\frac{x^2(1-a)}{2(1+\mathrm{a})}},$$

which is a PDF for random variable with normal distribution with $m = 0$ and $\sigma = \sqrt{\frac{1+a}{1-a}}$.

That trivial statement can be obtained if we consider the characteristic function of the normal distribution

$$\varphi(t) = e^{-\frac{t^2}{2}}.$$

Therefore

$$\frac{1}{b} f\left(\frac{x}{b}\right) = \frac{1}{2\pi} \int_{-\infty}^{\infty} e^{-itx} \frac{e^{-\frac{t^2}{2}}}{e^{-\frac{a^2t^2}{2}}} dt.$$

That gives us the mentioned before statement.

2.1.2. Uniform distribution is denoted $X \sim U[x_1; x_2]$ for $x_1 < x_2$ and has PDF.

$$f(x) = \begin{cases} \dfrac{1}{x_2 - x_1}, & \text{при } x \in [x_1; x_2], \\ 0, & \text{при } x \notin [x_1; x_2]. \end{cases}$$

Below we consider the standard uniform distribution with $x_1 = 0$ и $x_2 = 1$. It is considered to be a basic distribution in random variables modeling as it can be used in inverse transform sampling method.

If $a = 1/(n+1), n \in \mathbb{N}$ then the added random variable $X(t)$ in AR(1)-based algorithm for simulating random sequences with standard uniform distribution should have PDF

$$f(x) = a \sum_{k=0}^{n} \delta\left(x - \frac{k}{n}\right) = \frac{1}{n+1} \sum_{k=0}^{n} \delta\left(x - \frac{k}{n}\right),$$

where $\delta(x)$ is Dirac's delta function. That way each step of algorithm supposes to add an independent random variable with discrete uniform distribution in points $0, 1/n, 2/n, \ldots, (n-1)/n, 1$.

To find the given PDF we shall consider the characteristic function of standard uniform distribution.

$$\psi(t) = \frac{e^{it} - 1}{it}.$$

Therefore the characteristic function for $bX(t)$ is

$$\varphi(bt) = \frac{\frac{e^{it} - 1}{it}}{\frac{e^{iat} - 1}{it}} = \frac{e^{it} - 1}{e^{iat} - 1}.$$

As we only consider $a = 1/(n+1)$

$$\varphi(bt) = \frac{\left(e^{iat}\right)^{n+1} - 1}{e^{iat} - 1} = \left(e^{iat}\right)^n + \left(e^{iat}\right)^{n-1} + \cdots + e^{iat} + 1.$$

And PDF for $X(t)$ is

$$\frac{1}{b}\cdot f\left(\frac{x}{b}\right) = \delta(x-an) + \delta\big(x-a(n-1)\big) + \cdots + \delta(x-a) + \delta(x).$$

That relation leads to the PDF given above.

Joint PDF of adjacent sequence members $Y(t)$ and $Y(t+1)$ is

$$g(y_1, y_2) = \begin{cases} \sum_{k=0}^{n} \delta\left(\frac{y_2}{a} - y_1 - ka\right), y_1 \in [0;1] \text{ и } y_2 \in [0;1], \\ 0,\ y_1 \notin [0;1] \text{ или } y_2 \notin [0;1]. \end{cases}$$

As one can see, the support of the joint PDF is a set of line segments

$$y_2 = ay_1 + ka^2, k = 0..n,$$

therefore the simulated sequence might be not suitable for simulating some processes. In [4] the AR(1) algorithm was modified that it allowed simulating strictly stationary random sequences with standard uniform distribution and specified value of autocorrelation function $R(1)$ in $[-0.625, 0.625]$.

Example of support for $a = 0.2$ is shown on figure 1. Figure 2 shows 1000 point with the coordinates $\big(Y(t); Y(t+1)\big)$, where $Y(t)$ and $Y(t+1)$ are the members of the simulated sequence.

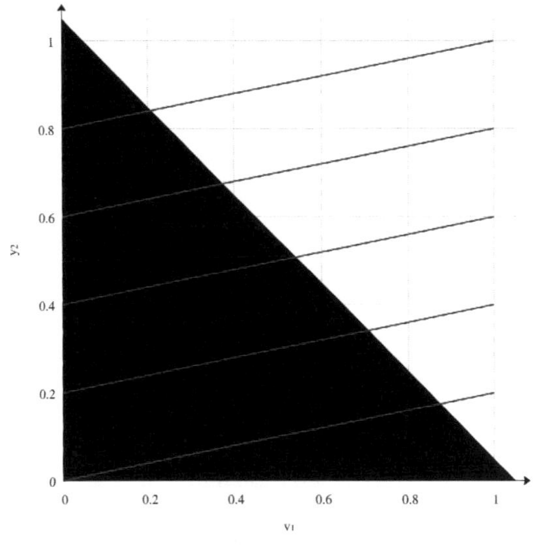

Figure 1. Support of joint PDF of $Y(t)$ and $Y(t+1)$ for $a = 0.2$

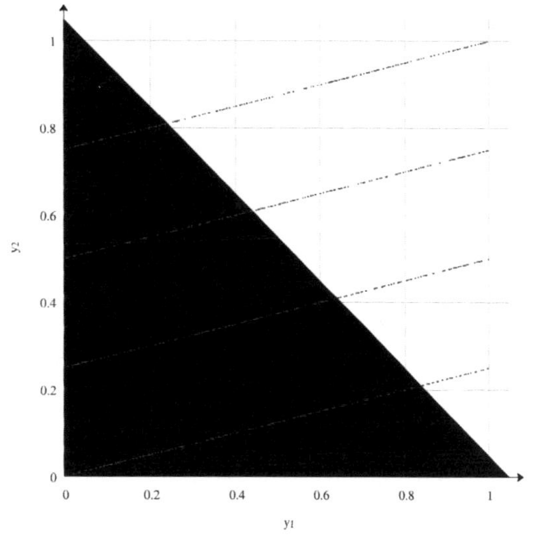

Figure 2. Location of 1000 members of simulated sequence as points with coordinates $\big(Y(t); Y(t+1)\big)$ for $a = 0.25$

2.1.3. *Exponential distribution.* The PDF has one parameter $\lambda > 0$.

$$g(y;\lambda) = \begin{cases} \lambda e^{-\lambda y}, & y \geq 0, \\ 0, & y < 0. \end{cases}$$

When we use AR(1) model to simulate random sequence with exponential distribution, the PDF of $X(t)$ is

$$f(x) = a\delta(x) + bg(x; \lambda b).$$

where $\delta(x)$ is Dirac's delta function, and $g(x; \lambda b)$ is a PDF of exponential distribution with parameter λb $(b > 0)$:

$$g(x; \lambda b) = \begin{cases} \lambda b e^{-\lambda b x}, & x \geq 0, \\ 0, & x < 0. \end{cases}$$

Therefore each step of simulating algorithm supposes to add an independent random variable, which is a mixture of exponential random variable and non-random value $X = 0$ with corresponding weights b and a.

That form of PDF can be obtained if we consider the characteristic function of exponential distribution

$$\psi(t) = \left(1 - \frac{it}{\lambda}\right)^{-1}.$$

Therefore

$$\frac{1}{b} f\left(\frac{x}{b}\right) = \frac{1}{2\pi} \int_{-\infty}^{+\infty} e^{-itx} \frac{1 - \frac{ita}{\lambda}}{1 - \frac{it}{\lambda}} dt = a \cdot \frac{1}{2\pi} \int_{-\infty}^{+\infty} e^{-itx}\, dt + \frac{1}{2\pi} \int_{-\infty}^{+\infty} b \left(1 - \frac{it}{\lambda}\right)^{-1} e^{-itx}\, dt.$$

Which leads to the given above PDF.

Joint PDF of adjacent sequence members $Y(t)$ and $Y(t+1)$ is

$$g(y_1, y_2) = \begin{cases} \lambda e^{-\lambda y_1}\left(a\delta(a y_1 - y_2) + \lambda\, b\, e^{\lambda(a y_1 - y_2)}\right), & y_1 \geq 0 \text{ and } y_2 \geq a y_1, \\ 0, & y_1 < 0 \text{ or } y_2 < a y_1. \end{cases}$$

2.1.4. Erlang distribution is a two parameter family of continuous probability distributions with the PDF

$$f(x; k, \lambda) = \begin{cases} \dfrac{\lambda^k x^{k-1} e^{-\lambda x}}{(k-1)!}, & x \geq 0, \\ 0, & x < 0. \end{cases}$$

Parameter $\lambda > 0$ is the "rate" and $k \in \mathbb{N}$ is the "shape.

The Erlang distribution with shape parameter $k = 1$ is the exponential distribution.

The chi-squared distribution with even number of degrees of freedom is a special case of the Erlang distribution.

It is a special case of the Gamma distribution. It is the distribution of a sum of k independent exponential variables with mean $1/\lambda$ each.

To simulate a random sequence with the Erlang distribution using AR(1) model-based algorithm, one should on each step add a random variable with PDF

$$f(x) = a^k \delta(x) + \sum_{i=1}^{k} \binom{k}{i} a^{k-i} b^i g(x; i, \lambda b).$$

where $\delta(x)$ is Dirac's delta function, $\binom{k}{i} = \frac{k!}{(k-i)!i!}$ are binomial coefficients, and $g(x; i, \lambda b)$ are PDFs of Erlang distribution with parameters i and λb:

$$g(x; i, \lambda b) = \begin{cases} \dfrac{(\lambda b)^i x^{i-1} e^{-\lambda b x}}{(i-1)!}, & x \geq 0, \\ 0, & x < 0. \end{cases}$$

Characteristic function of Erlang distribution is

$$\psi(t) = \left(1 - \frac{it}{\lambda}\right)^{-k}.$$

Therefore

$$\frac{1}{b} f\left(\frac{x}{b}\right) = \frac{1}{2\pi} \int_{-\infty}^{+\infty} e^{-itx} \left(\frac{1 - \frac{ita}{\lambda}}{1 - \frac{it}{\lambda}}\right)^k dt = \frac{1}{2\pi} \int_{-\infty}^{+\infty} e^{-itx} \left(\frac{b}{1 - \frac{it}{\lambda}} + a\right)^k dt.$$

As k is a positive integer, we can use the binomial formula

$$f\left(\frac{x}{b}\right) = b \cdot \frac{1}{2\pi} \left(\int_{-\infty}^{+\infty} a^k e^{-itx} dt + \int_{-\infty}^{+\infty} e^{-itx} \sum_{i=1}^{k} \frac{C_k^i b^i a^{k-i}}{\left(1 - \frac{it}{\lambda}\right)^i} dt \right).$$

After the Fourier transform we get the given above PDF, which is a mixture of non-random value and k Erlang-distributed random variables.

3. Results and discussion

Formula for finding the PDF of added random variable for AR(1) model-based algorithm was found. Examples the PDFs of added random variables for AR(1) model-based algorithm were examined. Joint PDFs of adjacent members were also found.

Discussed methods of simulation are easy to implement. Computer modeling confirms the theoretical calculations.

Further work may include using inverse transformation to the members of obtained sequences to get the sequences with different PDFs and different joint PDFs of adjacent members.

4. Acknowledgments

The reported research was funded by Russian Foundation for Basic Research and Novgorod region, grant №18-41-530001.

References

[1] Ogorodnikov V A and Prigarin S M 1996 Numerical Modelling of Random Processes and Fields *Algorithms and Applications* Utrecht VSP p 240

[2] Kir'yanov B F and Koznov A V 1994 Protsessy avtoregressii so sluchaynymi koeffitsi-entami i ikh primenenie pri modelirovanii radiotekhnicheskikh sistem *Prikladnaya matema-tika: Mezhvuz. cb. Novgorod* **1** pp 3–8 (In Russian)

[3] Garbar S V 2016 Manipulating intervals of probability density function in autoregression model for simulating strictly stationary random sequences *International Journal of Mathematical Models and Methods in Applied Sciences* **10** pp 270-273

[4] Garbar' S V 2012 Modelirovanie statsionarnykh markovskikh sluchaynykh protsessov s zadannoy plotnost'yu raspredeleniya metodom avtoregressii *Vestn. Novg gos un-ta* **67** pp 13-15 (In Russian)

Formation of spectral characteristic of silicon differential photoreceivers.

V.V. Gavrushko[1], A.S.Ionov[2], O.R. Kadriev[1], V.A. Lastkin[2]

[1] Novgorod State University,
ul. B. St. Peterburgskaya 41, 173003, Veliky Novgorod, Russia,
[2]PLC "Special Design Office Planeta" 173004, Veliky Novgorod, Russia

E-mail: Valery.Gavrushko@novsu.ru

Abstract. The article provides information about the structure and technology of manufacturing of silicon differential photoreceivers having a selective spectral characteristic. The possibility was established of displacement of the longwave boundary of sensitivity $\lambda_{0.5}$ from 0.37 to 0.47μm at change of a dose of implantation of arsenic in the range of 200-5000 $\mu C/cm^2$ at constant short-wave border ($\lambda_{0,5} = 0.28\mu m$). The effect of the implantation dose on the quantum yield of the differential channel was investigated.

1. Introduction.

Silicon-based photoreceivers are widely used in optoelectronics. This is due to the possibility of silicon devices to register optical radiation in wide spectral range, including ultraviolet, visible and near-infrared radiation. Besides, it is necessary to note the highly developed and comparatively low cost technology of silicon devices. However, the availability of wide spectral range is not always useful. If there is a problem of registration of radiation in the limited spectral range, for example in the UV area, the availability in IR and visible areas is parasitic and reduces noise immunity of the device. Using high-quality external filters for its deep suppression. increases the cost of devises and is not always effective. In addition, researchers note the tendency of filters to aging, especially when irradiated with UV light [1]. The method is of interest to form the spectral range of the photoreceiver by subtracting signals from a number of nearby photosensitive platforms with different spectral characteristics (differential photo receivers) [2.3]. This method allows to adjust the spectral characteristic by using only group technological operations of semiconductor manufacturing without application of the external filters. As a well-managed way to change the spectral characteristics, implantation can be applied to the surface layer, creating additional recombination centers and reduce sensitivity of photoreceiver in the shortwave range [4]. It was established, that it is effective for these purposes to use heavy ions, for example As [5]. The influence of the doping dose on the type of spectral characteristic of photoreceiver is shown. However, the known publications provide information for a limited interval of implantation doses. There is no information about the influence of the doping degree on the quantum yield of differential signal. The present article is devoted to solving these questions.

2. Photoreceiver structure.

IOP Conf. Series: Materials Science and Engineering **441** (2018) 012019 doi:10.1088/1757-899X/441/1/012019

The topology of differential photoreceiver is shown in Figure 1. Photoreceiver represents a system of 16 identical in size photodiodes divided into two groups. The technology of photodiodes of the first group provided high sensitivity in wide spectral range and the sensitivity of photodiodes of the second group was suppressed in the UV range by creating recombination centers in the surface area. The sensitive areas were positioned in checkered order, evenly filling the area of view of the photoreceiver. This provided improvement in the homogeneity of the sensitivity distribution on the area of photoreceiver.

For differential photoreceivers the necessary condition is the significant difference of spectral sensitivity in the working range of spectrum and at the same time nearness of spectral characteristics in the non-working area. The latter circumstance allows by subtracting the signals to practically eliminate the sensitivity in the nonworking part of spectrum. If one needs an ultraviolet photo receiver, the most important requirement for the main channel is to ensure maximum sensitivity in the shortwave range. Silicon photodiodes, obtained by the standard diffusion method, usually have a declining characteristic in the UV range. In Figure 2, the spectral sensitivity of the diffusion photodiode for equal flow of quanta with the depth of location of ***p-n*** barrier of 2.2 microns (curve 1) is given as an example. Decreasing the depth to submicron dimensions, such as 0.8-0.9μm is also insufficient (curve 2).

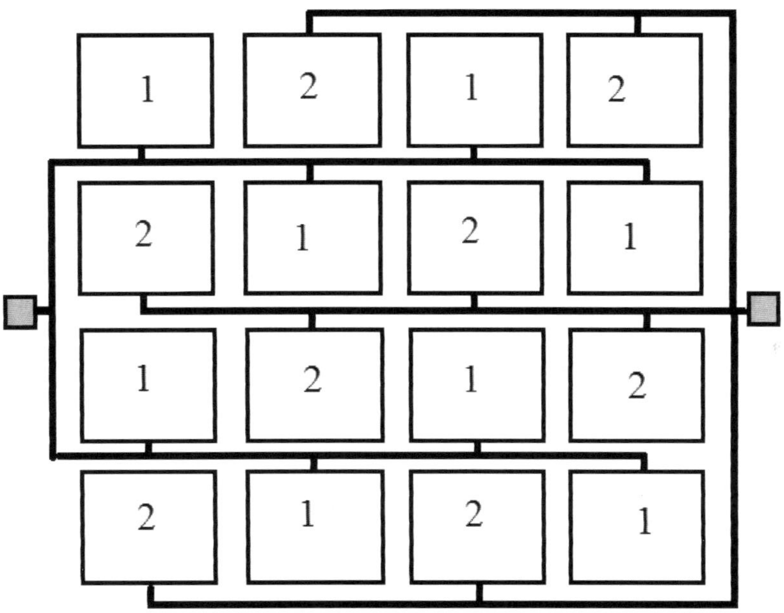

Figure 1. Photoreceiver topology.

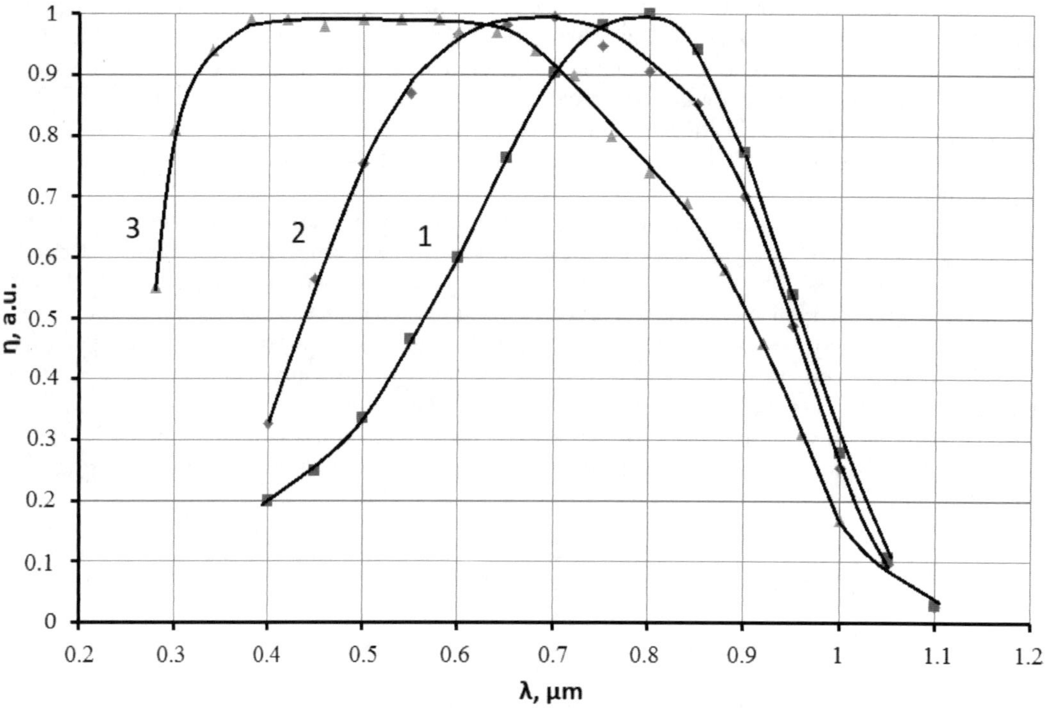

Figure 2. Spectral sensitivity of silicon photodiodes, obtained by deep (1) and shallow (2) diffusion and by implantation (3).

To increase the sensitivity of the main channel we used the photodiode structure created by implant doping of small dose of phosphorus of 30-70 $\mu C/cm^2$. After up-diffusion of the implanted layer necessary to form a barrier with good electrical characteristics, the ***p-n*** barrier with a sharp gradient of donor concentration at a depth of about 0.2μm was created. Using this method (curve 3) has significantly increased sensitivity in the shortwave range.

The second channel was formed simultaneously with the first, on the same wafer, with the use of combined technological operations, but with the addition at the final stage of the process of implantation of high dose of arsenic. This technique allowed to create additional recombination centers in the surface area of the second channel and reduce the sensitivity of the photodiode in the UV range. In addition, the presence of concentration maximum in the form of the impurities distribution during implantation created a decelerating electric field for photocarriers in the surface area. Thus, the formation of ion-implanted layer led to decreasing in the sensitivity of the auxiliary channel in the UV range, both due to recombination processes, and by creating a decelerating electric field. At the same time, the identity of the deep layers of both channels provided close spectral sensitivity in the longwave area, where the processes of collecting photocarriers were determined by the areas removed from the surface. Arsenic was chosen as intersitial impurity from the following considerations. This element has a small projected running, has low diffusion coefficient and high ultimate silicon solubility. Such properties allowed to form ***p-n*** barrier with high concentration of impurity at the minimum depth, poorly sensitive to the subsequent thermal treatments.

3. Spectral characteristic of photoreceiver.

One of the critical moments determining the type of spectral characteristic of the differential channel is the amount of the dose of arsenic implantation. In this connection series of photoreceivers with different degree of doping of auxiliary channel has been investigated. Figure 3 shows the normalized spectral characteristics of differential channels for the area of small doses, and in Figure 4 - for significant doses. As expected, with the increase in the implantation dose, expansion of the spectral range of sensitivity through the displacement of the longwave border was observed. Analysis of their relationship (Figure 5) showed that the infrared border on the level of $\lambda_{0.5}$ can be approximated by expression:

$$\lambda = \lambda_0 + k \ln \frac{Q}{Q_0},$$

where $\lambda_0 = 0.347\mu m$, $k = 3{,}2\ 10^{-2}\ \mu m$, $Q_0 = 100\ \mu C/cm^2$.

Thus, changing the doping dose within reasonable limits one can control the position of the longwave boundary sensitivity from 0.37 to 0.47μm. At that, the left boundary of the spectral range remained unchanged ($\lambda_{0,5} = 0.28\mu m$).

Another important characteristic is the level of the signal of the differential channel, the value of which was also dependent on the doping dose of the auxiliary channel. Figure 6 shows the dependence of quantum efficiency of differential channel for different doses of As. Values are given as a ratio of the quantum yield of the differential channel to the value of the quantum yield of the main channel. As the dose increased, the differential channel signals increased. Figure 7 presents the analysis of the dependence of the quantum efficiency of the differential channel from the doping dose of As for the fixed wavelength $\lambda = 0.34\mu m$ (in the range of maximum of spectral characteristic of the differential channel). As one can see, in the range of 500 to 2000 $\mu C/cm^2$ there was significant dose effect on the sensitivity of the differential channel. Starting from 2000 $\mu C/cm^2$ the quantum yield value slows.

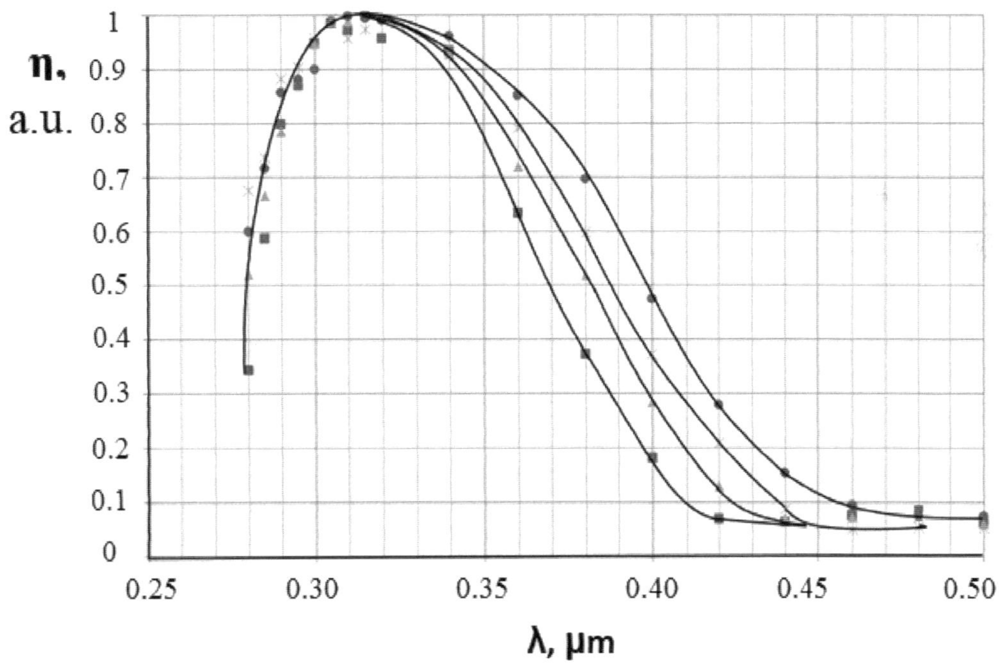

Figure 3. Dependence of spectral sensitivity of differential channel from doping dose in the area of small doses (200, 300, 400, 500 $\mu C/cm^2$).

IOP Conf. Series: Materials Science and Engineering **441** (2018) 012019 doi:10.1088/1757-899X/441/1/012019

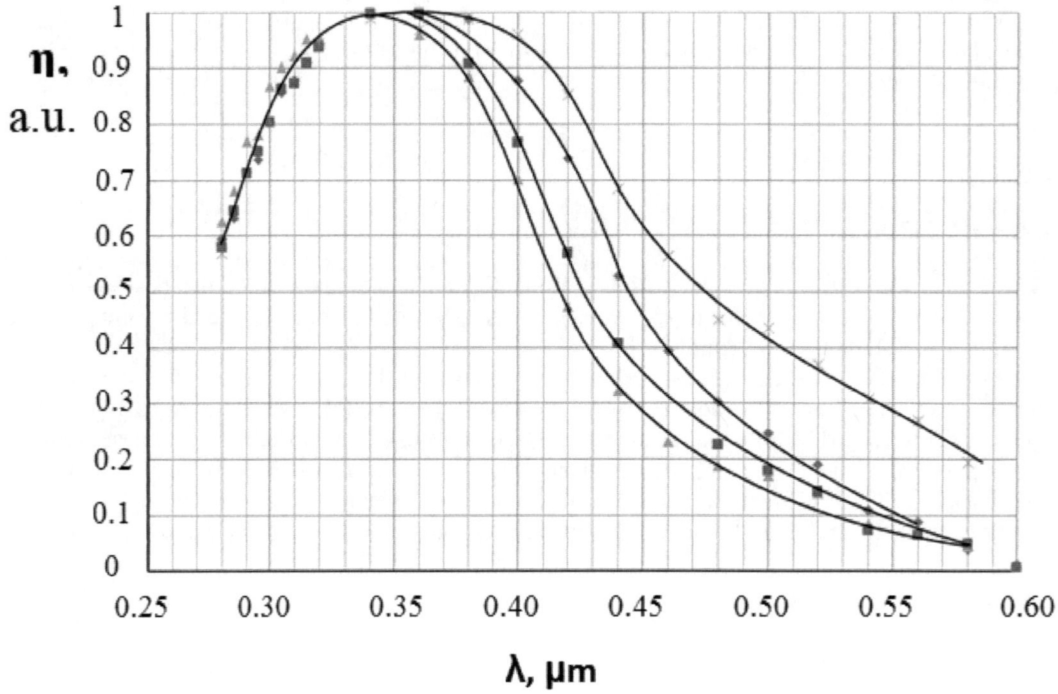

Figure 4. Dependence of spectral sensitivity of differential channel from doping dose in the area of large doses (700, 1000, 2000, 5000 μC/cm^2).

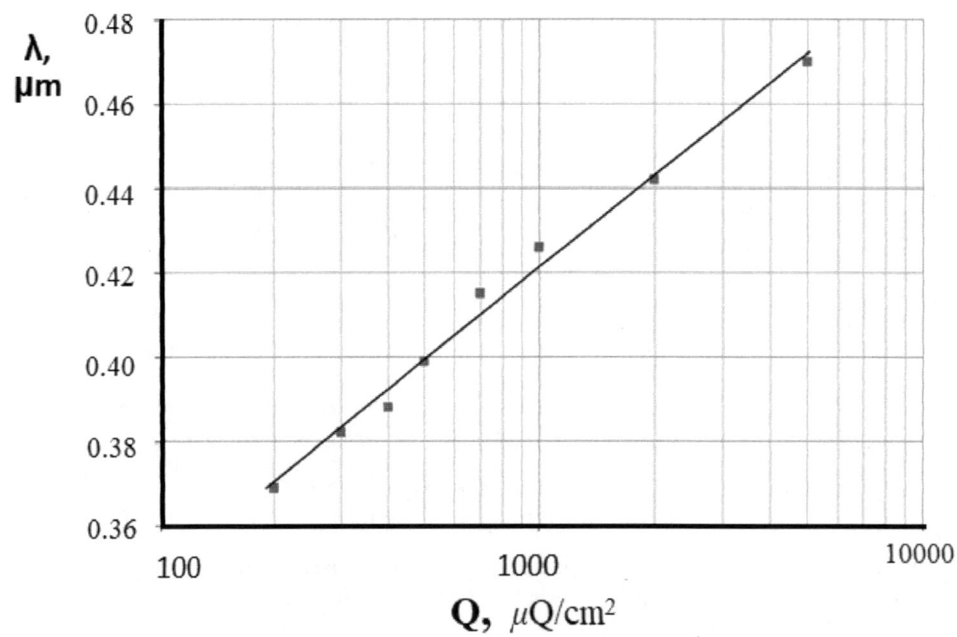

Figure 5. Longwave boundary dependence of the differential channel from As implantation dose.

IOP Conf. Series: Materials Science and Engineering **441** (2018) 012019 doi:10.1088/1757-899X/441/1/012019

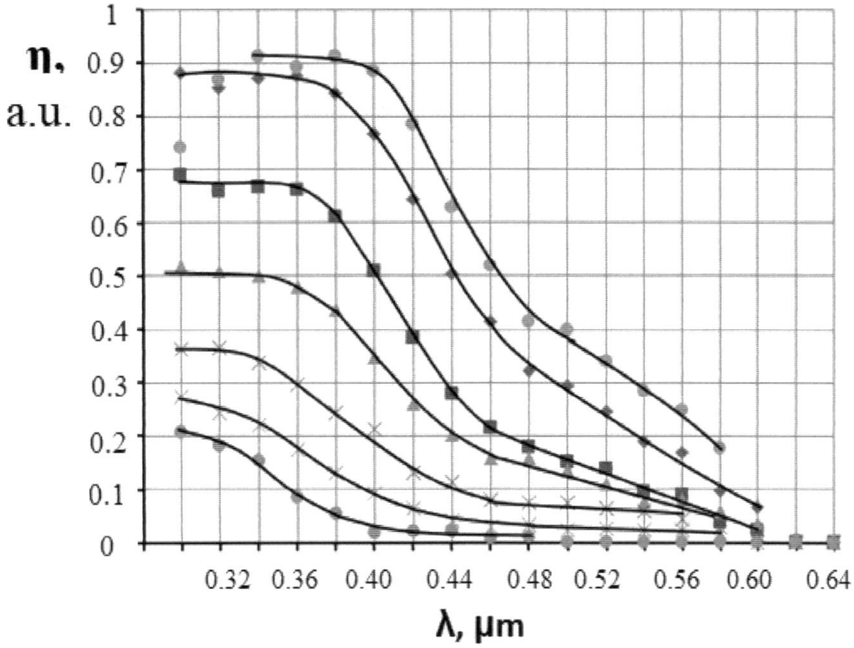

Figure 6. Spectral dependence of the ratio of quantum efficiency of the differential channel to the efficiency of the main channel from As implantation dose.

Doses: 200, 300, 500, 700, 1000, 2000, 5000 μC/cm^2.

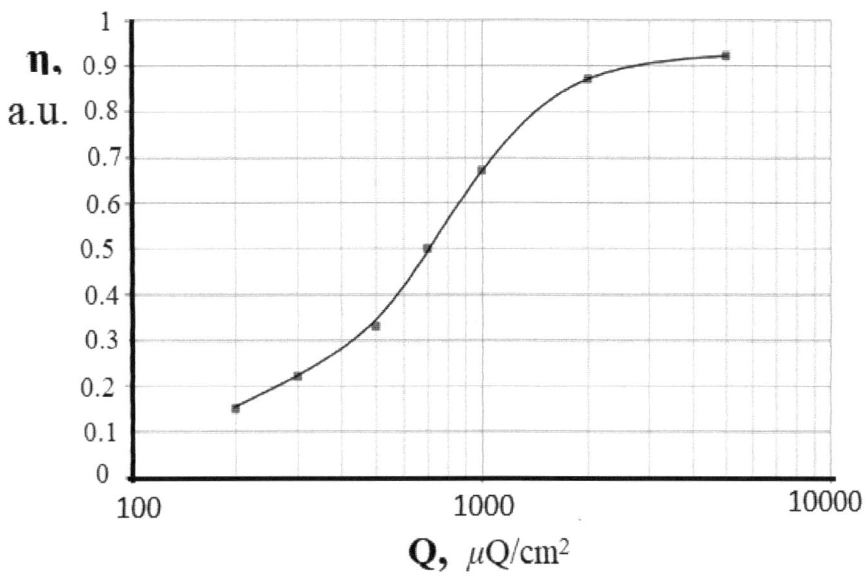

Figure 7. Quantum efficiency dependence of differential channel on As implantation dose for λ = 0.34 μm.

Approaching to 90% and further, increase in dose has little effect on sensitivity. Thus, if the problem of maximally possible sharpening spectral range is not necessary, the doping doses of 1000 – 2000 μC/cm^2 should be considered as close to optimum. At the same time, a good selectivity of spectral characteristic and the high sensitivity of the differential photoreceiver remains.

4. Conclusion.

The important advantages of silicon differential photoreceivers is the ability to control spectral sensitivity using methods of group technological operations of semiconductor manufacturing. On the basis of the relationships obtained during this work it is possible to form the position of the longwave boundary of the spectral range of photoreceivers in the range from 0.37 to 0.47μm by controlling the implantation dose. At that, the left boundary of the spectral range remained unchanged ($\lambda_{0,5}$ = 0.28 μm). The study of the dependence of quantum yield from implantation dose gives the reason to consider for such photoreceivers arsenic doping dose of 1000-2000 μC/cm^2 as close to optimum .

References

[1] Goltberg U.A. Tsarenkov B.V. 1996 *JTF* **66** Vol. 8 p. 195-97.

[2] Yukiko Takiba, Hiroshi Suzunaga. Pat. № 7196311b2 US, MkI h011 27/15 Semiconductor photosensor. Appl. 11.08.2005; Published. 27.03.2007.

[3] Blynskiy V.I. Malyshev S.A. Pat. 8532 U by, MkI h011 27/15 photoreceiver. Appl. 11.08.2005; Published. 27.03.2007.

[4] Gorshkov A.P., Karpovich I.A., Pavlova .D., Volkov N.S. 2012 *FTS* **46.** Vol. 12. p. 1542-45.

[5] Gavrushko V.V. IonovA.S., Kadriev O.P., Lastkin V.A. 2017 *JTF* **87** vol. 2 p. 310-11.

IOP Conf. Series: Materials Science and Engineering **441** (2018) 012020 doi:10.1088/1757-899X/441/1/012020

Research of the displacement transducer based on ferromagnetic shape memory alloy

N I Gorbatenko[1], V V Grechikhin[1], I S Kraevskiy[1], A V Kudrya[1] and J V Yufanova[1]

[1] The Platov South-Russia State Polytechnic University (NPI), 132 Prosvesheniya ul., Novocherkassk 346428, Russia

E-mail: vgrech@mail.ru

Abstract. The use of the ferromagnetic shape memory alloy as measuring transducers and actuators is a promising direction of development of positioning systems. However, the theoretical and experimental basis for designing such systems is not sufficiently developed. The article evaluates the possibility of using the ferromagnetic shape memory alloy as a linear-displacement transducer. The influence of various types of deformation of the active element from ferromagnetic shape memory alloy and temperature on the change in its inductance is investigated. The results of full-scale and model experiments are presented. Using the dependence of the deformation on the magnetic state element makes it possible to avoid the using of special displacement transducer and thereby simplify the positioning systems. However, it is necessary to provide for measures to reduce the effect of temperature on the measurement results to improve the positioning accuracy.

1. Introduction

The ferromagnetic shape memory alloy (FSMA) allows large reversible deformations to be realized by restructuring the martensitic domain structure in a magnetic field [1,2]. These properties of FSMA are used in positioning systems and electric drives with a wide range of linear displacements. The impulse field created in the distributed control windings is proposed to be used to ensure small overall dimensions and low energy consumption [3,4]. The development of such systems is hampered by non-linear dependencies between the mechanical, temperature, and magnetic properties of the FSMA, namely, phase transitions induced by change in the first derivative of thermodynamic potentials, such as temperature or magnetization. At the same time, the second derivatives of thermodynamic potentials undergo sharp changes. An example is the magnetic permeability, which in the chosen crystallographic direction can vary significantly during deformation of individual sections of the active element from the FSMA under mechanical loads without changing the phase of the material. Thus, the dependence between the deformation and the magnetic permeability allows obtaining information about the relative elongation of the active element of the drive.

The purpose of this work is to assess the possibility of using FSMA as a displacement transducer for positioning systems based on full-scale and model experiments.

2. Methodology

The research of the linear-displacement transducer is performed on the basis of the FSMA. The construction of the transducer is shown in Figure 1. In figure: *1* – the active element from the FSMA; *2* – measuring coil; *3* – support. The magnetic field is plane-parallel due to the design of the transducer. The Cartesian coordinate system is chosen for research. The geometrical parameters of the transducer are shown in Figure. 2, where X_{AE} – length of the sensitive element; X_0 – distance between the support and the measuring coil; X_C – length of the measuring coil; C_C – thickness of the winding of the measuring coil; Y_G – gap between the coil and the active element; Y_{AE} – active element width.

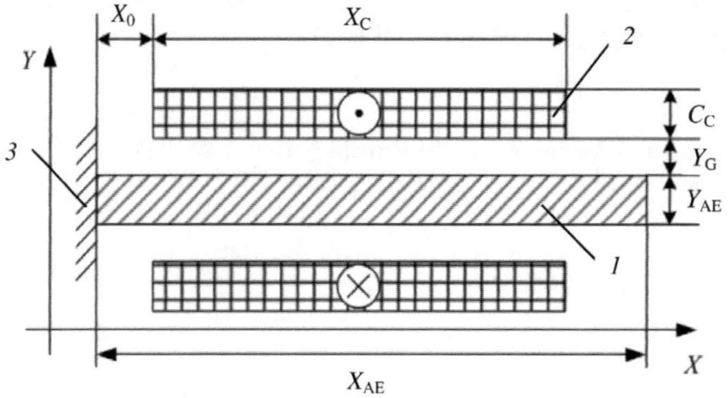

Figure 1. The construction and geometrical parameters of the displacement transducer of positioning systems.

The dependence of the magnetic induction *B* on the applied magnetic field strength *H* undergoes changes in the deformation of the active element from the FSMA [5]. The characteristics *B*(*H*) of the fully compressed (heavy axis) and the maximally elongated (light axis) active element are shown in Figure 2.

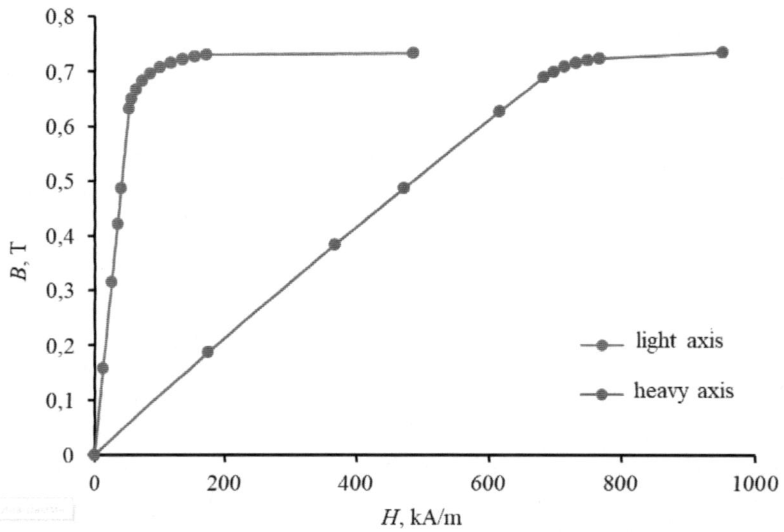

Figure 2. Magnetic characteristics of the fully compressed (heavy axis) and the maximally elongated (light axis) active element.

Assumptions are made: change the width of the active element is ignore; the magnetic state of the active element is taken into account by a linear dependence for two states (compressed and deformed) of the active element; the element is divided into p rectangles with a constant magnetic permeability within the rectangle; the vertices of the rectangles have some freedom (Figure 3). The cross-sectional area of a rectangle is determined by the formula $S = d_1 d_2 \sin\alpha/2$, where d_1, d_2 – diagonals of the quadrangles, α – the angle between them.

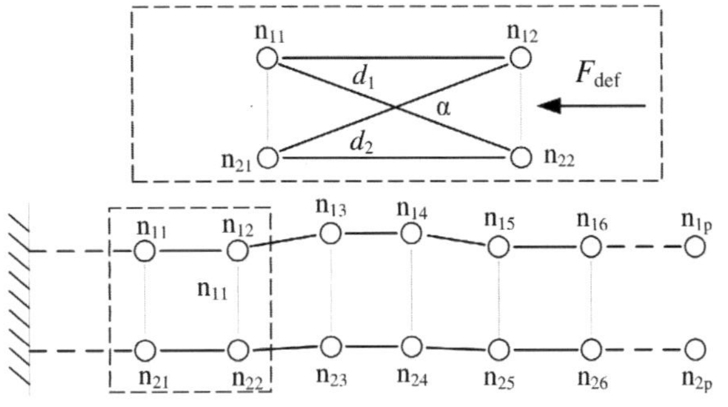

Figure 3. Deformation of the active element.

The control program, written in the complex of applied mathematical programs *Scilab* [6], and the FEMM program [7], which implements the numerical method for calculating physical fields (the finite element method) were used to conduct a model experiment. Application of finite element method provides high accuracy and short calculation time of magnetic field parameters. The joint use of programs allows implementing an algorithm in which the geometry and magnetic properties of a separate region of the active element were changed dynamically, and then the inductance of the measuring coil was calculated from the formula $L = \Phi/I$, where Φ – magnetic flux passing through the measuring coil; I – coil current. The change in the inductance of the measuring coil makes it possible to determine the change in the magnetic permeability of the active element from the FSMA.

3. Results and Discussion

Standard sample, manufactured by Adaptamat Inc. and which is a parallelepiped, was used as an active element from FSMA. Its geometric dimensions in the compressed state: length l = 20 mm; height h = 2.5 mm; width b = 1 mm.

The full-scale experiment was carried out on a model sample including a measuring transducer, a system for measuring and controlling the deformation of the active element. Various types of deformation of the active element were investigated (Figure 4).

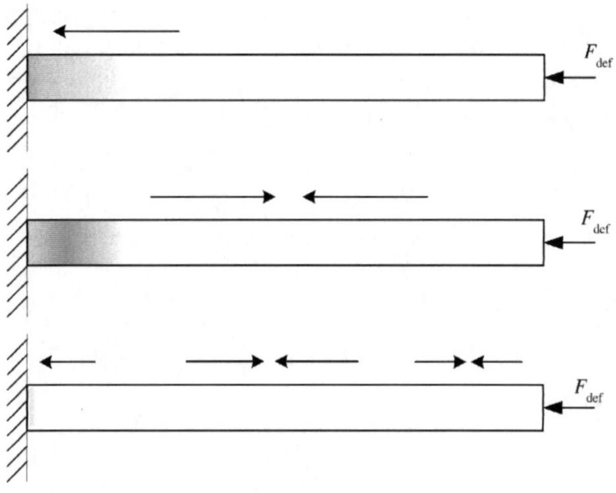

Figure 4. Types of deformation of the active element.

The set of characteristics of $L(X_{AE})$ (Figure 5), depending on the type of deformation, was obtained as a result of the experiment.

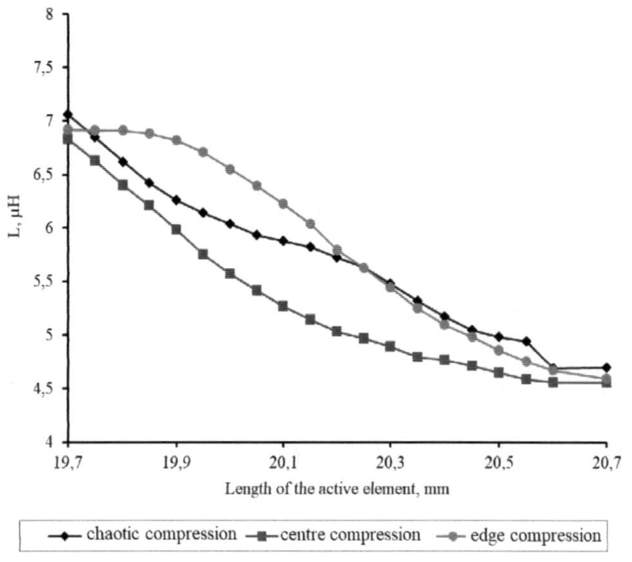

Figure 5. Experimental dependences of the inductance on different types of deformation of the active element

The model experiment was carried out on the basis of the proposed algorithm using the programs *Scilab* and FEMM. The types of deformations was modeled those that in the full-scale experiment. The results of the modeling are shown in Figure 6. They are in good agreement with the results of the full-scale experiment.

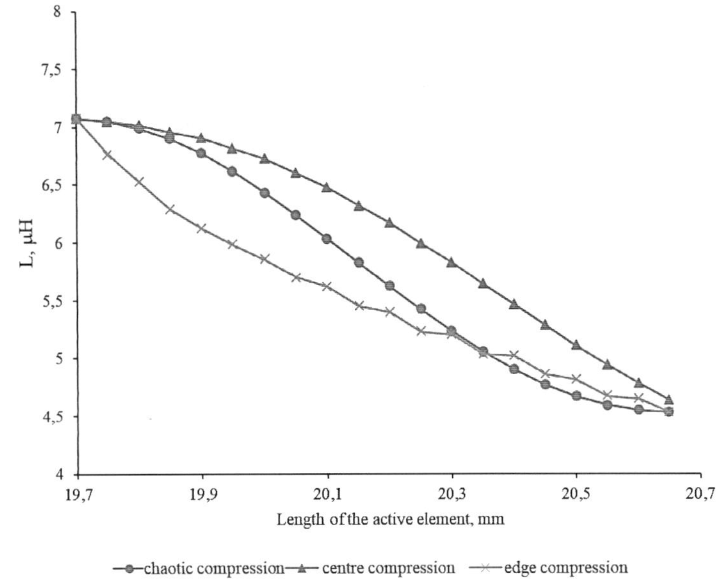

Figure 6. Model dependences of inductance on different types of deformation of the active element

The inductance of the measuring coil of the transducer changes 1.5 times under the influence of mechanical deformation. The effect of temperature on the magnetic state of the active element is a distinctive feature of the FSMA. Figure 7 shows the dependence between the inductance of the coil and the temperature deviation for the fully deformed (Figure 7*a*) and compressed state of the active element (Figure 7*b*). The obtained results show that the temperature of the active element is significantly correlated with the magnetic characteristics of the measuring transducer.

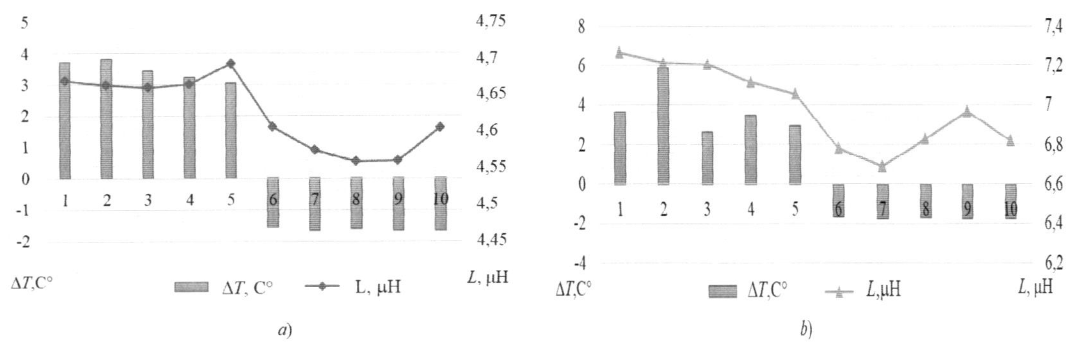

Figure 7. The effect of temperature on the inductance of the measuring transducer.

4. Conclusion

The dependence between the magnetic state and deformation of the FSMA opens up wide possibilities for using such materials not only as actuators, but also for obtaining measurement information in the implementation of automatic control systems for technological processes and industries. The use of this dependence allows us to abandon the use of special displacement transducer and thereby simplify the positioning systems and improve the positioning accuracy. The method of full-scale and model experiments for the study of FSMA as a displacement transducer is developed. It should be noted that it is necessary to provide for measures to reduce the effect of temperature on the measurement results

IOP Conf. Series: Materials Science and Engineering **441** (2018) 012020 doi:10.1088/1757-899X/441/1/012020

for the efficient use of FSMA in measuring transducers. The purpose of further research is the construction of mathematical models that take into account the mechanical, temperature and magnetic properties of the FSMA for the design of displacement transducers for high-precision positioning systems.

Acknowledgements
The study results are obtained with the support of the project #2.7193.2017/8.9 "Development of scientific bases of design, identification and diagnosis systems for highly accurate positioning with application of the methodology of inverse problems of electrical engineering", carried out within the framework of the base part of State job.

References
[1] Vasil'ev A N, Buchel'nikov V D, Takagi T, Khovailo V V and Estrin E I 2003 *Physics-Uspekhi* **46(6)** 559
[2] Wilson S A, Jourdain R P J, Zhang Q [ets.] 2007*Materials Science & Engineering R - Reports* **56** 1-129
[3] Gorbatenko N I, Grechikhin V V and Shaikhutdinov D V 2015 *Metal Science and Heat Treatment* **11** 609-613
[4] Grechikhin V V, Lozin O I, Kudrya A V, Kudrya N A, Shaikhutdinov D V and Yanvarev S G 2014 *Fundamental research* **11(4)** 744-748
[5] Schiepp T, Maier M, Pagounis E, Schlüter A and Laufenberg M 2014 *IEEE Transactions on Magnetics* **50(2)** 989-992
[6] Herman R. L. 2013 A Course in Mathematical Methods for Physicists *(CRC Press)*
[7] *Meeker D C 2010 Finite Element Method Magnetics.* **Version 4.2, 2nd Nov. Build.** Available at http://www.femm.info.Google Scholar

Development of Arc Welding Technology for Austenitic Steels in Reducing Media

E.A. Ivanaysky[1], V.V. Ivanaysky[2], A.A. Ivanaysky[3], E.V. Tokareva[1], A.I. Golubev[1]

[1] Altai State Technical University, No.46-Lenin Ave., Barnaul, Russia 656038
[2] Altai State Agrarian University No. 98-Krasnoarmeysky Ave., Barnaul, Russia 656067
[3] Emission Limited Liability Company, No.85-103 Baltiyskaya St., Barnaul, Russia 656049

E-mail: elena656012@mail.ru

Abstract. The possibility of performing the arc welding of austenitic 08Kh18N10 steel in a carbon monoxide medium is considered. The thermodynamic substantiation of the proceeding chemical reactions is presented. It was established that the use of reducing media prevents the oxidation of weld metal Comparative mechanical tests of welded joints made in various protective media were performed. The chemical and metallographic analysis of welded joints was done. It was established that the burnout of alloying elements during welding in a carbon monoxide medium does not occur. The studies conducted showed the possibility of using cheap carbon monoxide as a substitute for argon and helium in the welding of austenitic steels.

1. Introduction

Stainless austenitic steels are widely used in thermal power plants, pressure vessels, as well as in the automotive industry due to their high fracture toughness, good corrosion resistance, as well as the absence of the need for additional heat treatment [1]. Hygiene and environmental friendliness with a high level of elasticity and strength allow using steel data in the process of manufacturing equipment and various products for the chemical and food industries, medicine and household appliances.

Almost always in the manufacture of parts made of austenitic steels are used arc welding methods. The predominant welding with non-consumable electrode in argon or mixture of argon, carbon dioxide and hydrogen [2]. However, this welding method has a low productivity and high cost. The introduction of oxygen, carbon dioxide and others into the protective medium significantly affects the ongoing metallurgical reactions, mechanical properties, metal transfer, arc stability and other parameters [3]. The media used usually have oxidizing properties, or consist of inert gases. Quite often in the welding process, carbon monoxide is formed, which is considered a harmful impurity, due to its negative biological effects [4]. However, this gas has reducing properties, in relation to the oxides of most metals. It is therefore of interest to carry out welded joints in a reducing medium, mainly consisting of carbon monoxide.

The known reaction is the decomposition of carbon dioxide under the influence of a welding arc into carbon monoxide and oxygen, which then reacts with the weld bath metal:

$$CO_2 \rightarrow CO + O_2; \qquad (1)$$

$$2nMe + mO_2 \rightarrow 2Me_nO_m \qquad (2)$$

Therefore, the welding of austenitic steels in a carbon dioxide medium may be allowed only if the product is not subject to the requirements for resistance to intergranular corrosion. The formation of carbon monoxide leads to restrictions in the use of this welding method in closed volumes.
It is also known that carbon monoxide heated above a certain temperature enters into active chemical reactions with atmospheric oxygen, being oxidized to carbon dioxide [5]. Therefore, to neutralize the harmful effect of carbon monoxide, it was proposed to heat the gas heated to protect the welding arc above the ignition temperature [6].

2. Materials and methods

To investigate the possibility of using reducing protective media, the weld-jointing of Ø76 × 3 mm pipes was performed. 08X18H10 steel, the analog of AISI 304 steel, was used. Arc welding in argon and carbon monoxide media was used. Sv-12Kh18N10T filler metal wire, analog of the Autrod 308LSi wire, produced by ESAB Company, was used. Microstructural studies were carried out on an optical and scanning electron microscope with simultaneous microchemical analysis. The welded joints were tested for tensile strength to determine the breaking strength. Corrosion resistance tests were carried out in accordance with GOST 6032-2003.

3. Results and Discussion

In the investigated steel, the combination of basic technological properties is ensured by the content of chromium and nickel in them. The possibility of oxidation of these elements in the investigated medium was studied. The source of oxygen may be the dissociation of carbon monoxide in the arc column, as well as passing it into the weld bath from atmospheric air. Reverse metal-oxide reduction chemical reactions are also possible. The direction of the reactions was defined by determining the isobaric-isothermal potential (the Gibbs Criterion) using the method described in [7]. Data given on the site [8] was used.
We shall consider the various ways of oxidation of nickel and chromium alloying elements during their interaction with carbon monoxide.

$2CO \rightarrow 2C + O2.$ (3)

This reaction is reversible and proceeds with a noticeable release of oxygen at temperatures greater than 6000 K, that is, only in the anode and cathode zones of the arc. Accordingly, the amount of the decomposed monoxide will be very small. As will be shown below, this statement is confirmed by the results of chemical analysis

$Ni + CO \rightarrow NiO + C$ (4)

The isobaric & isothermal potential ΔG in the temperature range from 1100 to 3500 ° K decreases from 974 to 684 KJ, which makes this reaction practically impossible.

$2Cr + 3CO \rightarrow Cr2O3 + 3C$ (5)

ΔG in the above temperature range is even higher, and increases with increasing temperature from 1442 to 1560 KJ. It is believed that the films of chromium oxide inhibit oxygen diffusion, thus decreasing the oxidation rate of the metal [9].
Thus, the calculations performed showed that the oxidation of alloying elements during welding as a result of the reactions (3, 4, 5) is practically impossible.
Owing to poor weld shielding, the presence of oxide and organic films on the surface of the weld metal, oxygen can enter the arc zone, resulting in the formation of oxides of the alloying elements.

$2NiO + 2CO \Rightarrow 2Ni + CO2$ (6)

$Cr2O3 + 3CO \Rightarrow 2Cr+3CO2$ (7)

Consider the possibility of recovering the oxides of nickel and chromium under the action of carbon monoxide. The calculation results are presented in Table 1.

Table 1. Calculation result of the isobaric-isothermal potential

Temperature, oK	1100	1500	1900	2300	2700	3100	3500
Reaction	Isobar-isothermal potential, ΔG, KJ						

$NiO + CO \rightarrow Ni + CO_2$	-399	-338	-265	-211	-158	-93	-29
$Cr_2O_3 + 3CO \rightarrow 2Cr + 3CO_2$	247	253	258	264	270	276	283

As shown in Table 1, chromium oxides will not react with carbon monoxide, and nickel oxides will be actively reduced, giving up oxygen and oxidizing carbon monoxide to the dioxide. Similar calculations for other alloying elements are presented in [10]. Thus, preliminary calculations show that the proposed medium will not cause the oxidation of the investigated steels during welding.

It is established that when welding in a carbon monoxide medium, a qualitative weld seam was formed, without pores, micro-cracks or other defects. The micro-structure of the weld is shown in Figure 1.

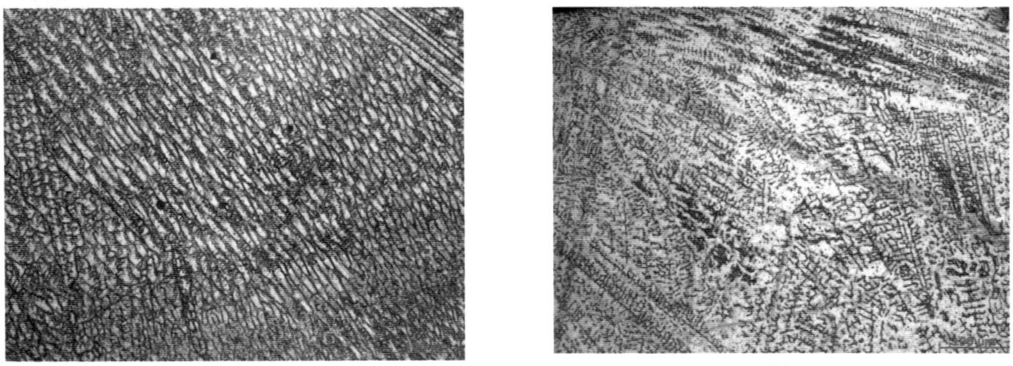

a) b)

Figure 1. Micro-structure of a welded joint made in a carbon monoxide (a) and argon (b) medium

In the weld metal deposited in the carbon monoxide medium was observed the formation of secondary walls that do not coincide with the direction of the dendritic growth (Figure 2).

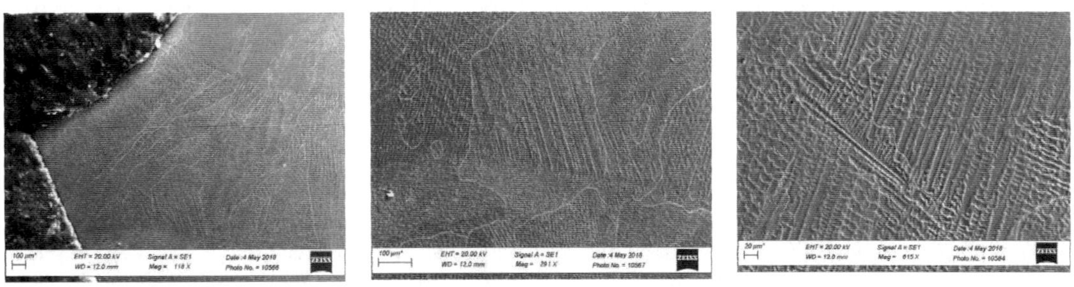

Figure 2. Secondary grains in the weld metal made in a carbon monoxide medium

In conventional argon-arc welding, the formation of these walls was not observed. It should be noted that in the listed steels was noted the formation of similar structures after recrystallization [11]. Additional research is required to establish the reasons for their occurrence in the welding process.

Tensile tests on the expansion of flat samples cut across the weld were performed. In all cases, the disintegration occurred over the heat-affected zone. Significant differences in the mechanical properties of samples welded using various protective media were found. The tensile strength varied from 558 to 579 MPa.

To determine the resistance of the welded joint to inter-crystalline corrosion, rings 50 mm long were cut from the samples, in which the weld was centered. 8 samples of welded joints were made in different protective media. Of these, 4 samples were subjected to boiling in an aqueous solution of copper sulphate and sulfuric acid in the presence of copper chips for 8 hours. All the produced samples

were flattened until a 13 mm-gap was obtained. After that, the curved samples were inspected with a magnifying glass with a sevenfold magnification. Cracks were not detected in the samples.
The results of the micro-chemical analysis of the weld metal obtained in argon and carbon monoxide media are presented in Figure 3.

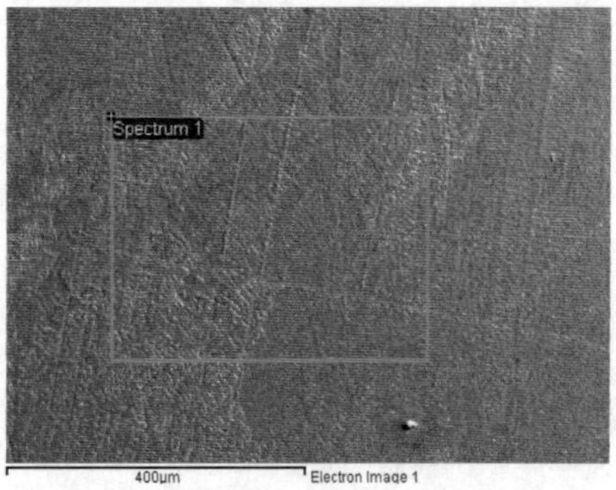

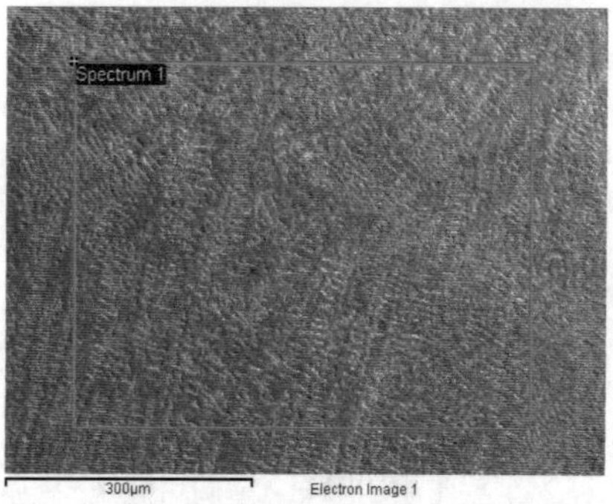

a) Element, Weight%				b) Element, Weight%			
Si	Cr	Mn	Ni	Si	Cr	Mn	Ni
0.31	21.16	1.30	9.25	0.68	21.99	1.43	10.00

Figure 3. Composition of weld metal made in carbon monoxide (a) and argon (b) media

As can be seen from the data presented in Figure 3, weld metal made using different media has approximately the same composition. Due to a significant error in measuring carbon during the microchemical analysis, an X-ray fluorescence analysis of the chemical composition was performed using the X-MET 7500 analyzer. In general, the results presented in Figure 3 were confirmed. However, an increase in the carbon content from 0.072 ... 0.106 to 0.144 ... 0.171% by weight was noted. The increase in the carbon content in the weld metal by 0.07% seems to be due to the dissociation of carbon monoxide in the near-electrode regions, which was mentioned above. Increasing the carbon content negatively affects the corrosion resistance of the metal. It is believed that for this class of steels, to ensure corrosion resistance, the carbon content limit in the steel should not exceed 0.2%. The specified level of carbon in the metal was not achieved when using carbon monoxide as a protective medium.

4. Conclusion

Based on the studies conducted, it was proved that when carbon monoxide is used as a protective medium for the welding of 08X18H10 steel, the chemical composition of the weld metal is close to the composition of the weld metal made in argon-arc welding. At the same time, the required mechanical properties and corrosion resistance of the welded joint are also provided.. The studies performed demonstrate the possibility of using carbon monoxide as a cheap alternative to the inert gases widely used in arc welding.

References

[1] Kumar R, Elangovan S, Siva N Shanmugam 2014 Parametric optimisation of pulsed – TIG welding process in butt joining of 304L austenitic stainless steel sheets, *International Journal of Research in Engineering and Technology* **6** 213-219
[2] Costanzaa G, Silib A, Tataa M E 2016 Weldability of austenitic stainless steel by metal arc welding with different shielding gas *21st European Conference on Fracture, ECF21* 20-24
[3] Mvola B, Kah P 2017 Effects of shielding gas control: welded joint properties in GMAW process optimization, *The International Journal of Advanced Manufacturing Technology* **88** 2369–2387

[4] [4] Ojima J, 2012 Generation rate of carbon monoxide from CO2 arc welding *J Occup Health.* **55(1)** 39-42
[5] Information on http://www.inchem.org/documents/icsc/icsc/eics0023.htm
[6] Ivanayskiy E A, Ivanayskiy A A 2014 Method of welding in a protective gas medium *Pat. of the Russian Federation2570609* Bul. **34**
[7] Ishkov A V 2014 Thermodynamic Substantiation of Chemical Reactions in the B4C-Borate Flux-Fe System under High-Temperature Heating *News of the Altai State University*. **3 (83)**. 199-203
[8] Information on https://www.sciencebysimulation.com/chemreax/Analyzer.aspx
[9] Soares A, Sabionia A, Huntzb E, Luza M, Mantel C 2005 Diffusion of iron in Cr2O3: polycrystals and thin films *Materials Science and Engineering* **392** 254–261
[10] Ivanayskiy E, Ishkov A, Ivanayskiy A, Malyshev B 2018 Thermodynamic Substantiation of Chemical Reactions Occurring during the Heating of Parts by High-Frequency Currents in Protective Media *Polzunovsky Herald* **1** 143-147
[11] Karcı F 2009 The effect of process parameter on the properties of spot welded cold deformed AISI304 grade austenitic stainless steel *Journal of Materials Processing Technology* **209**

Scientific and practical justification of haylage harvesting in APC "Levochsky" Novgorod region

S V Kartashov [1], S B Pavlov and M S Prokopov

[1]Yaroslav-the-Wise Novgorod State University
173003, Russia, Veliky Novgorod, Great St. Petersburg st., 41

E-mail: Sergey.Kartashov@novsu.ru

There are three options for harvesting haylage: haylage in the trench, haylage with separate packaging, haylage with the preparation of rolls in line. The comparative assessment above the listed technologies of harvesting haylage on the basis of APC "Levochsky" Khvoyninsky district of Novgorod region is given. According to the results of scientific and practical research, the lowest cost of finished products is achieved when using the option of harvesting of haylage with roll packing in line. A wrapper-wrapper is used for this technology.

Section 1

The most important condition for the creation of a solid fodder base in the farm is the preparation of high-quality vegetable fodders using modern progressive technologies that maximize the yield of nutrients from a unit of land area and preserve 85-90 % of them in the process of storing.

The wide spread of haylage in recent years is explained by the fact that harvesting makes it possible to obtain more nutrients from the same area, to reduce the number of components in the diet, to reduce the cost of livestock products in comparison with silage and hay.

The most important condition for obtaining high-quality feed is the creation of anaerobic storage conditions that exclude air access in the preparation of haylage from cured grasses. Such conditions are achieved by compaction of the hay mass and hermetic covering of the storage. If the mass is properly sealed, its temperature should not exceed 35÷37°C.

The use of the senile towers allows fully mechanizing the processes of laying the mass, its seizure and delivery to the places of feeding. And it also reduces storage losses.

The towers with the upper type of unloading mass are the most widespread among tower storages. Filling of such towers is carried out with the help of special equipment. Crushed mass from the field is delivered to the tower, where it is reloaded by a grader loader into the feeder-feeder, dosed and pneumatically loaded into the tower through hatches in the walls of the tower.

Trenches of ground, semi-buried and buried type are used for laying haylage because of the lack of tight towers. At present, ground trenches made of prefabricated reinforced concrete are widely used. Trenches of buried and semi-buried type should be lined with reinforced concrete slabs, the floors should be concreted [2].

The loss of dry matter during storage of haylage is 10% with proper observance of the technique trench filling with fodder and its coverage. If the technological requirements are not met, the losses increase to 40÷70% [1].

Section 2

The most promising technology of haylage currently is haylage in plastic packaging that avoids the difficulties and disadvantages of the conventional technologies (haylage in trenches and towers). According to the experience of the advanced economies, haylage in the package allows the increase of nutrient feeds by approximately 20% and provides a fully balanced food for animals, as well as its effective use. An increase in productivity of the animals by 20 ÷ 25% is achieved and the cost of dry concentrate feed and livestock production cost is reduced.

In recent years, high-speed packers are used in polymeric packaging, which pack rolls into a "line" and there is no need to wrap the ends of rolls, which saves 30-40% of the film, compared to the individual winding of each roll. It is recommended to put on special caps on the end sides of the formed lines for a high-quality sealing. This technology does not require the transfer of packed rolls to the storage location, which eliminates damage to the film. The line can have unlimited length, which allows using the storage area.

The main advantage of this packing method is high productivity of work - 100÷110 rolls per hour or 500 ÷600 tons per shift, which is comparable to the volume of the whole trench of haylage.

A comparative assessment of haylage harvesting technologies is given in the table1.

Table 1. A comparative assessment of haylage harvesting technologies.

Cost item	Haylage in the trench	Single packed haylage	Haylage with packing of rolls in a line
Mowing of herbs with squashing, rub/ha	622	622	622
Turning, rub/ha	261	261	261
Raking, rub/ha	226	226	226
Pressing, rub/ha	–	682	682
Loading of rolls, rub/ha	–	277	277
Transportation to the storage place, rub/ha	2834	946	946
Packing of rolls per packing, rub/ha	–	–	278
Packing of rolls (including the cost of the film), rub/ha	–	4327	2411
Stowage of packed rolls, rub/ha	–	285	–
Pickup of grass mass with grinding, rub/ha	1663	–	–
Leveling and tamping of the mass, rub/ha	3426	–	–
Sheltering with film, rub/ha	220	–	–
Amount of expenses, rub/ha	**9246**	**7636**	**5687**
Amount of expences per 1 ton, rub/ha	912	762	576
Average losses of the haylage mass (during harvesting, storage, feeding), %	15-30	0-5	0-05
Cost of finished products including losses, rub/ha	**1177**	**789**	**599**

Section 3

The agricultural production cooperative "Levochsky" is one of the leading enterprises of the Novgorod region. The own forage base available in the farm makes it possible to fully provide the livestock with quality fodders, with the exception of concentrated forages. It is practiced to prepare forage from the rolled grain, packed in a polymeric sleeve.

The main performance indicators of APC "Levochsky" are given in table 2.

Table 2. Indicators of the farm unit in 2014-2016.

№	Article	Unit of measurement	2014	2015	2016
1	Availability of agricultural land	ha	2312	2312	2312
2	Area of farm field	ha	2312	2312	2312
3	Crop acreage of cereals	ha	792	837	845
4	Gross harvest of cereals	t	1847,4	2224	1605,1
5	Hay making	t	1205,4	730	844
6	Haylage harvesting	t	1957	1942	1150
7	Silage harvesting	t	4548	4626	3086
8	Straw harvesting	t	316	170	190
9	Cereals yield	c/ha	23,3	26,6	19
10	Yield of perennial grasses	c/ha	105	86,6	135
11	Cattle stock	Heads	936	857	862
12	Milking herd	Heads	410	410	410
13	Cost of 1 ton of haylage	rub	727	866	1181

At present, the APC Levochsky stores haylage in ground trenches. Every year more than 1,000 tons of haylage are stored on the farm, which, in combination with hay and silage, makes it possible to completely provide the livestock with feed. The machine and tractor fleet is is able to produce all technological operations for harvesting plant fodder. The main problem in the preparation of haylage in the farm is the insufficient number and unsatisfactory condition of silage-haylage trenches, which require significant investments in the overhaul of structures. Due to the lack of concrete storage, part of the feed is stored in earth trenches for storage. This leads to a decrease in the quality of the resulting haylage and a significant increase in its losses during storage. Low quality of plant fodder when feeding a milking herd leads to a surplus of concentrates and sharply raises the cost price of meat and milk.

The technology of harvesting of haylage in APC "Levochsky" includes the following operations:

– - Mowing grass with squashing (tractor MTZ-826 + trailed mower-conditioner KRONE EasyCut 280, self-propelled mower-conditioner E-302);
– Turning (tractor MTZ-82,1 + rotary ripper Krone KW-6/72);
– - Shoveling grass mass into the roll (tractor MTZ-82,1 + Rotary swath former Krone swadro 807);
– - Pickup of mass from rolls, followed by grinding, application of a preservative and unloading into a vehicle (forage harvester DON-680M, forage harvester MARAL -125);
– - Transportation of crushed mass to the storage location (tractor T-150K + semitrailer PIM-40, tractor John Deere 730 + semitrailer PPS -15);
– - Distribution and tamping of the ground mass into a trench (John Deere 730 tractor, loader JCB 531-70AG);
– - Sheltering of a trench with a film;
– - Sheltering of the trench with ballast (loader JCB 531-70AG).

A technological map was developed for APC "Levochsky" for harvesting haylage by packing coils into polymer film, which is presented in table 3.

Polymer "Agrostreich" film is used for hermetic winding of rolls on the technology "haylage in packing". It is made from three layers of high-density polyethylene by casting method. The film has considerable resistance to tears and punctures, and also has the ability to resiliently stretch up to 75%. For winding rolls of haylage in a line, a wrapper is used, which is an autonomous machine and has its own gasoline internal combustion engine with a capacity of 12-15 hp, which ensures the operation of all the units of this wrapper. This machine moves independently on the haylage packing site, and the tractor is required only to move it along public roads. The working process is carried out by means of a hydraulic system, which includes a hydraulic pump, a hydraulic motor, a hydraulic distributor, control levers, pusher and rear wheel hydraulic cylinders, hoses and connectors [1].

Table 3. Technological map for harvesting haylage by packing rolls in polymer film.

The area of perennial grasses - 100 hectares
The yield of perennial grasses is 135 dt/ha
The volume of harvested sennage is 1150 tons
The average distance to the storage place is 5 km.

Volume works, ha. (t)	Terms of work		Composition of the unit		Personnel, people	Productivity		Number of machines	Labor costs, p,. p/ha	Fuel costs, kg/ha	Operating expenses				
	date	days	tractor	agricultural machinery		Per hour	Per shift				Salary	CLM	Amorti-sation	AM, OR storage	Total
	6 June	10	-	E 302	1	3,2	25,6	1	0,31	6,1	9,6	213	76	4,4	303
	6 June	10	MTZ 826	KRONE EasyCut 280	1	4	32	1	0,25	6,8	8,2	238	68	4,8	319
100	6 June	10	MTZ 826	KRONE KW 6/72	1	6	48	1	0,16	2,6	7,2	182	64	6	260
100	6 June	10	MTZ 82,1	KRONE Swadro 807	1	6	48	1	0,16	2,6	7	154,8	58	5,2	225
(1150)	6 June	10	MTZ 82,1	KRONE Round Pack 1250	1	(30) 2,2	(240) 18	1	0,45	7	9,2	225	87	8,1	330
(1150)	6 June	10	MTZ 82,1	PRF -180	1	(23) 1,7	(184) 14	1	0,58	7,1	9,2	257	75	9,4	351
(1150)	6 June	10	MTZ 82,1	PTM – 0,75 PMT – 0,1	1	(50)	(400)	1	0,26	3,3	7,1	195	69	7,2	276
(1150)	6 June	10	KamAZ 45143	Nefaz 8560	1	(48)	(384)	1	0,28	13	9,3	857	70	10,2	947
(1150)	6 June	10	JCB 531-70AG	PMT – 0,1	1	(50)	(400)	1	0,26	3	7,3	179	82	6,8	276
(1150)	6 June	10	-	Roll packer	1	(50)	(400)	1	0,26	1	9	56	61	2279	2405
									2,9	**52,7**					**5695**

IOP Conf. Series: Materials Science and Engineering **441** (2018) 012022 doi:10.1088/1757-899X/441/1/012022

Section 4

The calculation of the economic efficiency of the technology of harvesting of haylage with winding of rolls in a line is made in comparison with the technology of harvesting of haylage in concrete trenches used in APC "Levochsky". The initial data for the calculation are taken from the annual report of the farm for 2016.

The use of the proposed technology for harvesting haylage with packing of rolls in a line showed its advantages in comparison with the technology used previously. The results of economic efficiency are presented in table 4.

Table 4. Comparative technical and economic efficiency.

№	Indicators	Options	
		Basic	New
1	The volume of harvested fodder, t.	1150	1150
2	Costs for the production of fodder, rub.	1050065	672010
3	Capital expenditures, rub.	–	787000
4	Capital expenditures, rub.	–	378055
5	Payback period of capital investments, months.	–	24

Along with saving money spent on harvesting haylage, the quality of haylage will improve, which will effectively affect the productivity of animals.

References

[1] The patent of the Russian Federation for utility model №103439 The packer of rolls / M. F. Zakirov et.al., 2011.

[2] Ten A,G, Fodder production. -M.: Kolos, 1982. – 463p.

Analysis of reflected radiation from a semitransparent mirror of silicon carbide

M V Kazakova[1], V A Karachinov[2], I G Djerenov[1], D A Evstigneev[1]

[1] Novgorod State University, ul. B. St. Petersburgskaya, 41 173003 Veliky Novgorod, Russia

[1]E-mail: mv_kazakova@inbox.ru
[2]E-mail: Polnovo@yandex.ru

Abstract: Within the adapted McMahon theory optical properties of flat mirror based single crystal SiC is researched. The method of assessment share of tinted radiation in total reflected from mirror flow with effect thermal radiation wide spectrum in condition high temperature is offered and tested for polytypes mirror on based 6H-SiC<N> for temperature 293K and 800K.

Introduction

Interest in the research of heterogeneous systems of the reflective type based on silicon carbide (SiC) is at present predominantly stimulated by the further development of power optics, for example, within the development and use of powerful special-purpose lasers, as well as television industrial safety systems resistant to a number of aggressive factors [1-5]. Being a typical representative of refractory semiconductor materials with optical transparency, SiC at the same time stands out among them a whole set of such competing properties as abrasive, chemical and radiation resistance, high thermal conductivity, etc. [6,7]. The listed properties are technologically manageable, both due to the possibility of doping the starting material with various impurities (basic diffusion and ion implantation processes), and due to the well-known phenomenon of polytypism caused by the layered structure of SiC at the level of nanosize and realized in crystallization processes [8,9]. The noted advantages of SiC most fully meet the criteria of optical quality when creating input windows and objective-lens and especially mirrors of a wide range of applications [10-12]. Thus, in [3] the possibility of using a "hot" SiC mirror of a small thickness in the television endoscope for obtaining images of objects under extreme conditions is shown. For a number of technical applications of such endoscopic systems, narrow-band heat-resistant mirrors are of particular interest. Within the framework of this approach, the problem arises of determining the degree of the influence of the color of a semitransparent SiC mirror on the color of the reflected radiation when a wide spectrum of radiation is exposed to it under high-temperature conditions. At present, some works are known, partially addressing this problem for some optical materials, for example [13].

The present work is devoted to the solution of this problem.

Research methodology

When developing the research methodology, plate-like single crystals of silicon carbide, whose color properties are presented in Table 1, were considered as prototypes of mirrors.

Pure silicon carbide with hexagonal structure is colorless crystals with diamond shine. The color of the material depends on the raw material and technology of obtaining crystals and is determined by the type and amount of impurity, and the degree of deviation of the composition from the stoichiometric. SiC can have a variety of colors: white, gray, yellow, green and black [6]. Studies of SiC heterogeneous reflecting systems were carried out within the framework of the adapted theory of MacMagon [14, 15]. In this case the geometric model shown in Figure 1 was used. In this model, a semitransparent mirror, which is in thermodynamic equilibrium with the source of thermal radiation at high temperature t, has the following properties: $R(\lambda,t)$ - reflection coefficient from the mirror surface, $T(\lambda,t)$ - transmittancy. The mirror is colored in a certain color, that is, the intensity of its transmission is maximal at a certain wavelength λ. At it at an angle, the radiation from the object with intensity I_λ falls. Quantitative estimation of the color content in the reflected radiation was carried out by determining the proportion of the colored radiation $k_{col}(\lambda,t)$ in the total reflected. At the same time, the possibility of using a color translucent material as a narrow-band mirror was justified by the criterion $k_{col}(\lambda,t) \rightarrow 100\%$.

Table 1. Color properties of some silicon carbide polytypes [6].

Polytype	Dopant impurity	Color
3C	pure	yellow
3C	nitrogen (N)	yellow-green
8H	nitrogen (N)	yellow-orange
6H	pure	colorless
6H	nitrogen (N)	green
6H	aluminum	blue - purple
15R	nitrogen (N)	pale green
4H	pure	colorless
4H	nitrogen (N)	light brown
24R	nitrogen (N)	purple
27R	nitrogen (N)	red

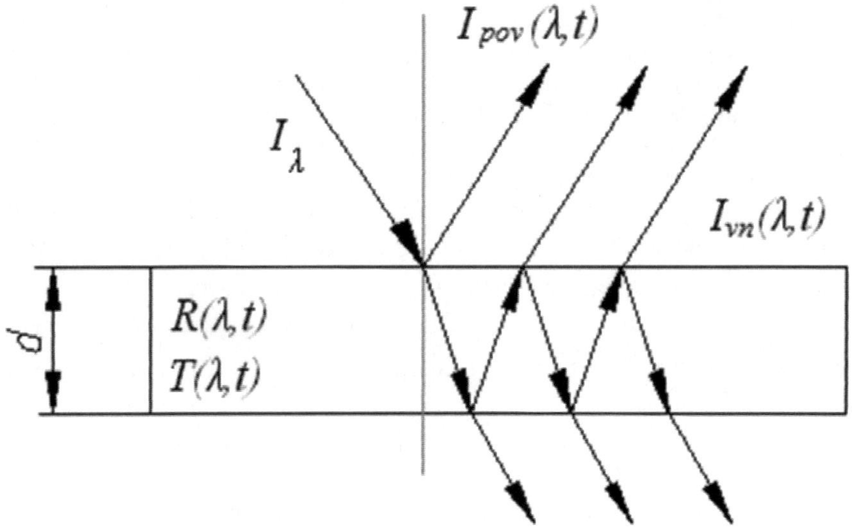

Figure 1. Optical-geometric model of light interaction with a semitransparent SiC mirror.

Then the fraction of the colored radiation is determined from the expression:

$$k_{col}(\lambda,t)=\frac{I_o(\lambda,t)}{I_{otr}(\lambda,t)}\cdot 100\% \tag{1}$$

Where: $I_{otr}(\lambda,t)$ - the intensity of the reflected radiation from the semitransparent mirror, $I_o(\lambda,t)$ - the intensity of the colored radiation. In general, the intensity of the reflected radiation of the translucent material $I_{otr}(\lambda,t)$ will be composed of the intensity of the mirror reflected by the surface $I_{pov}(\lambda,t)$ and transmitted through the mirror as a result of internal reflection $I_{vn}(\lambda,t)$:

$$I_{otr}(\lambda,t)=I_{pov}(\lambda,t)+I_{vn}(\lambda,t) \tag{2}$$

Proceeding from the MacMahon theory, equality (2) in a more complete form was given by the expression [14]:

$$I_{otr}(\lambda,t)=I_{\lambda}\cdot R(\lambda,t)+\sum_{m=1}^{m=\infty}\frac{I_{\lambda}\cdot(1-R(\lambda,t))^2}{R(\lambda,t)}\cdot(R^2(\lambda,t)\cdot T^2(\lambda,t))^m \tag{3}$$

where: m - the number of reflections, $T(\lambda,t)$ - the transmittance is:

$$T(\lambda,t)=\exp(-k(\lambda,t)\cdot d) \tag{4}$$

d - plate thickness, $k(\lambda,t)$ - crystal absorption coefficient.

The intensity of the radiation reflected from the surface of the crystal was considered a constant value, independent of the color of the material. Since within the visible spectrum the change in wavelength slightly affects the refractive index, the monochromatic reflection coefficient $R(\lambda,t)$ remains practically unchanged, and the color of the rays reflected by the transparent colored materials does not change.

Expression (3) was used in the calculation as follows:

$$I_{otr}(\lambda,t) = I_{\lambda} \cdot R(\lambda,t) + \frac{I_{\lambda} \cdot (1 - R(\lambda,t))^2 \cdot R(\lambda,t) \cdot T^2(\lambda,t)}{1 - R^2(\lambda,t) \cdot T^2(\lambda,t)} \quad (5)$$

The radiation intensity after passing through the material layer due to internal reflection will be maximum at wavelength λ corresponding to the color range of the material, while radiation of other wavelengths will be absorbed rather intensively, that is, it can be written:

$$I_o(\lambda,t) = I_{vn}(\lambda,t) \quad (6)$$

Due to the fact, that the intensity of radiation weakens, passing through the material, it will be less than the intensity of radiation reflected from the surface. To determine the radiation fraction, we substitute the values of $I_{otr}(\lambda,t)$ and $I_{vn}(\lambda,t)$ from (5) in (1):

$$k_{col}(\lambda,t) = \frac{x}{x+1} \cdot 100\% \quad (7)$$

where

$$x = \frac{(1 - R(\lambda,t))^2 \cdot T^2(\lambda,t)}{1 - R^2(\lambda,t) \cdot T^2(\lambda,t)} \quad (8)$$

It should be noted that this technique is adapted to the influence of only thermal radiation, while using a narrow-band light source, as is well known, can lead to the imposition of spectra. To determine the color in such cases, special diagrams are used [13].
The results of the research and their discussion
Consider the effect of the color of a heat-resistant mirror made of semitransparent single-crystal SiC on the color of the reflected radiation. Figure 2 shows graphs of the theoretical dependences of the transmission coefficient $T(\lambda,t)$ for SiC crystals on the wavelength of the radiation: colorless, green and blue colors.

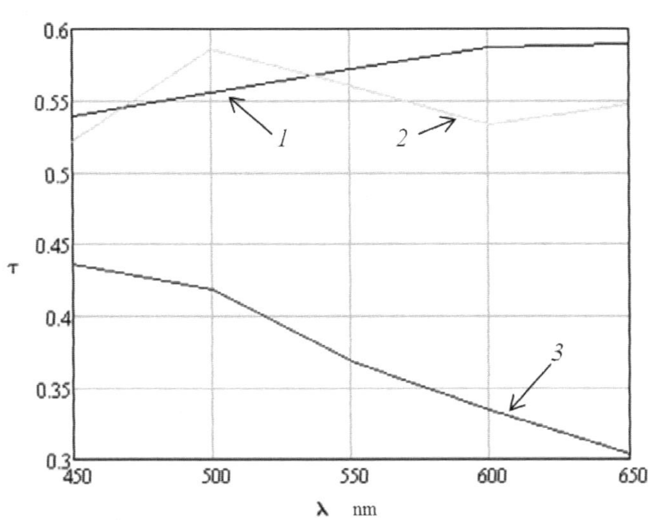

Figure 2. Spectral dependence of the transmittance of a flat SiC mirror of colorless (1), green (2), and blue (3) color. The thickness of the plate crystal is d = 0.1 mm.

To calculate the numerical values $T(\lambda,t)$, the well-known technique was used [15], taking into account the SiC absorption coefficients according to [16]. It is well known that due to the doping of the most widespread and sufficiently studied 6H polytype with nitrogen during the growth process, the color of the SiC crystal acquires a green tint (Table 1). That is, the maximum of its transmission falls on the green region of the spectrum. The remaining radiation with wavelengths of the visible spectrum is absorbed quite intensively. The studies carried out for this case using the Fresnel formula [13] showed (Fig. 3) that the intensity of the reflected radiation from the surface of the flat mirror 6H-SiC <N> in a wide range of wavelengths remains practically constant.

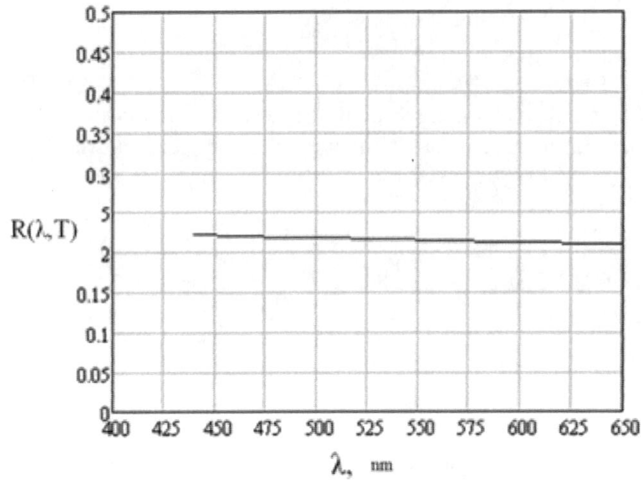

Figure 3. Dependence of the reflectivity of the surface of a planar 6H-SiC <N> mirror on the wavelength of the radiation. The thickness of the mirror is d = 0.1 mm, N_d- N_a = 4•10^{18}cм$^{-3}$.

The scientific and practical interest in the problem being solved is the influence of the angle of incidence of radiation relative to the reference surface of the 6H-SiC <N> mirror on the spectral composition of the reflected radiation. As follows from the geometric model, Figure 1), the radiation reflected from the inner surface will be colored and spatially separated from the non-color radiation reflected by the outer surface. Since the reflection coefficient of the surface is constant in the visible range and the transmittance has a maximum at a wavelength of 0.5 μm, only the intensity of the light of the inside of the crystal affects the increase in the fraction of the colored reflection. Due to the shift of the absorption coefficient to the long-wave part of the spectrum with increasing temperature, an analogous shift in the dependence of the fraction of the reflected radiation is observed. Examples of the results of the study are shown in Figure 4.

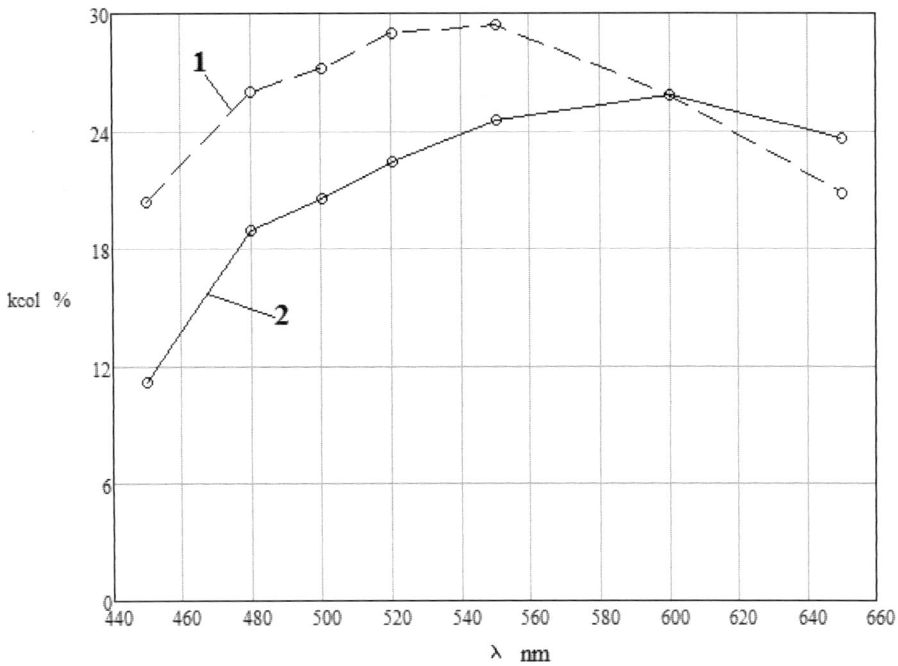

Figure 4. Dependence of the proportion of the colored radiation in the total reflection from the radiation wavelength at different temperatures for a planar 6H SiC <N> mirror. The thickness of the mirror is d = 0.1 мм, $N_d - N_a = 4 \cdot 10^{18} см^{-3}$. The angle of incidence of radiation is 0 grad. *1* - T=293K. *2* – T=800K.

Conclusions

The proposed analytical expression allows one to adequately calculate the fraction of the reflected radiation of a partially transparent SiC heterogeneous system at the wavelength characterizing its transmission maximum at high temperatures. The fraction of the colored radiation in general reflected for flat mirrors based on 6H-SiC <N> crystals of green color with thickness d = 0.1 mm is less than <30%, therefore the color of the reflected radiation will depend almost completely on the color of the incident radiation. With increasing temperature of a plane mirror based on a 6H-SiC <N> crystal, the spectral dependence of the reflected radiation shifts to the long-wavelength region of the spectrum.

References

[1] Apollonov V V 2014 Power Optics. *Quantum Electronics* No **2** 102-21

[2] Karachinov V A , Toritsin S B and Karachinov D V 2010 A System for Brightness Pyrometry of Objects via Water Streak. *Instruments and Experimental Techniques* No **2** 305-6

[3] Karachinov V A , Kazakova M V and Karachinov D V 2014 A Television Endoscopic System with a Heat_Resistant Silicon Carbide Mirror, *Instruments and Experimental Techniques* No **3** 336-9

[4] Riza N A 2015 Hybrid Optical Sensor for Extreme Temperature Measurement Greater Power Plants AMA Conf. (Nuremberg Germany) pp 511-3

[5] Gruzintsev A N 2013 Study of photomodulated reflection of single crystals of 6H-SiC *FIT* No **4** pp 442-6

[6] Tairov Y M, Tsvetkov V F 1988 Semiconductor compounds AIVBIV. In the book: Handbook of electrotechnical materials, ed Y V Koritskogo, V V Pasynkova, et al (L .: Energoatomizdat) pp 446-72

[7] Samsonov G V and Vinitsky I M 1976 Refractory compounds (M: Metallurgy) p 560
[8] Karachinov V A 2005 Preparation of profiled silicon carbide single crystals by sublimation and electric erosion methods: Abstract of thesis. dis. doc. technological sciences. (St. Petersburg) p 32
[9] Verma A R and Krishna P 1969 Politipism and Polymorphism in Crystals (Moscow: Foreign Literature) p 270
[10] Karachinov V A , Karachinov D V 2013 Pat. 2484569 RF IPK H02G 15/22 Cable with blind BI 16.
[11] Karachinov V A , Karachinov D V and Petrov A V 2013 Pat. 2486559 RF, MKI G03B11 / 04 Protective hood for optical instruments RF 18
[12] Khimich Y P , Evteev G V and Nikitin D B 2007 Optical shaping of a large-sized aspherical mirror made of silicon carbide, *Optical Journal* №**2** pp 70-2
[13] Tilley R .J D 2011 Colour and the optical properties of materials: an investigation of the relationship between light, the optical properties of materials and colour (WILEY) p 510
[14] Mc Mahon H O 1950 Thermal radiation from partially transparent reflecting bodies. ,*J. Optical Soc. of America* No **6** pp 376-80
[15] Karachinov V A, Karachinov D V and Kazakova M V 2012 Thermophysical and Optical Properties of Silicon Carbide-Based Microsystems with a Dimple Relief, *Technical Physics* No **8** pp 1167–71.

Development of SLM quality system for gas turbines engines parts production

V V Kokareva[1, 3], V G Smelov[1], A V Agapovichev[1], A V Sotov[1], V S Sufiiarov[2],

[1] Samara University, 34, Moskovskoe sh., Samara, 443086 Russia
[2] National Technology Initiative Center of Excellence in Advanced Manufacturing Technologies at Peter the Great St. Petersburg Polytechnic University, 29, Polytechnicheskaya, St. Petersburg, 195251 Russia

[3]E-mail: victoriakokareva@gmail.com

Abstract. The work represents the approaches for designing the quality system for gas turbines engines parts additive manufacturing. For gas turbines engines applications, the quality parameters obtained by the selective laser melting are insufficient and require the subsequent surface treatment. To reduce the cost of additional manufacturing and improve the quality parameters, it is necessary to predict the surface structure that could be formed by selecting process parameters. The paper deals with each element of the selective laser melting technological process which effects on the manufactured part quality and suggests appropriate measures to achieve and maintain a high level of quality.

1. Introduction

Selective laser melting (SLM) is the powder bed fusion process, which are occurred at a high melting temperature. High local temperature gradients and briefly cooling cause the residual stresses and part deformations [1]. The consequences are additional surface treatment and reduced productivity.

To understand how to control the formation of the residual stresses and forms deformation, a reliable method is investigation of influences between the technological parameters and quality behaviours. There are some basic physics mechanisms of selective laser melting process they lead to part distortion and cracks: high temperature gradient, high viscosity and surface tension of molten powder zone, un-melting powder and oxidized particles [2].

The following variables of SLM process can be established [3,4]:

- Powder, composition, size distribution, shape, thickness of melting layer;
- Laser, power, spot size, beam spatial distribution, scanning velocity and protective gas atmosphere;
- Strategy of manufacturing

The main target of the research is to find and control the optimum SLM process parameters to minimize the roughness, residual stresses and part deformations. The SLM quality system for gas turbines engines parts production is based on interaction model of the technological factors affecting the quality of fabricated parts.

There are many methods for predicting the temperature distribution and residual stress during SLM process. These methods can be classified into three groups. The first group is the simulation method [5]. The second group is the researches which focus on the experimental works. The last group is the comparison of simulation and experimental results [6]. It is difficult to predict part distortion due to enormous computational layer analysis. Li et al. proposed to divide a SLM process for a practical part into three scales such as micro scale, meso scale and macro scale [7]. With this approach, the temperature history and residual stress fields during the SLM process were predicted. Thermal information has been transferred through micro scale laser scanning, meso scale layer hatching, and macro scale part build-up [8].

2. Description of the model

We developed the model of influences of SLM process parameters on quality, Ishikawa diagram.
The quality of the final part is decided by powder properties, process parameters, SLM equipment characteristics, finishing and detail behaviours as shown in Figure 1.

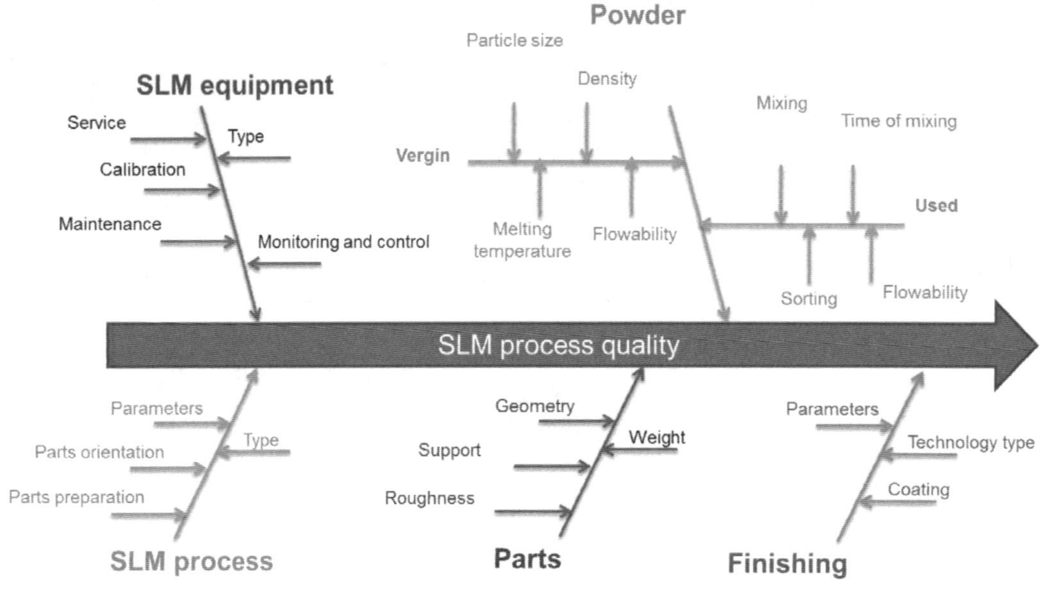

Figure 1. Ishikawa diagram of SLM process quality.

SLM equipment characteristics are determined by type of printers, monitoring system, kind of technologies, frequencies of service and maintenance. In order to ensure the technological accuracy it is recommended to calibrate production system and built every month test samples as benchmark of complex shape. Then it is necessary to check weight (density), dimensions, tolerances, and surface roughness under different parts orientation. The quality maintenance requires keeping the equipment daybook, where all actions are recognised: powder changing, cleaning, stopping, optic system controlling, parts replacement [9].
The specific behaviour of the molten material depends on the building conditions, where the temperature distribution in the meltpool is important for instance intermetallic phases, oxides, carbides, sulphides etc. The local cooling gradients and rates, which are dependent on the set of the main building parameters laser power and scan speed drive the microstructure formation and precipitation behaviour.
The investigating the quality of virgin powder is a first step of material controlling for achieving acceptable accuracy of part distortion and residual. Powder analyse previous to every manufacturing process regarding particle size distribution and particle shape using scanning electron microscope. Furthermore, it is necessary to evaluation powder flowability and apparent density [10]. The SLM

quality system should include the registration the qualitative and quantitative parameters of powders especially the proportion of mixed powders. In addition, the main material quality parameter is a rate of sieved and reused in a subsequent process powder.

It is clear that the part quality is dependent from SLM process parameters, which should be controlled and managed. In order to determine the optimal built parameters with an aspired objectives and technical requirements there is a need to consider many factors, such as cost, time, part quality, batch quantity together. For simplicity this task we developed a data base of SLM technological parameters for domestic powders: aluminium, titanium, heat resistant, stainless steel. We realised this data base in PDM system Teamcenter Manufacturing. The input technological parameters wish influence on part quality: scan speed and laser power, the layer thickness, the hatching distance, the hatching angle.

A list of corresponding minimal technological data to be controlled and stored is specified in Table 1.

Table 1. The quality ensuring activities

Quality parameters	Challenges
Input/output powder	Atmosphere controlled, optical inspection
Accuracy	Deviation prediction model, layer thickness
Surface	Post treatment, part orientation

The process of predicting and controlling the building parameters for part quality ensuring is promising method for manufacturing cost decreasing such expensive and poor machinability powder like titanium.

3. Development of SLM quality system

In this study, an effort to better understand the factors influencing on the part quality an evaluation method was applied.

Technological parameters were divided into two types: controlled by operator of additive machine (input) and defined by parts functional use (limiting conditions).

The input parameters:

- gas atmosphere concentration (percentage of oxygen);
- layer thickness;
- set of process conditions: scan speed, laser power, hatching.

The limiting conditions: powder behaviours, geometry accuracy, grain size.

The input parameters influence on a SLM process, the layer thickness increasing effort on bed fusion and density of melting material is decreasing. Another example, if we increase the oxygen concentration in the building camera a melting material becomes more crack-sensitive.

We propose to use the SLM quality system which is based on managing and controlling of input parameters taking into account the limiting conditions.

The main blocks of the SLM quality system is shown in Figure 2. In order to select the appropriate set of technological parameters the system uses a making decision algorithm, selection of input parameters depend on the link between part requirements (accuracy, geometry, surface) and building regimes for corresponding material and mechanical behaviors. The main idea of this system, is that decision and denoting of SLM parameters are based on experience, statistic database included in making decision algorithm. After each successfully building the data base records input parameters by certain limiting conditions. Successfully is mean that all quality requires are satisfied.

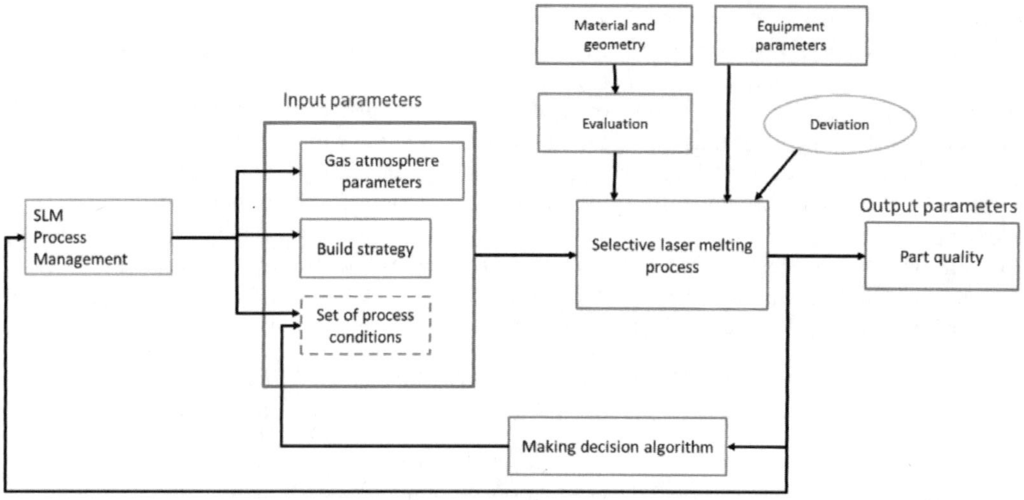

Figure 2. The SLM quality system.

The making decision algorithm should include not only statistic database, but a method of quality prediction. The prediction of accuracy and surfaces behaviors found on physical process during SLM: temperature gradient and distortion, internal stress and deformations. For this approach we need monitoring SLM process and have ability to manage this process. Such system is the step to digital manufacturing transformation according the Industry 4.0 concept.
The Figure 3 demonstrates the developed perspective SLM quality system. It should be noted that we need the simulation model of SLM process for better understanding the link between input and output parameters under different limiting conditions.

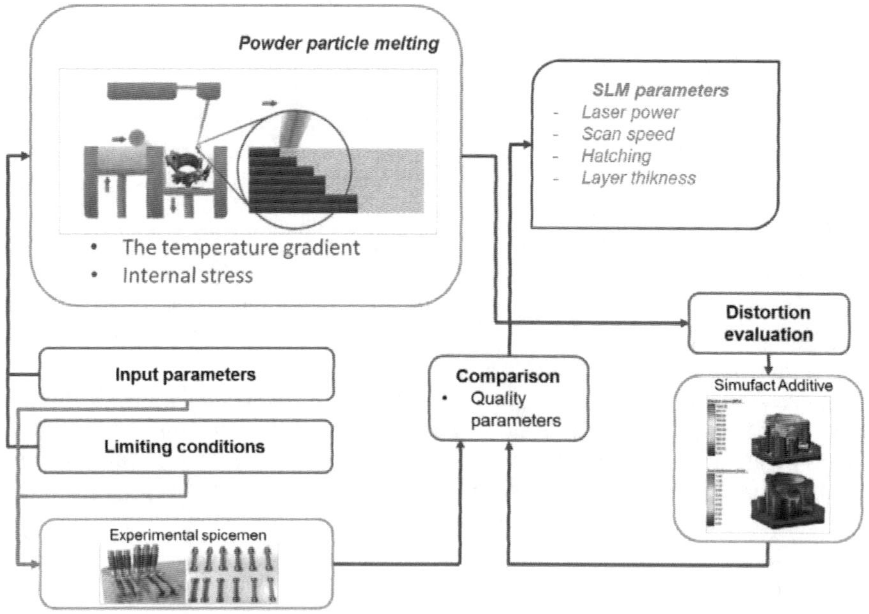

Figure 3. The perspective SLM quality system.

Simulation methods have been widely used to predict residual stress and part distortion of the SLM processes. But it is only suitable for analyzing the thermal-mechanical model to predict residual stress and distortion of sintered specimen [11]. For original SLM part, it is difficult to predict part distortion due to requiring millions of micro-scale laser scans which will increases the computational ability of the computer hardware. We suggest using MSC Software Additive Simufact and compare numerical and experimental results.

A multi scale approach is highly needed to achieve acceptable accuracy of part distortion and internal stress. Li et al. proposed to divide a SLM process for a practical part into three scales such as micro scale, meso scale and macro scale [11]. With this approach, the temperature history and residual stress fields during the SLM process were predicted. Thermal information has been transferred through micro scale laser scanning, meso scale layer hatching, and macro scale part build-up

The aim of this research is to develop a perspective quality system for SLM process based on making decision algorithm and predicting the part quality by SLM process simulation in consideration of the temperature distribution and internal stress in the workpiece. For carrying out the system, concept model for SLM quality system was established.

4. Conclusion

We developed the model of influences of SLM process parameters on quality and built Ishikawa diagram. The SLM quality system includes technical-organizational method of managing and controlling SLM process.

For getting the requirement part quality influence factors must be considered such as limiting conditions (material properties, equipment specifications), input parameters (building conditions and process parameters) [10]. However, during SLM process, the localized compression and tension caused by the large temperature gradients and fast cooling will increase that lead to the significant internal stresses in workpiece and shape deformation.

Further study concern of investigation the influences of physical SLM parameters on part quality and development mathematical model of part melting according with requirement part quality.

References

[1] Popovich A, Sufiiarov V 2016 Metal Powder Additive Manufacturing. Chapter: **10** *New Trends in 3D Printing. InTech.* 215–236

[2] Manfred Schmid, Gideon Levy 2012 Quality Management and Estimation of Quality Costs for Additive Manufacturing with SLS *Fraunhofer Direct Digital Manufacturing Conference*

[3] Sufiiarov V Sh, Popovich A A, Borisov E V, Polozov I A, Masaylo D V, Orlov A V 2017 The Effect of Layer Thickness at Selective Laser Melting *Procedia Engineering* **174** 126-134

[4] Grigoriev A, Polozov I, Sufiiarov V, Popovich A 2017 In-situ synthesis of Ti2AlNb-based intermetallic alloy by selective laser melting *Journal of Alloys and Compounds* **704** 434-442

[5] Tom Craeghs, Stijn Clijsters, Evren Yasa, Jean-Pierre Kruth Online quality control of selective laser melting *Proceedings of the 20th Solid Freeform Fabrication (SFF) symposium, Austin (Texas)*, **8-10** 212-226.

[6] Agapovichev A V, Sotov A V, Kokareva V V, Smelov V G, Kyarimov R R 2017 Study of the structure and mechanical characteristics of samples obtained by selective laser melting technology from VT6 alloy metal powder *International Journal of Nanomechanics Science and Technology* **8(4)** 323-330

[7] Li C, Liu J F and Guo Y B 2016 Efficient multiscale prediction of cantilever distortion by selective laser melting Annual *International Solid Freeform Fabrication Symposium*

[8] Smelov V G, Sotov A V, Agapovichev A V 2016 Research on the possibility of restoring blades while repairing gas turbine engines parts by selective laser melting *IOP Conf. Series: Materials Science and Engineering* **140**

[9] Agapovichev A V, Balaykin A V, Smelov V G 2015 Production technology of the internal combustion engine crankcase using additive technologies *Modern Applied Science* **9** 335-343
[10] Park H. S, Tran N H, Nguyen D S 2017 Development of a predictive system for SLM product quality *IOP Publishing IOP Conf. Series: Materials Science and Engineering* **227**
[11] Li C, Fu C H, Guo Y B 2016 A multiscale modeling approach for fast prediction of part
[12] distortion in selective laser melting *J Materials Processing Technology* **229** 703

IOP Conf. Series: Materials Science and Engineering **441** (2018) 012025 doi:10.1088/1757-899X/441/1/012025

Experience of convergence of methodologically various decisive rules for improvements of quality of identification of complex objects on analog signals

R A Krupchatnikov[1], M V Artemenko[2], N M Kalugina[2], A F Rybochkin[2]
[1]Kursk State Agricultural I.I. Ivanov Academy 305021, 70, Karla Marksa ave, Kursk, 305021, Russia
[2]Southwest state university, 94, 50 years of October ave, Kursk, 305040, Russia

E-mail: artem1962@mail.ru

Abstract
The article deals with the author's experience of solving the problem of improving the quality of identification of complex objects by the recorded analog signals using recognition systems that implement the convergence of decision rules synthesized on the same samples using methodologically different approaches. The mathematical tools are considered: the statistical approach implemented by calculating the author's indicators of the system organization – analogues of statistical moments of the third and fourth orders; the synergetic approach-indicator latent variables are proposed to form on the basis of matrix transformations of the bi-spectral autocorrelation function of the recorded signal, the parameters of the time shift of which are adjusted in accordance with the Fourier spectrum of the reference classes of the objects. The results of testing the proposed methods on the example of solving a medical problem-preventive diagnosis of destructive lung conditions in asthmatic diseases are reduced. It is shown that the convergence of different rules allows to obtain identification solutions more correlated with independent external expertise than the results of their application separately.

1. Introduction

Quality control and correction of the states of complex technical biological and biotechnical objects determines the application of a system approach and specific methods of its identification as an element of a certain class, for which the content of the bases of shares of the acting agents is known [1]. According to the classical definition by Rastrigin L. A. the complex objects, first and foremost, characterized by: a lack of sufficient for the formalization of the mathematical description; a large number of feedbacks and non-target object processes that lead to additive superposition and modulation of the registered signals; the presence of auto control [2]; the stationary nature and temporal uniqueness (the phenomenon of the arrow of time by Prigogine I.).

These objects include: open-type technical systems, the elements which structures are in constant interaction with the external environment (vehicles, computer networks); physiological and functional systems that ensure the normal functioning and execution of the target functions of the body as a whole (cardio, respiratory, musculoskeletal, etc. system); economic and social systems of different hierarchical levels; population; socio-technical landscapes, etc.

Since the functioning of complex objects (systems) systemically involved in the provision of the basic processes of human life and determines largely its safety, and to date in various information databases and sources accumulated experience in the management of various, more or less similar to each other objects, improving the quality of identification of the object, as a representative of a certain cluster, is a permanent current scientific and technical task [3, 4].

The transparency of the complex objects functioning causes a high degree of uncertainty which requires the development of adequate models and methods of decision what have got synergetic properties and based on the achievements of artificial intelligence and computer-information technologies. The behavior of a complex object from a cybernetic point of view, in general, is described by the change in the position in the space of the recorded features [5]. Features are represented as discrete and-or analog signals, direct and latent, scalar or matrix forms. The variety of forms leads to the need to apply different analytical paradigms and theories, methodologically and epistemologically, different from each other. Various expert councils, who designed to implement the optimal convergence, are involved to coordinate the final management conclusions [6]. Councils can be organized: by combining specialists (in this case, classical well-developed methods of expert analysis are used, by combining the means of "computer intelligence", by combining "representatives" of artificial and natural intelligence. The latter option is most preferable, because it allows to combine the advantages of objective and subjective [19] analyses in a reasoned and targeted way, especially when it is necessary to make insightful management decisions in the conditions of the situations of "gray" and "black" swans (Taleb N.).

According to the work [7], in general, the impact on the object is represented as: $FacInf = \{A, FI(t), TI, Rez\}$, where A - unambiguously determining the power of the impact factor, , $FI(t)$ - effect function, TI - period of exposure, Rez – resonance coefficient, which assesses the possibility of resonance phenomena under the influence where $Rez \in [0,1]$, $Rez = 0$ – resonance phenomena are impossible, $Rez = 1$ – « surely arising ». As an indicator of Rez it is proposed to apply the function of belonging to the base variable of the square of the ratio of the object's natural frequency to the difference between the frequencies of its own and external influence (external or internal, integral or particular, of different physical nature).

If the registration of the values of signs, defined as informative, is carried out continuously, we can talk about the analysis of some analog signals for the classification identification of both the object and its state, by calculating the values of the indicator variables and the application of certain decision rules. In most cases, these rules are well formalized in the form of products.

The specific signal from the object is characterized by spatiotemporal components reflected by spectral parameters (obtained, for example, by Fourier or Walsh transformations) and autocorrelation and autoregressive dependences [8], of different orders. Statistical analysis involves the use of values of particular and integral indicators (for example, indicators of the system organization in the analysis of the functioning of certain physiological systems [9]). Synergetic methodology involves the calculation of the indicators values which systematically reflecting the self-organization processes occurring in the object during the registration of the signal, in the form of changes in the structures and parameters of autocorrelation functions.

Thereby taking actuality of the problem by qualitative identification of the control object and the possibility of its solution by existing means of artificial intelligence, the purpose of the study is to increase the quality of the diagnostic classification of the object in the process of studying the structural components of the generated specific signal, by developing a convergence method of the application of statistical and synergetic results what approaches to the analysis of its space-time structure.

2. Results and Discussion

The classical (statistical) approach to the synthesis of decisive rules of the production type is to execution the following steps:

1. Formation of a set of features *{Ω}* what are characterizing the state and behavior of the object of study, the values of which can be observed and recorded, by expert analysis ways.
2. The selection of *{Ω}* the set of informative features *{ΩI},* which allows to use them as indicator variables of the crucial antecedents of production rules for the implementation of consequential that are certained by specialists.
3. Synthesis of antecedents, by determining on training samples specific ranges of values of the indicator variables included in the antecedents in the case of numerical values, it is recommended, after evaluating the distribution law, to calculate the confidence intervals with acceptable errors of the first and second types.
4. Evaluation on the examination sample of values (or functional definition) of confidence coefficients for decision rules.

The formation of *{Ω}* is specific for analog signals recorded by the sensors from the object, presupposes:

- first, the synthesis of additional, latent signals сигналовд *S(t)* and iS(t): $dS(t) = \frac{\partial S(t)}{\partial t}$. $iS(t) = \int S(t)dt$, *S(t)* – the recorded analog signal is represented in discrete form;
- secondly, the choice of informative frequency ranges-a lot of *{DFR}* (for *S(t), dS(t)* and *iS(t)*);
- third, the definition of the composition of features for each of the ranges *{DFR}* (for *S(t), dS(t)* and *iS(t)*).

Classical and proven methods and algorithms for distinguishing characteristics from *S(t)* are presented in a number of foreign and domestic sources, for example, in the process of "blind" signal processing [8]. Exploration spectral analysis allows to select elements of the set *{DFR}* (lower and upper bounds) using the following algorithm [7].

1. Registration of the signal from the object. Formation of training and control samples. Set of diagnostic frequency bands *{DFR}=Ø.*
2. Setting the frequency range randomly - *df.*
3. Determination of spectral characteristics in *df* (maximum and minimum values of the spectral function, the spectral area in the range) – $\{SV_{df}\}$.
4. Definition of diagnostic capabilities $\{SV_{df}\}$ by DSM method [30] with the fixation of the criterion of diagnostic possibilities (the criterion of diagnostic opportunities - CDO_{df}).
5. If $CDO_{df} \subset tr_{df}$ then df are included to *{DFR}* and transition to item 7 is carried out. Otherwise, the check is performed - if the process of formation of *{DFR}* is completed, then the transition to item 7 , otherwise – the transition to item 3. (Here - tr_{df} , – user-defined designer acceptable threshold range for in *df.*)
6. If *{DFR}* does not significantly improve the values of the diagnostic criterion on the control sample, then the transition to item 3.
7. Ordering *{DFR}* in ascending order of frequencies.
8. If there are restrictions on the number of ranges in *{DFR}*, then *{DFR}* - either decreases - by eliminating the ranges according to the selected criterion, or increases - by going to item 2.

(Note that the algorithm allows for intersecting, but not matching, *{DFR}* elements.)

As elements of the set *{Ω}* is proposed to use calculated in each band characteristics [10]: the minimum value of the spectrum minDFR, the maximum value of the spectrum *maxDFR*, the modal value of the spectrum *modDFR*, average power of signal *PDFR*, the mean and standard deviation - *msdDFR*, entropy - *entDFR*, negentropy - *nentDER*, the area of the spectrum, *SDFR*, pairwise relations – *maxDFR/minDFR, PDDFR/modDFR, msdDFR/modDFR, (maxDFR-minDFR)/modDFR.*

The formation of a set of informative features {ΩI} is recommended to be carried out depending on the data structure and volumes by the methods described in [11] using the Kendell concordance coefficient.

To determine statistically significant differences between objects belonging to different classes, it is proposed to use the following algorithm:

1. On the training sample, according to the recommendations [12], the discriminant function is identified as $Y(\omega_k)=Y(\omega_{k-1})+(-1)^k\cdot H_{\omega_{k-1}}$, where $Y(\omega_k)$ - value of the identifier for a class ω_k, $H_{\omega_{k-1}}$ - step "inter-class differences" (it is recomended: $H_{\omega_i}=H_{\omega_j}=10$), k=0,1,...K, $Y(\omega_{-1})=0$, $H_{\omega_{-1}}=0$. On examination - thresholds are determined $Y(p)$ for decision rules and confidence coefficients equal to the values of diagnostic efficiencies in each alternative class ω_k.

2. Values of indicators of the system organization of object are defined (indicators of the system organization – *ISOas, ISOex*) by formuls [23]:

$$ISOas=\frac{\sum_{i=1}^{Qr}(p_i-\bar{p})^3}{Q_r\cdot\sigma^3},\quad ISOas=\frac{\sum_{i=1}^{Qr}(p_i-\bar{p})^4}{Q_r\cdot\sigma^4}, \tag{1}$$

where: Qr – number of comparison implementations; σ^2, $\bar{p}$ – variance and mean, respectively, of the characteristic p_i from $\{\Omega I\}$.

3. Built membership function $K2_k$ on the media:

$$dISO_k^2=(\mathrm{ISOas}_k-\mathrm{ISOas}_0)^2+(\mathrm{ISOex}_k-\mathrm{ISOex}_0)^2 \tag{2}$$

(index k – corresponds to the k-th sign).

Functions $K2_k(dISO^2)$ determined by expert or analytical means. For the latter case, the authors propose empirically obtained formula:

$$K2(dISO_k)=\begin{cases} th(\pi\cdot\frac{dISO_k}{db_k})\cdot\cos(\varphi),\text{for a standard } Se_k \\ h(\pi\cdot\frac{dISO_k}{db_k}),\text{in } other\ \ case\end{cases}, \tag{3}$$

where: $\overrightarrow{(\mathrm{ISOEas}_0,\mathrm{ISOEex}_0)}$ - reference state vector S_k, db_k - module $\overrightarrow{(\mathrm{ISOEas}_k,\mathrm{ISOEex}_k)}$, φ- the angle between the vectors $\overrightarrow{(\mathrm{ISOEas}_k,\mathrm{ISOEex}_k)}$ and $\overrightarrow{(\mathrm{ISOEas}_0,\mathrm{ISOEex}_0)}$.

4. Confidence is determined $K3_k$:

$$K3_k=(\sqrt{K1_k}+\sqrt{K2(\mathrm{dISO}_k)}-\sqrt{K1_k\cdot K2(\mathrm{dISO}_k)})^2 \tag{4}$$

5. The production decision rules of the form are synthesized:

"IF $K2(dISO_k)>0.62$, then the object is in the S_k state with confidence $K3_k$.»

In general, the form of recording such decisive rules is generalized to the form:

$$\text{"IF } Ans(\{ISO\})_{\omega_k}\text{, then the object} \in\omega_k \text{ with confidence } U_{\omega_k}\text{"} \qquad \text{(ISO rule)}$$

Where: $Ans(\{ISO\})_{\omega_k}$ - the antecedent for the class ω_k whose arguments are the indicators of the ISO system organization.

To analyze the synergetic components of the analog signal, it is proposed to use the bispectral autocorrelation function, the values of which are calculated by the formula [8]:

$$R_{x,x+k,x+l}=\frac{1}{M}\sum_{i=0}^{M-1}x(i)\cdot x(i+k)\cdot x(i+l), \tag{5}$$

Where: $x(i)$ the i-th discrete of vector X; M – the number of the discrete vector X; *k, l* – «time shift »; *k=0,1, ...M/2, l=0,1, ...M/2;* X – a volume-constrained vector of values $S(t)$ or $dS(t)$ or $iS(t)$.

In order to optimize the process of obtaining the matrix R and its further processing, it is proposed to limit its dimension to ten rows and columns (time shifts), and to make the shift step self-organizing and dependent on the spectrum of the analyzed signal. In this way from formula (5), it is proposed to move to the formula:

$$R_{x,x+k_j,x+l_j} = \frac{1}{10}\sum_{i=0}^{9} x(i)\cdot x(i+k_j)\cdot x(i+l_j) \quad , \tag{6}$$

where: k_j and l_j are elements of an array of integers (j=0,1,...9) that are harmonic numbers of 10 local maxima of the spectral function.

The synthesis of decision rules is realized by the following method:

1. Spectral function is identified and are formed a vector of values of k_j and l_j.
2. Randomly selected from the training sample is one of the signals and a discrete vector X is formed.
3. The matrix of the bi-spectral autocorrelation function R_o is determined by the formula (6). The inverse matrix R^{-1} of the matrix R_o is calculated. The obtained R^{-1} matrix is defined as a support for the synthesis of the conditions of the decision rules in the future.
4. On the training sample of object-specific signals are formed by the formula (6) of the matrix B_l ($l=0,1...K,K$ - number of classes) which is a bi-spectral autocorrelation function. Indicator matrices C_i are calculated as a product of: $C_i=R^{-1}*B_i$. For each matrix C_i, the average values of the matrix squares are determined:

$$IP_i = \sqrt{\frac{1}{100}\sum_{i=1}^{10}\sum_{j=1}^{10}(C_{i,j})^2} \; . \tag{7}$$

5. On the training sample for IP, confidence intervals are formed, which are typical for each alternative object class.
6. For each of the classes the decisive rules of the form are formed:

$$\text{"}IF \;\; Ans(\{\mathrm{R}\})_{\omega_k}, \; then\ the\ object \in \omega_k \;\; with\ confidence\ U_{\omega_k}\text{"} \quad , \qquad \text{(rule R)}$$

where: $Ans(\{\mathrm{R}\})_{\omega_k}$ - the antecedent for the class ω_k formed by the bi-spectral autocorrelation function and shows the belonging (or non-belonging) of the value of the indicator variable IP to the corresponding confidence interval.

Since the obtained "ISO rules" and "R rules" reflect various methodological aspects of the analysis, it is proposed to synthesize convergent decision rules as follows. Let's denote by $Rul_{r,l}$ some decisive rule for class l. The general decision rule, in this case, is as follows:

$$\text{"}IF \;\; (\wedge_1^{Rr}(Ans_{\omega_k}(\mathrm{II})_{Rul_r})) \;\; that\ object \in \omega_k \;\; with\ confidence \;\; (U\wedge)_{\omega_k}\text{"} \qquad \text{(rule convergent)}$$

where: $(\wedge_1^{Rr}(Ans_{\omega_k}(\mathrm{II})_{Rul_r}))$ - the conjunction of the ancestor of the decision rules Rul_r obtained on the indicator variables II; $(U\wedge)_{\omega_k}$ - confidence in the correlation (identification) of the object to the class ω_k, calculated by the formula:

$$(U\wedge)_{\omega_k} = 1-\max(\mathrm{U}_{\omega_k,\mathrm{Rul}_{r,l}})\cdot\prod_{r=1}^{R}(1-a_r)-\prod_{r=1}^{R}(a_r\cdot(1-\mathrm{U}_{\omega_k,\mathrm{Rul}_{r,l}}))/ \quad a_r = \begin{cases} 1, \text{if the rule was applied } Rul_{r,l} \\ 0, \text{if the rule was not applied } Rul_{r,l} \end{cases} .$$

(8)

Application (8) improves the quality of object identification by increasing the value of the confidence function in case of convergence of decision rules. To eliminate the influence of "weakly-confident" decision rules, in the formula (8) of the binary variable a_r assign a zero value if the confidence of the rule is below a certain level defined by the user.

As an experimental basis for checking the effectiveness of the proposed method of convergence of decision rules, the medical problem of identification of various destructive asthmatic forms in a patient was chosen [11] (I — stage of compensation, II — stage of increasing ventilation disorders or stage of decompensation, III stage — hypercapnic coma). The experimental material representing sound signals was formed on the basis of international medical information databases.Various decision rules are obtained, for example:

- If $(Ip \subset 1 \pm 0.2)$, then the state of the respiratory system are normal and asthma is absent with a confidence of 0.96;
- If $(Ip \subset 2{,}1 \pm 0.6)$, then the state of the respiratory system are normal and asthma is absent with confidence of 0,92 (stage I);
- If the Ip values do not belong to the specified intervals, additional examination with a confidence of 0.95 is required to determine the state of the respiratory system.

To calculate these indicators as informative, $\{\Omega I\}=\{modDFR, PDFR, msdDFR, entDFR, SDFR, maxDFR/minDFR, PDDFR/modDFR, msdDFR/modDFR\}_k$. ($k$=2 – two frequency bands are selected) were selected. The calculated values of ISO_{as} and ISO_{ex} system organization indicators for the classes "healthy" and "asthma" were: $ISO_{as\_0}=0.38\pm0.15$, $ISO_{ex\_0}=1.9\pm0.32$, $ISO_{as\_1}=-0.1\pm0.2$, $ISO_{ex\_1}=0.98\pm0.41$. According to the given data, the corresponding production rules are constructed and the values of the confidence functions are determined.

Options for convergent application of the rules are presented in table 1 - the values of diagnostic efficiency. The values of the indicators set by an expert doctor when listening to sound signals are highlighted in brackets.

Table 1. Confidence in the correct diagnostic solution.

Apply rules		Patient condition			
(rule ISO)	(rule R)	«healthy»	I stage	II stage	III stage
no	no	0.04	0.06	0.16	0.05
no	yes	0.96	0.92	0.86	0.95
yes	no	0.88	0.94	0.84	0.94
yes	yes	0.995	0.995	0.978	0.997
Medical expert		(0.97)	(0.96)	(0.95)	(0.98)

Analysis of the data in the table allows us to draw the following conclusions:

1. The results of the application of the obtained decisive rules are mutually consistent, which emphasizes their acceptability for operation in real conditions;
2. Convergence of decision rules allows to improve the quality of preventive diagnostic identification of the object of study;
3. The convergence results of several decision rules have a stronger correlation with external, independent expertise than the application of each of them separately.

3. Conclusion

The results of the research allowed to develop a method of convergence of the decision rules of identification of the object of study on the specific analog signals issued by it, synthesized methodologically different approaches. The basis of the method is the combination of antecedent one-way consequents and the modification of the values of the confidence coefficient, which is more correlated with the objective data of expert identification.

For the static approach of the synthesis of adequate decision rules it is recommended to apply the indicators of the system organization, similar to the particular moments of 3 and 4 orders. In the analysis of the synergetic properties of the object as an indicator variable is recommended to use the average sum of the squares of the elements of the diagnostic matrix, which is the product of two matrices: the bi-spectral autocorrelation function of the object and the inverse matrix of this function for some object belonging to the reference class. Self-organization of matrix elements calculation, expressed in the determination of time shifts when calculating the values of the autocorrelation function, by analyzing the spectrum of the signal of the reference class, allows to reduce the resources of the computational process and increase its efficiency.

The developed methods are well formalized for use in the design of decision support systems.

Decision-making systems - areas of theoretical and practical application of research results, automatic control and pattern recognition, diagnostics of complex objects.

4. Acknowledgments

The work was carried out within the framework of the federal target program "Research and development in priority areas for the development of the scientific and technological complex of Russia for 2014-2020" and in accordance with the scientific direction of the South-West State University "Development of medical and environmental information technologies" and financial support for the grant No.11643GU/2017 as part of the contest "UMNIK"(cod 0033037).

References

[1] Rassel S, Norvig P 2018 *Iskusstvennyj intellekt: sovremennyj podhod* (Moskva: Vil'jams)1408.

[2] Zhdanov A A 2009 *Avtonomnyj iskusstvennyj intellekt* (Moskva: BINOM)

[3] Guchuk V V 2018 Applied aspects of the organization of an interactive mode for increasing tye security of functioning of control systems of complicated objects *European Science*. **No 1 (33).** 18-21

[4] Semenov S S , Voronov E M , Poltavskij A V, Krjanev A V 2016 *Metody prinjatija reshenij v zadachah ocenki kachestva i tehnicheskogo urovnja slozhnyh tehnicheskih sistem* (URSS)

[5] Viner N 2001 Chelovek upravljajushhij. (1. Kibernetika i obshhestvo. 2. Tvorec i robot. 3. Stat'ja "Kibernetika i obshhestvo" URSS)

[6] Ushakov V N, Lavrov N G, Ushakov A V 2014 Konstruirovanie reshenij v zadache o sblizhenii stacionarnoj upravljaemoj sistemy *Tr. IMM UrO RAN*, **20, No 4** 277–286

[7] Artjomenko M V, Borodin M V, Kalugina N M, Kurochkin A G 2017 Identifikacija faktorov vozdejstvija na obekt samoorganizacionnym analizom spektra harakterizujushhego signala *Informacionno-izmeritel'nye i upravljajushhie sistemy* **T. 15. No 5** 8-12.

[8] Kravchenko V F 2007 *Cifrovaja obrabotka signalov i izobrazhenij v radiofizicheskih prilozhenijah* (Moskva: FIZMATLIT)

[9] Artemenko M V, Dronova T A, Kalugina N M, Bojcov A A 2017 Primenenie pokazatelej sistemnoj organizacii v avtomatizacii diagnostiko-terapevticheskogo processa *Izvestija Jugo-Zapadnogo gosudarstvennogo universiteta. Serija: Upravlenie, vychislitel'naja tehnika, informatika. Medicinskoe priborostroenie.* **T 7 No 2 (23)** 122-134.

[10] Artemenko M V, Pisarev M V 2016 Synthesis and study of matimatical models of biomedical signals methods of selforganization modeling by the example of the photoplethysmogram *International Journal of Applied and Fundamental Research.* **5** 31-35.

[11] Artemenko M V, Chernetskaia I E, Kalugina N M, Shekina E N 2018 Bootstrap and counter-Bootstrap approach for formation of the cortege of informative indicators be results of measurements *Jurnal of Physics: Conference Series* **998** 120-126

[12] Voroncov I M, Shapovalov V V, Sherstjuk Ju M 2006 Zdorov'e. Opyt razrabotki i obosnovanie primenenija avtomatizirovannyh sistem dlja monitoringa i skrinirujushhej diagnostiki narushenij zdorov'ja (SPb.: «IPK «Kosta»)

IOP Conf. Series: Materials Science and Engineering **441** (2018) 012026 doi:10.1088/1757-899X/441/1/012026

Analysis of forming properties during the isothermal upsetting of cylindrical workpieces in the viscous-plasticity mode

S N Larin [1], **A A Pasynkov** [1]

[1] Tula State University, 92, Lenina ave., Tula, 300012, Russia

E-mail: mpf-tula@mail.ru

Abstract. In the manufacturing of axisymmetric parts made of high-strength alloys, slow hot deformation is widely used, which makes it possible to increase significantly the plasticity properties of the material and to reduce the deformation force, as well as to achieve greater deformation. However, currently there are few studies discussing the issues of high-strength materials formation theory. This article considers the isothermal upsetting of cylindrical workpieces made of hard-to-deform alloys in the viscous-plasticity mode. The article introduces isothermal upsetting models regarding the short-term creep mode. The article reveals the influence of technological parameters and friction conditions existing on the contact surfaces of the working tool and workpiece on the material flow kinematics, stressed and deformed states, and the force conditions. The results obtained during their development can be used to create design methods for innovative, science-intensive, competitive technological processes of next generation for manufacturing axisymmetric geometrically-complex parts by means of isothermal deformation methods.

1. Introduction

The main problem facing engineering nowadays is to increase the efficiency and competitiveness of products manufacturing of high-strength metals and alloys by pressure-processing methods that provide the maximum possible performance characteristics. In the manufacturing of axisymmetric parts from high-strength alloys, slow hot deformation is widely used, which makes it possible to increase significantly the plasticity properties of the material and to reduce the deformation force, as well as to achieve greater deformation. The pressure processing of hard-to-deform alloys in the short-term creep mode opens great prospects for the creation of new technological processes. Despite the large number of works devoted to theoretical and experimental studies of isothermal die forging of bar stocks, the problems of high-strength materials forming theory in short-term creep mode have not been fully discussed at present.

2. Materials and methods

Upsetting is a typical operation for the processing of metals by pressure [1]. The upsetting of high-strength materials workpiece is carried out under isothermal conditions on hydro-press equipment. The forging modes are determined by the temperature-speed conditions, as along as with deformation hardening of the material, there is a creep condition, depending on the processing speed [2,3]. In addition, the upsetting is non-stationary, which also affects the technology in terms of operating

pressure and the limiting degree of deformation. Thereupon let's consider a variant of upsetting properties calculation. We will use the energy calculation method based on the upper bound theorem of plasticity theory with respect to the discontinuous axisymmetric displacement velocity field [4].
Figure 1 schematically shows the upsetting process. The scheme also demonstrates the displacements velocity field, consisting of hard blocks "0" and a block of deformations "1". The blocks represent the rotation factors and the limited surfaces of discontinuity velocity.

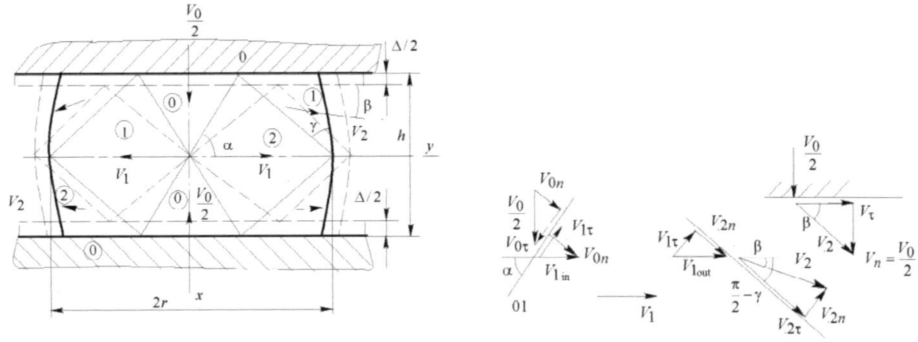

a) b)

Figure 1. Upsetting scheme, field (a) and velocity plan (b)

3. Results and Discussion

The velocity field is non-stationary, because it changes during the upsetting (shown in dotted lines). The line "01" rotates relative to "no". The lines "01", "12" move along the coordinate axes "x", "y" respectively, which causes additional normal velocities on discontinuity velocity surfaces. The deformations occur on these surfaces and in the deformation block.
Equivalent stresses are expressed on the basis of the equality of the state of the material being deformed. On the surface of the discontinuity velocity is

$$\sigma_э = A\varepsilon_\partial^m \xi_n^n = A\left(\frac{V_0}{\Delta h}\right)^n \varepsilon_э^{m+n}; \tag{1}$$

In the deformation block

$$\sigma_\partial = A\left(\frac{\Delta h}{V_0}\right)^m \cdot \xi_э^{m+n}. \tag{2}$$

Expressions (1) and (2) take into account the stress relaxation in connection with creep. The power of the internal forces on the surface of the velocity discontinuity is expressed using $N_p = \frac{1}{V_3}\sigma_э V_\tau S$, where $\sigma_э$ are the equivalent stresses (9); $V\tau$ - tangential velocities; S - velocity discontinuity areas.
We obtain the following expressions:

$$N_{01} = \frac{\pi A h^2}{2\sqrt{3}\sin\alpha}\left(\frac{1}{\Delta h}\right)^n V_0^{1+n}\left(1+ctg^2\alpha\right)\varepsilon_{э01}^{m+n}; \tag{3}$$

$$N_{12} = \frac{\pi A h^2}{2\sqrt{3}\cos^2\gamma}\left(\frac{2r}{h}\cos\gamma - 1\right)\left(\frac{1}{\Delta h}\right)^n V_0^{1+n}\frac{\cos(\alpha+\beta)}{\sin\beta}\left[tg(\beta+\gamma) - tg\gamma\right]\varepsilon_{э12}^{m+n}; \tag{4}$$

$$N_{02} = \frac{\pi A h^2 tg^2\gamma}{2\sqrt{3}tg\beta}\left(\frac{1}{\Delta h}\right)^n V_0^{1+n}\varepsilon_{э02}^{m+n}. \tag{5}$$

In the deformation block, the powers are determined using expressions (4) and (5) by the following equation:

$$N_1 = \int_w \sigma_э \xi_э dw = \pi A r h y_{H.T.} \left(\frac{\Delta h}{V_0} \right)^m \int_{y01}^{y12} dy \int_0^h \xi_э^{m+n} dx \qquad (6)$$

Here w- is the power of the deformation block; $y_{H.T}$ is the ordinate of block's longitudinal section on the velocities' plan (triangular ones with sides x=0, y_{01}, y_{12}). Upsetting pressure at any stage follows from the balance equation for the powers of the upper and internal forces. Using the expressions (3) - (6), we obtain

$$q \le 2(N_{01} + N_{12} + N_{02} + N_1)/\pi r^2 V_0. \qquad (7)$$

The pressure depends on the initial workpiece dimensions, upsetting degree, its speed and deformation's non-stationarity. It can be calculated for any stage of the operation for a given Δh. The lower value of the upper bounded pressure is determined by minimizing expression (7) with respect to angle α. When forging on crank presses, creep does not appear in formula (7) n = 0 as well.

The calculated and technological results are obtained with reference to the upsetting of cylindrical workpieces with dimensions 2r=50 mm, h0 – 40 mm made of aluminum alloys at 450 ° C and titanium alloys at 950 ° C. Upsetting pressure, as follows from the table and figure 2, decreases with increasing operation time.

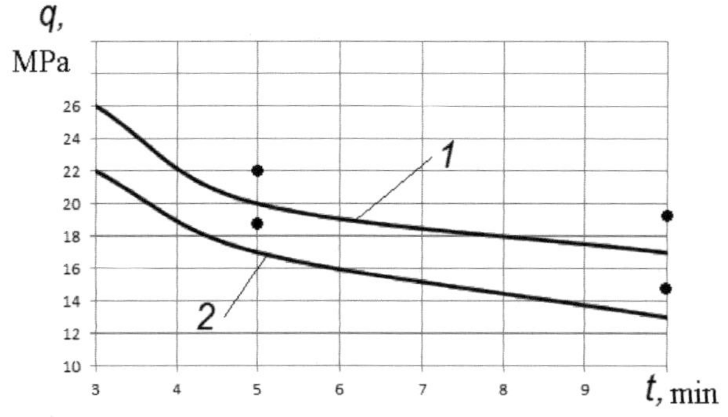

Figure 2. Dependence of the specific upsetting pressure from the deformation time: 1 - AMg6 alloy; 2 - alloy VT14

Therefore, for one group of materials the upsetting degree can be increased by increasing the duration of the operation. For another group, there is no such dependence. Taking into account the non-stationary nature of the process during the calculations corrects the deformation values, surface stresses and, consequently, affects the pressure estimates and the critical degree of deformation.

In order to take into account the influence of the deformation rate, friction and workpiece dimensions on the quality of the products obtained, strain and deformation forces, this process was studied with the QFORM software. Its use is justified by the qualitative compliance of the results obtained in the course of the study based on the upper boundary plasticity theorem with the results obtained in the QFORM studies. Studies were carried out for the titanium alloy VT14, since its deformation is the most difficult one. For a more complete picture of the study, workpieces with a length of 40; 60; 80 mm and with a diameter of 40 mm were used. The proportions of these sizes are 1; 1.5 and 2, respectively. Friction varied in the range between 0.1 and 0.3, which corresponds to the real type of lubrication, that is to the liquid glass (with friction coefficient value being 0.1), as well as a mixture of graphite and oil (in this case the friction coefficient is 0.3).

Figure 3 shows the dependence of the change in the force during the upsetting on the deformation rate for different dimensions of workpieces at $\mu = 0{,}1$.

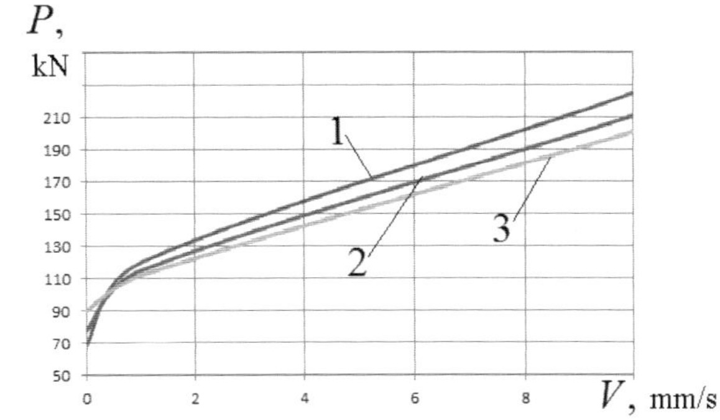

Figure 3. Dependence of forces under isothermal upsetting from deformation rate ($D_b = 40\ mm$; $\mu = 0{,}1$): 1 - $H_b = 40\ mm$; 2 - $H_b = 60\ mm$; 3 - $H_b = 80\ mm$

Figure 4 shows the dependence of the change in the force during the upsetting from the deformation rate for different dimensions of the workpieces at $\mu = 0{,}3$.

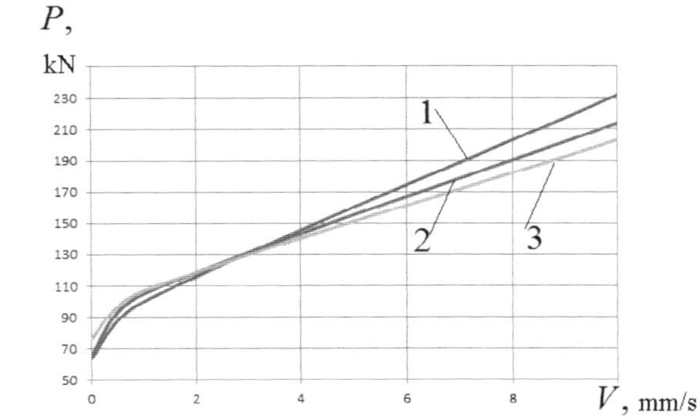

Figure 4. Dependence of forces during isothermal upsetting from deformation rate ($D_b = 40\ mm$; $\mu = 0{,}3$): 1 - $H_b = 40\ mm$; 2 - $H_b = 60\ mm$; 3 - $H_b = 80\ mm$

According to the analysis of the dependencies presented in fig. 3 and 4 we can see that together with the increase of the deformation rate the force is also increasing, but the greatest intensity of force growth is achieved in the velocity intervals between 0 and 1 mm/s. Changes in the height of the workpiece with its constant diameter do not affect the force growth greatly. However, we can say that the smaller the ratio of the height of the workpiece to its diameter, with the same stroke of the punch, the greater the force is. An increase in the friction value from 0.1 to 0.3 evenly increases the force values by 10%.
Figure 5 shows the dependence of the change in the force during the upsetting from the deformation rate for different workpiece dimensions.

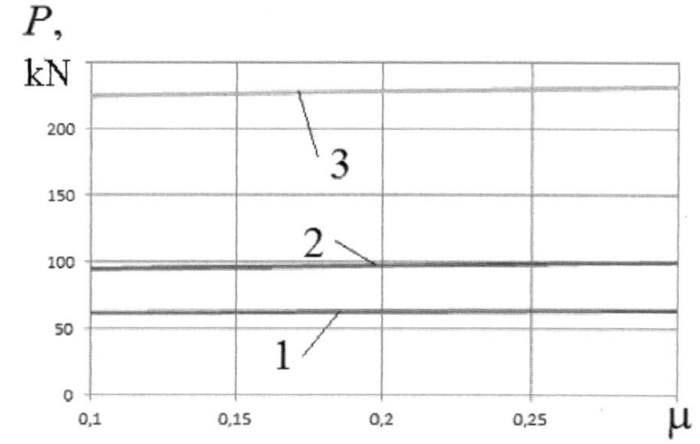

Figure 5. Dependence of forces during the isothermal upsetting from the friction coefficient ($D_b = 40\ mm; H_b = 40\ mm$):1 - $V = 0{,}01\ mm/\sec$; 2 - $V = 0{,}1\ mm/\sec$; 3 - $V = 10\ mm/\sec$

According to the analysis of the dependence shown in fig. 5 it is seen that the force increases with increasing deformation rate. The increase in the friction value leads to an even growth of the force value.

It has been established that during the upsetting, a defect arises for all workpieces, depending on the deformation rate, which manifests itself in the form of barrel or cavity formation in the central section of the part. In this connection, we carried out a study, the purpose of which was to determine the rate of deformation, at which such a defect does not arise. It can be seen that for workpieces made of titanium alloy at the deformation temperature $900°C$ the best value of the deformation rates for the height $H_b = 40\ mm$ is $V = 0{,}2\ mm/\sec$; for height $H_b = 60\ mm$ is $V = 0{,}5 mm/\sec$; for height $H_b = 80\ mm$ is $V = 0{,}7 mm/\sec$. According to the obtained data, we can construct the diagram shown in Fig. 6.

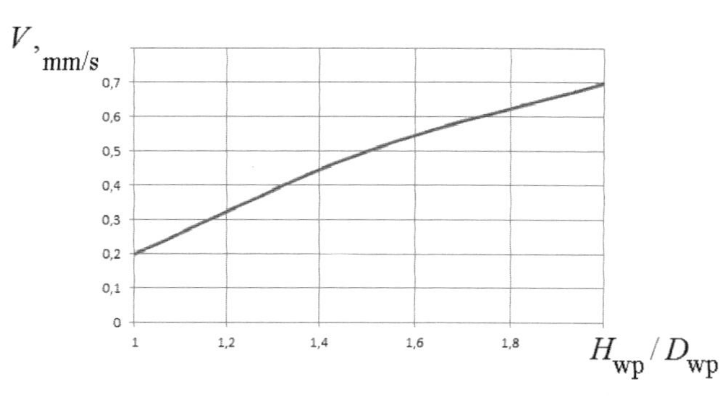

Figure 6. The diagram, estimating the rational deformation rate

4. Conclusion

It can be concluded that for the design of technological processes of isothermal forging, with upsetting being the main deformation scheme, the isothermal upsetting occurs under strain hardening conditions and stress relaxation in connection with the development of creep of the material being processed. This fact affects the punching mode; the operation pressure decreases as the upsetting rate decreases. At the same time, workpiece material damage is accumulated; for some materials the less the speed of the operation is, the less is the final damageability for a given upsetting degree, for other materials, the dependence from speed (time) does not appear; taking into account the non-stationarity of the upsetting approximates the pressure estimation and the critical degrees of deformation to the actual ones. As a result, recommendations to ensure a defect-free upsetting in a viscous-plastic flow mode without barrel or cavity formation were provided.

5. Acknowledgments

The results of the research project are published with the financial support of Tula State University within the framework of the scientific project.

References

[1] Unksov E P Theory of plastic deformations of metals 1983 *M.: Mechanical Engineering* 598.
[2] Romanov K I Mechanics of hot metal forming 1993 *Moscow: Mashinostroenie* 240.
[3] Zhang H, Li X, Tang W, Deng X, , Reynolds A P, Sutton M A Heat transfer modeling of the friction extrusion process 2015 *Journal of Materials Processing Technolog.* **Volume 221** 21–30
[4] Matthias Haase, Erman A, Tekkaya Cold extrusion of hot extruded aluminum chips 2015 *Journal of Materials Processing Technology* **Volume 217** 356–367
[5] Peter Krizika, Martin Baloga, Emilia Illekovab, Peter Svec Sr.b, Igor Matkob, Matej Stepaneka, Martin Noskoa, Frantisek Simancika The oxidation behavior of gas-atomized Al and Al alloy powder green compacts during heating before hot extrusion and the suggested heating process 2014 *Journal of Materials Processing Technology* **Volume 214 Issue 6** 1165–1172.
[6] Hong-ye Li, Min Xia, Zhen Zhao, Yu Zheng An experimental study on local thickening laws based on multi-plate extrusion 2013 Journal of Materials Processing Technology **Volume 213 Issue 7** 1213–1220
[7] Marcus Engelhardt, Norbert Grittner, Hermann von Senden genannt Haverkamp, Wilfried Reimche, Dirk Bormann, Friedrich-Wilhelm Bach Extrusion of hybrid sheet metals 2012 *Journal of Materials Processing Technology* **Volume 212 Issue 5** 1030–1038.
[8] Wenxiong He, Erde Wang, Lianxi Hu, Yang Yu, Hongfei Sun Effect of extrusion on microstructure and properties of a submicron crystalline Cu–5 wt.Cr alloy 2008 *Journal of Materials Processing Technology* **Volume 208, Issues 1–3** 205–210.
[9] Nicolas Fiétiera, Yves Krähenbühlb, Matthieu Vialardb New methods for the fast simulations of the extrusion process of hot metals 2009 *Journal of Materials Processing Technology* **Volume 209, Issue 5** 2244–2259.
[10] Malinin N N Creep in metal processing 1986 *M.: Mechanical Engineering* 216.
[11] Yakovlev S S, Yakovlev S P, Chudin V N, Tregubov V I, Chernyaev A V Isothermal shaping of anisotropic materials by a rigid tool in the regime of short - term creep 2009 *M.: Mechanical Engineering* 412.

Magnetoelectric current sensor: miniaturization and perspective

Leontiev V.S., Petrov R.V., Solovev A.I., Bichurin M.I.

Novgorod State University, Veliky Novgorod, Russia, 173003
E-mail: viktorsergeevich.novsu@gmail.com

Abstract. One of the main tasks of microelectronics is the miniaturization of devices with the keeping of previous functions and technical characteristics. In this paper, we present a magnetoelectric current sensor using a microprocessor. Compared with the previously presented magnetoelectric current sensor, we reduced the overall dimensions by almost 2 times and saved technical characteristics. Integration of the microprocessor opens up wide prospects for the application of the sensor in various industries, for example in robotics.

1. Introduction

An analysis of the world trends of current sensors indicates that such devices are in great demand today in various areas of industry. There is a massive replacement of traditionally used current sensors: shunts, current transformers, sensors on the Hall effect to modern miniature current sensors. There are searches for current sensors based on fundamentally new effects. In most modern sensors, current measurement is performed using magnetic fields, which provides their main advantages: galvanic isolation, high sensitivity, wide operating temperature range. Low price, small dimensions and at the same time high sensitivity contribute to the fact that they are beginning to be applied in measuring systems, security systems, the automotive industry, the railway industry, space technology, robotics and etc.

In most multiferroics (Cr_2O_3 and etc.) the magnetoelectric (ME) effect is not significant and its magnitude does not exceed α_E= E/H ~ 20 mV/(cm·Oe), and observed at low temperatures or in large magnetic fields, which limits their practical application. A significantly larger ME effect was observed in composite structures containing a magnetostrictive and piezoelectric phase. Thus, the use of composite structures opens up wide possibilities for practical application [1]. One of the main applications is high-sensitivity magnetic field sensors is magnetometers [2,3]. The current sensors are also a perspective trend. The previously presented magnetoelectric current sensor [4-6] has a significant disadvantage that is no digital output. ME current sensor is based on the magnetoelectric effect and designed to measure the current in electrical circuits of a direct current. The use of the ME effect allows us to solve a lot of problems and tasks facing us. The main applications of the sensor are measuring systems, safety systems, automotive industry, space technology, robotics.

The paper discusses the principle of operation, three-dimensional modeling of the case, a printed circuit board using a microprocessor, a layered model of a printed circuit board for a magnetoelectric current sensor, and also a discussion of possible digital outputs.

IOP Conf. Series: Materials Science and Engineering **441** (2018) 012027 doi:10.1088/1757-899X/441/1/012027

2. Principle of Operation

The magnetoelectric current sensor works as follows. When measuring current flows through the current coil, a constant magnetic field H_0 (displacement field) is created. At the input of the solenoid coil, an alternating signal of the generator is applied, which creates an alternating magnetic field $H_\sim$. The alternating magnetic field $H_\sim$ and the permanent magnetic field H_0 are co-directed and perpendicular to the polarization vector P of the piezoelectric composite layer. As a result of the interaction of magnetic and electric fields in the ME element, we observed the transverse magnetoelectric effect. As a result of the transverse magnetoelectric effect, an alternating electrical signal appears, proportional to the strength of the measured current, which is taken from the electrodes.

The sensitive element of the ME sensor is a magnetoelectric composite (Figure 1), made in the form of a piezoelectric plate 1, on whose side of surfaces the plates of Metglas 2 are glued on. The device also includes measuring circuit, consisting of low-frequency signal generator, solenoid coil and current coil. The magnetoelectric element made of composite laminate, for example, the composition Metglas - piezoceramic PZT- Metglas, with a volume fraction v = 0.5.

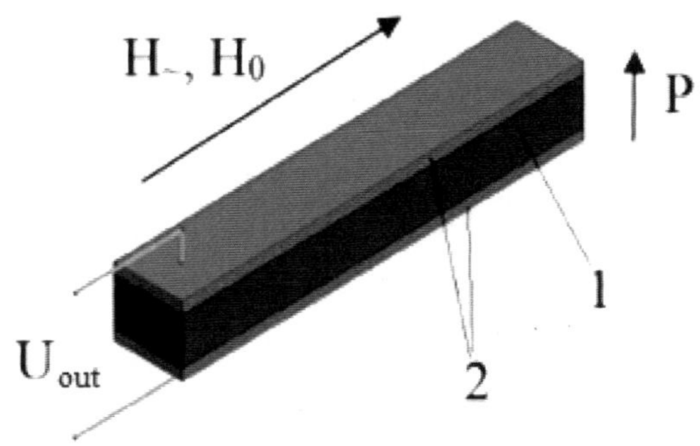

Figure 1. Magnetoelectirc element: 1) piezoceramic plate of PZT; 2) magnetostrictive layers of Metlgas.

Also, the main advantage of the sensor over analogues is the use of the main working mechanism of the magnetoelectric effect. ME composite structures have high sensitivity values, approaching ultra-sensitive and costly SQUID systems, and exceed the sensors of current on the Hall effect in terms of sensitivity.

3. Modeling of the PCB and the case

In Figure 2 the 3D printed circuit board model of ME current sensor is presented. Simulation of the two-layer PCB was made in an Altium Design. In the simulation of PCB, the problems of the optimal position of electronic components and the miniaturization of the sensor were set. The use of modern materials and electronic components can significantly reduce and simplify the sensor PCB. The layered structure of the PCB of the sensor is shown in Figure 3.

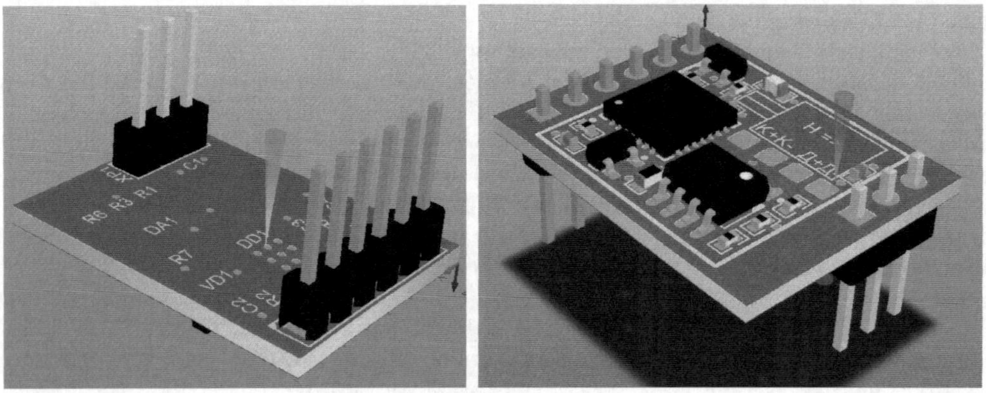

Figure 2. Three-dimensional model of the printed circuit board ME of the current sensor (ME element and coil not shown.

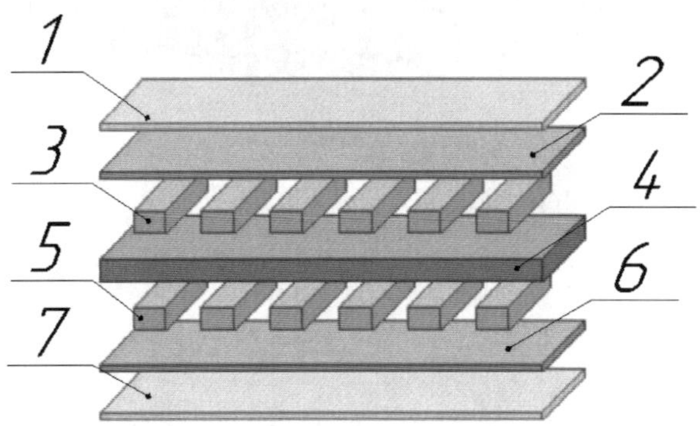

Figure 3. The layered 3D model of the PCB: 1 - Top Overlay, 2 - Top Solder, 3 - Top Layer, 4 - Dielectric, 5 - Bottom Layer, 6 - Bottom Solder, 7 - Bottom Overlay.

Figure 4 shows a three-dimensional model of magnetoelectric current sensor case. Simulation of the case was carried out in CAD Compass 3D V14 and printed on a 3d printer. This case has smaller overall dimensions.

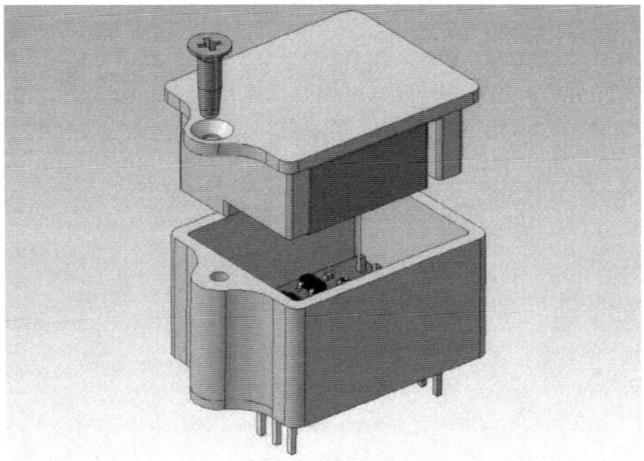

Figure 4. Three-dimensional model of ME sensor current case.

4. Discussion

Microprocessors are indispensable in modern technology and its use at the heart of sensor control is a big advantage. The microprocessor enclosed in a miniature case combines functions of the processor and peripheral devices such as ADCs, comparators, timers, etc., on one chip, and also contains RAM and / or ROM. In fact, it is a single-chip computer capable of performing relatively simple tasks.

Of the many and varied applications of microprocessors, microprocessor systems, object-oriented computing systems, for example, for control, diagnostics, digital signal processing and imaging, occupy one of the first places in terms of volume and use. The microprocessor will work with various interfaces. The following types of interfaces exist: 1-wire; RS-232; RS-485; SPI; I2C; CAN and etc. Each of which has its own peculiarities, advantages and disadvantages. For example, in the 1-wire protocol, only two wires are required to communicate with the device: for data and grounding, the data transmission distance is up to 300 m, the configuration of any 1-Wire network is changing in the process of its operation. Advantages of the RS-232 interface are a large fleet of operating equipment using this standard; simplicity and cheapness of the connecting cable; simplicity and availability of software for working with the interface. The RS-485 interface has the following advantages: convenience of system debugging and simplicity of synchronization of parcels, the possibility of organizing a protocol with direct transmission of binary data and the possibility of being "pure" binary, i.e. without the allocation of special control characters. The SPI protocol has a number of advantages: full-duplex data transfer by default, the ability to arbitrarily select the length of the packet, the length of the packet is not limited to eight bits, the simplicity of the hardware implementation, only four outputs are used, which is much less than for parallel interfaces, the unidirectional nature of the signals allows, if necessary it is easy to organize a galvanic isolation between the master and slave devices, the maximum clock frequency is limited only by the speed of the devices participating in the exchange e data. Among the advantages of the I2C protocol are the following: you need only one microcontroller to manage the set of devices; Use only two conductors to connect multiple devices; simultaneous operation of several master devices connected to the same I2C bus; the standard provides for "hot" connection and disconnection of devices during the operation of the system; A built-in microchip filter suppresses bursts, ensuring data integrity. Car standard CAN has the following advantages: the ability to work in hard real-time mode; simplicity of implementation and minimum costs of use; high resistance to interference; arbitration of access to the network without loss of bandwidth; reliable control of transmission and reception errors; wide range of work speeds.

5. Conclusions

The presented magnetoelectric current sensor is designed for all modern world standards. The sensor is based on the magnetoelectric effect. The sensitive element of the sensor is a magnetostriction-piezoelectric composite structure consisting of the amorphous material Metglas and the plate of piezoceramic PZT-19.

The developed ME sensor PCB using a microcontroller allows the sensor to be used in the variety of modern devices, such as: current monitoring systems, safety systems, etc.

Advantages of such a sensor are: digital output of the device, high sensitivity, low power consumption, small overall dimensions, high linearity of the output characteristic.

The reported study was funded by RFBR according to the research project № 18-42-530001.

References

[1] Bichurin M I et al 1990 Magnetoelectric materials: features of technology and prospects of application *Segnetomagnitnye matter* Science pp 118-133 (in Russian)

[2] Bichurin M I, Petrov V M, Petrov R V, Kiliba Y V, Bukashev F I, Smirnov Yu and Eliseev D N 2002 Magnetoelectric Sensor of Magnetic Field *Ferroelectrics* **280** pp 199-202

[3] Petrov R V and Leontiev V S 2013 Magnetoelectric magnetometer *Vestnik of Yaroslav the Wise Novgorod state university* **75** pp 29-32

[4] Solovyov I N and Bichurin M I 2013 Low-frequency magnetoelectric current sensor *Vestnik of Yaroslav the Wise Novgorod state university* **75** pp 35 -36

[5] Petrov R V, Solovyev I N, Soloviev A N and Bichurin M I 2013 Magnetoelectic current sensor *PIERS Proceedings* Stockholm Sweden pp 105-108

[6] Bichurin M I, Petrov R V, Leontiev V S, Semenov G A and O.V. Sokolov 2017 Magnetoelectric Current Sensors *Sensors* **17** 1271

Additive based on aluminosilicates for lime dry mortar mixes

Loganina V.I. [1,a] , Skachkov Yu.P[2,b], Ryzhov A. D.[3,c]

[1] Department "Quality management and technology of construction production" Penza State University of Architecture and Construction,Russia, 440028, Penza, street Germana Titova, 28

[2] Rectorate Penza State University of Architecture and Construction,Russia, 440028, Penza, street Germana Titova, 28

[3] Department "Quality management and technology of construction production" Penza State University of Architecture and Construction,Russia, 440028, Penza, street Germana Titova, 28

[a]email: loganin@mail.ru. [b]email: rector@pguas.ru [c]email: penza48@yandex.ru

Keywords: coating, additive, amorphous aluminosilicate, lime, adsorption centers
Abstract. Information on the properties of the additive based on amorphous aluminosilicates is given. The technology of synthesis of the additive is described. The influence of the additive on the process o2f the formation of a calcareous mixture is shown. The information on the properties of the developed dry construction mix are given.

Intorduction

In the process of operation, protective and decorative coatings of the outer enclosing structures of buildings are exposed to moisture due to the action of rain, high relative humidity of the air, and also due to the diffusion of water vapor through the enclosing structure from the inner surface to the outer one during the cold period of the year [1, 2]. In the conditions of external influences on the coating, in some cases there is a partial destruction of the plaster coating, which manifests itself in the form of detachments or a grid of small hair cracks. One of the possible causes of these defects may be the condensation of moisture at the interface between the enclosing structure and the plaster coating and the subsequent destruction of the coating due to the increase in the volume of water in the pores of the material,when the water freezes. In practice, such cases occur more often when using plaster coatings with low values of vapor permeability (Figure.1, 2).

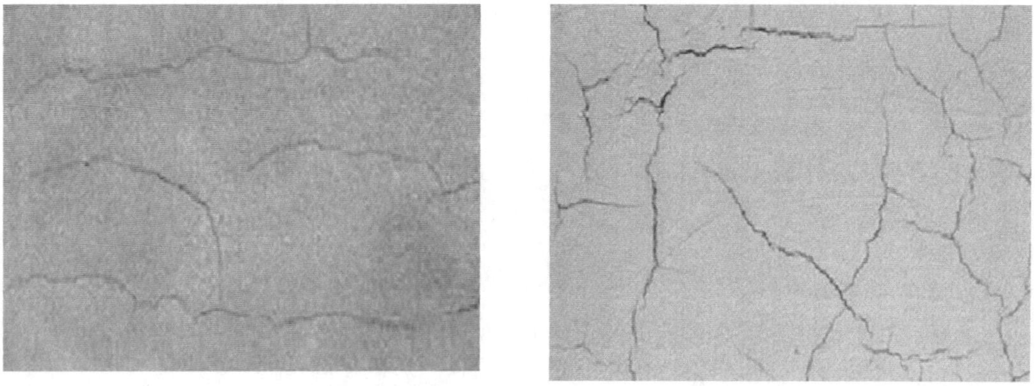

a b

Figure 1. Photo of the facade of the building

a- on Tsiolkovsky Street after three years of operation (plaster on cement basis), Penza, Russia

b - on Kalinina Street after three years of operation, (plaster on cement basis) Penza, Russia

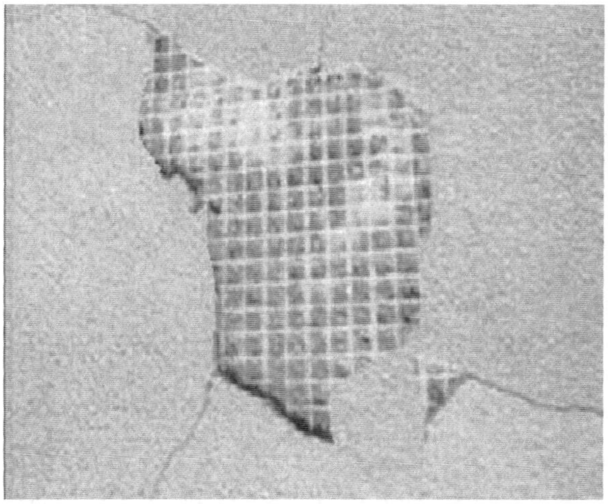

Figure 2. Photo of the facade of the building on Mozhaisky Street after six years of operation, (plaster on cement basis). Penza, Russia

It is established, that the main operational defects are cracks in the coating at the end of the building, as well as peeling of the coating.

As a rule, dry mixtures for outdoor work are made on a cement binder. Calcareous building mixtures are used less often because of the low resistance of coatings based on them. However, coatings based on lime compositions are characterized by greater crack resistance, better perceive tensile forces, have good vapor permeability and resistance to biodestruction [3,4].

Disadvantages of such compositions are low strength and water resistance. As is known, calcareous binder very slowly hardens, which complicates the production of finishing works. To accelerate the curing of lime and increase the durability of lime coatings, various additives are introduced [5,6,7]. There are additives based on amorphous aluminum oxides (γAl_2O_3) and aluminum hydroxide

$Al(OH)_3$, (Axilat SA_50_2 (Rhodia, France) and Amga (OJSC RUSAL Boksitogorsky alumina), Alumina CTC-20 (Germany). Additives are characterized by high dispersity and activity with respect to alkalis [8].

It is proposed in [9] to introduce additives based on synthesized calcium hydrosilicates (CHS) into the formulation of calcareous finishing compositions. Introduction to the lime additives based on CHS leads to an acceleration of the hardening process and increase the strength of calcareous samples.

It is of interest to investigate the possibility of using in calcareous mixtures of additives, containing amorphous aluminosilicates . The interaction of the calcareous binder with amorphous aluminosilicates will promote the acceleration of curing and increase the water resistance, as well as the mechanical strength of the coating.

At present, materials containing amorphous aluminosilicates are used as ion exchangers (for example, for water purification), as adsorbents in chromatography, during purification, drying and separation of gases, as well as catalytic cracking of petroleum products [10,11].

The analysis of patent and scientific and technical literature shows that the questions of their use in the formulations of calcareous mortar require careful consideration. The mechanism of interaction of synthetic adsorbents with astringent is not revealed, the patterns of structure formation of the composite have not been established. This indicates testifies to the expediency the feasibility of investigating the possibility of using additives based on amorphous aluminosilicates in calcareous dry mixtures.

Results and Discussion

The following synthesis technology of the aluminosilicate additive was used in the work. Microdispersed aluminum powders ПАП -1were added to sodium liquid glass at a temperature of 60-90 ° C for 30-120 minutes.

To reveal the distribution and concentration of acid-base centers (active sites) on the surface of the additive particles, an indicator method for adsorption of indicators with different dissociation constant values pK_a^x was used. . The quantitative determination of adsorption sites ($q_{pK_a^x}$, mg-eq / g or mg-eq / m^2) of a given acid strength was carried out by a photometric method. The solutions were photometricized in cuvettes relative to the solvent on a KFK-3KM photoelectrocolorimeter at the wavelength corresponding to the maximum absorption of each indicator (). The following indicators were used for the tests: diamond green (pK_a = + 1.5), fuchsin (pK_a = + 2.1), methyl orange (pK_a = + 3.46), bramfenol blue pK_a (= + 4.1), bromocresol purple (pK_a = + 6.4), bromothymol blue (pK_a = + 7.3), timol blue (pK_a = + 8.8), nil blue (pK_a = + 10.5) and ethylene glycol (pK_a = + 14.2).

Studies were carried out in the field of Brönsted acid centers (pK_a from 0 to 7) and basic centers (pK_a from 7 to 13), and Lewis acid centers(pK_a > 13). The value pK_a = +7 correspond neutral centers.

In the composition of the additive revealed a high content of oxides of Al_2O_3, SiO_2, Na_2O, constituting, respectively, 51.03%, 36.36%, 11.89%. The amorphous phase is represented by sodium aluminate. The content of the amorphous phase is 83%. Using a scanning electron microscope Tescan VEGA 3, it was found, that the structure of the synthesized additive is represented by particles of different shapes, (lamellar and acicular) in size from 0.11 to 10.49 μм

It is established, that the surface of the sample of the additive is characterized by the predominance of the centers $pK_a = + 4.7$, $pK_a = + 6.4$. The total amount of acid centers is $\Sigma q_{pKa} = 53{,}72$ mmol / g and the base centers is $\Sigma q_{pKa} = 7{,}70$ mmol / g (Fig.3).
To study formation of structure of calcareous composites, samples were made on the basis of hydrated lime 1 grade with an activity of 84.4%. The maximum strength of calcareous samples is achieved by adding an aluminosilicate additive in an amount of 10% of the weight of the lime. The increase in the strength of calcareous composites when adding an additive based on amorphous aluminosilicates is due to the presence of additional chemical formations.
It was found, that the mineralogical composition of calcareous composites is represented by portlandite, calcite, calcium hydrosilicates, calcium hydroalumates. To confirm the results, a differential thermal analysis (DTA) was carried out with the help of the "Thermoscan-2" installation.
It was found, that in control samples in the age of 28 days of air-dry hardening, the amount of free lime is 47.67%, and in samples with using an additive based on amorphous aluminosilicates - 31.41%.

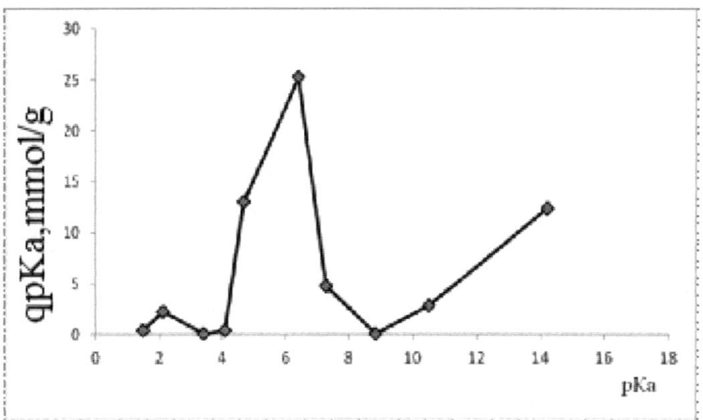

Figure 3. The distribution curve of acid-base centers on the surface of the particles of the additive

The introduction of an aluminosilicate additive into the calcareous system results in a slight decrease in the pH of the liquid phase. So, after one and a half hours from the moment of start-up, the pH of the control compounds (without additive) is 13.43, and at additive content of 10% - pH = 13.31.
It is found, that when the aluminosilicate additive is introduced in an amount of 10% of the weight of the lime, the maximum temperature of the structure formation is 24 ° C and is reached after 60 min
Additive based on amorphous aluminosilicates is expected to be used in the manufacture of heat-insulating dry construction mixtures.
Table 1. shows the main technological and operational properties of the thermal insulation composition based on the developed formulation using an additive based on amorphous aluminosilicates and coatings based on it.

Table 1. Technological and operational properties of the developed calcareous heat-insulating DBM

Indicator name	The value of the indicator for the composition

	design	prototype 1	prototype 2	prototype 3
Average density DBM, [kg/m^3]	700	300	550-600	1100
Consumption of the composition when applied with a layer thickness of 10 mm, [kg/m^2]	8,4	3,2	5,5	12
Water retention ability, [%]	98,3	96,0	97,0	98
Coefficient of vapor permeability, [mg/(mhPa]	0,18	0,05	0,11	0,1
Compressive strength, [MPa]	4,7	>0,5	3,5	3,5
Strength of adhesion to the base, [MPa]	0,38	0,5	0,3	0,4
Frost resistance	F35	F25	F50	F35
Coefficient of thermal conductivity, [Watt/(m^{o}C]	0,18	0,07	0,13	0,35
Warranty period of storage, [month]	12	12	12	12

As a prototype 1, was chosen a dry cement-pearlitic plaster mix "Perlutka SHT-4" , as the prototype 2 - a heat-insulating plaster mixture UF-2 TM "UMKA" (produced by OOO "Ecotermogroup"), as a prototype 3 - plaster cement heat-insulating facade Knauf-Grünband (the firm "Knauf").

Conclusions

The developed DBM composition with an additive based on amorphous aluminosilicates possesses a number of advantages in comparison with analogs: fast setting time, high water retention ability, high adhesion strength to the surface under different operating conditions and low value of shrinkage deformations.

References

[1] Grachev I.A.1995 Causes and occurrence of high humidity in the walls of the building, *Housing construction*, **9** p.18-20.
[2] Vasilik P.G., Golubev I.V.2003 Cracks in plasters, *Building Materials*.**4** p.14-16.
[3] Loganina V.I., Makarova L.V., Tarasov R.V., Zhegera K.V.2014 Improving the Quality of Lime Composites, *International Journal of Research in Engineering & Technology*,**vol.2, 4** p.25-30
[4] Loganina V.I., Petuhova N.A., Akzhigitova E.R.2011 Development of an organomineral additive for dry construction mixtures, *Bulletin of the Belgorod State Technological University V.G. Shukhov*.**3** p. 8-12.
[5] Loganina V.I., Makarova L.V., Tarasov R.V.2013 Dry construction mixstures with cjmpozite lime astringent. *International Journal of Resefch in Engeneering and Technology* p.7-12

[6] Zakharov S.A., Kalachik B.S.2007 Highly active metakaolin - a modern active mineral modifier of cement systems, *Building materials*, **5** p.56-57.
[7] Kozlov V.V. 2000 Dry building mixtures, Moscow, ASV.
[8] Liner A.I., Eremin N.I., Yu.A. Liner et al Alumina production, Moscow, *Metallurgy*, 1978 .
[9] Loganina V.I., Makarova L.V., Tarasov R.V., Sergeeva K.A.2014 Improving Reliability of Covers Based on Lime Finishing Compositions *American Journal of Materials Science and Application*, **vol.2, 1** p. 1-5
[10] Krupennikova A.Yu, Tsitsishvili G.V., Burdzhanadze M.N. and others 1978 Adsorbents, their preparation, properties and application, Leningrad, Science .
[11] Milinchuk V.K., Shilina A.S., **RU** Patent. 2402486 (2000)

Application of an eddy-current method to measure electrical conductivity of thin films

S F Dmitriev[1], V N Malikov[1], A V Ishkov[2], A O Katasonov[1], A M Sagalakov[1]

[1] Altai State University, 61, Lenina ave., Barnaul, 656057, Russia
[2] Altai State Agracultural University, 98, Krasnoarmeisky ave., Barnaul, 656057, Russia

E-mail: osys11@gmail.com

Abstract. The research objective was the development of subminiature eddy-current transducer(ECT) designed for studying thin films. The possibility of studying the electrical conductivity of thin films of metals as per the amplitude of the eddy-current transducer signal. A subminiature surface eddy-current transducer of the transformer type was developed, and the optimum shape and size of the core and number of inductance coils and turns in them were defined, this provides efficient localization of electromagnetic field. On the basis of eddy-current transducer, a software and hardware complex was developed to control the eddy-current transducer (to generate alternating current of various frequencies, to provide its supply to the eddy-current transducer, to receive a useful signal from transducers to ensure its convenient visualization). The signal working frequency on the energizing winding of the eddy-current transducer was 3 MHz, the signal working amplitude on the energizing winding was 1.1 V. The hardware and software signal processing made it possible to significantly improve the accuracy of measurements. In the developed system, the parameter with information about the monitoring object is the averaged amplitude of the voltage applied by the eddy currents into the measuring winding of the ECT. The algorithm to determine the averaged signal amplitude is given. The dependence between the averaged amplitude of the eddy-current transducer signal and the electrical conductivity of thin films was analyzed and studied. The results of testing of Al, Cu and Zn films were presented, and the electrical conductivity values of the samples were obtained.

1. Introduction

The scientific and technical direction associated with the obtainment and use of thin metal films has remarkably developed over the past decades and occupied key positions in many branches of modern production. At present, the use of thin films in microelectronics, microwave equipment, optics and many other branches of science and technology opens the prospects for the creation and improvement of not only new devices but also entire technological areas. Thin-film conducting structures based on various special conductive films and coatings are used in the electronics industry for the production of integrated circuits, in aviation and space technology, as well as reflective and antireflective coatings in optics and microwave technology.

The variety of structures and specific properties associated with small thickness of such objects leads to the fact that their physical characteristics can differ significantly from the same materials but in the bulk condition.

For the full use of thin films in various areas of electronics, complete information about their material constants and physical parameters, in particular, electrical conductivity, is necessary. Consequently, new methods are needed to determine the characteristics of the materials being studied in the form of thin films.

The various existing methods in the field of resistometry based, as a rule, on creating a good electrical contact between the film surface and measuring elements (electrodes, probes, etc.) can often not be acceptable, since their use can change the properties of the film.

The most significant effect on the quality of the produced conductive films and coatings is provided by reliable monitoring of such important parameters as structural defects, pores, electrical conductivity, thickness, etc. In this case, in view of the considerable technological dispersion of these parameters, it is necessary to carry out 100% outgoing inspection. To monitor these parameters, the following non-destructive testing methods are widely used: visual-optical, capillary, magnetic, ultrasonic, eddy current.

The analysis of works [1-4] showed that eddy current method (ECM) is the simplest, environmentally friendly, and at the same time efficient and high-performance at 100% continuous automatic non-destructive testing of thin films. It is based on the analysis of the interaction of an external electromagnetic field with the electromagnetic field of eddy currents induced by an energizing coil in an electrically conductive film.

From the general considerations developed in work [5], it follows that the efficient technical implementation of the eddy current method of controlling the parameters of thin metallic films is feasible with the help of attachable and screen-type ECTs It is necessary to develop methods for reliable and accurate control of such parameters of thin non-ferromagnetic films as thickness, specific electric conductivity, the presence of imperfections and structural defects.

The choice of the eddy current method is due to the fact that it has significant advantages in comparison with other methods. It provides the ability to make measurements easily and accurately without the direct contact with the object under investigation, and also provides to make measurements directly in a vacuum, and this allows to more precisely control the process of films growth. In addition, the method makes it possible to carry out a lot of measurements of a wide range of thicknesses of thin films, because of the possibility of a quick change in the penetration depth of eddy currents depending on frequency. [6, 7]

In works [8-13], an eddy current method based on a single-frequency signal was used to study thin films. The use of a single frequency makes it possible to simplify the design of the instrument, but does not make it possible to study thin films at different depths simultaneously. To eliminate this problem, a multifrequency signal is used to control thin films [14-15].

Among a number of models of eddy current defectoscopes currently available on the market, only one-of-a-kind models are good for obtainment of thin films. This is due to the thickness of these objects (100-500 nm), which determines a number of requirements to ECT and other components of the system. In particular, the operating frequency range required for studying thin films is 1-10 MHz.

The company Suragus GmbH produces eddy current devices for various purposes, including those designed for the study of thin films. For example, EddyCus® TF lab 2020 allows point-by-point measurements of the electrical conductivity of samples with dimensions from 10×10 mm to 200×200 mm in the resistance range from 0.001 to 3000 Ohm/m^2 [16].

A significant drawback of this class of instruments is the use of eddy current transducers with a measuring winding radius of 5 to 25 mm, which does not allow achieving high resolution.

An eddy current system is known for studying thin non-ferromagnetic films using the method of resonant screen eddy current nondestructive testing. This system has a high efficiency in detecting defects of thin non-ferromagnetic films of 0.01 μm thickness at the frequency of the excitation current in the range of 10-30 MHz. The disadvantage of this system is the screen-type ECT and as a result the object under investigation should be located in the gap between the measuring and compensation coils, which leads to a significant complication of the design. This is due to the fact that the objects under control can have a base layer of different thickness, so that it is necessary to vary the width of the gap

between the measuring and compensation coils during the studies.
The purpose of this work was the use of an eddy current method to measure the electrical conductivity of thin metal films using a developed measuring system based on a subminiature eddy current transducer that effectively localizes the electromagnetic field in small areas, with subsequent hardware and software processing that allows to process mathematically the received signal.

2. Materials and methods

As the analysis of work [4] has shown, to control the defects of sheets, foils, thin metal films, it is efficiently to use transformer-type attachable ECTs. Based on the results obtained in other works, for example, in [17-18], several designs of a transformer-type attachable ECTs, differing in shape of the core, location of windings, etc., were developed. All ECTs had energizing, measuring and compensating windings.
The energizing and measuring windings contained 200 turns. To wind the turns, copper wire having a thickness of 5 μm was used. The compensating winding contained 140÷170 turns. The cores were made of ferrite 2,000 Nm3 with an initial magnetic permeability value of 500.
The core was a truncated cone with a height of 4.3 mm with a base diameter of 1.5 mm (Figure 1).

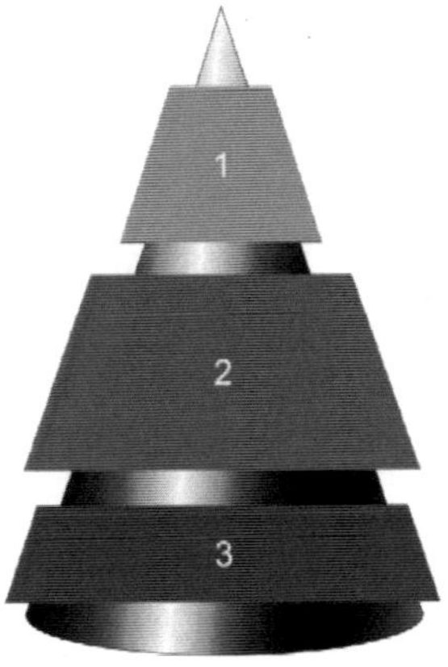

Figure 1. Scheme of ECT windings, 1 — measuring winding, 2 — generator winding, 3 — compensating winding.

The measuring winding 1 was located on the tip of the cone (200 turns), the generator winding 2 was centrally located (200 turns) and wound so that the radius was as small as possible to achieve maximum localization of the field.
To eliminate the influence of the generator winding on the measuring winding, a compensating winding 3 was added to ECT. The windings were made of copper wire with a diameter of 15 μm. Windings 1 and 3 were connected according to differential circuit.
The measuring system, which is based on a miniature eddy-current transducer, operates as follows. The software of the personal computer controls the operation of the generator, which produces a train of rectangular voltage pulses with the repetition rate that is necessary for the operation of the eddy-current transducers. The voltage pulses are transmitted from the generator output to two series integrators. They are then directed to the input of the power amplifier. From the amplifier output the voltage pulses arrive at the exciting inductance coils of the eddy-current transducers. The difference of the output voltages of the measuring coils of the transducers contains information on the structural heterogeneities of the tested object that is located in the effective area of the eddy-current transducers. It is detected and amplified in a special microphone amplifier. The signal arrives at the amplitude detector after the transmission through two series high-quality low frequency filters and two series selective amplifiers. The signal is then transmitted through an analog-to-digital converter to a personal computer. Due to the simultaneous control of the generated signal frequency at the exciting coil and

the cutoff frequency of the filtering system and the selective amplification, the useful signal, which contains information on the electric conductivity distribution inside the object is detected.
Special software written in the C++ language for the Windows operating system allows controlling the signal on the exciting coil and receiving the signal from the measuring winding. By means of a special mixer, voltage is applied to the exciting winding of the transducer and the level and frequency of the sinusoidal signal of the generator are set.
Figure 2 shows the relative position of ECT and the object under control. The transducer is perpendicular to the plane OK, so that the measuring winding is at a minimum distance from the OK, but the ETP does not contact the OK.

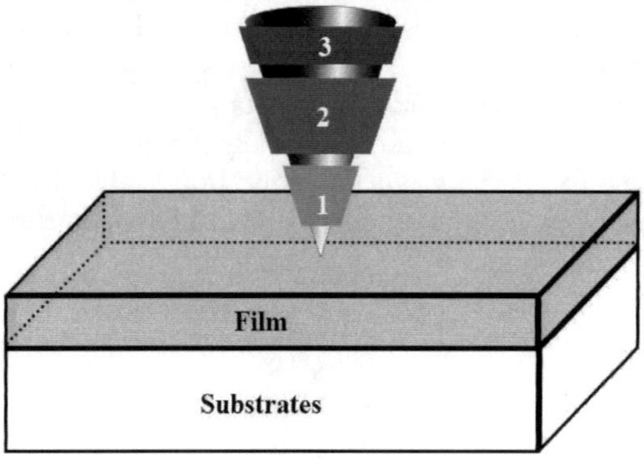

Figure 2. ECT over the object to control.

3. The result and discussion

As objects for studying the operation of the developed system, thin films of metals Al, Cu and Zn, obtained by condensation during the gas phase in a vacuum onto glass substrates, were chosen. The dimensions of the substrates were 23×23 mm. Dimensions of films under investigation were 23×21 mm.
After testing the operation of the system with different amplitudes of the exciting signal and at different frequencies, the amplitude was 1.1 V and the frequency was 3 MHz as the working frequency.
In the developed system, the parameter with information about the object under control is the average amplitude of the voltage introduced by the eddy currents into the measuring winding of ECT. The average voltage amplitude was determined as follows: at the initial stage, the maximum and minimum signal values were measured separately. At the next stage, the average maximum and minimum values of the signal were determined. The difference between the average values $<A_1>$ corresponded to twice the average signal amplitude $<U_1>$.
Figure 3 shows the initial data — the measured values of voltage on the measuring winding of ECT, in the absence and in the presence of an Al film. Area 1 (N = 150-600) corresponds to the signal in the absence of the object under control — the decompensation of ECT. Area 2 (N = 620-1190) corresponds to the signal in the presence of an Al film. Figure 4 shows the results of using the algorithm for finding the average amplitude.

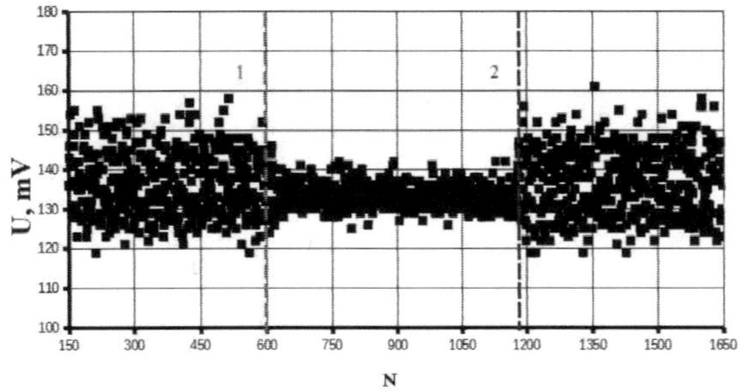

Figure 3. Voltage on the measuring winding of ECT (Al film). 1 — signal without object of control, 2 — signal in the presence of the film under study.

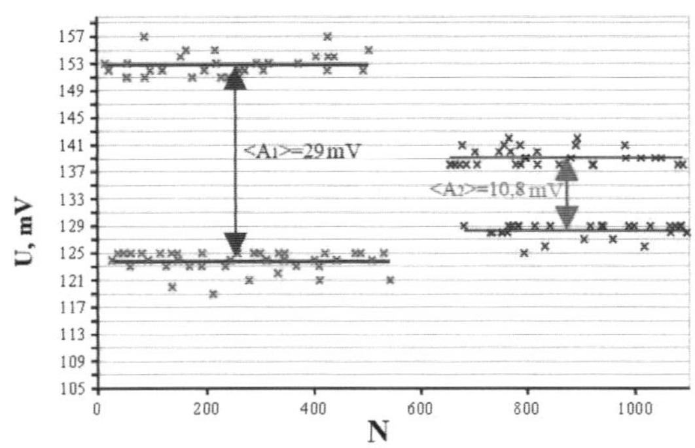

Figure 4. Determination of the averaged signal amplitude, $\langle A_1 \rangle$ — the signal span without the object of control, $\langle A_2 \rangle$ — in the presence of the Al film on the glass base surface..

In area 1, the average signal amplitude was 14.53 mV, in the area 2-5.41 mV. The difference between the amplitude in area 1 and the amplitude in area 2 ($\Delta \langle U \rangle$) is 9.12 mV.
$\Delta \langle U \rangle$ for the Zn film was 6.34 mV and for the Cu film was 37.5 mV.
Figures 5(a,b) show the results of measuring the response from Zn and Cu films, respectively.

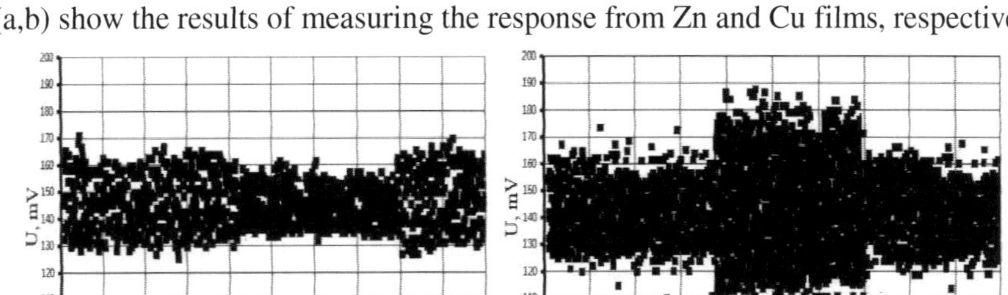

a b

Figure 5. Voltage on the measuring winding of ECT: a – Zn film; b – Cu film.

For Al and Zn films, the voltage applied by the eddy currents is less than the decompensation, and consequently in Figure 3 and Figure 5.a we see a decrease in the signal amplitude. For a Cu film, the contribution of eddy currents is much greater than the decompensation, as a result of which we see an increase in the signal amplitude in Figure 5.b.
To obtain a connection between the values of the average signal amplitude and the electrical conductivity of the material, samples with known electrical conductivity were used.

Based on the results of approximation of the data obtained from measurements of samples with known electrical conductivity, we found the electrical conductivity of Al, Zn, Cu films (Table 1).

Table 1. The results of calculation electrical conductivity of films Zn, Al, Cu.

	<ΔU>, mV	σ, MS/m
Zn	6.34	0.1433
Al	10.89	0.5114
Cu	18.27	1.1084

4. Conclusion

Eddy current transducer designed to effectively localize an electromagnetic field in small areas of electrically conductive objects has been developed. On the basis of a subminiature eddy current transducer, a measuring system was developed to determine the electrical conductivity of thin non-ferromagnetic films. The measuring system includes: software, hardware, ECT and allows to produce hardware and software processing of the received signal in order to improve the accuracy of measuring the electrical conductivity of thin films. In the course of the work, it was possible to create an algorithm to find the change in the average amplitude of the output signal, the magnitude of which makes it possible to infer the parameters of the film, in particular the electrical conductivity. The samples of thin films were examined using the developed measuring system, the electrical conductivity values of Al, Cu, and Zn films were obtained. For an aluminum film, the electrical conductivity value was 0.5114 MS/m, for a copper film — 1.1084 MS/m, for a zinc film — 0.1433 MS/m. The inaccuracy of the obtained values did not exceed 5%.

References

[1] Yurkov V, Ryzhii V 2008 Effect of Coulomb scattering on graphene conductivity *JETP Lett* **vol. 88, 5** 370-373

[2] Abramchuk S 2007 Novel highly elastic magnetic materials for dampers and seals: Part I. *Preparation and characterization of the elastic materials. Polym. Adv. Technol* **vol. 18, 11** 883-890

[3] Li W A 2017 Thickness Measurement System for Metal Films Based on Eddy-Current Method With Phase Detection. *IEEE Transactions on Industrial Electronics* **vol. 64,.5** 3940-3949.

[4] He D F, Tang J 2017 A contactless method to measure the electrical conductivity, *Proc. 3rd "IEEE International Conference on Control Science and Systems Engineering"* 305-308.

[5] Bakunov A S, Yefimov A G 2009 Eddy current non-destructive testing in defectoscopy of metal products. *Testing. Diagnostics* **4** 21-22

[6] Angani S, Park D G, Kim G D 2010 Differential pulsed eddy current sensor for the detection of wall thinning in an insulated stainless steel pipe *J. Appl. Phys* **vol. 107, 9**

[7] Arez J C 2012 Liftoff insensitive thickness measurement of aluminum plates using harmonic eddy current excitation and a GMR sensor *Measurement* **vol. 45, 9** 2246-2253.

[8] Qu Z L 2014 Improvement of sensitivity of eddy current sensors for nano-scale thickness measurement of Cu films. *Nondestructive Testing & Evaluation International* **61** 53-57

[9] Qu Z L, Zhao Q, Meng Y G 2013 In-situ measurement of Cu film thickness during the CMP process by using eddy current method alone *Microelectron. Eng* **108** 66-70.

[10] Li W, Wang H B, Feng Z H 2016 Non-contact online thickness measurement system for metal films based on eddy current sensing with distance tracking technique *Rev. Sci. Instrum* **vol. 87, 4**

[11] Kral J, Smid R, Ribeiro A L 2013 The Lift-Off Effect in Eddy Currents on Thickness Modeling and Measurement. *IEEE Trans. Instrum* Meas **vol. 62, 7** 2043-3047

[12] Yin W L, Xu K A 2016 Novel Triple-Coil Electromagnetic Sensor for Thickness Measurement Immune to Lift-Off Variations *IEEE Trans. Instrum. Meas* **vol. 65,1** 164-169

[13] Wang H B 2015 Noncontact thickness measurement of metal films using eddy-current sensors immune to distance variation *IEEE Trans. Instrum. Meas* **vol. 64, 9** 2557-2564
[14] Pinotti E., Puppin E. Simple 2014 Lock-In Technique for Thickness Measurement of Metallic Plates *IEEE Trans. Instrum. Meas* **vol. 63, 2**, 479-484
[15] Yin W, Peyton A J 2007 Thickness measurement of non-magnetic plates using multi-frequency eddy current sensors *Nondestructive Testing & Evaluation International* **vol. 40, 1**. 43-48.
[16] Busch M 2015 Measurements of the properties of thin films by non-contact (Dresden, SURAGUS GmbH)
[17] Dmitriev S F, Sagalakov A M, Malikov V N 2017 Flaw inspection of welded joints in titanium alloys by the eddy current method. Welding International **vol. 31, no.8** 46-49.
[18] Dmitriev S F, Sagalakov A M, Malikov V N 2017 Scanning the Welded Seams of Titanium Alloys by Using Subminiature Eddy Current Transducers *Materials Science Forum* **vol. 906** 147-152.

Development of methods of structural-analytical mesomechanics that take into account the statistical properties of martensitic transformations in materials with shape memory effect

V G Malinin[1], **N A Malinina**[1], **G V Malinin**[1], **O A Malukhina**[2]

[1] Orel state agricultural institute, 69, General Rodin ul., Orel, 302019, Russia
[2] Yaroslav-the-Wise Novgorod State University, 41, ul. B. St. Petersburgskaya, 41, 173003 Veliky Novgorod, Russia

E-mail: malinin.mvg@yandex.ru

Abstract. Methods for structural-analytical mesomechanics for materials with shape memory effect are developing, in these methods a pronounced synergetic effect is manifested, which is characterized by the appearance of self-accomodating groups of martensitic crystals, forming during the initiation of martensitic reactions extended domains oriented towards acting effective stresses.

Introduction
The calculation of the structural and mechanical behavior of polycrystals undergoing martensitic transformation under thermomechanical action is a complex task. This is due to the fact that under real deformation conditions, elements of different scale and structural levels are involved in its process, the behavior of which is determined by their individual characteristics, ensemble properties and self-consistent evolution of the structure at all interacting scale levels. A significant role is played by the statistical properties of the kinetics of martensitic phase transformations [1]. In this paper, on the basis of the results of the experimental studies carried out on thin-walled tubes of an alloy Cu–12 %Al–4 %Mn, using the methods of structural-analytical mesomechanics [2-4], we propose a model for calculating the deformation properties of materials, which takes into account the influence of the state of stress and the statistical features of phase transformations on the shape memory effects and plasticity of transformation. The model of a material with an SME can be used to solve an important problem of creating methods for calculating the stability of spherical segments of alloys with an SME under thermocyclic action [5].

1. The main hypotheses
The concentration hypothesis of the onset of martensite transformation in all structural elements of the orientation space [2] is assumed, provided that the critical nucleus of the new phase is formed. The criterion for the formation of a critical nucleus is formulated at the macroscale level in terms of effective temperature and effective stresses [1, 2] as a condition for the formation of a new phase at the macroscale level. The generalized Clausius-Clapeyron principle at the macroscale level is postulated and experimentally substantiated, which takes into account the different nature of chemical

reactions in the forward and reverse martensitic transformation, depending on the initial phase composition. The formulation of the Clausius-Clapeyron equation uses concepts of the effective temperature $T*$ [2]. It is taken into account that the direct conversion of austenite to martensite (AM) occurs with the release of internal sources of heat, ie, an exothermic reaction takes place, and in the reverse martensitic transformation of MA, an endothermic reaction occurs, that is, heat absorption in local structural elements, where the inverse reaction.

It is assumed that at the structural level self-accumulating groups of martensitic crystals arise and the plasticity of the direct martensitic transformation is determined by the disturbing effect of the deviator of local stresses [2, 6-7].

A physically grounded hypothesis is used, which asserts that the mechanism of phase reactions under the reverse martensitic transformation at the structural level obeys the "exactly backward" principle [2].

The hypothesis of an isotropic continuum at the macro-scale level and a quasicontinuum at the structural level is accepted. The influence of the anisotropy factor of martensitic crystals is reflected by taking into account the structural-static properties of the kinetics of phase transformations.

The hypotheses formulated have received good experimental substantiation both in experiments at the physical level [1] and in experimental studies on macroscale samples [2].

2. Theoretical and experimental investigation of the influence of the form of the stressed state on the thermomechanical hysteresis of the alloy Cu-12% Al-4% Mn

In materials with reversible martensitic reactions, a pronounced reversible thermomechanical hysteresis is formed during thermal cycling under voltage through the temperature range of phase transformations. An analysis of this phenomenon for simple strained states such as pure stretching and pure shear is performed within the framework of the structural-analytic theory of strength [2].

This article presents the results of a theoretical and experimental study of the influence of the form of the stressed state and the statistical properties of the kinetics of phase transformations on the thermomechanical hysteresis of an alloy Cu–12 %Al–4 %Mn.

Using the formulated hypotheses and applying the methods of structural-analytical mesomechanics for constructing models of processes of martensitic transformations and deformations initiated by them at the structural and macroscopic levels [2-8], after corresponding transformations we obtain the determining equations for the prediction of thermomechanical hysteresis in the form of formulas (1) and (2).

$$
\begin{aligned}
\varepsilon_{ik}^{\Pi\Pi} = B_{\phi}\sigma_{ik}^{'}\Bigg\{ & \frac{1}{2\Delta(M_{\text{н}}-M_{\text{к}})}\left(M_{\text{н}}-T+\kappa\sigma_i+\frac{\Delta}{2}\right)^2 H(M_{\text{н}}-M_{\text{к}}-\Delta) \\
& \times H\left(T-\kappa\sigma_i-M_{\text{н}}-\frac{\Delta}{2}\right) * H\left(M_{\text{н}}-T+\kappa\sigma_i+\frac{\Delta}{2}\right)+\frac{1}{(M_{\text{н}}-M_{\text{к}})} \\
& \times (M_{\text{н}}-T+\kappa\sigma_i)H(M_{\text{н}}-M_{\text{к}}-\Delta)H\left(T-\kappa\sigma_i-M_{\text{к}}-\frac{\Delta}{2}\right)\times H\left(M_{\text{н}}-T+\kappa\sigma_i-\frac{\Delta}{2}\right) \\
& +\frac{1}{\Delta}\left[M_{\text{к}}-T+\kappa\sigma_i+\frac{\Delta}{2}+\frac{1}{2(M_{\text{н}}-M_{\text{к}})}\left\langle (M_{\text{н}}-M_{\text{к}})^2+\left(M_{\text{н}}-T+\kappa\sigma_i-\frac{\Delta}{2}\right)^2\right\rangle\right] \\
& \times H(M_{\text{н}}-M_{\text{к}}-\Delta)H\left(T-\kappa\sigma_i-M_{\text{к}}+\frac{\Delta}{2}\right)H\left(M_{\text{к}}-T+\kappa\sigma_i+\frac{\Delta}{2}\right)\Bigg\}
\end{aligned} \tag{1}
$$

Here $\varepsilon_{ik}^{\Pi\Pi}$ – deviation tensor of the transformation plasticity deformations, σ_{ik} – stress tensor deviator, σ_i – stress intensity, T – temperature.

To take into account statistical properties, the width of the hysteresis Δ by the following rule: $2\Delta = A_к + A_н - M_н - M_к$, where $M_н$, $M_к$ – characteristic temperatures of the beginning and the end under direct martensitic transformation, $A_н$, $A_к$ – corresponding characteristic temperatures during reverse martensitic transformation.

Equation (1), as the analysis shows, quite satisfactorily describes the kinetics of the increase in the components of the strain tensor at the cooling stage.

In the derivation of formula (2), it was assumed that the beginning of the reverse phase transition was preceded by the preliminary deformation ε_{ik}^0 due to the active loading of martensite or as a result of plasticity of the direct martensitic transformation, which the mathematical object stably maintained at temperature $T < M_к - \Delta/2$. Here, as the heating progressed, this deformation returned, which was classified as a shape memory effect. Calculations on details, which do not stop because of their cumbersome nature, made it possible to obtain the following general expression for the components of the tensor of restoring deformation upon heating $\varepsilon_{ik}^{ПФ}$:

$$\varepsilon_{ik}^{ПФ} = \varepsilon_{ik}^0 \cdot H(A_к - A_н - \Delta) \cdot \left\{ \frac{1}{\Delta}\left(A_н - T + \kappa\sigma_i + \frac{\Delta}{2} \right) + \frac{1}{2\Delta(A_к - A_н)} \right\}$$
$$\times \left[(A_к - A_н)^2 - \left(A_к - T + \kappa\sigma_i - \frac{\Delta}{2} \right) \right] \cdot H\left(T - \kappa\sigma_i - A_н + \frac{\Delta}{2} \right) \cdot H\left(A_н - T + \kappa\sigma_i + \frac{\Delta}{2} \right)$$
$$+ \frac{1}{(A_к - A_н)} (A_к - T + \kappa\sigma_i) \cdot H\left(T - \kappa\sigma_i - A_н - \frac{\Delta}{2} \right) \times H(A_к - T + \kappa\sigma_i)$$
$$+ \frac{1}{2\Delta(A_к - A_н)} \left(A_к - T + \kappa\sigma_i - \frac{\Delta}{2} \right)^2 \cdot H\left(T - \kappa\sigma_i - A_к + \frac{\Delta}{2} \right) H(A_к - T + \kappa\sigma_i) \quad (2)$$

Having experimental data on thermomechanical hysteresis, for example, in pure torsion, for two stress intensity values, it is possible to determine the model parameters that retain their values for predicting shape memory effects and plasticity of transformation for any kind of stress state.

$$K = \frac{T_н^{A \to M}\left[\sigma_i^{(2)}\right] - T_н^{A \to M}\left[\sigma_i^{(1)}\right]}{\sigma_i^{(2)} - \sigma_i^{(1)}}, \; B_ф = \frac{2}{3}\frac{\varepsilon_i^0}{\sigma_i^0}.$$

For alloy Cu–12 %Al–4 %Mn model parameters $B_ф$ and K have the following values: $B_ф = 3{,}218 \cdot 10^{-6}$ [МПа$^{-1}$], $K = 0{,}161$ [°C·МПа$^{-1}$].

Formula (2) is valid for materials with a narrow hysteresis of the phase transformation, when $\Delta = A_к - M_н \le A_к - A_н$, which corresponds to the alloy Cu–12 %Al–4 % Mn. This formula leads to a rather simple and consistent with the kinetics of the recovery of deformation upon heating. It contains in the middle part of the temperature interval of mold recovery a section linear with respect to the temperature of the phase transition. At the beginning and at the end of the strain return schedule, quadratic segments with positive and negative curvature are observed, respectively.

The methods of structural-analytical mesomechanics naturally describe thermomechanical hysteresis under a complex stress state, which makes it possible to use it within the framework of solving boundary value problems of the mechanics of materials.

As an example, Picture 1 shows the results of experiments performed on thin-walled tubular samples of an alloy Cu–12 %Al–4 %Mn for various stress states, including a program of fifteen loading schemes, which have been described with good agreement between theoretical (solid lines) and experimental (point) results. The deviation of the theoretical data, with respect to the experimental data, does not exceed the spread (up to 5%) observed in the experiments. To calibrate the parameters of the model, it is sufficient to perform the experiment at a single stressed state, for example, under pure shear.

IOP Conf. Series: Materials Science and Engineering **441** (2018) 012030 doi:10.1088/1757-899X/441/1/012030

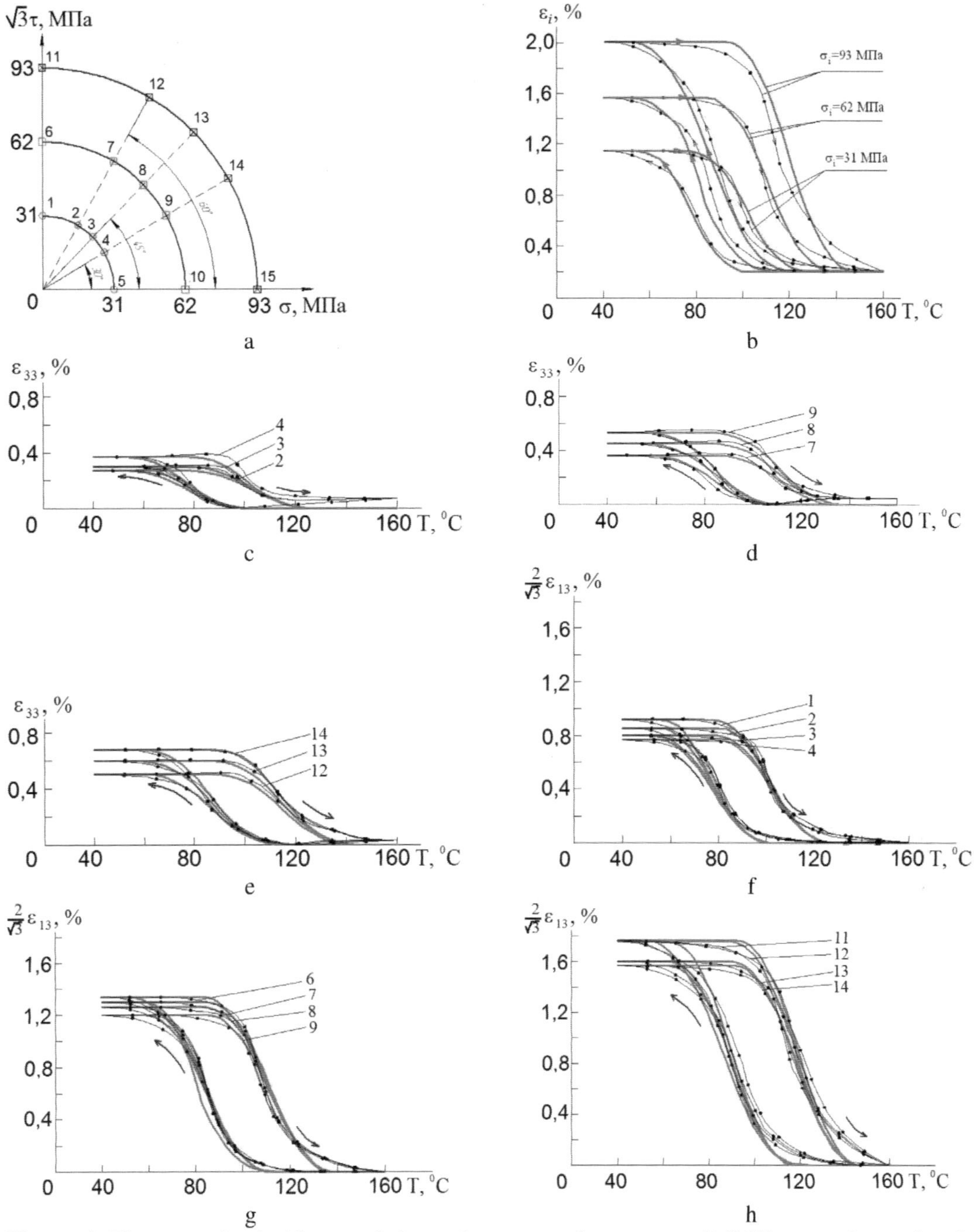

Picture 1. Thermomechanical hysteresis for various types of stress state. Solid lines are the results of a theoretical calculation; points are experimental data.

The stress space (a); the dependence of the strain intensity on the temperature (b); dependence of the component ε_{33} on T in points 2,3,4 (c); dependence of the component ε_{33} on T in points 7,8,9 (d); dependence of the component ε_{33} on T in points 12,13,14 (e); dependence of the component $(2/\sqrt{3})\varepsilon_{13}$ on T in points 1, 2,3,4 (f); dependence of the component $(2/\sqrt{3})\varepsilon_{13}$ on T in poinys 6, 7,8,9 (g); dependence of the component $(2/\sqrt{3})\varepsilon_{13}$ on T in points 11, 12,13,14 (h).

Picture 1a shows the scheme of the program for experimental and theoretical studies of thermomechanical hysteresis. The coordinates are used - the intensity of the tension at pure tension (σ), intensity of shear stress ($\sqrt{3}\tau$). Circles characterize the coordinates of points of different stressed states with the same intensity of stresses σ_i. Three levels are considered σ_i =31; 62; 93 МПа. Picture 1b shows the graphs of the thermomechanical hysteresis in the coordinates: the strain intensity ε_i – temperature T. Pictures 1c, d, e illustrate the kinetics of the thermomechanical hysteresis of the relative linear component of the strain tensor ε_{33} for various types of stress state. The graphs shown in picture 1f, g, h illustrate the thermomechanical hysteresis in the coordinates of the shear component of the strain tensor $(2/\sqrt{3})\varepsilon_{13}$ on temperature T. The figures in Pic.1a characterize the corresponding form of the stressed state. Comparing the obtained theoretical and experimental results, it can be noted that the proposed version of the theory makes it possible to create effective and simple methods for calculating the nontrivial deformation properties of the plasticity of transformation, shape memory, and thermomechanical hysteresis under conditions of a complex stress state in order to perform calculations of products deformed under conditions of initiation of reversible martensitic reactions.

Summing up and conclusions

Experimental studies of the processes of transformation plasticity and shape memory of an alloy Cu–12 %Al–4 %Mn in the case of a complex stressed state are fulfilled. Analytical relations at the macroscale level in the form of equations of the deformation type (1) and (2) are obtained within the framework of structural-analytical mesomechanics methods, which is caused by the use of the concentration hypothesis and the corresponding method of integration over the variables of the microscale level of the orientation space [2]. In this case the Clausius-Clapeyron equation is formulated through the invariant of the stress tensor σ_i. The criterion for the formation of a critical nucleus at the structural level that triggers the martensitic reaction in all elements of the orientational space is formulated at the macro-scale level in the form of the Heaviside function operator $H\left[M_н - \Phi(M_н - M_к) - T^*\right]H\left(\dot{T}^*\right) = 1$ upon cooling during the transformation plasticity step and the operator $H\left[T^* + \Phi(A_к - A_н) - A_к\right]H\left(\dot{T}^*\right) = 1$ when heated, under conditions of initiating the effect of shape memory. The presence of the designated Heaviside operators in this model takes into account the history of thermomechanical action, which is fundamentally different from the approach of variants of deformation theory of plasticity.

References

[1] Likhachev V A, Kuz'min S A and Kamentseva Z P 1987 Effekt pamiati formy (Shape memory effect) Leningrad LGU Publ 216 p

[2] Likhachev V A and Malinin V G 1993 Strukturno-analiticheskaia teoriia prochnosti *(Structural analytic theory of strength)* Saint-Petersburg "Nauka" Publ 471 p

[3] Saprykin A and Saprykina N 2012 Engineering support for improving quality of layer-by-layer laser sintering *7th International Forum on Strategic Technology IFOST 2012* 6357719

[4] Sharkeev Yu P, Eroshenko A Yu, Kovalevskaya Zh G, Saprykin A A, Ibragimov E A, Glukhov I A, Khimich M A, Uvarkin P V and Babakova E V 2016 Structural and Phase State of Ti–Nb Alloy at Selective Laser Melting of the Composite Powder *Russian Physics Journal* **59(3)** pp 430-434

[5] Malukhina O A and Filippov D A 2017 Rabotosposobnost' sfericheskikh segmentov iz splavov s effektom pamiati formy pri termomekhanicheskom tsiklirovanii cherez interval martensitnykh prevrashchenii *(International Research Journal)* **7(61)** p 3 pp 50-53

[6] Malinina N A, Malinin V G and Malinin G V 1999 Razvitie teorii fizicheskoi mezomekhaniki dlia sred s martensitnym mekhanizmom massoperenosa *(Vestnik Novgorodskogo gosudarstvennogo universiteta)* **13** pp 39-49

[7] Malushin N N, Kovalev A P, Valuev D V, Shats E A, Borovikov I F 2016 Development and

study of hard-facing materials on the base of heat-resisting high-hardness steels for plasma-jet hard- facing in shielding-doping nitrogen atmosphere *IOP Conference Series: Materials Science and Engineering* **142**
[8] Malushin N N, Valuev D V, Il'yaschenko D P, Trifonov V A, Nekrasov R Yu 2018 Technological Improvement of Surfacing of Parts of Hammer Crushers Used in Coke-Chemical Industry *J. Materials Science Forum* **927** 168-175

Mathematical model, the solution of boundary value problems for alloys with shape memory effect by analytical, numerical-analytical and numerical methods

G V Malinin[1], V G Malinin[1], N A Malinina[1], O A Malukhina[2]

[1] Orel state agricultural institute, 69, General Rodin ul., Orel, 302019, Russia
[2] Yaroslav-the-Wise Novgorod State University, 41, ul. B. St. Petersburgskaya, 41, 173003 Veliky Novgorod, Russia

malinina2006@yandex.ru

Abstract. In this paper we consider the application of analytical, numerical-analytical and numerical methods for solving boundary value problems for structures made from alloys with shape memory. As the defining relations, the equations of structural-analytic mesomechanics were applied.

Introduction

Materials with the shape memory effect (SME) have unique properties that are absent in most traditional materials used in engineering applications. They are called intellectual materials of the XXI century, as well as smart materials. Accordingly, their use provides new design capabilities, which makes it possible both to improve the characteristics of devices and to offer innovative solutions. Alloys with shape memory (ASM) are widely used in medicine, engineering, aircraft building, construction, etc. To calculate the stress-strain state and the engineering solution of problems, it is necessary to develop an appropriate mathematical apparatus that allows to reflect the real mechanical properties of ASM. The development of a mathematical model of ASM that takes into account the features of their behavior in simple and complex deformations, as well as the creation of methods for the analytical and numerical solution of boundary value problems for structures and products from SPF are relevant.

The basic equations of structural-analytic mesomechanics are given in [1-3]. They make it possible to perform a forecast of the basic deformation phenomena in alloys possessing shape memory effects. In this paper, the defining relations of the deformation type are presented, and therefore it becomes possible to apply the mathematical apparatus of the theory of plasticity for solving boundary value problems with elements made of ASM. Within the framework of this approach, some boundary-value problems that admit an analytical solution are solved-the problem of deformation of a long thick-walled tube under internal pressure; numerical-analytical solution by the method of boundary elements - calculation of the truss structure; numerical methods - rod system, bending of the beam, and also plates with different support of the ends. Only isothermal martensitic transformations are considered. A detailed derivation of the defining relations is given in [3].

1.Determined relations for an alloy with shape memory effect

at the load stage:

$$\varepsilon_i^{\Sigma}=\frac{\sigma_i}{E}+\left(\frac{2}{3}B_{\phi}\sigma_i\frac{k\left(\sigma_i-\sigma_{н\uparrow}^{A\to M}\right)}{M_н-M_к}\right)H\left(\sigma_i-\sigma_{н\uparrow}^{A\to M}\right)H\left(\sigma_{к\uparrow}^{A\to M}-\sigma_i\right)H\left(\dot{\sigma}_i\right) \qquad (1)$$

At the unloading stage:

$$\varepsilon_i^{\Sigma} = \frac{\sigma_i}{E} + \left(\frac{2}{3} B_{\phi} \sigma_i^0 \frac{k\left(\sigma_i + \sigma_{\Pi\gamma} - \sigma_{н\uparrow}^{A\to M}\right)}{M_н - M_к} \right) H\left(\sigma_i - \sigma_{к\uparrow}^{A\to M}\right) H\left(\sigma_{н\uparrow}^{A\to M} - \sigma_i\right) H\left(\dot{\sigma}_i\right) \qquad (2)$$

where σ_i – voltage intensity; E – elastic modulus; B_{Φ} – material constant; σ_i^0 – intensity of voltage accumulated during the loading phase; $k = \frac{T_0}{q_0} D_i$; T_0 – thermodynamic equilibrium temperature; q_0 – thermal reaction effect; D_i – distortion of phase transformation; $\sigma_{н\uparrow}^{A\to M} = \frac{T_D - M_н}{k}$ – the voltage of the onset of direct martensitic transformation; T_D – deformation start temperature; $M_н$, $M_к$ – the temperature of the beginning and the end of the direct martensitic transformation, respectively; $\sigma_{к\uparrow}^{A\to M} = \frac{T_D - M_к}{k}$ – endurance of martensitic transformation, $\sigma_{\Pi\gamma} = \frac{\Delta}{k}$ – limit of pseudoelasticity at unloading; Δ – width of hysteresis of the martensitic transformation; $\dot{\sigma}_i = \frac{d\sigma_i}{dt}$; $H(\ldots)$ – Heaviside function.

The diagram at the stage of loading in the regime of austenitic pseudoelasticity (ferroelasticity), obtained by the formula (1) is shown in Fig. 1.

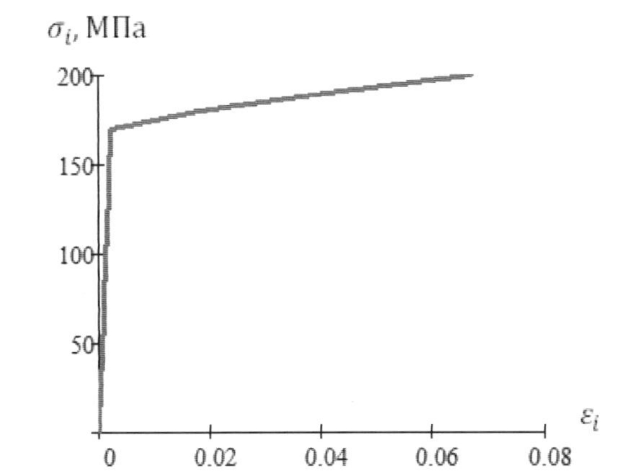

Figure 1. Deformation diagram $\sigma_i - \varepsilon_i$ under active loading.

2. Boundary value problems for ASM

Initial data for the alloy TiNi are presented in Table 1.

Table 1. The initial data for Ti-Ni alloy.

Name	Formula
Characteristic temperatures of forward and reverse martensitic transformation	$M_н$ =330K, $M_к$ = 320K $A_н$ =370K, $A_к$ = 380K
Crystallographic constant k	k = 0,29 KMPa^{-1}
Constant B_{Φ}	B_{Φ} = 0,06·10^{-2} MPa^{-1}
Elastic modulus E	E = 7,42·10^{4} MPa

Deformation temperature T_{D1}	$T_{D1} = A_к = 380К$
Deformation temperature T_{D2}	$T_{D2} = (A_к + A_н)/2 = 375К$
Voltage of the beginning of a direct martensitic transformation at deformation temperature T_{D1} (phase yield strength) $\sigma_{н1\uparrow}^{A\to M} = \sigma_{T1}$	$\sigma_{н1\uparrow}^{A\to M} = \sigma_{T1} = 172\text{MPa}$
Voltage of the beginning of a direct martensitic transformation at deformation temperature T_{D2} (phase yield strength) $\sigma_{н2\uparrow}^{A\to M} = \sigma_{T2}$	$\sigma_{н2\uparrow}^{A\to M} = \sigma_{T2} = 155\text{MPa}$
Voltage of the end of a direct martensitic transformation at deformation temperature T_{D1} $\sigma_{к1\uparrow}^{A\to M} = \sigma_{\max 1}$	$\sigma_{к1\uparrow}^{A\to M} = \sigma_{\max 1} = 207\text{MPa}$
Voltage of the end of a direct martensitic transformation at deformation temperature T_{D2} $\sigma_{к2\uparrow}^{A\to M} = \sigma_{\max 2}$	$\sigma_{к2\uparrow}^{A\to M} = \sigma_{\max 2} = 190\text{MPa}$
The module of ferroelasticity $E_{\Phi\gamma 1} = E_{к1}$ (tangent module) at the deformation temperature T_{D1}	$E_{\Phi\gamma 1} = E_{к1} = 414\text{MPa}$
The module of ferroelasticity $E_{\Phi\gamma 2} = E_{к2}$ (tangent module) at the deformation temperature T_{D2}	$E_{\Phi\gamma 2} = E_{к2} = 452\text{MPa}$

2.1. Analytical calculation of a thick-walled pipe from ASm under internal pressure, under conditions of direct martensitic transformation

Formulation of the problem. Determine the voltage distribution in the cross section of a thick-walled pipe r_1 = 10 мм, r_2 = 20 мм under internal pressure, made of alloy *TiNi*, having a shape memory effect, under conditions of direct martensitic transformation.

Assumptions:

1) the material characteristics are calculated at different but constant temperatures;
2) the material undergoes elastic and inelastic (phase) deformations;
3) we take a bilinear model of material deformation.

The results of the calculation for the deformation temperature $T_D = A_к$ are presented in table 2.

Table 2. Results of calculation of pipe yield radius depending on internal pressure.

N	Yield radius r_T , мм	Load *p*, MPa
1	10	75
2	13	110
3	15	125
4	18	137
5	20	139

Graphs of voltage distribution normalized to $\sigma_{н1\downarrow}^{A\to M}$ for different pressure values are shown in the figure 2.

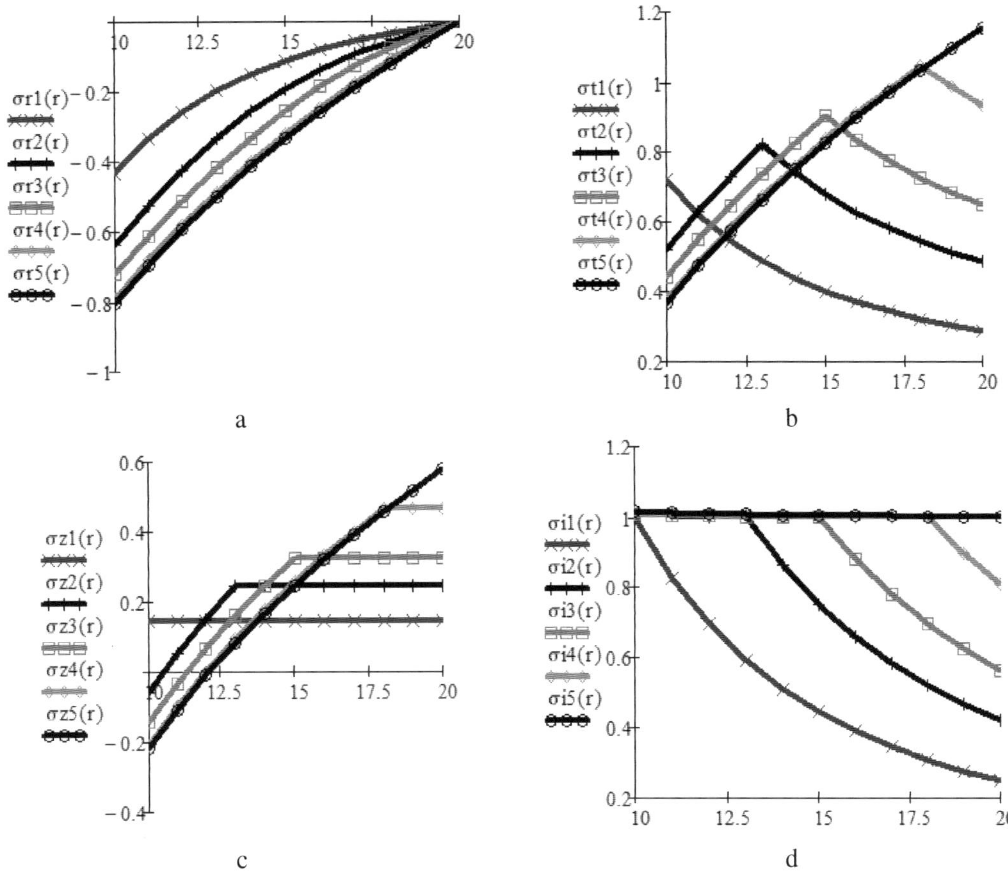

Fogure 2. The distribution of voltage normalized to $\sigma_{н\downarrow}^{A\to M}$ by pipe radius:
a) diagram of radial voltage; b) the diagram of the circumferential voltage;
c) diagram of axial voltage; d) an equivalent voltage diagram.

2.2. Core systems

Calculation of a statically indeterminate rod system.

Formulation of the problem. The rod system consists of two rods made of ASM with different mechanical characteristics (Figure 3). Both materials have the same elastic moduli E. Tangent modules and voltage of the onset of direct martensitic transformation $\sigma_{н\uparrow}^{A\to M}$ are different. The external horizontal force is $F = 2\cdot A\cdot\sigma_{н\uparrow}^{A\to M}$ and is applied at the junction of two rods, where A is the cross-sectional area of both rods. Deformation temperature of the first rod T_{D1}, of the second T_{D2}. It is required to determine the stresses in the rods by the finite element method, using the iterative Newton-Raphson algorithm with the following initial data:

Table 3. Initial data.

Cross-sectional areas of rods *A*	$A = 314{,}159$ мм^2
Length of each rod *L*	$L = 50$ мм
Load	$R = 2\cdot A\cdot\sigma_{T1} = 1{,}083\cdot 10^5$ Н

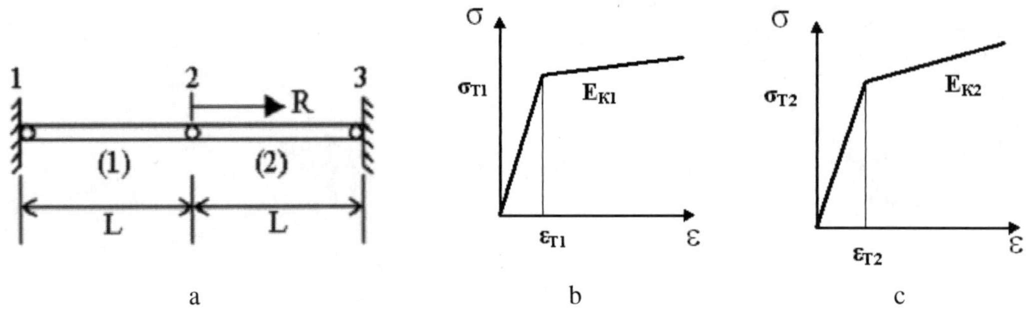

Figure 3. The calculation scheme. Deformation diagrams.

Table 4. Results of the calculation of the rod system.

Voltage in the first rod:	Voltage in the second rod:
$\sigma_1 = \sigma_{T1} + E_{k1}(\Delta_1 \cdot \varepsilon_{T1}) = 180\text{MPa}$	$\sigma_2 = -\sigma_{T2} - E_{k2}(\Delta_2 \cdot \varepsilon_{T2}) = -164\text{MPa}$

Stresses in the first rod are tensile, in the second - compressive.

2.3. Calculation of a statically indeterminate farm

For calculation, the numerical-analytical method of boundary elements [4] is used in conjunction with the methods of elastic solutions and variable elasticity parameters.

Formulation of the problem. Determine the internal forces in the rods of a statically indeterminate shape made from ASM, figure 4.

Table 5. Initial data.

Name	Formula
Tangent module, Pa	$E_к = \dfrac{15 q_0}{T_D \cdot D_{31}^2}(M_н - M_к) = 2{,}469 \cdot 10^{11}$
Phase yield strength, Pa	$\sigma_T = 6 \cdot 10^8$
Deformation temperature, K	$T_D = 300$
Cross-sectional areas of rods	$A = 3{,}142 \cdot 10^{-4}$ м^2
Load	$F = EA \cdot 0{,}002$
Geometric parameter	$a = 1$ м

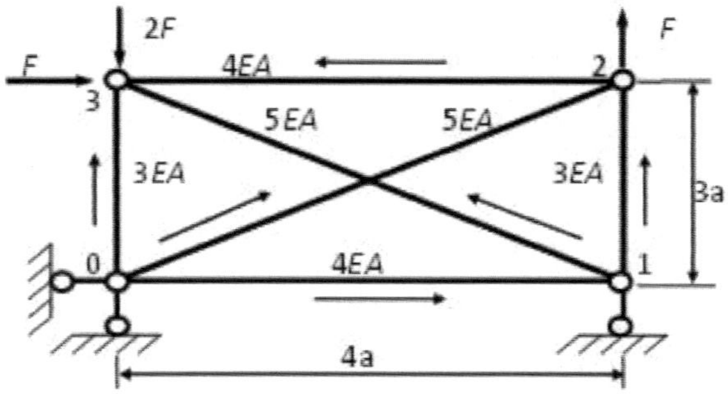

Figure 4. The calculation scheme. Boundary element grid.

Table 6. Results of the calculation of the shape.

Elastic solution	The solution of the nonlinear problem
$N^{0-1}=2{,}727\cdot 10^5\,H$; $N^{0-2}=2{,}089\cdot 10^5\,H$; $N^{0-3}=-6{,}751\cdot 10^5\,H$; $N^{1-2}=3{,}145\cdot 10^5\,H$; $N^{1-3}=-3{,}409\cdot 10^5\,H$; $N^{2-3}=-1{,}671\cdot 10^5\,H$.	$N^{0-1}=1{,}827\cdot 10^5\,H$; $N^{0-2}=3{,}214\cdot 10^5\,H$; $N^{0-3}=7{,}307\cdot 10^3\,H$; $N^{1-2}=2{,}470\cdot 10^5\,H$; $N^{1-3}=-2{,}284\cdot 10^5\,H$; $N^{2-3}=-2{,}571\cdot 10^5\,H$.

2.4. Calculation of a beam on two supports loaded with a distributed load

A single-span beam of square cross-section, made of ASM with the following parameters, presented in Table 7, was calculated.

Table 7. Initial data.

Load, MPa	$q = 10$
Deformation temperature, K	$T_D = A_к = 380$
Length, мм	$a = 10$
Thickness, мм	$h = 1$
Width, мм	$b = 1$
The axial moment of inertia, мм^4	$J_z = b\cdot h^3/12 = 0{,}083$

The calculation was carried out by three methods: Ritz, elastic solutions, variable elastic moduli. For approximation of the deformed beam axis, the Vlasov method was used. The problem was solved with two boundary conditions: 1) both edges are hinged-supported, 2) both edges are rigidly embedded.

Table 8. Results of the calculation of the beam by the Ritz method.

The bending moment diagram *M(x)*	
Hinged-supported beam	*Rigged beam*
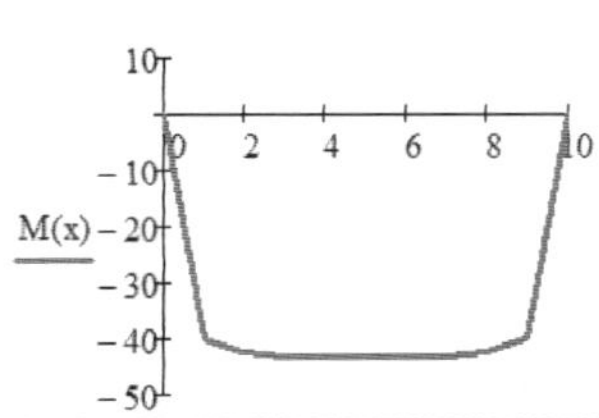	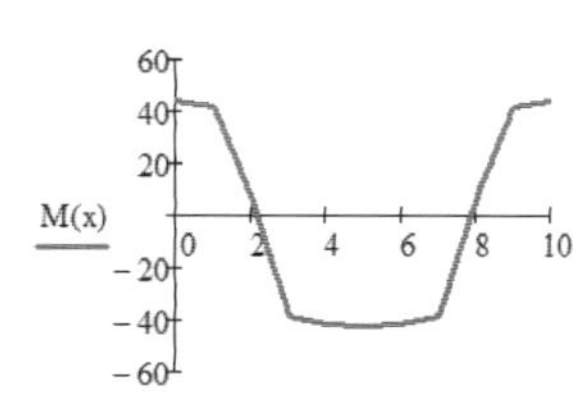

Table 9. Maximum voltage.

The Ritz method	The method of elastic solutions	The method of variable elasticity parameters
Hinged-supported beam cross-section $\sigma\,(a/2;h/2)$, MPa		
177	176	176
Rigged beam cross-section $\sigma\,(0;h/2)$ and $\sigma\,(a;h/2)$, MPa		
178	174	174

2.5. Plates

Formulation of the problem. A square plate of ASM loaded with a distributed load with two anchors along the contour: articulated and rigidly fixed. The calculation was carried out by three methods: Ritz, Vlasov-Kantorovich together with the method of elastic solutions and the finite difference method together with the method of elastic solutions.

Table 10. Initial data.

Load, MPa	$q = 10$
Deformation temperature, K	$T_D = A_к = 380$
Length, мм	$a = 10$
Thickness, мм	$h = 1$

Table 11. The results of calculating the plate by the Ritz method.

Voltage intensity σ_i, MPa	
Hinged-supported plate	*Rigged plate*
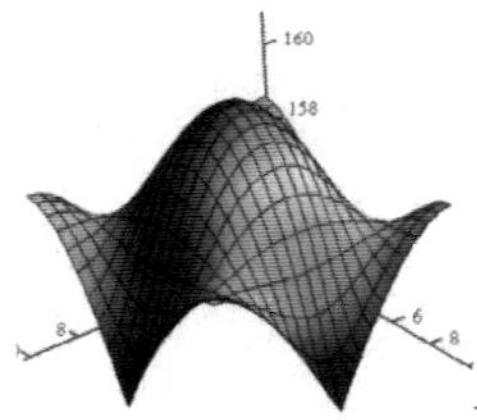	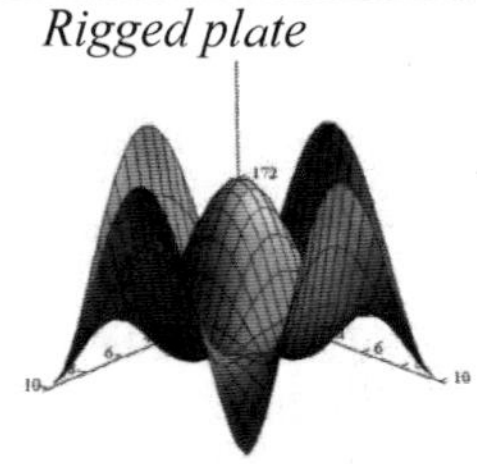
In the center of the plate $\sigma_{i,0} = 179$	In the center of the plate and in the middle of the sides $\sigma_{i,0} = \sigma_{i,C} = 173$

Table 12. Results of calculating the maximum voltage intensity for a hinged-supported plate by various methods, MPa

The Ritz method	The Vlasov – Kantorovich method together with the method of elastic solutions	The finite difference method together with the method of elastic solutions
178	173	174

References

[1] Malinin G.V. Strukturno-analiticheskaia mezomekhanika nanostrukturnykh sostoianii sredy s obratimymi martensitnymi prevrashcheniiami [Structural analytic mesomechanics of nanostractured conditions of the medium with reversible martensitic transformations]. Nanoinzheneriia, 2012, no. 5, pp. 22-29.

[2] Malinin V.G., Malinina N.A., Malinin G.V., Perelygin A.I. Metod prognoza effekta plastichnosti prevrashcheniia pri slozhnom napriazhennom sostoianii na osnove razvitiia strukturno-analiticheskoi mezomekhaniki [Method of forecasting the conversion plasticity effect in complex stress state on the basis of structural-analytical mesomechanics]. Fundamental'nye i prikladnye problemy tekhniki i tekhnologii – Fundamental and Applied Problems of Engineering and Technology, 2012, no. 2(292), pp. 18-23.
[3] Malinin G.V., Malinina N.A. Vyvod opredeliaiushchikh sootnoshenii dlia materiala s effektom pamiati formy [Finding the constitutive relations for the shape memory material]. X Vserossiiskaia nauchno-prakticheskaia konferentsiia s mezhdunarodnym uchastiem «Sovremennye problemy gorno-metallurgicheskogo kompleksa. Energosberezhenie. Ekologiia. Novye tekhnologii» [Proc. 10th All-Russian Sci. & Pract. Conf. with Int. Participation "Current Issues of Mining and Metals Sector. Energy conservation. Ecology. Innovative technologies"]. Stary Oskol, 2013, pp. 57-61.
[4] Bazhenov V.A., Dashchenko A.F., Orobei V.F., Sur'ianinov N.G. Chislennye metody v mekhanike [Numerical methods in mechanics]. Moscow, "Mir" Publ., 2004. 564 p.

ANISOTROPY OF POISSON'S RATIO OF DENTIN AND ENAMEL

S A Muslov

Evdokimov Moscow State Medical Stomatological University (MSMSU), Delegatskaya St. 20-1, 127473, Moscow, Russia

E-mail: muslov@mail.ru

Abstract. In this paper, the orientation dependence of the Poisson's ratio of teeth's dentin and enamel on the basis of matrices of elastic constants and the compliance coefficients of hexagonal crystals, such as crystals of dentine hydroxyapatite, was obtained for the first time. The results of calculating the Poisson's ratios of dentin and enamel as a crystalline system with a hexagonal structure are presented in the form of tables and diagrams in the polar and Cartesian coordinate systems. It is shown that the maximum value of the Poisson's ratio of dentin (0.54) is greater than the upper limit for the Poisson's ratio of isotropic materials, including known restoration materials, which in some cases may reduce the quality of restorations. The effective elastic characteristics of dentine and enamel, including Poisson's ratio using different averaging methods, are calculated.
Key words: enamel, dentin, Poisson's ratio, crystals of hydroxyapatite.

1. Introduction

Elastic properties of hard tissues of tooth and hydroxyapatite, as their mineral component, quite often cause interest among researchers [4-7, 12-15, etc.]. However, a number of aspects of this subject, in particular the value of the Poisson's ratio (coefficient of transverse deformation) of dentine and enamel and as anisotropic media with hexagonal symmetry, remain relevant to the end. The study was conducted on the basis of two different methods of calculation [2, 16].

According to modern concepts, the range of possible values of the Poisson's ratio [1, 3] can be negative $\mu<0$ (auxetics), and more than 0.5. For dentin and enamel the "technical" coefficient according to the literature data lies in the range of 0.29-0.33 [9], for example, according to the results of ultrasonic measurements for dentin, it is equal to 0.32, in enamel – 0.28 [15].

2. The study of anisotropy Poisson ratio of dentin and enamel

It is believed that dentin is a biocomposite and consists of approximately 45-70 % of inorganic material [10, 14] in the form of hydroxyapatite crystals. Crystals of hydroxyapatite are located between the collagen fibers and the class of symmetry belong to hexagonal crystal structure. The mineral base of enamel is represented by hexagonal crystals of hydroxide, carbonate, chlorine, fluorapatite, and enamel prisms are the main structural formation. Therefore, there is every reason to consider dentine and enamel as anisotropic media, which means that their elastic properties are described by the matrix of elastic constants c_{ij} or compliance coefficients s_{ij} .

The Poisson ratios of such a medium in the general form can be defined as

$$\mu_{kl} = -\frac{\varepsilon_{ll}}{\varepsilon_{kk}}, \quad (1)$$

where ε_{kl} – components of the strain tensor. As a result, hexagonal systems, dentin and enamel are also described by two Poisson's ratios μ'_{31} and μ'_{32} [16]. According to the literature data, the values of the elastic constants of dentin and enamel are well known [12, 13], so knowing them, we can calculate the coefficients of compliance and the Poisson's ratio of dentin and enamel by the corresponding formulas [8, 16] for hexagonal crystals.

$$c_{11} + c_{12} = \frac{s_{33}}{s}, \quad c_{11} - c_{12} = \frac{1}{s_{11} - s_{12}}, \quad c_{13} = -\frac{s_{13}}{s}$$
$$c_{33} = \frac{s_{11} + s_{12}}{s}, \quad c_{44} = \frac{1}{s_{44}}, \quad c_{66} = \frac{1}{s_{66}}, \quad s = s_{33}(s_{11} + s_{12}) - 2s_{13}^2, \quad (2)$$

$$\mu'_{31} = -[\sin^2\psi\sin^2\theta\cos^2\theta(s_{11} + s_{33} - s_{44}) + (\cos^2\theta - \sin^2\theta\cos 2\theta\sin^2\psi)s_{13} + \sin^2\theta\cos^2\psi s_{12}] / s'_{33} \quad (3)$$

and

$$\mu'_{32} = -[\cos^2\psi\sin^2\theta\cos^2\theta(s_{11} + s_{33} - s_{44}) + (\cos^2\theta - \sin^2\theta\cos 2\theta\cos^2\psi)s_{13} + \sin^2\theta\sin^2\psi s_{12}] / s'_{33}, \quad (4)$$

where

$$s'_{33} = \cos^4\theta s_{33} + \sin^2\theta\cos^2\theta(2s_{13} + s_{44}) + \sin^4\theta s_{11}, \quad (5)$$

θ и ψ – Eulerian angles (describing the rotation of an absolutely rigid body in a three-dimensional Euclidean space). Orthogonal system x_i has the Ox_3 axis parallel to the <c>-axis of the crystal.
Equation (4) gives Poisson's ratio for an extension along Ox'_3 and a contraction along Ox'_2, equation (3) gives Poisson's ratio for a contraction in the perpendicular direction Ox'_1.
For an isotropic solid $s_{11} = s_{22} = s_{33}$, $s_{44} = s_{55} = s_{66} = 2(s_{11} - s_{12})$, $s_{12} = s_{13} = s_{23}$ and equation (3), (4) and (5) are reduced to $s'_{33} = s_{11}$, $\mu'_{32} = \mu'_{31} = -s_{12}/s_{11}$, so that Poisson's ratio is independent of orientation, what we wanted to prove.
Values of elastic constants c_{ij} and compliance coefficients s_{ij} of dentin and enamel are presented in Table 1 and 2.

Table 1. Elastic constants of dentin and enamel (GPa)

Material	c_{11}	c_{12}	c_{13}	c_{33}	c_{44}
Dentin	37.00	16.60	8.70	39.00	5.70
Enamel	115.00	42.40	30.00	125.00	22.80

Table 2. Compliance coefficients of dentin and enamel (GPa^{-1})

Material	s_{11}	s_{12}	s_{13}	s_{33}	s_{44}
Dentin	0.0346	-0.0145	-0.0045	0.0276	0.1754
Enamel	0.0104	-0.0034	-0.0017	0.0088	0.0439

The results of calculating the Poisson coefficients (directional dependence) of the dentin and enamel as crystal systems with a hexagonal structure are shown in Figure 1 and Figure 2.

Cartesian coordinates Polar coordinates

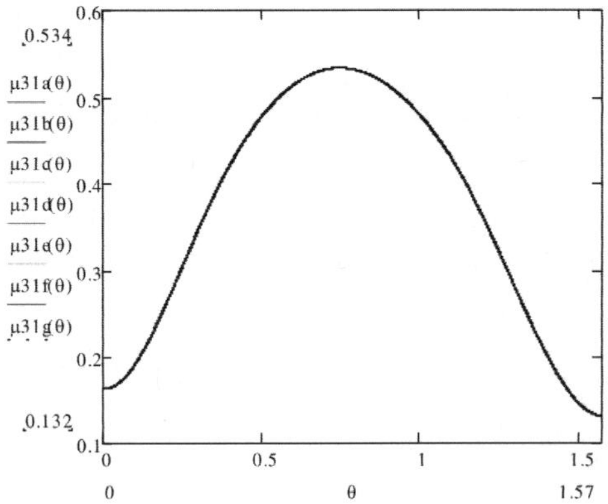

1

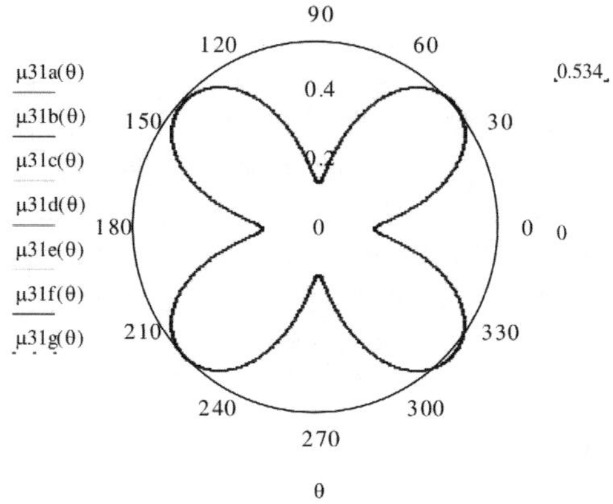

2

$\mu'_{31}(\theta)$ (a-g: $\psi = 0, 0.25, 0.5, 0.75, 1, 1.25, 1.5$)

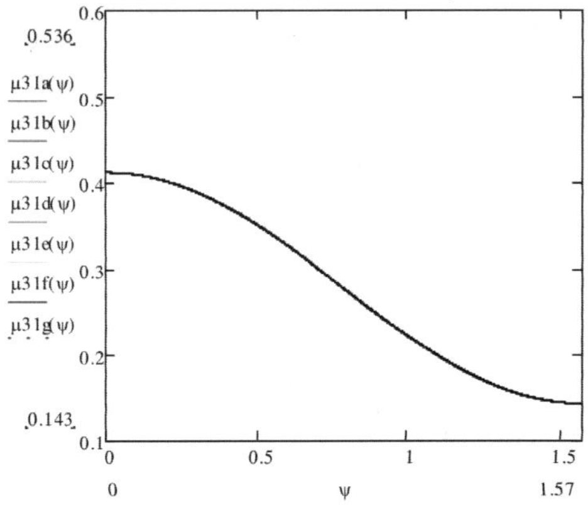

3

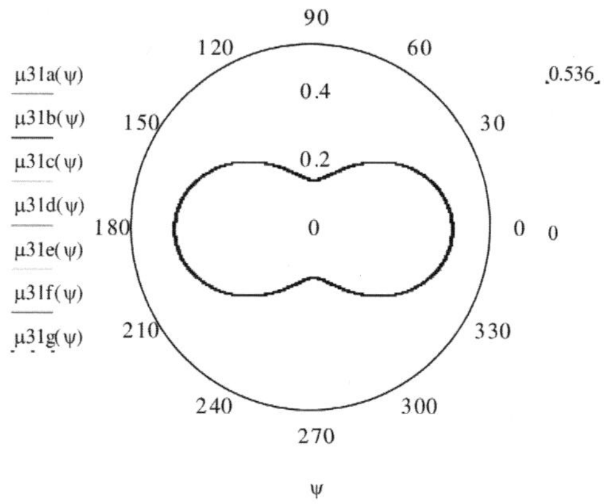

4

$\mu'_{31}(\psi)$ (a-g: $\theta = 0, 0.25, 0.5, 0.75, 1, 1.25, 1.5$)

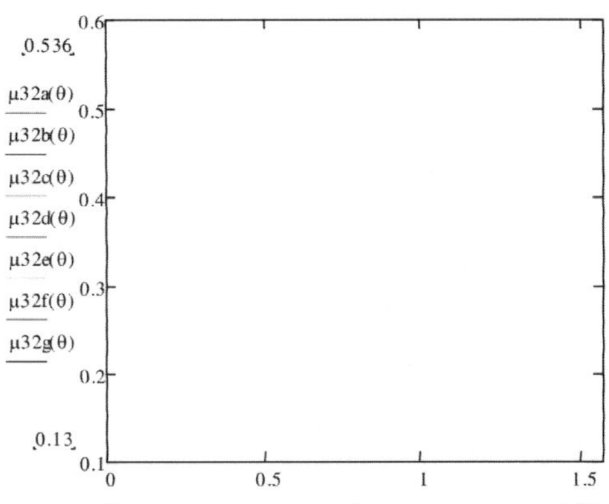

5

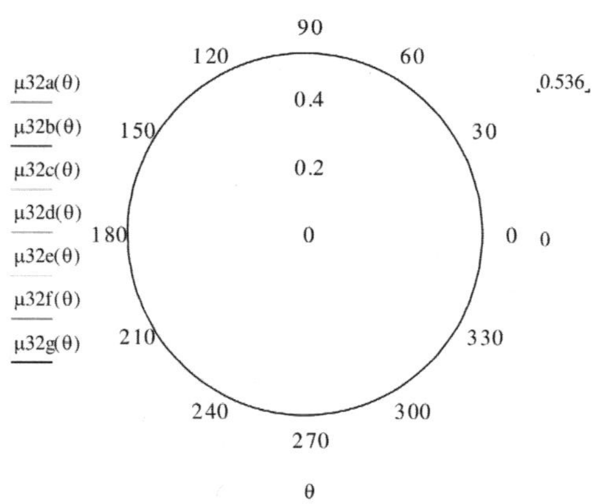

6

$\mu'_{32}(\theta)$ (a-g: $\psi = 0, 0.25, 0.5, 0.75, 1, 1.25, 1.5$)

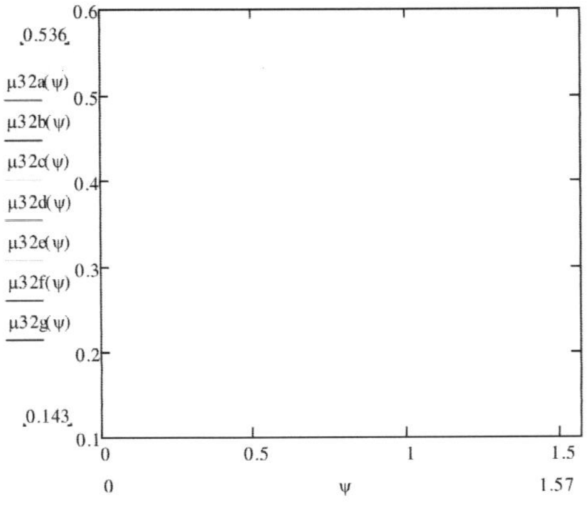

7

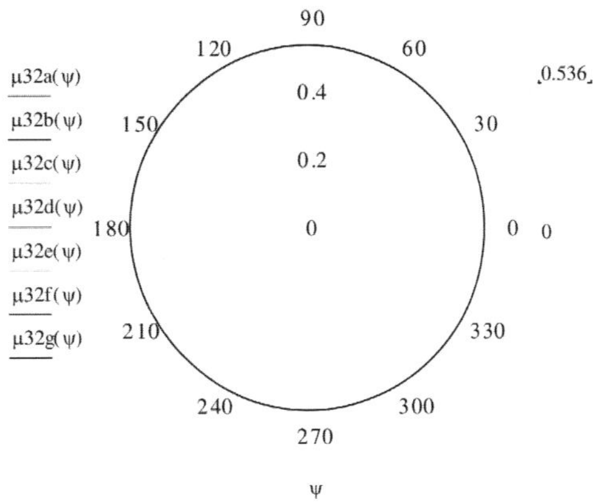

8

$\mu'_{32}(\psi)$ (a-g: $\theta = 0, 0.25, 0.5, 0.75, 1, 1.25, 1.5$)

Figure 1. The Poisson's ratio μ'_{31} (1-4) and μ'_{32} (5-8) of dentin for different directions θ and ψ.

Cartesian coordinates

Polar coordinates

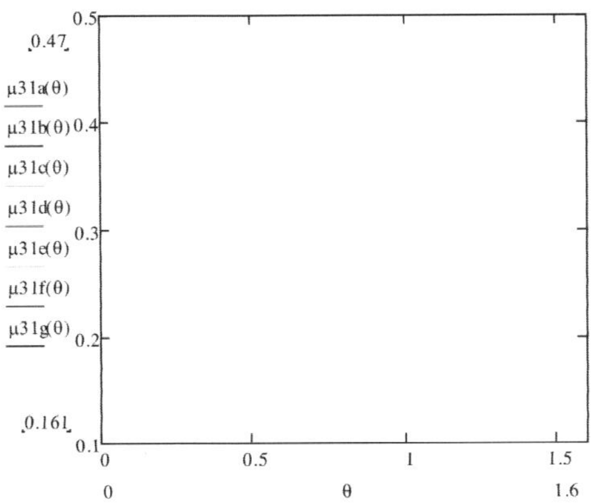

1

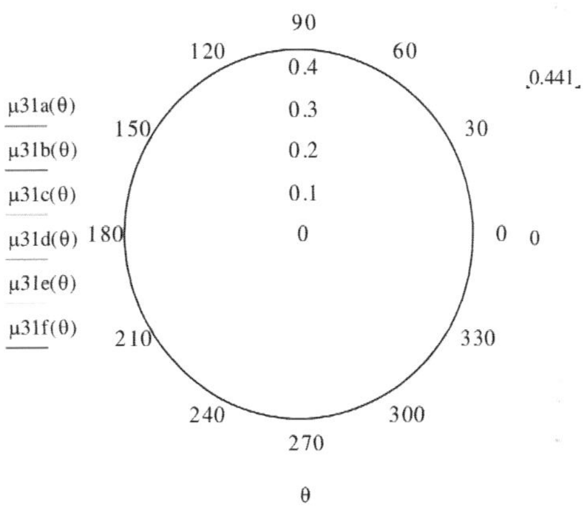

2

$\mu'_{31}(\theta)$ (a-g: $\psi = 0, 0.25, 0.5, 0.75, 1, 1.25, 1.5$)

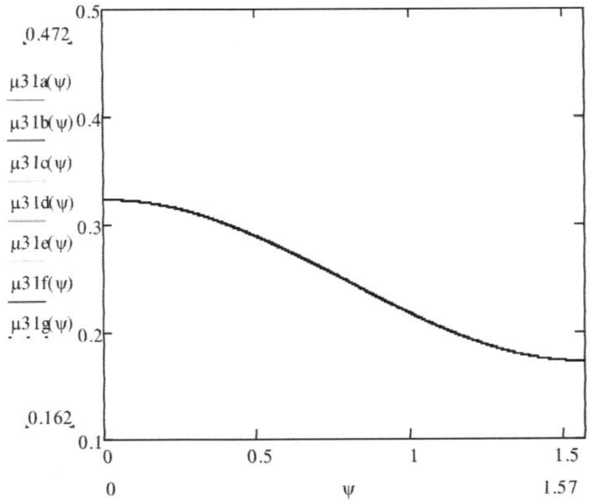

3

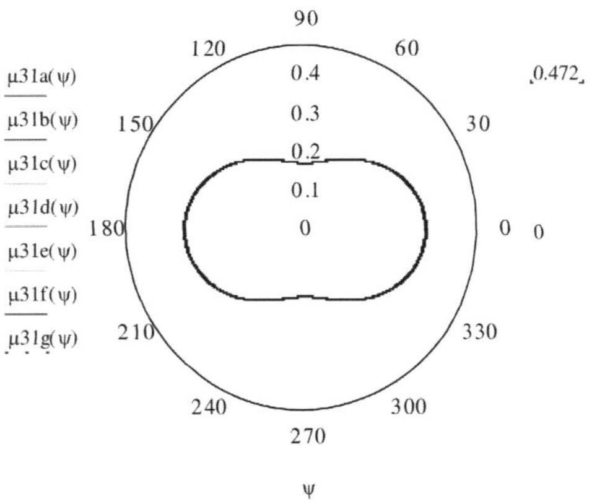

4

$\mu'_{31}(\psi)$ (a-g: $\theta = 0, 0.25, 0.5, 0.75, 1, 1.25, 1.5$)

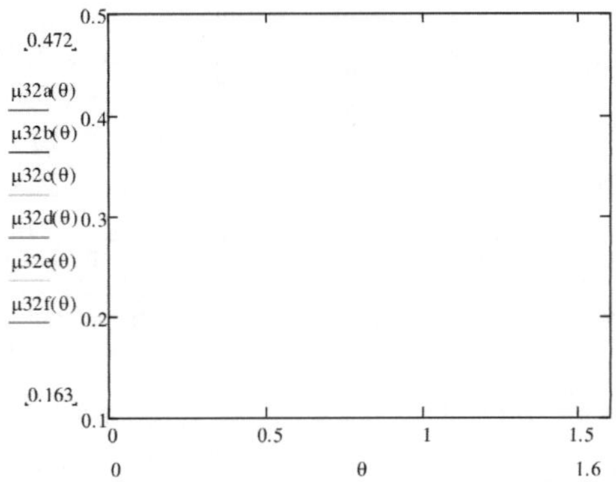

5

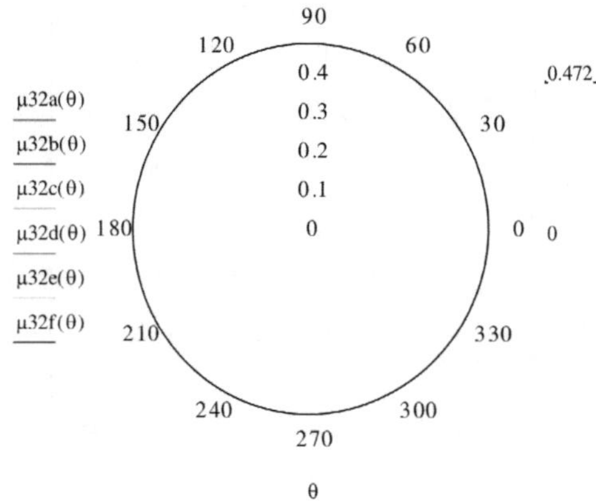

6

$\mu'_{32}(\theta)$ (a-g: $\psi = 0, 0.25, 0.5, 0.75, 1, 1.25, 1.5$)

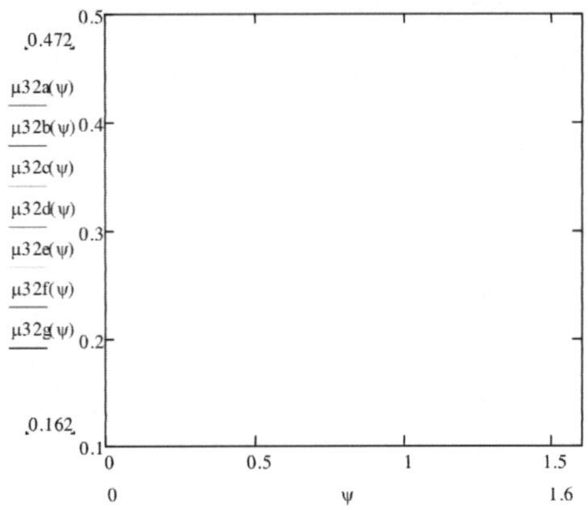

7

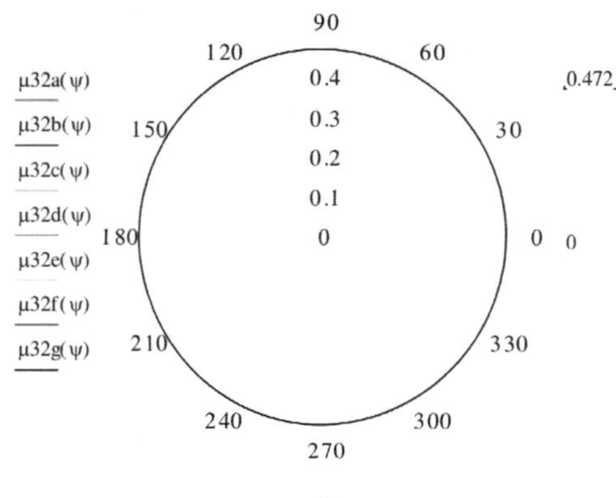

8

$\mu'_{32}(\psi)$ (a-g: $\theta = 0, 0.25, 0.5, 0.75, 1, 1.25, 1.5$)

Figure 2. The Poisson's ratio μ'_{31} (1-4) and μ'_{32} (5-8) of enamel for different directions θ and ψ.

Worthy of attention approach based on calculation of extreme values of the Poisson ratio of crystalline media and create isosurfaces of the coefficient in space [2, 11]. The results of calculating the coefficients of the dentin and enamel Poisson as a crystal hexagonal structure are presented in Table 3 and Figure 3.

Table. 3. Extreme values of Poisson's ratio (μ_{min}, μ_{max}) for dentine and enamel, values of Poisson's ratio in particular orientations and "anisotropy coefficient" μ_{max} / μ_{min}

Material	μ_{max}	μ_{min}	$\mu_{[2\bar{1}\bar{1}0],[0001]}$	$\mu_{[01\bar{1}0],[2\bar{1}\bar{1}0]}$	$\mu_{[0001],[2\bar{1}\bar{1}0]}$	μ_{max} / μ_{min}
Dentin	0.54	0.13	0.16	0.13	0.42	4.15
Enamel	0.47	0.16	0.19	0.16	0.33	2.94

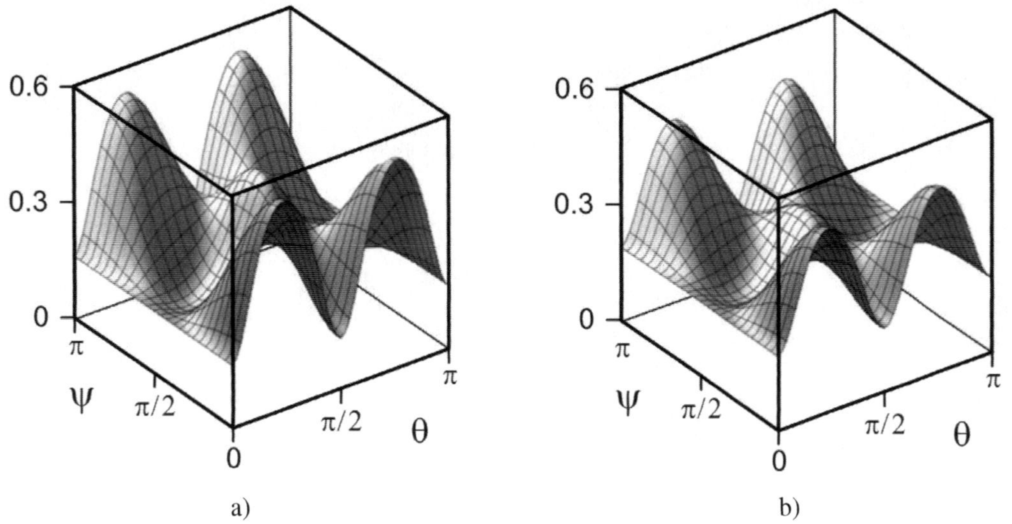

Figure 3. Variability of the Poisson's ratio of dentin (a) and enamel (b) in space.

All graphics (Figure 1-3) clearly show the pronounced anisotropic behavior of the Poisson's ratio of dentin and enamel crystals. Its values, depending on the direction in space, vary within a very wide range (4.15 times in dentin and 2.94 times in enamel). This number of times can be called the anisotropy coefficient for the Poisson ratio.

Note the anomaly high value of the maximum value of the Poisson's ratio (0.534-0.54) along a number of directions, which is extremely unusual for materials. This means that the compression of the local areas of the dentin along these directions will increase their volume, and when stretched – on the contrary, decrease. The elastic behavior under load of the restoration material is fundamentally different (Poisson's ratio of the filling material is significantly less than 0.5, for example, according to [17] it is between the values of 0.24 to 0.35 for dental composite materials), so its volume decreases when compressed and increases when stretched. This discrepancy in the deformation behavior at the boundary of heterogeneous media of the filling material and dentin can lead to the formation of overstress domains at this boundary, to the weakening of adhesion of the restoration material with dentin and, as a negative result, the degradation of fixation and the often encountered situation of destruction of direct and sometimes indirect restoration, especially composite materials.

3. The average of the Poisson's ratio of dentine and enamel

The average value of the Poisson's ratio of dentine and enamel can be obtained by applying the formulas (3-5) and the expression [16]:

$$<\mu_{ij}^{'}>=\frac{1}{2\pi}\int_0^{\pi}\int_0^{\pi}\mu_{ij}^{'}(\theta,\psi)\sin\theta d\theta d\psi . \qquad (6)$$

Numerical calculations of this integral in Mathcad software give the results: 0.312 for dentine and 0.286 for enamel, which is very well consistent with the data of ultrasonic measurements [15].

4. Calculation of the effective value of the Poisson's ratio of dentine and enamel on the basis of their elastic constants and compliance coefficients

On the basis of data obtained on single crystals we can predict the elastic properties of dentin and enamel as microinhomogeneous materials [18].

According to the Voigt averaging, the module of all-round compression and the shear modulus for hexagonal crystals are equal respectively

$$K_V = \frac{2c_{11} + c_{33} + 2(c_{12} + 2c_{13})}{9}$$
$$G_V = \frac{7c_{11} + 2c_{33} - 5c_{12} - 4c_{13} + 12c_{44}}{30}. \quad (7)$$

In the approach proposed by Reuss, the averaging of the compliance tensors is performed, as a result of which for the hexagonal lattice we obtain

$$\frac{1}{K_R} = 2s_{11} + s_{33} + 2(s_{12} + 2s_{13})$$
$$\frac{1}{G_R} = \frac{2(7s_{11} + 2s_{33} + 3s_{44} - 5s_{12} - 42s_{13}}{15}. \quad (8)$$

In many cases, a good agreement with the experimental data gives the proposed by Hill arithmetic mean value, found by the averaging for Voigt and Reuss, namely

$$K_H = \frac{1}{2}(K_V + K_R),\ G_H = \frac{1}{2}(G_V + G_R). \quad (9)$$

Keeping in mind that two modules are independent in the isotropic approximation, we can write expressions for E_H and μ_H as functions K_H and G_H:

$$\langle E_H \rangle = \frac{9K_H G_H}{3K_H + G_H},\ <\mu_H> = \frac{3K_H - 2G_H}{6K_H + 2G_H}. \quad (10)$$

Table 4 shows the values of the averaged elastic modules and the Poisson's ratio of dentine and enamel, calculated from the values of elastic constants and compliance coefficients using the described averaging methods used for single-phase hexagonal polycrystals.

Table 4. Averaged elastic modules and Poisson's ratio of dentine and enamel

Material	K_V	G_V	K_R	G_R	K_H	G_H	E_H	μ_H
Dentin	20,11	9,59	20,03	8,21	20,07	8,90	23,26	0,31
Enamel	62,20	33,22	62,20	30,54	62,20	31,88	81,69	0,28

Note that the calculated values of μ_H are in good agreement with the known experimental data [15] and results of section 3 for $<\mu'_{ij}>$.

Conclusion

1. Dentin and tooth enamel are not isotropic media due to the symmetry of their mineral component – hydroxyapatite crystals.
2. The expressed anisotropy of the Poisson's ratio of dentine and enamel is established on the basis of calculations by formulas of elastic constants and compliance coefficients for hexagonal symmetry. The maximum value of the Poisson's ratio of dentine (0.534-0.54) is more than 4 times (4.15) higher than the minimum (0.13).
3. The maximum value of the Poisson's ratio of dentine is higher than the upper limit for the Poisson's ratio of isotropic, including restoration materials used in dentistry, which can locally affect the

quality of restorations. In this context, it is suggested that the established elastic anisotropy of the dentin model with crystal hexagonal symmetry is a clinically undesirable factor.

4. A more careful analysis of the elastic anisotropy of the hard tissues of the tooth as a mineral-organic complex and a microinhomogeneous heterophase system is possible with the involvement of the theory of anisotropic media with a crystallographic texture (despite the fact that all the prisms of minerals in the dentine and enamel have the same or similar crystal structure, they differ in the mutual orientation of the crystallographic axes). Further analysis is based on the knowledge of the spatial distribution of the crystallographic axes of individual mineral prisms.
5. The study of the anisotropy of dentine and enamel as an anisotropic inhomogeneous medium is of practical importance in the study of the strength of tooth tissues and the quality of restorations.
6. The values of the Poisson's ratio of enamel are not fundamentally different from those of dentin. They are still quite large and indicate anisotropy of elastic properties of tooth enamel, although to a lesser extent: $\mu_{max} / \mu_{min} = 2.94$.
7. Calculation of effective Poisson's ratios of hard tooth tissues based on elastic constants and compliance coefficients and different averaging methods gives results that are in good agreement with empirical data.

References

[1] Belomestnykh V N, Tesleva E P 2003 Poisson's ratio and the Grüneisen parameter of solids *Bulletin of the Tomsk Polytechnic University* **306(5)** 8-12
[2] Goldstein R V, Gorodtsov A, Lisovenko D S 2011 Variability of elastic properties of hexagonal auxetics *DAN* **441(4)** 468-471
[3] Ivanov G P, Lebedev T A 1964 On the physical sense of the Poisson's ratio *Proceedings of Leningr. polytechnic. in-ta named M.I. Kalinina* **236** 38-46
[4] Lebedenko I Yu, Arutyunov S D, Muslov S A, Useinov A S 2009 Nanohardness and Young's modulus of tooth enamel *Vestnik RUDN, series Medicine* **4** 637-638
[5] Lebedenko I Yu, Arutyunov S D, Muslov S A, Useinov A S 2009-2010, Winter Study of nanomechanical properties of tooth enamel *Chair* **32** 24-28
[6] Lebedenko I Yu, Arutyunov S D, Muslov S A, Bolataev Z B 2010 Elastic anisotropy of the hard tissues of the tooth *Proceedings of the Eurasian Congress of medical physics and engineering "MEDICAL PHYSICS-2010", June 21-25, 2010. Moscow State University* **1** 279-280
[7] Lebedenko I Yu, Arutyunov S D, Useinov A S, Muslov S A, Brandt N N, Apresyan S V Study of the effect of ultralow (4.2 K) temperatures on the nanomechanical properties of tooth enamel 2010 *Proceedings of the Eurasian Congress of medical physics and engineering "MEDICAL PHYSICS-2010", June 21-25, 2010. Moscow State University* **4** Part I 323-324
[8] Nye D 1960 *Physical properties of crystals* M: IL 385 p
[9] Pertsov S S, Stureva G M, Muslov S A, Sinitsyn A A, Korneev A A, Zaitseva N V 2017 *The basics of biomechanics for the dentists* M: MSMSU 115 p
[10] Khyshiktuyev B S, Khyshiktueva N 2004 *Biochemistry of the oral cavity* Chita 84 p
[11] Goldstein R V, GorodtsovV A, Komarov A M, Lisovenko D S 2017 Extreme values of the shear modulus for hexagonal crystals *Scripta Mater.* **140** 55-58 Doi: 10.1016 / j.scriptamat.2017.07.002
[12] Katz J L, Ukraincik K 1971 *J. Biomech.* **4** 221. Doi: 10.1016 / 0021-9290 (71)90007-8
[13] Katz J L 1971 *J. Biomech.* **4** 455. Doi: 10.1016 / 0021-9290 (71)90064-9
[14] Lees S, Davidson C L 1977 The role of collagen in the elastic properties of calcified tissues *J. Biomechanics* **10(8)** 473-86
[15] Lees S et al 1972 Anisotropy in hard dental tissues *J. Biomech.* **5(6)** 557-64
[16] Povolo F et al 1983 Poisson's ratio in zirconium single crystals. *J. Nuclear Materials* **118(1)** 78-82
[17] Sew Meng Chung et al 2004 Measurement of Poisson's ratio of dental composite restorative materials. *Biomaterials* **25** 2455–2460

[18] Shermergor T D 1977 *Theory of elasticity of microinhomogeneous mediums* Moscow Science 399 p

IOP Conf. Series: Materials Science and Engineering **441** (2018) 012033 doi:10.1088/1757-899X/441/1/012033

Digital substation conceptual model for the complex full-scale-model diagnostics

R G Oganyan, N D Narakidze, D V Shaykhutdinov, E V Kirievskiy, S S Kostinskiy

Platov South-Russia State Polytechnic University (NPI), 132, Prosveschenia Str., Novocherkassk, 346428, Russia

E-mail: roman.work18@gmail.com

Abstract. This article describes the development of the detailed conceptual model of digital substation and to a research of opportunities of digital substation on the basis of the comparative characteristic of digital and traditional substations. Accurate accounting of all components of system and their interaction is necessary for efficient development of methodology of modelling of similar complex technical systems. It will give a powerful spur to development of systems of diagnostics, the considered complex technical system. A review of existing projects related to the implementation of the "Digital Substation" Technology has been conducted with a view to generalizing the approaches to the realization of this technology.

1. Introduction

A well-known fact is increases the production of electricity from year to year, which leads to an increase in the load on the energy system. The existing energy infrastructure, in most cases is overloaded, which exacerbates the problem of ensuring the reliability of energy supply. In addition to this, the term of operation of the equipment used to date has long since expired, which indicates a high probability of failure of the equipment operated. From the economic and technical point of view, it is impossible to replace equipment at the same time, which aggravates the problem of the normal functioning of the energy sector.

Based on the above, it is necessary to build new substations, and taking into account the development of modern technologies, the new substations are digital substations. Digital substations have a number of advantages and differences in comparison with traditional substations. According to the November conference "Digital Network", transition to digital substation is inevitable. However, for a detailed study of a new facility, a thorough analysis of its structure and differences from the traditional substation is necessary in order to use them effectively. Besides, there is no most common structure of digital substation for realization of systems of monitoring and diagnostics on their basis. In this article, it is proposed to develop such a structure on the basis of an analysis of existing projects for the implementation of a digital substation.

2. Materials and methods

When developing the conceptual model, the first step is to model the model to identify the properties of the system in accordance with the purpose of modeling - the development of a unified structural description of a digital substation. The next step in the development of the conceptual model is the choice of the level of detail of the model (stratification). The system model is represented as a set of parts (subsystems, elements). The model must have all the elements that implement the execution of all technological processes. In addition, the model includes elements that serve to manage resources and processes, as well as to store the information needed for management (detailing). The next step in creating a conceptual model is its localization, which is carried out by representing the external environment in the form of generators of external influences included in the model as elements.

3. Results and Discussion

Application of optical measuring transformers belongs to the first main difference of digital substations from traditional:

Table 1. Comparative characteristics based on measuring transformers of digital and traditional substations

Digital substation (Optical measuring transformers)	Traditional substation (Electromagnetic measuring transformers)
Ecological compatibility	Oil-filled transformers - explosion and fire hazardous
Increased measurement accuracy, wide frequency range	Inadequate frequency range, which imposes a restriction on the development of relay protection, diagnostics, the error due to imbalance in the load
Absence of ferroresonance	The emergence of ferro-resonance, which leads to improper operation of electric power equipment, its failure, the development of major accidents
Reduced weight and dimensions	High weight and dimensions
Safe current measuring transformers	Danger when opening secondary circuits of the current transformer
Not exposed to electromagnetic interference	Secondary circuits of analog voltage transformers are subject to electromagnetic interference
Changing the transformation ratio	The need for additional equipment to change the ratio of transformation
Elimination of the effect of "overload" on secondary circuits	The presence of the effect of "overload"

However, there is a problem of optical measuring transformers, which is their cost.
The second major difference between the digital substation and the traditional one is the use of the IEC 61850:

Table 2. Comparative characteristics based on the used protocols of digital and traditional substations

Digital substation (IEC 61850)	Traditional substation (MEK 60870-101/103/104, Modbus, DNP3 etc.)
Ethernet with speed 100 Mbit/s	The speed of information exchange based on the use of serial interfaces is not more than 0.0192 Mbit / s
Transmission of instantaneous current and voltage values	Lack of real-time data transmission (delay)
The ability to transfer a large amount of information	Limiting the transfer of large amounts of information
Built-in ability to work with other protocols	Reassignment of signals in firm protocols and vice versa leads to loss of information

IOP Conf. Series: Materials Science and Engineering **441** (2018) 012033 doi:10.1088/1757-899X/441/1/012033

Digital substation (IEC 61850)	Traditional substation (MEK 60870-101/103/104, Modbus, DNP3 etc.)
Horizontal data exchange between devices (GOOSE)	There is no horizontal data exchange
Self-diagnosis of communication channels	Low level of diagnostics of communication channels

The use of digital data channels makes it possible to simplify secondary connections, due to the use of optical fiber. In addition, the transfer of information in digital form is safe, in contrast to analog transmission of data. With these differences in mind, the IEC 61850 protocol certainly has advantages, however, this protocol also has drawbacks [1].

To take into account all components of this complex technical system, a model approach is proposed. To simulate such complex technical systems, the most accurate scheme is needed, allowing to evaluate the interaction of all elements. At the current time, there are many structural diagrams of the digital substation, however, each of the schemes is oriented for a particular object, but the most detailed and general scheme is not presented. For the development of the scheme, it is necessary to analyze existing projects related to the implementation of the digital substation.

One of the pilot projects "Digital substation" is a substation "Sosnovaya" 110/10 kV located in Snezhinsk, Chelyabinsk region. The whole complex "Digital substation" (optical current and voltage transformer and digital information and measuring system (DIMS) was put into trial operation in February 2013. This project is implemented in compliance with the requirements of IEC 61850 [2].

The architecture of the digital substation according to the pilot project is as follows:

- The process level includes a primary current and voltage transducer NXVCT-230
- The bay level is represented by an intelligent electronic device with data reception according to the IEC 61850 protocol 9-2LE: "SysteLLogic Emergency Events Recorder", "SysteLLogic device for measuring parameters" and "SysteLLogic Unified Power Quality Index"
- The station level is represented by a server of digital substation, an automated workstation (AWP), a time synchronization server.

The physical implementation of the station bus and process bus is performed on Ethernet switches. To synchronize time, the SSV-1G time server is used, which provides 1PPS reference signals and supports NTP and SNTP protocols. A three-phase set of fiber-optic current and voltage converters was placed on an open switchgear parallel to the basic traditional measuring equipment. The architecture of this substation is shown in Figure 1.

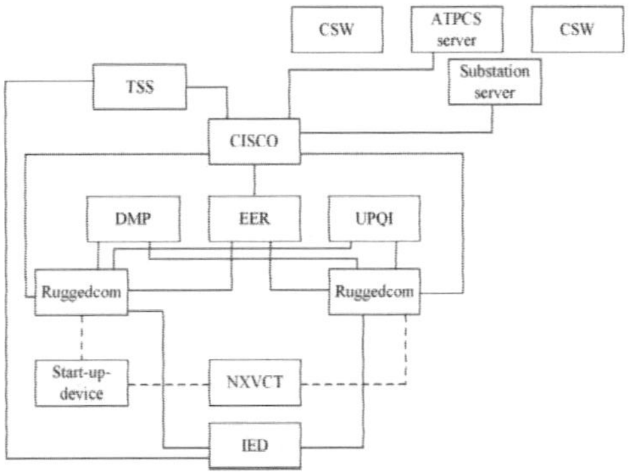

Figure 1. The architecture of the digital substation at 110/10 kV Substation "Sosnovaya"

The secondary measuring equipment is the IED:

a) "SysteLLogic device for measuring parameters (DMP)" - the device for measuring electrical energy parameters. Transmission of operational and archival data to the upper level is carried out according to IEC 61850-8-1 protocol or by intra-company protocol SystelNet in XML format.

IOP Conf. Series: Materials Science and Engineering **441** (2018) 012033 doi:10.1088/1757-899X/441/1/012033

b) "SysteLLogic Unified Power Quality Index" - a device for changing the power quality indicators and their comparison with regulatory values (in accordance with GOST R 54149-2010). Transmission of operational and archive data to the upper level is identical to "SysteLLogic DMP".
c) "SysteLLogic Emergency Events Recorder" - the device provides recording and storage of oscillograms of stationary and transient processes preceding and accompanying emergency modes in electrical networks.
During the commissioning of the DIMS for the digital substation, the mobile verification complex was used. The results of metrological studies showed that the digital substation and traditional telemechanics are in good agreement within the limits of the declared errors.
Another pilot project for the implementation of the technology "Digital substation" is the automation of a 10 kV Distribution Transformer Substation located in Moscow [3]. The contractor proposed a solution for the implementation of the concept by installing on its own a software and hardware complex called iSAS [4].
The elements of the digital substation at the substation "Emelino" in the Sverdlovsk region have been successfully introduced. The integrated system is based on the automated process control system implemented through the modern IEC 61850 data exchange standard [5].
The implementation of the "Digital substation" at the Nizhniy Novgorod HPP of JSC RusHydro [6,7]. In 2013, JSC RusHydro decided to implement the pilot implementation of equipment that operates using the technology of a "digital substation", based on the Nizhny Novgorod HPP. The hydroelectric power station is equipped with 8 hydraulic units with a total installed capacity of 520 MW. Within the framework of the pilot project, a digital automatic control system was implemented in one block № 6 of the generator-transformer of the Nizhny Novgorod hydroelectric power station.
The pilot-industrial sample of the software and hardware complex consists of three levels.

- Process level

The cabinet of field controllers based on the NPT microRTU controller, optical current transformers installed at 110 kV voltage from the higher voltage side of the block transformer, optical flexible current transformers at the main terminals of the 13.8 kV voltage generator, and also generator neutral, electronic voltage transformers, controllers with built-in multiplexer NPT MU board.

- Bay level

Microprocessor relay protection and automation devices, NPT BAY connection controller (9.2), electricity meter.

- Station level

Server hardware and software SCADA NPT Expert.
The process bus and the station bus are made in the form of a single optical ring and are separated using VLAN technology. The discrete signals at the field level are collected by NPT microRTU devices and, in the form of GOOSE messages, are transmitted via the station bus to the level of connections. The digitized measured values are transferred to process bus switches in the form of Sampled Values (SV) messages. The system of single time is built on the basis of the Meinberg M400 / PTP exact time server, which provides high-precision synchronization of devices over the LAN.
The important result was the provision of joint work of intelligent electronic devices from 7 manufacturers (Energoprom Automation, PROFOTEK, EKRA, Brestler, ABB, Alstom, Prosoft Systems). At the moment, according to the November conference "Digital Network", work on this facility continues, with the task of further expanding the level of accession and cybersecurity of the substation.
In addition, at present, work is underway on digital substation projects implemented on energy clusters "Elgaugol" [8] and "Vanino" [9].
August 15, 2017 was put into operation substation of the Rostov region " Sportivnaya" [10]. The automated control system allows employees to monitor and control technological and operational parameters with their visualization, to receive information about the state of the substation's electrical circuits in real time, to signal the output of process parameters for permissible norms. All the equipment used at the substation "Sportivnaya" is equipped with a full amount of locks and

protections, including locks from erroneous actions, modern design solutions provide a high degree of safety for maintenance personnel. Operative control of the substation can be carried out in several ways - traditional dispatching control, remote control of switching devices from the workplace of the dispatcher, remote telecontrol from the dispatching office of the production department. The software complex conducts continuous diagnostics of the temperature of the active part, the state of isolation of high-voltage bushings, the parameters of transformer oil, and the content of dissolved gases.
June 4, 2018 opened the first in the Moscow region digital substation 110 kV, "Medvedevskaya" [11]. Using the latest digital technology, it allows you to carry out online monitoring of electrical equipment. All probable malfunctions will be eliminated upon their occurrence. Measurement of electrical and technological parameters of the primary equipment, control of its condition and control is carried out using digital technologies in accordance with the international standard IEC 61850. Two power transformers with a capacity of 80 MVA are installed at the substation; Complete switchgear with gas-insulated insulation (SF6) 110 kV; 4-section switchgear 20 kV with vacuum switches, designed for 20 linear cells; complex relay protection and automation and automated process control system. The loss reduction for the new substation for the first time in Russia will exceed 30%.
There are also polygons "Digital substation", namely:

- JSC "Scientific and Technical Center of FGC UES" Center "Experimental polygon Digital substation" [12]
- JSC "ChEAZ" (Test Center LLC) - Testing area of the software and hardware complex of the automated process control system in the Information Storm mode [13]
- JSC "NTC UES" a test stand of JSC "NTC UES" (LLC “Energoprom Automatization”) [14]
- "NRU MEI" "Laboratory for the study of the interoperability of devices operating in accordance with the IEC 61850 standard" [15]

4. Conclusion

Combining information from various structural diagrams, we obtain the scheme of a digital substation, which is depicted in Figure 2.

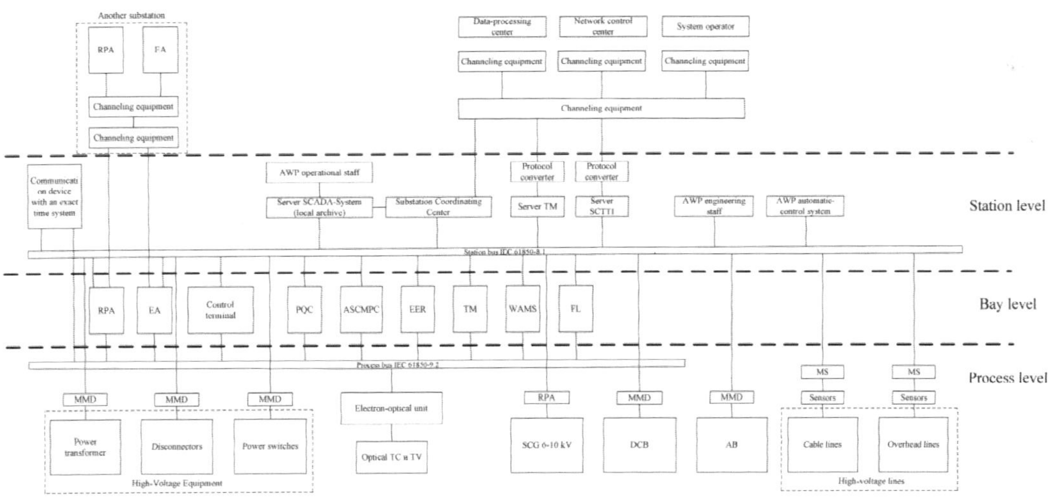

Figure 2. Structural diagram of the digital substation

The notation in Figure 3: MMD – man-machine device; DCB – direct-current board; AB – auxiliaries board; MS – monitoring system; RPA – relay protection and automation; EA – emergency automation; PQC – power quality control; ASCMPC – automatic system for commercial measurement of power consumption; EER – emergency event recorder; TM – telemechanics; WAMS – wide-area measurement system; FL – fault localization; AWP – automated workstation; SCTTI – system for the collection and transmission of technological information; SCG – switchgear and control gear

From the above described it follows that at present there is a mass introduction and approbation of solutions of the class "Digital substation" based on the standards of the IEC 61850 series. The application of the technology of the "Digital substation" should allow in the future to significantly reduce the costs for the design, commissioning, operation and maintenance of power facilities. However, at the moment, efforts are focused on implementing this technology, forgetting about the diagnosis of such a complex technical system. Repeatedly the energy community stressed that it is impossible to transfer the methodology of diagnostics from a traditional substation to a digital substation, which is an obstacle to the introduction of innovative technology. It is necessary to improve the methodological basis on the basis of taking advantage of the digital substation, and given the complexity of the system under consideration, the model approach is the best option. When developing the model, the following stages of the description are distinguished: conceptual, mathematical, program. In this article, the first step is the development of a conceptual model. Based on this model, together with the use of substation data, a mathematical model can be developed.
Thus, for full-scale-model diagnostics implementation of the "Digital substation" technology in the Russian energy sector, it is necessary to overcome the obstacle associated with the development of diagnostic bases. The application of fundamentally new approaches to the evaluation of the state of the equipment of digital substations will allow accelerating the introduction of innovative technology and will also solve one of the problems of IEC 61850 - the development of full-fledged diagnostic tools.

5. Acknowledgments
The reported study was funded by RFBR according to the research project № 18-38-00950.

References
[1] IEC-61850 Available at: http://cons-systems.ru/m-k-61850 (In Russ.)
[2] Rykovanov S N, Ukhov V I, Slabospitssky S R, Kovtsova I O 2014 *J. Releyshchik* **1** 32-39
[3] Implementation of the "Digital substation" technology in a pilot project for automation of a 10 kV DTS. Available at: http://digitalsubstation.ru/releyshik/2016/02/18/realizatsiya-tehnologii-tsifrovaya-podstantsiya-v-pilotnom-proekte-po-avtomatizatsii-rtp-10-kv/ (In Russ.)
[4] integrated Substation Automation System (iSAS). Available at: http://mail.lysis.su/index.php/ru/katalog-produktsii/8-katalog-produktsii/40-obshchee-opisanie-integrirovannoj-innovatsionnoj-sistema-zashchity-i-upravleniya-elektricheskoj-podstantsiej-isas (In Russ.)
[5] In the Urals region, the launch of a new generation "Emelino" substation 500 kV. Available at: http://newsdesk.pcmag.ru/node/13393 (In Russ.)
[6] Innovative complex based on the technology of the "Digital substation" at the operating hydroelectric power station. Available at: http://energy-polis.ru/energo/2774-innovacionnyy-kompleks-na-baze-tehnologii-cifrovoy-podstancii-na-deystvuyuschey-ges.html (In Russ.)
[7] Drozdova T V, Elov N E and Morozov A P 2016 *J. Energy of a Uniform Network* **3** 54-61
[8] Yanchenko A S, Kozlov A N 2014 *News of the Amur State University: Natural. and econ. Scien.* **67** 80-84
[9] The energy cluster "Vanino". Available at: http://www.ntc-power.ru/ies-aas/pilot_projects/energoklaster_vanino/ (In Russ.)
[10] A new substation "Sportivnaya" was put into operation in Rostov-on-Don. Available at: https://big-rostov.ru/v-rostove-na-donu-vvedena-v-ekspluataciyu-novaya-podstanciya-sportivnaya/ (In Russ.)
[11] In Skolkovo opened digital substation 110 kV Medvedevsky. Available at: http://www.en-prom.ru/2018/06/05/v-skolkovo-otkry-ta-tsifrovaya-ps-110-kv-medvedevskaya/ (In Russ.)

[12] Department of technology implementation "Digital substation". Available at: http://www.ntc-power.ru/field_of_activity/research_and_development/center_experienced_a_polygon_digital_substation/ (In Russ.)
[13] Testing ground for HSS ATPCS in "Information storm" mode. Available at: http://digitalsubstation.ru/blog/2013/07/01/poligon-ispytanijj-ptk-asutp-v-rezhime-informacionnyjj-shtorm/ (In Russ.)
[14] Test stand of relay protection devices and automation systems. Available at: https://www.ntcees.ru/departments/sections/test_stand.php (In Russ.)
[15] Testing relay protection and automation devices for interoperability according to IEC 61850. Available at: http://www.combienergy.ru/stat1202.html (In Russ.)

Study of technological factors' effects on deforming of thin-wall tubes deformation forces in short-term creep mode

A A Pasynkov [1], **S N Larin** [1]

[1] Tula State University, 92, Lenina ave., Tula, 300012, Russia

E-mail: mpf-tula@mail.ru

Abstract. In the modern world, slow isothermal deformation, which leads to a significant increase of the deformed metal's plasticity, have become more widely used in the manufacturing of ax-isymmetric products made of hard-to-deform light alloys. Despite a large number of works devoted to theoretical and experimental research of distribution processes, the problems of the forming theory have not been fully discussed at the present time. This paper presents the results of studies of the pipe workpiece expansion under isothermal conditions, performed in the QFORM 3D software. The workpiece was made of high-strength alloys AMg6 and BT6. It was assumed that the behavior of the workpiece material during deformation corresponds to the following equation, which characterizes the transition of the workpiece material from viscous to viscoplastic state. The study reveals the influence of a wide range of technological factors on the stress-strain state and operational force parameters. The results can be used to create scientifically based methodologies for the design of high technology processes for manufacturing of axisymmetric geometrically-complex parts using isothermal deformation methods.

1. Introduction

In the modern world, slow isothermal deformation, which leads to a significant increase of the deformed metal's plasticity, have become more widely used in the manufacturing of axisymmetric products from hard-to-deform light alloys. When developing the technology of isothermal expansion, it is necessary to take into account that the deformation values become extremely high, and, accordingly, the mechanical characteristics of the deformed materials change significantly. In connection with the foregoing, there is a need to apply the plasticity and creep theory to analyze isothermal forging and expansion in particular.

Despite a large number of works devoted to theoretical and experimental studies of crimping and expansion, the issues of the forming theory for anisotropic materials with cylindrical anisotropy of mechanical properties have not been practically discussed at present. A wide range of problem related to the design of technological processes of crimping and expansion of tube workpieces and the search for rational conditions for managing these processes, ensuring the manufacturing of products of specified quality, has not been resolved.

2. Materials and methods

The paper presents the results of studies of tube workpiece expansion under isothermal conditions, performed in the QFORM 3D software. A complex based on the finite element method was used due to the analysis of the data obtained in the course of theoretical studies based on the upper boundary plasticity theorem and the results of experiments. The scheme of the process is shown in Fig. 1. The investigations were carried out with the following tool and equipment parameters: the radius of the conical sections' transformation into the cylindrical ones is 1 ... 7 mm; the taper angle of the tool is 10 ... 40 degrees; workpiece thickness: 1 ... 3 mm; friction coefficient: 0.1 ... 0.8; deformation rate: 0.01 ... 1 mm / s, expansion coefficient is 1 ... 1.5.

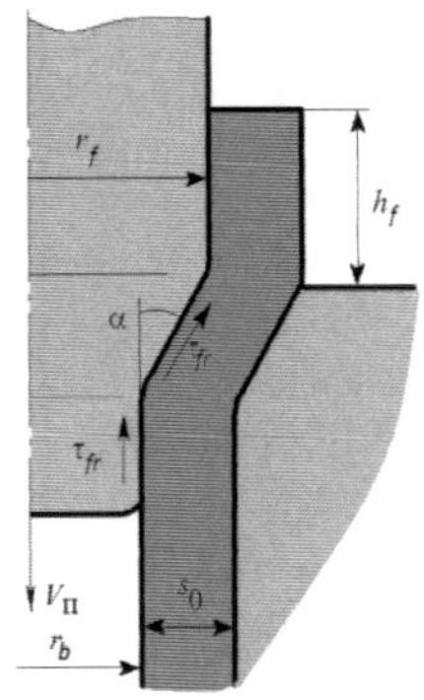

Figure 1. Expansion scheme

3. Results and Discussion

The workpiece was made of high-strength alloys AMg6 and BT6. Theoretical studies, carried out according to the model developed earlier, are based on the upper boundary plasticity theorem, the solution of which will be based on energy calculation methods [1-15]. It was assumed that the behavior of the workpiece material during deformation corresponds to the following equation, which characterizes the transition of the workpiece material from viscous to viscoplastic state

$$\sigma_e = A\varepsilon_e^m \xi_e^n, \tag{1}$$

where σ_e, ε_e, ξ_e - equivalent stress, strain and deformation rate; A, m, n - material properties. For our case, the following energy inequality is valid:

$$N_{\text{в}} \le N_{\text{вн}}, \tag{2}$$

where $N_{\text{в}} = qSV_n$; $N_{\text{in}} = N_f + N_{\text{fr}}$; $N_f = \int\limits_{V_{\text{f}}} \sigma_e \xi_e dV_{\text{f}}$; $N_{\text{тр}} = \int\limits_{S_{fr}} \tau_{fr} V_k dS_{fr}$.

In this inequality q is the forging pressure; τ_{fr} is the stress arising from friction between the tool and the workpiece; V_n - speed of the working tool; V_k - workpiece speed; S, S_{fr}, W - the cross-sectional area of the tube workpiece, the area at the point of contact of the part and the tool, and the volume of the deformation zone.

The expansion pressure is expressed by the formula [1-3]

$$q \le \frac{Ak^{m+n}}{2\sin\alpha} r_n^{\left(\frac{R_z n-1}{1+R_z}\right)} V_{\text{п}}^n \left(\sqrt{\frac{2(2+R_z)}{3(1+R_z)}} + \mu \operatorname{ctg}\alpha\right) \int\limits_{r_0}^{r_1} r^{1-(1+n)\frac{1+2R_z}{1+R_z}} \left(\ln K_p\right)^m dr_{\text{т}}. \tag{3}$$

On the basis of the formulas presented earlier, the optimum shape and dimensions of the tool were revealed, providing the best quality properties of the produced parts and the minimum deformation force. The calculations were made with the following geometric dimensions of the workpieces: $r_b = 10...20$ mm; $r_f = 20...30$ mm; $s_0 = 1...3$ mm; $h_{\text{f}} = 5...10$ mm; $\alpha = 15...45^\circ$.

Figure 2 shows the graphical dependencies between the speed of the deforming tool and the relative pressure value of expansion with different friction values, which characterize the lubricants used in the

production. Their analysis shows that the most significant effect on the distribution force has the change in the punch speed in the interval from 0.01 to 0.1 mm/s. Then the pressure increases, but not so intensively. The increase of the friction coefficient gives a uniform and linear increase in pressure during the expansion, i.e. with the increase of friction from 0.1 to 0.4, the pressure increases by 1.5.
The results of the studies, based on mathematical and computer modeling, and on the experimental data were compared. According to the results it can be seen that the results obtained in the computer modeling have a small difference (no more than 10%) from the results obtained in the course of mathematical modeling and experiments and qualitatively repeat the patterns of influence of various factors on force. This allows us to carry out more complete studies of this process, depending on a greater number of factors, using computer modeling.

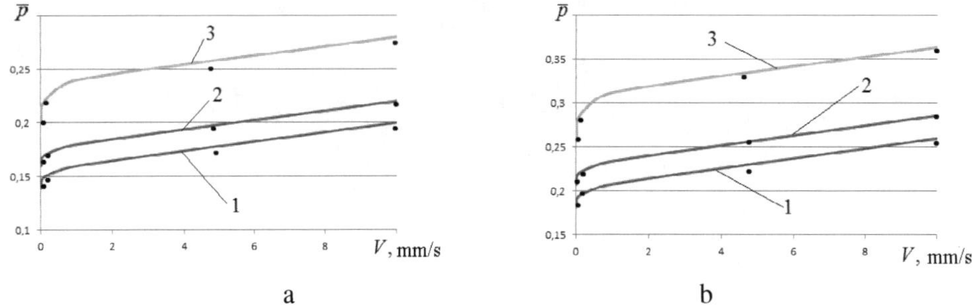

Figure 2. Dependence of change in $\overline{p}$ from V during expansion of alloys AMg6 (a) and BT6C (b): 1 - $\mu = 0{,}1$; 2 - $\mu = 0{,}2$; 3 - $\mu = 0{,}4$

Figure 3 shows the dependence of the isothermal expansion force on the taper angle of the tool and friction on the tool and workpiece.

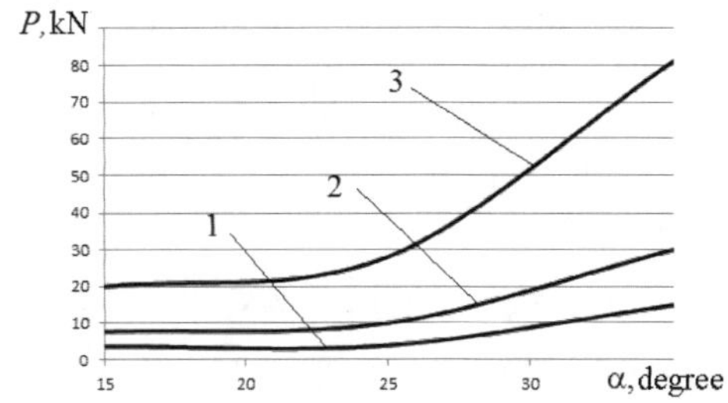

Figure 3. Dependence of the expansion force on the taper angle of the tool: 1 - $\mu = 0{,}1$; 2 - $\mu = 0{,}4$; 3 - $\mu = 0{,}8$

This dependence demonstrates that with increasing tool taper angle and friction the distribution force is also increasing. The greatest growth of strength occurs after increasing the taper angle of the tool higher than 25 degrees. The increase in the friction coefficient linearly affects the growth of the distribution force. Analysis of this dependence makes it possible to establish rational parameters of friction and the geometry of the expansion tool, i.e. the angle at a single transition should not exceed 30 degrees when friction is the lowest.
Figure 4 shows the dependence of isothermal expansion force from the isothermal expansion coefficient and the radius of tool curvature for the zones of transition from the cylindrical into tapered sections.

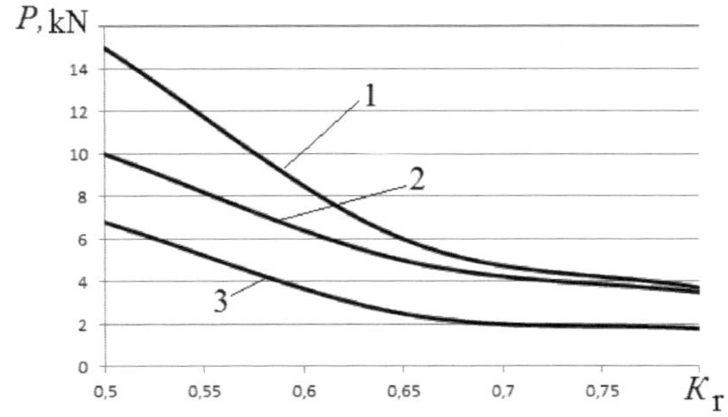

Figure 4. The dependence of the force from the expansion coefficient: 1 - $R=1$ mm; 2 - $R=3$ mm; 3 - $R=7$ mm

This dependence demonstrates that with the increase in the expansion coefficient and the tool curvature radius, the expansion force decreases. The greatest, more than double, increase in strength occurs at the value of the expansion ratio of 0.65 and below. The increase in the radius of the transition zone from cylindrical to tapered sections does not affect the force value so intensively, but at radii less than 3 mm there is a noticeable increase in the force value. Thus, when the expansion coefficient equals 0.8, the growth of cylindrical-tapered transition radius from 1 mm to 3 mm leads only reduction force by 15%, while at the same time when the expansion coefficient is 0.5, the same growth radius reduces the force by 60%.

Figure 5 shows the dependence of isothermal expansion force from the relative thickness of the workpiece and the taper angle of the tool.

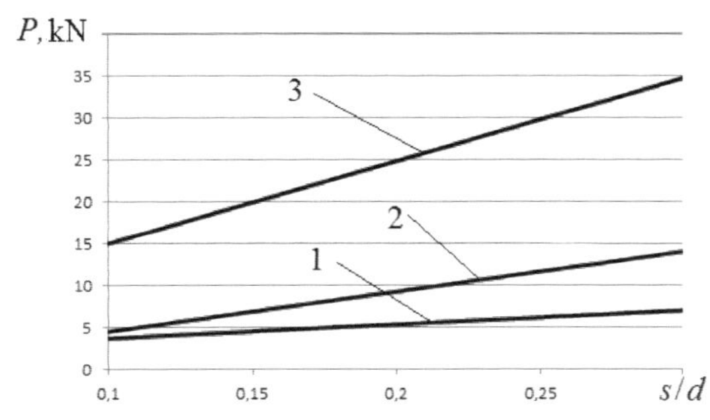

Figure 5. The dependence of the expansion force from the relative thickness of the workpiece: 1 - $\alpha=15°$; 2 - $\alpha=25°$; 3 - $\alpha=35°$

This dependence demonstrates that with the increase of tool's taper angle and relative thickness of the workpiece, the expansion force increases. It should be noted that the increase in the thickness of the workpiece in parallel with the increase of the taper angle of the punch leads to a more and more intensive increase in the force, that is, with simultaneous increase in both the thickness and the angle of expansion. With the given parameters of the taper angle growth of more than 25 degrees, the force increases by more than three times.

4. Conclusion

This paper presents the research of expansion of tube ends made of high-strength materials in the short-term creep mode. Experiments on the expansion and upsetting of the tube workpiece end under isothermal conditions were carried out using the QFORM 3d software. The results reveal the influence of a wide range of technological factors on the stress-deformation state and the force parameters of the operations under research.

By analyzing and generalizing the obtained theoretical results, we formulated the recommendations for the design of manufacturing technology for "adapter" parts. The results can be used to create scientifically valid, innovative, high-tech and competitive processes of the next generation for manufacturing of axisymmetric geometrically-complex parts by means of isothermal deformation.

5. Acknowledgments

The results of the research project are published with the financial support of Tula State University within the framework of the scientific project № _________.

References

[1] Unksov E P Theory of plastic deformations of metals 1983 *M.: Mechanical Engineering* 598.

[2] Romanov K I Mechanics of hot metal forming 1993 *Moscow: Mashinostroenie* 240.

[3] Springer P, Prahl U Characterisation of mechanical behavior of 18CrNiMo7-6 steel with and without nb under warm forging conditions through processing maps analysis 2016 *Journal of Materials Processing Technology* **Volume 237** 216-234

[4] Netoa D M, Martins M P, Cunha M, Alves L M, Oliveira C, Laurent H, Menezes L F Thermo-mechanical finite element analysis of the AA5086 alloy under warm forming conditions 2017 *International Journal of Solids and Structures Available* online.

[5] Zhengyang Cai, Min Wan, Zhigang Liu, Xiangdong Wu, Bolin Ma, Cheng Cheng Thermal-mechanical behaviors of dual-phase steel sheet under warm-forming conditions 2017 *International Journal of Mechanical Sciences* **Volume 126** 79-94.

[6] Verena Kräusel, Peter Birnbaum, Andreas Kunke, Rafael Wertheim Metastable material conditions for forming of sheet metal parts combined with thermomechanical treatment 2016 *CIRP Annals - Manufacturing Technology* **Volume 65 Issue 1** 301-304.

[7] Aksenov S A, Chumachenko E N, Kolesnikov A V, Osipov S A Determination of optimal gas forming conditions from free bulging tests at constant pressure 2015 J*ournal of Materials Processing Technology* **Volume 217** 158-164.

[8] Kyung-Hun Leea, Byung-Min Kim Advanced feasible forming condition for reducing ring spreads in radial–axial ring rolling 2013 *International Journal of Mechanical Sciences* **Volume 76**, 21-32.

[9] Malinin N N Creep in metal processing 1986 *M.: Mechanical Engineering* 216.

[10] Yakovlev S S, Yakovlev S P, Chudin V N, Tregubov V I, Chernyaev A V Isothermal shaping of anisotropic materials by a rigid tool in the regime of short - term creep 2009 *M.: Mechanical Engineering* 412.

IOP Conf. Series: Materials Science and Engineering **441** (2018) 012035 doi:10.1088/1757-899X/441/1/012035

Magnetoelectric sensor for measuring weak magnetic biological fields

V.N. Lobekin, R.V. Petrov, M.I. Bichurin, A.V. Rebinok, R.A. Sulimanov

[1]Novgorod State University, B.St.-Peterburgskaya str., 41, 173003, Veliky Novgorod, Russia

E-mail: roman.petrov@novsu.ru

Abstract. The paper is devoted to the sensors used for measuring biological magnetic field. The possibilities and sensitivities of magnetoelectric sensors for determining weak magnetic fields were investigated. The importance of human biological magnetic fields measurement for disease diagnosis and treatment was discussed. Recommendations on the use of magnetoelectric sensors for studying the weak biological magnetic fields of living organisms were given.

1. Introduction

To date, high-sensitivity magnetic field sensors are widely used in various areas of human activity, for example, in the automotive industry, space vehicles, archaeological research, robotics, measurement technology, and medical diagnostics. Classical old methods are constantly being improved, mainly with the use of new materials and technologies, such as miniaturization. These new sensors bring revolutionary changes in technology and, in some cases, demonstrate new opportunities. Very promising are hybrids of various effects, for example, magneto-optical, magnetoacoustic or magnetoelectric (ME) effect. In the ME effect, the magnetization can be caused by an electric field or the reverse electric polarization can be caused by a magnetic field.

To measure weak magnetic fields generated by living organisms, high sensitivity is necessary. Generation of weak magnetic fields can be caused by constant currents or magnetic particles that are very effectively used in biomedicine, therapy and diagnostics of a number of serious diseases such as coronary heart disease, cerebral ischemia, brain tumor, epilepsy, gastrointestinal tract diseases that makes measurement and diagnostics of biological magnetic fields very relevant for this study.

Another topic for discussion is about the methods for spatial matching of magnetic localization data (which, naturally, is carried out with respect to the measurement system) with anatomical structures of the heart. The advantages and disadvantages of various methods of topographic mapping, applied at present, are analyzed in detail. These are magnetic resonance imaging (MRI), computed tomography, X-ray tomography and echocardiography. It is concluded that, despite the high cost of the equipment, it is more appropriate to still use MRI and, possibly, a CT scan. Other methods do not provide enough detailed information, but it is hoped that the use of modern software that allows 3D reconstruction of echocardiography data will make this cheaper mapping method sufficiently accurate [1].

The aim of this work is to study the influence of weak biological magnetic fields on living organisms, in particular humans, and to review and compare the characteristics of magnetic field sensors.

2. Human magnetic fields and their importance for the human life

Studies of the biological effects of permanent magnetic fields are concentrated on large fields of field level in MRI devices (magnetic resonance imaging), usually several Tesla (several tens of thousands of Gauss). Unfortunately, studies of the effect of fields typical for magnetic therapy products, most of which are limited to a few hundred Gauss even on the surface of a magnet, are very few. Nevertheless, the main mechanisms of the influence of magnetic fields on biological organisms that allow developing magnetic therapy are already known. These mechanisms include: 1) increased blood flow as a result of increased oxygen content (both of which are the basis of the body's ability to self-repair); 2) a change in the rate of migration of calcium ions, as a result of which, on the one hand, calcium enters the broken bone faster and it fuses faster, and on the other hand, calcium is washed out of the affected with arthritis joint more quickly; 3) change in the acid-base balance (pH) of various fluids in the human body and animals (imbalance is often a consequence of the disease); 4) a change in the production of hormones by the endocrine glands; 5) change in enzyme activity and rates of various biochemical processes, 6) change in blood viscosity.

Magnetic fields of a living organism are caused by three causes. 1) Ionic points arising from the electrical activity of cell membranes (mainly muscle and nerve cells). 2) The smallest ferromagnetic particles, trapped or specially introduced into the body. These two sources create their own magnetic fields. 3) When an external magnetic field is applied, the inhomogeneities of the magnetic susceptibility of various organs appear and distort the superimposed external field [2].

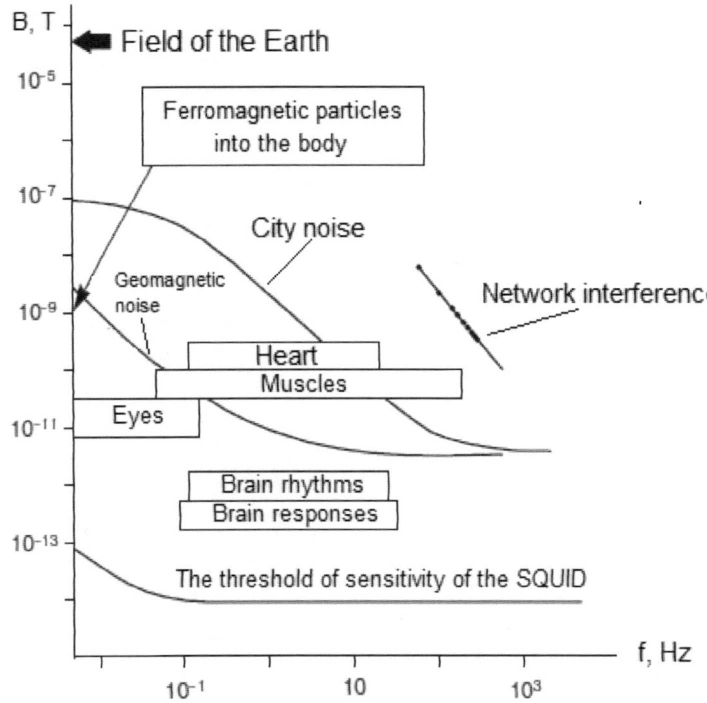

Figure 1. Place of human biomagnetic signals in the magnetic field scale. Specific interference levels and frequency ranges of signals are indicated [1].

Registration of these potentials is used in research and clinical practice: electrocardiography, electroencephalography, and the like. Also there are magnetic analogues, i.e. magnetocardiography and magnetic-encephalography, recording the changes in time of the magnetic component of the same electrical processes in the body.

Magnetography is used to study the heart, fetus, skeletal muscles, eye, retina, brain, magnetic contamination of the lungs, permanent currents in human skin, etc.

Magnetic fields are rapidly weakened when removed from the source of activity, since they are a consequence of relatively strong currents in the operating body, while surface potentials are determined by weaker and "smeared" currents in the skin. Therefore, magnetography is more convenient for the precise determination (localization) of the site of bioelectric activity [3].

It is possible to single out the main directions in the clinical application of the MCG:

• MCG of healthy volunteers;

• MCG conduction system of the heart;

• localization of additional ways of conducting irriotmogenic foci and evaluating the effectiveness of their radiofrequency ablation;

• Diagnosis of cardiac hypertrophy;

• Myocardial ischemia and its viability (especially in patients with a small or unchanged ECG, evaluation of the effectiveness of antianginal therapy, etc.);

• diagnosis of cicatricial changes;

• study of repolarization (prolongation and dispersion of QT), stratification of the risk of arrhythmias and evaluation of the effectiveness of antiarrhythmic therapy;

• monitoring of rejection after cardiac transplantation.

To identify patients with a high risk of sudden cardiac death, there is no single diagnostic method [4]. As effective diagnostic markers of the approaching structural adjustment of the electrophysiological properties of the myocardium, it is possible to use the characteristics of low-amplitude oscillations of the detected parameters, which, when approaching the points of loss of structural stability, begin to change earlier than it appears in the value of the average values of the parameters recorded. That is, the sensitivity of the MCG to local currents, the possibility of detecting small changes should allow us to evaluate the electrical homogeneity of the myocardium and to separate the pathological zones of the myocardium with fast and slow conductivity.

The advantage of magnetocardiography remains in its absolute non-invasiveness (passivity), that is, the absence of any energy that is induced on the body of the subject. However, recently there are publications that the successes related to the solution of the inverse problem of electrocardiology in terms of epicardial potentials and modern methods of ECG analysis on the basis of appropriate biophysical models allow us to approach a sufficiently high accuracy in the diagnosis of myocardial ischemia. Here the registration of multiple surface ECGs is supplemented with geometric information about the surfaces of the heart and the body and their mutual arrangement, obtained by computed tomography. But in magnetocardiography, the same task of anatomical binding arises in the solution of the inverse problem of magnetocardiology [5].

2.1 Importance of the magnetic field

Reducing the level of the external magnetic field leads to a violation of the magnetic field in the circulatory system, as a result of which blood circulation is disturbed, oxygen and nutrients are transported to organs and tissues, which can eventually lead to the development of the disease. Thus, the insufficient level of external magnetic effect in terms of the degree of harm caused to the body can fully compete with the deficiency of minerals and vitamins.

Under the influence of magnetic fields there is an increase in the permeability of blood vessels and epithelial tissues, so that it is possible to accelerate the absorption of edemas and the dissolution of medicinal substances. This effect is the basis of magnetotherapy and is widely used for various types of injuries, injuries and their consequences.

2.2 Effect of magnetic field on various systems of the body

Weak magnetic fields of technogenic and natural origin affect the circadian rhythms and physiological functions of a person, which ultimately affects the general state. In natural conditions, a person is exposed only to natural electromagnetic fields, which he tuned throughout the entire evolution process on planet Earth. When artificial sources of magnetic, electric and electromagnetic fields intervene in this process of interaction, a synchronization violation occurs. On the average, the Earth's magnetic field varies with an average frequency of 8 Hz, although this value can fluctuate significantly. Our body is already tuned to perceive this frequency and considers it a natural background. Our cells are thus sensitive to a given frequency of magnetic field exposure.

Various scientific studies have shown that a low-frequency (2-8 Hz) electromagnetic field affects the rate at which a person responds to an optical signal. The magnetic field in the range of 5-10 Hz changes the time of the human brain reaction to many other external influences.

Studies have shown that when a short-term alternating magnetic field with a frequency of 0.01-5 Hz is applied to the human body, the character of the human brain's electroencephalogram changes abruptly. Under the influence of weak variable magnetic fields, the person's pulse rate increases, the head starts to hurt, the state of health worsens and weakness in the entire body is felt. In this case, there is a strong change in brain electrical activity [6].

3. Magnetic field sensors used in medicine

Various converters of non-electrical quantities in electrical have firmly taken their place in many areas of human knowledge, and even more so in medicine. It is difficult to imagine a modern doctor dealing with the diagnosis of various diseases and their treatment, not based on a huge number of achievements of such sciences as radio electronics, microelectronics, metrology, materials science. And although the sensors are one of the slowest growing areas of medical electronics, and indeed of all electronics in general, the vast majority of diagnostic and therapeutic devices and systems directly or indirectly contain a variety of different transducers and electrodes, without which, sometimes the operation of this system is unthinkable.

To measure weak constant and slowly changing magnetic fields, magnetoresistive transducers, nuclear magnetic resonance converters, SQUIDs and ferroprobe converters are currently used. Due to the compactness, low cost, wide operating temperature range, low noise level, the possibility of measuring the direction of the magnetic field, the flux-gate transducers and magnetometers based on them are the most preferable.

Magnetic fields are usually divided into superstrong (over 100 T), strong (from 4 to 100 T), medium (from 0.05 to 4 T), and weak (less than 0.05 T). In medical diagnostic systems, high sensitivity is required, since many living organisms generate weak but measurable magnetic fields of $B \leq 10$ nT. Modern medicine practices numerous passive (for example, bone replacement) and active implants (apparatuses for auxiliary circulation, artificial heart, various stimulants, etc.). Non-invasive testing of functional characteristics, resources and other properties is possible using high-sensitivity magnetic field sensors.

The magnetic field sensor is the main element of any magnetometer and is designed to convert the magnetic induction B into an electrical signal, most often into a voltage U. When creating magnetic field sensors, various physical effects are used, for example, Hall, Gauss, Suhl effect, and others.

The main characteristics of magnetic field sensors can be identified as follows: 1) measuring range, 2) linearity, 3) hysteresis, 4) inaccuracy in the conversion of magnetic induction (including temperature dependence), 5) displacement, 6) long-term stability, 7) noise, 8) directional pattern, 9) sensitivity to the transverse field; 10) frequency range of measurements, 11) geometric dimensions of the sensor, 12) power consumption; 13) heat generation, 14) operating temperature range.

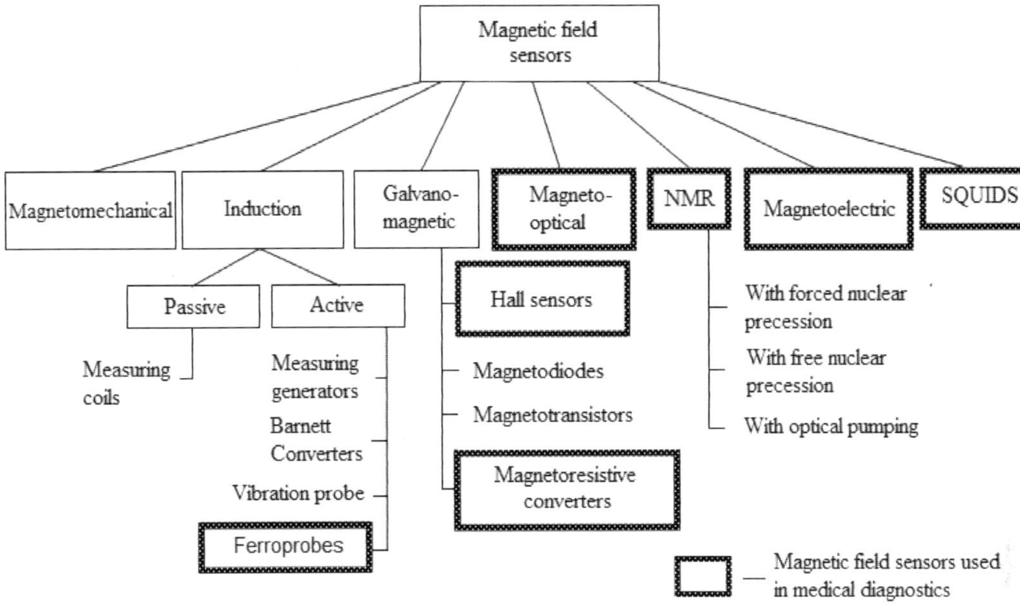

Figure 2. Types of magnetic field sensors.

To measure weak magnetic fields over a wide temperature range, magnetoresistive sensors are most widely used. The principle of operation of magnetoresistive sensors is based on the effect of changing the electrical resistance of a material in a magnetic field. Magnetoresistive sensors can be manufactured using integrated technologies, which significantly reduces their size and cost.

The temperature response of the magnetoresistors depends on the material of the doping impurities. As a rule, with increasing temperature, the resistance decreases, and after applying the magnetic field, the decrease occurs faster. Magnetoresistors have low noise, are not affected by surface effects and is characterized by very small aging.

Depending on the physical effect and the material, the following magnetoresistive sensors are distinguished:

• magnetoresistors based on the Gauss effect;
• anisotropic magnetoresistive sensors (AMP);
• giant magnetoresistive sensors (GMR);
• Spin-tunnel magnetoresistive sensors (STMS). [7]

SQUID (Superconducting Quantum Interference Device) is a supersensitive magnetometer used to measure very weak magnetic fields. SQUID magnetometers have a record high sensitivity, reaching 5×10^{-33} J / Hz (sensitivity to the magnetic field - 10^{-13} T). For long-term measurements of averaged values over several days, sensitivity values of 5×10^{-18} T can be achieved [8].

The Hall sensor (position sensor) is a magnetic field sensor. The operation of the device is based on the Hall effect. This effect is based on the following principle: if a certain conductor is placed with a direct current in a magnetic field, then a transverse potential difference (Hall voltage) appears in such a conductor.

Nuclear magnetic resonance (NMR) tomography, better known as magnetic resonance imaging (MRI), is an important medical diagnostic tool used to study the functions and structure of the human body. It allows obtaining detailed images of any organ, especially soft tissue, in all possible planes. It is used in areas of cardiovascular, neurological, musculoskeletal and oncologic imaging.

Table 1. Types of sensors for quasistationary magnetic fields.

Sensor type	Hall Effect	AMC, ГMC	Ferroprobe	SQUID	ME sensors
Current/Field	Field	Field	Current	Current	Field
Max sensitivity	1000 nT	0,5 nT	100 pT	5 fT	5pT
Frequency range	30 kHz	100 MHz	10 kHz	10 MHz	10 MHz
Strengths	Low cost, linearity of characteristics, dynamic range of measurements ≤70 dB	Low cost, sensitivity, dynamic range of measurements ≤100 dB	High sensitivity, dynamic range of measurements ≤70 dB	Very high sensitivity, wide range of frequencies, dynamic range of measurements ≥120 dB	High sensitivity, small size, dynamic range of measurements ≤70 dB
Weaknesses	Low sensitivity	Low dynamic range, 1/f noise	Large size	Low operating temperature (4 K), fragile, high cost	Complexity of technology and calculations

Researches of the influence of a strong static magnetic field on human blood were carried out numerous times using methods such as nuclear magnetic resonance (NMR), magnetic tomography (MRI). As early as 1936, the scientists Powling and Koryel reported the diamagnetic susceptibility of oxyhemoglobin (ie, oxygen-enriched blood) and the paramagnetic susceptibility of deoxyhemoglobin (ie, oxygen-poor blood). In the course of these studies it was possible to estimate, in particular, the magnitude of the effective magnetic moments of the complex Fe + 2, which is part of the hemoglobin of human blood. 10 years ago (1993) Higashi and co-authors studied the orientation of normal red blood cells in a strong permanent magnetic field with a maximum value of up to 8 Tesla. It was found that the red blood cells are oriented in such a way that the plane of their disc is parallel to the direction of the applied field. Finally, in 1997, American researchers Hike and Chen from the University of Florida studied various aspects of the effect of strong permanent magnetic fields on human blood, namely: on magnetic susceptibility, magnetomotive force and viscosity [9].

The magnetic susceptibility of blood was measured using a SQUID magnetometer. It was found that the blood behaves as a diamagnetic liquid when it is enriched in oxygen (in the arteries) and as a paramagnetic material when it is deoxygenated (in the veins). Figures 3 and 4 show the results of measuring the magnetization of blood in the arteries (1) and veins.

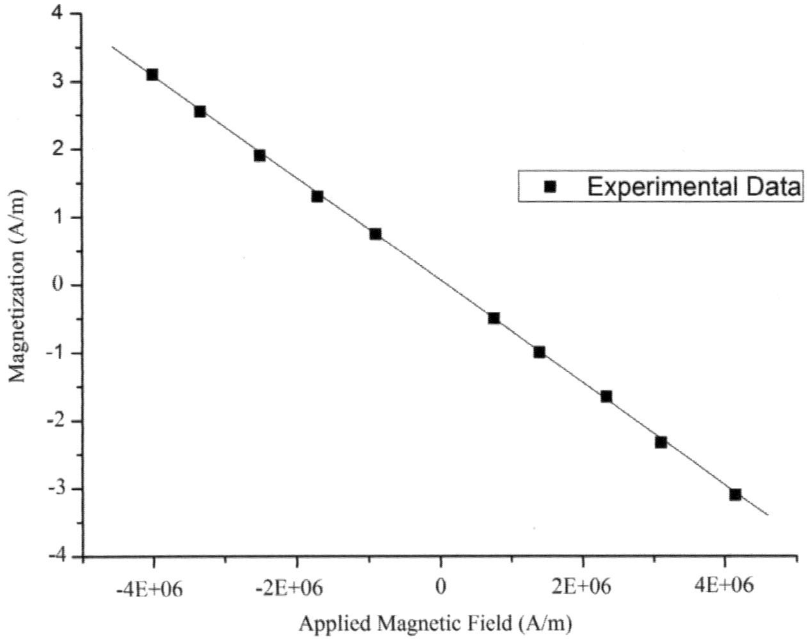

Figure 3. Oxygen-enriched blood magnetization.

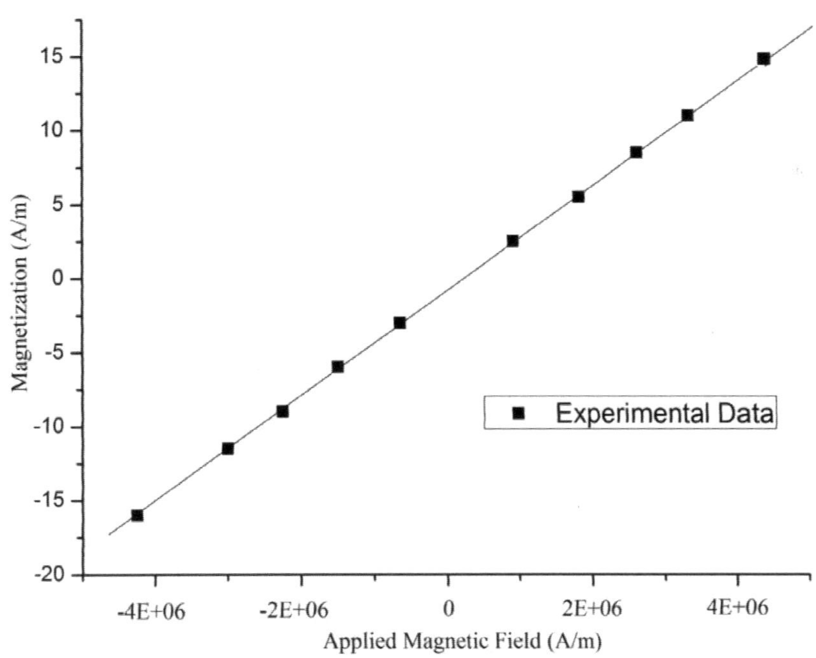

Figure 4. Oxygen-poor blood magnetization.

4. ME sensors, comparison

Sensors on ME materials are designed for operation in DC and AC circuits and fix the presence of both an alternating electromagnetic field and a constant magnetic field. Analogues of ME sensors have been widely used: Hall sensors, induction coils, including double induction coils (Helmholtz coils), SQUID sensors, ferroprobes, magnetotransistors, magnetodiodes, magnetoresistors, magneto-optical and fiber optic systems, etc. In turn, ME sensors can be widely used in medical technology as measuring devices for human magnetic fields and living beings, for measuring magnetobiological reactions, for electrical signals of the heart, for searching for ferromagnetic inclusions signals of skeletal muscle, eye, background and evoked brain activity, eye tissue, may also be used for magnetic sensors in ME tomography. In security systems, ME sensors can be used as motion detectors, in metal detectors; for the automotive industry - in ABS systems, engine management systems; in robotics – in control of angular and linear movements; in measuring equipment for the production of magnetometers, instruments for measuring the characteristics of the magnetic field and the magnetic properties of materials. Depending on the value being determined, it is possible to use field strengths (oerstedmeters), field directions (inclinators and deklinators), field gradients (gradiometers), magnetic induction (teslameters), magnetic flux (webmeters, or fluxometers), coercivity (coercimeters), magnetic permeability (mu-meters), magnetic susceptibility (kappa-meters), magnetic moment; in devices of automation and electronics as non-contact current sensors. Other possible applications are in such areas as geology, mineral search; archeology, during archaeological excavations; in astrophysics, in the study of the orbits of planets; in navigation on the sea, space and aviation; in seismology [10, 11, 12].

To measure weak constant and slowly changing magnetic fields, magnetoresistive transducers, nuclear magnetic resonance converters, SQUIDs and ferroprobe converters are currently used. Due to the small size, low cost, wide operating temperature range, low noise level, the possibility of measuring the direction of the magnetic field, the flux-gate transducers and magnetometers based on them are the most preferable.

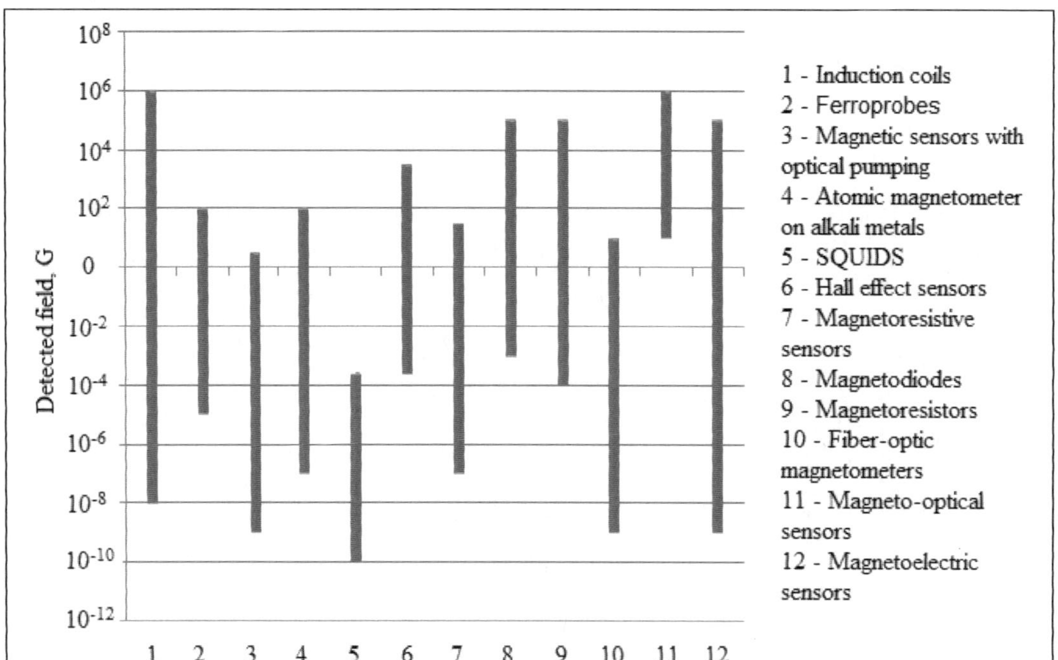

Figure 5. Magnetic sensor comparison.

As noted, the ME effect is the induction of the electric field by the magnetic field, or the inducing of the magnetization by the electric field, i.e. is a cross-effect that allows simultaneous use of dual control of magnetic and electric fields, which leads to a number of advantages over sensors operating according to other principles [12]. The study of the ME effect, materials and sensor construction has paid much attention to the work of foreign authors, which proves their relevance and demand. ME sensors have great prospects for measuring weak magnetic fields, since they have high sensitivity, low cost, and also small dimensions, for ease of use.

5. Conclusion

In this paper an overview of magnetic field sensors is presented, comparative characteristics are given, and the advantages of ME sensors are indicated. The current application of magnetic field sensors in various fields of human activity is indicated, especially for recording biological signals caused by electric fields or magnetic particles in medical diagnostics. The influence of weak magnetic fields on living organisms, their importance for human health and the influence of biological fields on various body systems are also considered. Recommendations are given on the use of ME sensors for studying the weak magnetic biological fields of living organisms.

6. Acknowledgements

The article was prepared with the financial support of the Ministry of Education and Science of Russia in the framework of the state task, project No. 8. 11100. 2018/11. 12.

References

[1] Polyakova I.P., 2011 Magnetocardiography: historical information, current status and perspectives of clinical application. *Magazine "Creative Cardiology"* **#2**
[2] Nikiforov V.N., 2013, Medical Applications of Magnetic Nanoparticles, *"News of the Academy of Engineering Sciences named after A.M. Prokhorov"*, Periodical **#1**
[3] Vvedensky V.L., Ozhogin V, **II** 1986, *Supersensitive Magnetometry and Biomagnetism (Moscow: Nauka)*
[4] Bokeria L.A., Golukhova E.Z., 2011 Clinical cardiology: diagnosis and treatment. *(Moscow: A.N. Bakulev, Scientific Center of the Russian Academy of Medical Sciences)*
[5] Wang Y Rudy Y 2006 Application of the method of fundamental solutions to potential-based inverse electrocardiography Ann. *Biomed. Eng.* vol **34** pp 1272-1288
[6] Mizun Yu G Mizun PG 1984 *Cosmos and health* p 144 *(Moscow: Znanie)*
[7] Baranova V.E., 2015 Measurement of a weak magnetic field based on a flux-probe sensor. dis. Ph.D. 05.11.01 Baranova Vitaliya Evgenevna, *Tomsk* p 134
[8] Ran Shannon K'doah 2004 Gravity Probe B: *Exploring Einstein's Universe with Gyroscopes-einstein NASA*, P 26
[9] Chaykovsky I.A., Magnetocardiography in an unshielded location in coronary artery disease: Doktorgrades der Medizin Chaykovsky I A. - Duisburg-Essen Universität, 2005.
[10] Bichurin M.I., Petrov R.V., Solovyov I.N., Soloviev A.N., 2012 Research of magnetoelectric sensors based on piezoceramics of TSS and metglas. *Modern problems of science and education* No. **1**
[11] Petrov R.V., Bichurin M.I., Kiliba Y. V., 1996 Sensors of the electromagnetic field Tez. doc. *8th All-Russia. scientific-techn. Conf. "Sensors and converters of information systems for measurement, control and management"* T.**2**. p 324 M .: MGIEM
[12] [12] Petrov R.V., Leontev V. S., 2013 Magnetoelectric Magnetometer, *Bulletin of Novgorod State University* No. **75** p 4

Development of compact Schottky diode model on GaN

M N Petrov[1]

[1] Yaroslav-the-Wise Novgorod State University, ul. B. St. Peterburgskaya, 41, 173004, Veliky Novgorod, Russia

E-mail: Mikhail.Petrov@novsu.ru

Abstract. In the article the technique of formation of compact (SPICE) model of Schottky diode on GaN, manufactured by heteroepitaxial growth on a sapphire substrate (Al_2O_3) is considered. As a compact model, the modified model of a conventional diode on the p-n junction is used. The procedure is proposed for extraction of Spice parameters of the Schottky diode model on the basis of experimental volt-ampere (VAC) and volt-farad characteristics (VFC) using the Matlab computer mathematics system. The procedure for extracting the parameters of the compact model and the electrophysical characteristics of the device can be useful for commissioning the technological process for Schottky power diodes production.

1. Inroduction

A *compact model* is a virtual analog of a real device.

The purpose of the compact model [1] is to provide users with reference material on the parameters of the manufactured instrument.

In addition, this model is used at the stage of the circuit design of microelectronic devices, including integrated circuits.

The synthesis of the compact Schottky diode model can be carried out in two ways:

1. On the basis of the theoretical calculation of the complete model (VAC and VFC) by numerical solution of the system of fundamental equations (in the diffusion-drift and / or hydrodynamic approximation) with the help of one of the programs of device-technological modeling (for example, Sentaurus TCAD) with subsequent extraction the parameters of compact model from it.
2. By extracting the parameters of the compact model from the experimental data: VAC, VFC, and temperature dependences of the Schottky diode current.

The first approach gives satisfactory results only for silicon technology, which provides rather high accuracy of simulation due to the presence of a large number of perfect models of electrophysical parameters.

In the case of fabrication of devices on complex semiconductors, the second approach based on experimental data is more effective.

An additional advantage of this approach is the possibility of extraction of a number of electrophysical parameters: the height of the Schottky barrier, the concentration of impurities in the substrate, the density of surface states at the metal-semiconductor interface, the Richardson constant.

Although these parameters are not parameters of the compact model, nevertheless, they are extremely important from the point of view of evaluating the quality of the diode manufacturing process.

2. Compact (SPICE) Schottky diode model

The compact model (Figure 1) includes a set of parameters used to simulate the device in different modes of operation: static (DC), dynamic (for transient analysis, TR) and frequency-response (in small signal mode, AC).
Each of the listed modes of analysis has its own set of parameters.

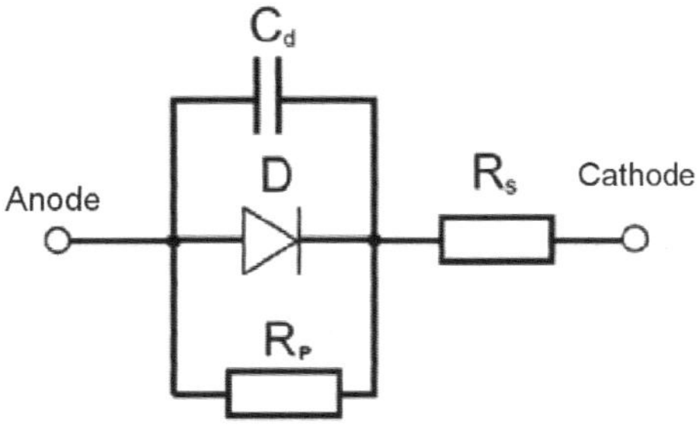

Figure 1. Full equivalent circuit of Schottky diode.

In most SPICE-like programs for circuit simulation, the compact (SPICE) Schottky diode model is a variation of the conventional diode model on p-n junction. The main difference between them is that the equivalent circuit of the Schottky diode (Figure 1) does not contain the diffusion capacitance, which is responsible for the processes of accumulation and discharge of minority charge carriers. Therefore, for the transition from conventional diode model to Schottky diode model it is sufficient to equate to zero the transit time of charge carriers of the diode base. In the SPICE model, this time corresponds to parameter **TT** = 0, used to simulate the diffusion capacity.

In this article, in order to obtain the compact Schottky diode model, three types of parameters were extracted, used for:

- simulation of static operation mode in direct switching: ***Is***, ***n*** and ***Rs***. These parameters are extracted from the experimental results of measuring the direct branch of VAC;
- simulation of static operation in reverse mode: ***Rp***, ***BV*** and ***IBV***. These parameters are extracted from the experimental results of measuring the reverse branch of VAC;
- simulation of the dynamic (transient) mode of operation in the process of switching, for example, from one static mode to another. This process is related to the presence of a parasitic barrier capacitance ***Cd*** in Schottky diode, which is modeled by a set of corresponding parameters (***CJO***, ***VJ***, ***M***), for extraction of which the results of measurements of diode VFC are used.

3. The method of extraction of parameters of compact model

Two types of numerical analysis methods were used to extract the parameters of the compact Schottky diode model from experimental data [2]: polynomial approximation (**polyfit**) by the method of least squares (OLS) and the technique based on the mathematical apparatus of optimization theory. Both of these tools are included in the tool kit of the *Curve Fitting ToolBox* of the computer mathematics system MATLAB.

3.1. Extraction of static parameters

To extract the static parameters of Schottky diode, simplified version (Figure 2) of the above mentioned equivalent diode circuit is used-without the barrier capacitance ***Cd***.

3.1.1 VAC of diode in direct switching. For the diode forward-bias region of VAC, the parallel leakage resistance can be neglected, which corresponds to the equivalent circuit shown in Figure 2.

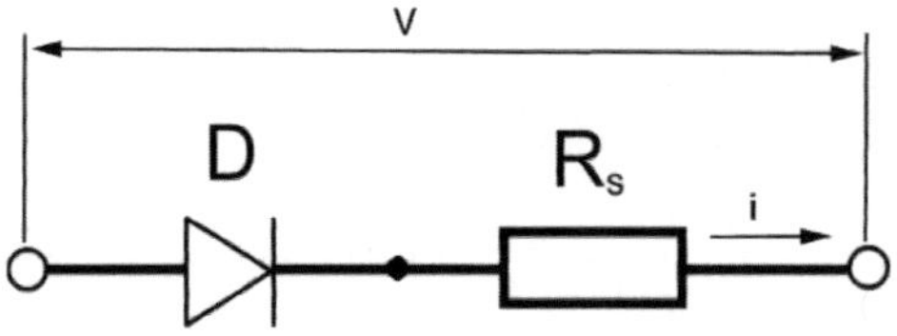

Figure 2. Static model of diode in direct switching.

This equivalent circuit corresponds to the mathematical expression:

$$i = Is * (\exp((v - i * Rs)/(n * \varphi_T)) - 1), \quad (1)$$

which includes three parameters:
Is – saturation current (***IS*** in compact model);
n – coefficient of non-ideality (***N*** in compact model);
Rs – resistance of the base region (quasi-neutral region of diode). It corresponds to ***RS*** in compact model.

In the case where the input voltage exceeds 0.1 V, expression (1) can be simplified, resulting in the form:

$$i = Is * \exp((v - i * Rs)/(n * \varphi_T))). \quad (2)$$

This is a transcendental equation, which can only be solved by a numerical method.

To estimate the values of the static parameters of the diode (***Is***, ***Rs*** and ***n***), they must be extracted from the experimental data (the direct branch of VAC of the diode). In numerical mathematics, this procedure is called the *approximation* of tabular data with the help of appropriate analytic function with a set of necessary parameters. In our case, this is the exponential function (2) with parameters ***Is***, ***Rs*** and ***n***.

To extract the static parameters of the Schottky diode, we used an experimentally-removed VAC in the direct switching (Figure 3).

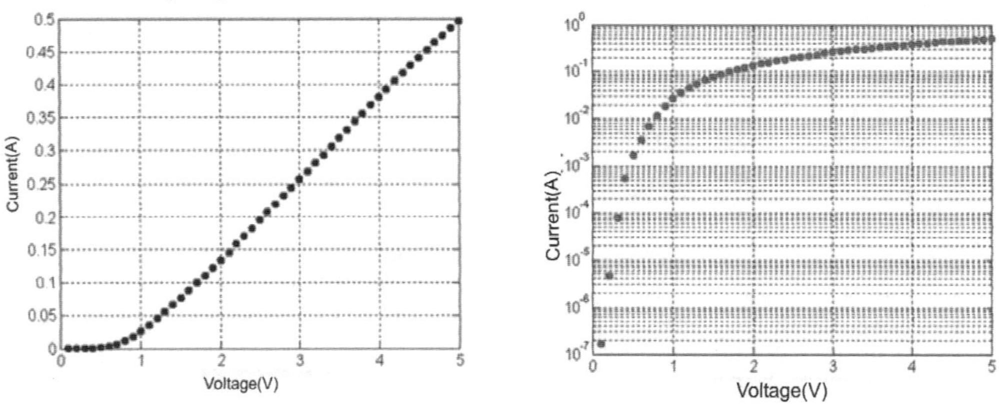

Figure 3. The direct VAC of Schottky diode in the ordinary (left) and semilogarithmic (right) scales.

Due to the complexity of the task, the extraction procedure for three parameters was performed in two stages:

1. In the region of low voltages 0.1-0.3 V, the resistance ***Rs*** can be neglected. As follows from the published data and is confirmed by the experimental data shown in Figure 4, the thermionic charge transfer mechanism dominates in this section of VAC. This allows us to use it to estimate the values of parameters ***Is*** and ***n***.

2. Then, using expression (2) and the obtained values of the two parameters, the third parameter, the quasineutral resistance ***Rs***, is estimated for the entire range of currents and voltages of the direct branch of VAC.

Extraction of ***Is*** *and* ***n*** *parameters.* In the region of small values of forward-bias (in the range of 0.1-0.3 V), the influence of the series resistance ***Rs*** can be neglected, which makes it possible to convert expression (2) to the form:

$$\mathrm{i} = \mathrm{Is} * \exp(\mathrm{v}/(\mathrm{n} * \varphi_\mathrm{T}))). \quad (3)$$

To simplify the procedure of extracting ***Is*** and ***n*** parameters included into this expression, the original expression should be logarithmized, transforming to the form:

$$\ln(\mathrm{i}) = \mathrm{v}/(\mathrm{n} * \varphi_\mathrm{T}) + \ln(\mathrm{Is}). \quad (4)$$

The resulting expression describes a straight line in coordinates **v** and $\mathbf{ln}(\boldsymbol{Is})$ with a slope $\mathbf{1/(}\boldsymbol{n} * \boldsymbol{\varphi_T})$ and the point of intersection with the coordinate y corresponding to the saturation current $\boldsymbol{I_S}$. Thus, the static parameters of diode ***n*** and ***Is*** can be extracted from the experimental data after the approximation of the latter by expression (4).

In MATLAB program for the approximation of experimental data, the straight line provides **polyfit** function that finds the coefficients of the polynomial ***P (x)*** of *n* degree. The general expression for this function is:

$$coeff\_xy = polyfit(x, y, n)), \quad (5)$$

where

x and ***y*** are the vectors of the values of the argument and the functions in the columns of the table (in our case they correspond to the voltage vectors (***v***) and current (***i***)).

n is the degree of the polynomial used to approximate the table data.

In our case, to approximate a straight line, we must specify a polynomial of the first degree (***n*** = 1) containing two coefficients $\mathrm{c}_\mathbf{0}$ and $\mathbf{c_1}$:

$$coeff\_xy = \mathrm{c}_0 + \mathrm{c}_1\mathrm{x}. \quad (6)$$

Comparison of expressions (4) and (5), indicates that

$\mathrm{c}_0 = ln(Is)$ и $\mathrm{c}_1 = 1/(n * \varphi_\mathrm{T})$.

After estimating the values of coefficients $\mathbf{c_0}$ and $\mathbf{c_1}$ using expression (5), we determine the values of two required parameters $\boldsymbol{Is} = \mathbf{8.2153}\boldsymbol{e} - \mathbf{9}$: and ***n*** = 1.2509.

The curve shown in Figure 4 indicates a good agreement of the experimental data with the results of modeling of VAC at the initial section.

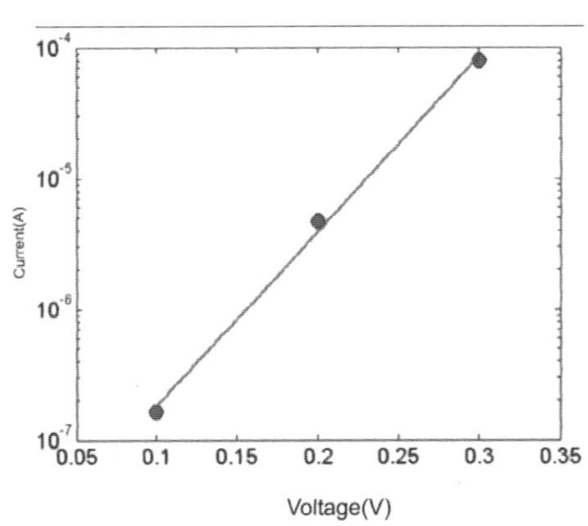

Figure 4. Fragment of the initial section of the experimental VAC (red circles) and the model (solid) obtained from formula (4) for two values of the extracted parameters: *Is* = 8.2153e-9, *n* = 1.2509.

Estimation of the height of Schottky barrier. Since the value of the coefficient of coefficient nonideality is different from one, it can be concluded that in this section of VAC there is a combined mechanism for transfer of charge carriers with the dominance of the thermionic mechanism [3]. The latter allows us to make an approximate estimate of the Schottky barrier height (for accurate estimate, it is necessary that $\boldsymbol{n} = 1$).

$$\mathrm{Is} = (\mathrm{A} * \mathrm{R} * \mathrm{T}^2) * \exp(-\phi_{\mathrm{bo}}/\varphi_{\mathrm{T}}) \tag{7}$$

Solving this equation with respect to the height of Schottky barrier $\boldsymbol{\phi_{bo}}$, we obtain: $\boldsymbol{\phi_{bo}} = \mathbf{0.6957}$ V

Analyzing the result obtained, it is worth noting that the estimated value of the barrier is understated, in fact representing the value of the effective Schottky barrier. This is due to the fact that the value of non-ideality coefficient is different from one, which is characteristic of ideal Schottky diode, which indicates the combined carrier transfer mechanism. This is most likely a variant of *thermionic-field emission* (Thermionic-field emission, TFE), as a result of which electrons with thermal energy of less than Schottky barrier height can tunnel through the depletion region at the interface.

*Extraction of the resistance of the quasineutral region of diode **Rs**.* As already noted, the real Schottky diode, in addition to the space-charge region, also contains a quasineutral region, which, in combination with the contact resistance, is modeled by the resistance ***Rs*** (see Figure 2).

This resistance begins to affect the characteristics of the diode in the region of increased current values, reducing the effective voltage loss across the space charge region of Schottky diode by amount $\boldsymbol{i * Rs}$.

Here, for extraction from experimental data of parameter ***Rs***, we use the functions **fittipe** and **fit**, which are part of the tool kit of the MATLAB computer mathematic system *Curve Fitting ToolBox*. The first of the mentioned functions forms the objective function, the second starts the process of extraction of the required parameters.

Formation of the objective function. At the first stage, it is necessary to specify the objective function, which is the expression used to approximate the experimental VAC. In this case, as the objective function, we use expression (2), solved with respect to the voltage on the diode ***v***:

$$\mathrm{v} = \mathrm{i} * \mathrm{Rs} + \mathrm{n} * \varphi_{\mathrm{T}} * \ln(\mathrm{i}/\mathrm{Is}) \tag{8}$$

After substituting in it the values of the parameters extracted at the previous stage $\boldsymbol{Is} = \mathbf{8.2153}\boldsymbol{e} - \mathbf{9}$ and $\boldsymbol{n} = \mathbf{1.2509}$, and also the value of the thermal potential $\boldsymbol{\varphi_T} = \mathbf{0.0259}$ V we obtain the final expression for the objective function:

$$\mathrm{v} = i * Rs + 0.0324 * ln(i/8.2153e - 9). \tag{9}$$

We form the objective function using expression:

$$ftype \;=\; fittype('x * Rs + 0.0324 * ln(x/8.2153e - 9)').$$

Here, x is the vector of current values of VAC.

Then we start the procedure for extracting the parameter ***Rs*** using operators ***fresult*** and ***fit***:

$$fresult \;=\; fit(x, \mathrm{v}, ftype, 'StartPoint', [8]).$$

Here, x and v are current and voltage vectors of experimental VAC;

ftype - expression of the objective function (9);

StartPoint - initial approximation of the desired parameter, from which the search for unknown parameter begins (in this case, the value of $\boldsymbol{R_S^{(0)}} = \mathbf{8}\,\Omega$ is set).

As a result of the calculation, we get parameter value $\boldsymbol{RS} = \mathbf{9,173}$.

Thus, the equation used to approximate the direct branch of VAC takes form:

$$\mathrm{v} = i * 9.173 + 0.0324 * ln(i/8.2153e - 9) \tag{10}$$

Below in Figure 5, a comparison of the experimental VAC with that described by expression (10) is given. It can be seen that the greatest difference for the two curves is observed in the voltage range 0.5-1.5 V. This is due to the fact that in this range the value of the non-ideality coefficient is a function of the applied voltage (Figure 6), which is associated with an increase in the role of the field transfer mechanism carriers (tunneling) with increasing voltage.

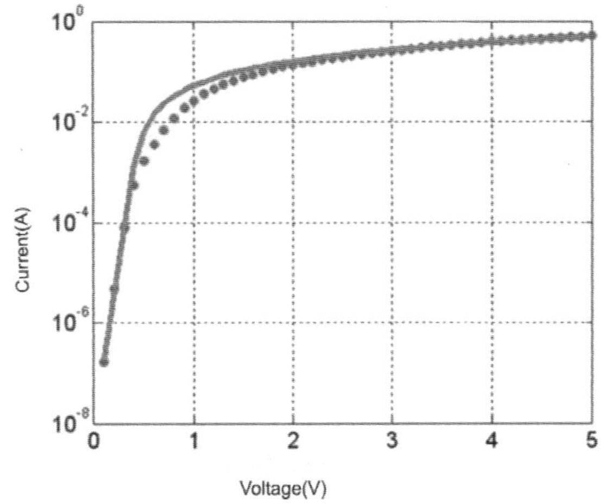

Figure 5. VAC of Schottky diode in direct switching: points - experiment; The solid line is the compact model ($\boldsymbol{IS} = \mathbf{8.2153}\boldsymbol{e} - \mathbf{9}$ $\boldsymbol{N} = \mathbf{1.2509}\boldsymbol{RS} = \mathbf{9.173}$).

In the framework of the built-in SPICE model of diode a more accurate simulation of the experimental curve is almost impossible due to the use of only one averaged value of non-ideality coefficient. Therefore, in order to obtain more accurate model, it is necessary to develop a special macromodel of diode that takes into account various mechanisms of charge carrier transfer.

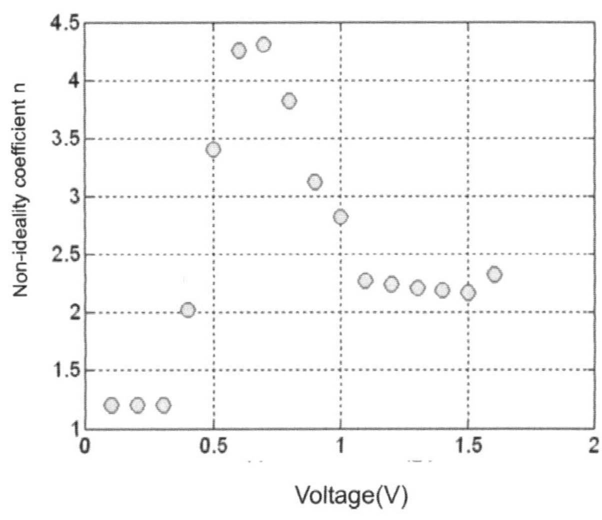

Figure 6. Dependence of non-ideality coefficient **n** on the voltage.

Note. Function (10) represents a nonlinear equation. Therefore, in order to construct a model VAC on the basis of this model, a multiple solution of the given equation for various voltage values is necessary. In MATLAB program, **solve()** function is provided for this purpose.

To solve this problem, this function is represented as:

$$\mathrm{solve('f(x)', x)}$$

where:

'f (x)' is the solvable equation (in this case (10)) written in single quotes;

x is the sought unknown value.

Below is an example of the implementation of this procedure when solving equation (10) for the voltage value v = 0.5 V:

$solve('x * 8.043 + 2.622 * 0.0259 * log(x/2.794e - 7) - 0.5 = 0', x)$,

the start of which allows one to estimate the value of the current flowing through the diode:

$x = 0.41904029609134856432949733350643e - 3.$

Defining additional parameters. In addition to extracting SPICE parameters of Schottky diode model from the direct branch of VAC, two additional parameters can also be estimated: *the voltage drop for the direct-on-turn* $\boldsymbol{V}_{\text{пр}}$ and *the current density at which the self-heating of the diode begins,* $\boldsymbol{J_h}$.For the estimation of both parameters, the above VAC (see Figure 3) is presented in new coordinates (Figure 7): current density-voltage [4]. From this curve, we can estimate the value of $V_{\text{пр}}$ ~2.4B (for the current density of 100A/cm^2). The second parameter is that the current density ***Jh*** should have a value of ~ 103A/cm^2. For considered diode, the maximum current density is 312.5A/cm^2, which is much lower than current density at which diode self-heating occurs. The latter is most likely due to the large magnitude of the series resistance limiting the magnitude of the current flowing through diode.

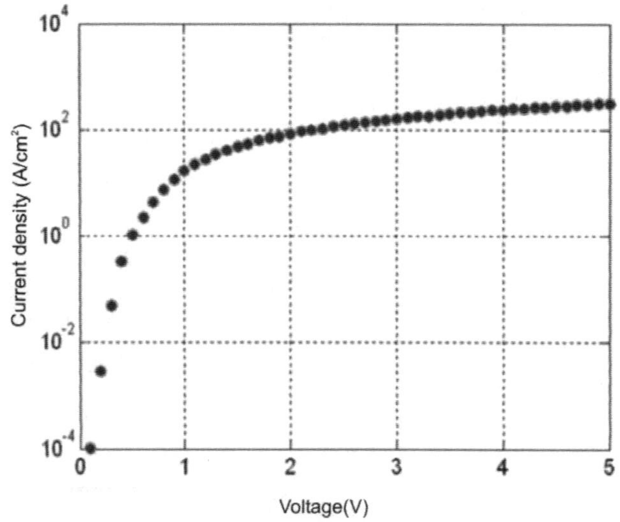

Figure 7. Dependence of the current density on the voltage for the direct branch of VAC.

3.1.2. VAC of the diode in reverse switching. For the reverse branch of VAC, the dominant contribution in VAC introduces the leakage resistance (at least in the initial stage), so the equivalent circuit will take the form shown in Figure 8.

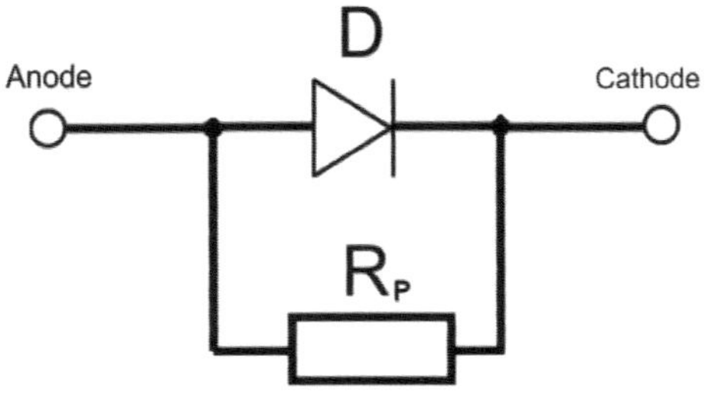

Figure 8. Static model of diode in reverse switching.

In the region of small reverse voltages, this scheme corresponds to the mathematical expression:

$$i = v/R_p \ , \tag{11}$$

which includes one parameter: $\boldsymbol{R_p}$- *leakage resistance.*

As in the previous case, we will use VAC to extract this parameter (Figure 9).

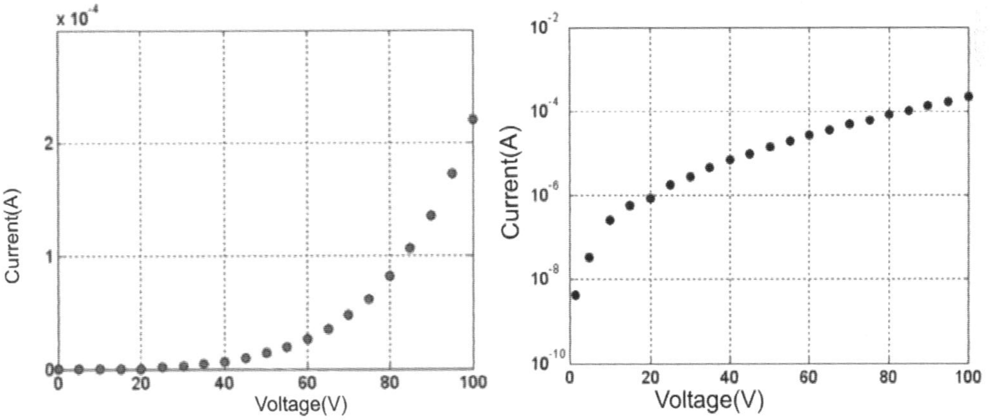

Figure 9. The reverse branch of VAC of Schottky diode in the ordinary (left) and semilogarithmic (right) scale.

Analysis of VAC indicates that, in the voltage range [0, 4 V], the diode is shunted by the leakage resistance, as evidenced by Figure 10, where a fragment of the initial section of the reverse VAC with a higher resolution is shown (in 0.1 V increments).

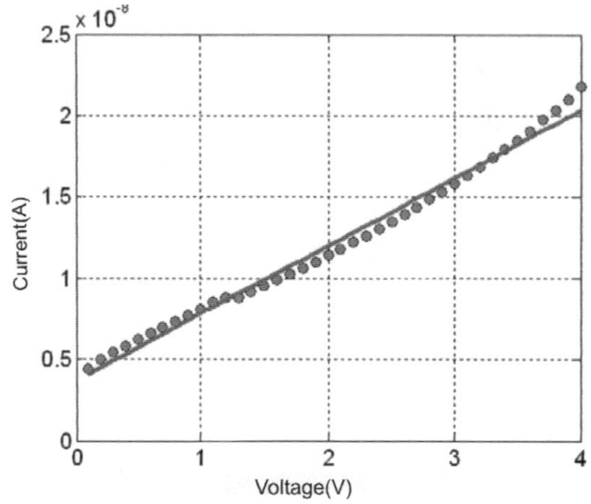

Figure 10. Fragment of the initial section of the reverse branch of the VAC (points) and its approximation with leakage resistance $\mathbf{R_p=2.4e8\Omega}$ (solid line).

In this section of the BAX, the reverse diode saturation current is masked by the leakage current, the components of which are surface leakage currents and leakage currents of dislocation nature. As noted, in SPICE, to simulate the presence of leakage current in an equivalent diode circuit, leakage resistance $\boldsymbol{R_p}$ is introduced.

To estimate the magnitude of this resistance, let us perform approximation of the initial portion of the VAC by straight line. As with the extraction of ***Is*** and ***n*** parameters for the direct branch of VAC, we use the polynomial function **polyfit**, which approximates experimental data by a polynomial of the first degree (6).

After setting the values of the voltage vectors **(v)** and current **(x),** we start the procedure of extraction of two coefficients of the polynomial $\boldsymbol{c_0}$ и $\boldsymbol{c_1}$.

$$p1\_fit = polyfit(\mathrm{v}, x, 1)$$

As a result of the calculation we obtain $\boldsymbol{c_1 = 0.4169e - 8}$, $\boldsymbol{c_0 = 0.366e - 8}$. So far as $\boldsymbol{c_1 = 1/R_p}$, we find $\boldsymbol{R_p = 1/c_1 = 2.4e8}\ \Omega$.

Comparison of the experimental data with the results of the calculation is shown in Figure 10.

Estimation of the breakdown voltage. Analyzing the behavior of the reverse branch of VAC for voltages exceeding 4V, one can see (Figure 9) that change in the voltage within two orders of magnitude (~ 95V) leads to increase in current by four orders of magnitude. This indicates a nonlinear dependence of the current on the voltage, which may be due to the action of complex of causes, including thermionic current, generation-recombination current of depleted region, and thermal-field (tunnel) current. Identifying the role of each of these components is the independent task that requires additional research. Nevertheless, from a formal point of view, the presence of this section on the reverse branch of VAC can be modeled by introducing an additional nonlinear resistance into the equivalent circuit, which would require abandoning the standard diode model and switching to the synthesis of its macromodel.

One of the main parameters of diodes intended for power electronics is breakdown voltage $\mathbf{V_B}$, which in compact model is associated with two parameters: actual breakdown voltage **VB** and **IBV** current at which breakdown is observed. As in the case of the earlier estimate of the voltage loss for the direct displacement $\boldsymbol{V_{np}}$, it is also necessary to present inverse branch of VAC given in Figure 9 in new coordinates: current density-voltage.

When analyzing the results shown in Figure 11, we use the criterion for estimating breakdown voltage value proposed in [5]. It corresponds to current density of 10 A/cm2. As a result, we get

$\boldsymbol{BV} = \mathbf{97.5}$B and $\boldsymbol{IBV} = \mathbf{1.96e-4}$A. It should be noted, that the obtained value of breakdown voltage does not reach the theoretical values predicted at 700V, which is most likely due to the imperfection of the heteroepitaxy of gallium nitride due to the presence of high dislocation density.

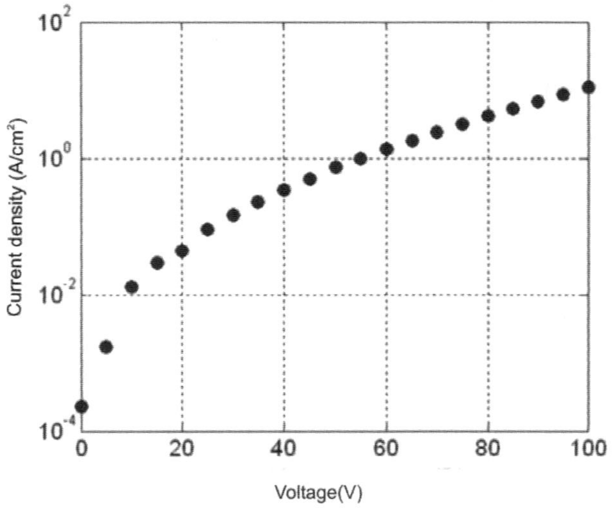

Figure 11. Dependence of current density on voltage for the reverse branch of VAC.

3.2. Extraction of dynamic parameters

To design circuits using Schottky diodes in the transient mode or for frequency-response analysis, an equivalent diode circuit (see Figure 1) should contain parasitic barrier capacitance $\mathbf{C_d}$. For the diode based on p-n junction, the SPICE capacitance model is approximated by expression:

$$\mathrm{Cd} = \mathrm{CJO} * \left(1 - \frac{\mathrm{Vdi}}{\mathrm{VJ}}\right)^{-\mathrm{M}}, \quad (12)$$

where

$\boldsymbol{V_{di}}$ is the voltage loss on the space charge region of Schottky diode (since the barrier capacitance is measured at reverse bias, this is actually the voltage **v** applied to the diode);

$\boldsymbol{CJO}$ is the barrier capacitance at zero bias at the p-n junction;

$\boldsymbol{VJ}$ is the value of the contact potential difference;

$\boldsymbol{M}$ is the coefficient that depends on the type of impurity distribution profile in the region of p-n junction.

In the case of Schottky diode, parameter $\boldsymbol{VJ}$ is understood as the barrier height at the semiconductor-metal interface. Usually for it the coefficient $\boldsymbol{M} = \mathbf{0.5}$, which is analogous to the case of sharp p-n-junction with constant impurity concentration in the base region.

Therefore, for ideal Schottky diode, formula (12) takes form:

$$\mathrm{Cd} = \mathrm{CJO} * \left(1 - \frac{\mathrm{Vdi}}{\mathrm{VJ}}\right)^{-0.5}. \quad (13)$$

To extract the SPICE parameters of the barrier capacitance, we use experimental VFC (see Figure 14).

Before proceeding with the procedure of extracting the barrier capacitance parameters, it is necessary to determine the expression for calculating the barrier capacitance. As already noted, this can be either (12) or (13). The criterion of choice can be the form of the VFC, constructed in coordinates, the graph of which is shown in Figure 12.

For a sharp jump p-n-junction ($\boldsymbol{M} = \mathbf{0.5}$), this graph is a straight line. In our case, there are two linear sections on the graph, which indicates the presence of two regions with different levels of doping [6]. Therefore, as the approximating function, we use expression (12).

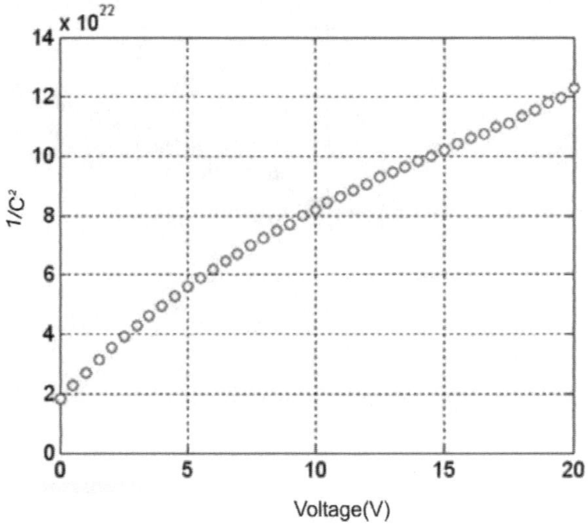

Figure 12. The graph of dependence $1 / C^2$ = **f (v).**

As in the case of parameter $\boldsymbol{Rs}$ extraction, we use the set of functions fittipe and fit to extract the parameters of the barrier capacitance (($\boldsymbol{CJO}$, $\boldsymbol{VJ}$ and $\boldsymbol{M}$).

First, we define objective function using expression (9):

$$ftype = fittype('(CJO/(1+x/VJ)^{\wedge}M)'), \quad (14)$$

where in quotes approximating function is specified, **x** is the voltage vector.

Note. We note that in the last expression, in comparison with expression (12), the minus sign is replaced by a plus sign. This allows you to enter positive values for reverse bias.

Further, to extract these parameters, we use function **fit**, which in this case has form:

$$fresult = fit(x, cr, ftype, 'StartPoint', [CJO\ \ VJ\ M]).$$

where
x is the voltage vector (table voltage values);
cr is the vector of barrier capacitance values (tabular values of barrier capacitance);
ftype is approximating function with extracted parameters;
StartPoint is the operator used to specify the initial values of extracted parameters;
$[\boldsymbol{CJO}\ \boldsymbol{VJ}\ \boldsymbol{M}]$ is the vector of initial values of extracted parameters.

Substituting in (11) the values of the components of the vector of initial values, we obtain:

$$fresult = fit(x, cr\ , ftype, 'StartPoint', [7.5, 1.0, 0.5]).$$

Starting the calculation, we get the estimate of the extracted parameters:

$\boldsymbol{CJO} = 7.484\ (7.438, 7.53)$;

$\boldsymbol{VJ} = 1.027\ (0.9672, 1.087)$;

$\boldsymbol{M} = 0.3194\ (0.3138, 0.3251)$,

where the values of confidence intervals at 95% for each of the extracted parameters are indicated in parentheses.

As a result, we obtain an expression approximating the experimental VFC:

$$Cmod = 7.484/(1 + x/1.027)^{0.3251}. \quad (15)$$

Comparison of the experimental and calculated VFC shown in Fig. 13, indicates a good agreement between calculation and experiment.

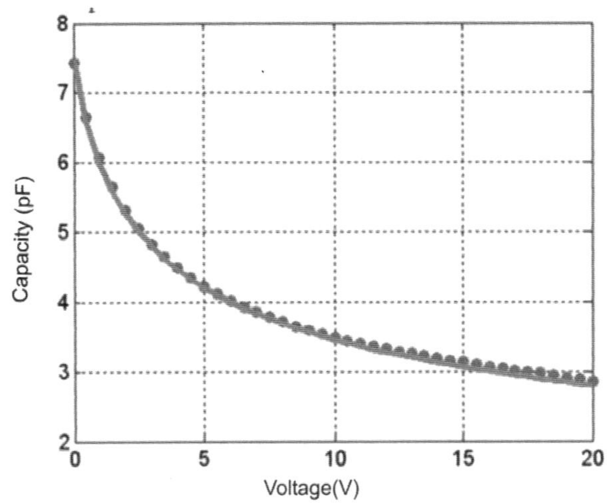

Figure 13. Comparison of VFC: experimental (circles) and model (solid line), calculated by formula (15).

4. Conclusion

In this work, extraction from experimental data of nine SPICE parameters of Schottky diode model on GaN was performed. Eight of them (**IS, n, RS, BV, IBV, CJO, VJ and M**) are directly represented in the compact Schottky diode model described by the listing below:

Model DSGAN D Is=8.215e-9 n=1.2509 Rs=9.173 BV=97.5 IBV=1.96e-4 + CJO=7.484 VJ=1.027 M=0.3194

The ninth parameter ***RP*** (leakage resistance) is not included directly in the list of parameters of compact Schottky diode model (see Figure 14). It is used as tuning parameter that replaces the default value of the similar parameter in SPICE diode model.

Due to the fact that the characteristics of a real Schottky diode on gallium nitride are quite different from the characteristics of ideal diode, the diode model built into the SPICE program requires perfection. This can be done by synthesizing the original model, implemented as a macromodel. However, at this stage of the development of production technology of such diodes, this is hardly advisable because of imperfect production technology itself. Nevertheless, the method of extraction of parameters of compact model and electrophysical characteristics of the device, given here, is useful for commissioning the technological process for the production of Schottky power diodes.

References

[1] Gildenblat G 2010 *Compact Modeling. Principles, Techniques and Applications*. Ed. (London: Springer)

[2] John O Attia 2010 *PSPICE and MATLAB for Electronics. An Integrated Approach*. 2nd ed. (London: CRC Press)

[3] Liliental-Weber Z, Jasinski J and Zakharov D N 2004 GaN grown in polar and non-polar directions. *Opto-Electronics Review* 12 p 339
[4] Umesh S Mishra K 2008 *Semiconductor Device Physics and Design*. (Berlin: Springer)
[5] Tian Y Chua S J and Wang H 2003 Theoretical study of characteristics in GaN metal-semiconductor-metal photodetectors. *Solid State Electron* 47 p 1863
[6] Schroder D K 2006 *Semiconductor Material and Device Characterization*, Third Edition (*John Wiley & Sons*)

IOP Conf. Series: Materials Science and Engineering **441** (2018) 012037 doi:10.1088/1757-899X/441/1/012037

Magnetoelectric effect in self-bias gradient structure $CoFe_2O_4/Ni/BaTiO_3$ with 0-3 connectivity

O V Sokolov[1], M I Bichurin[1], R V Petrov[1*] and Su-Chul Yang[2]

[1]Novgorod State University, ul. B. St. Petersburgskaya, 41 173003 Veliky Novgorod, Russia
[2]Department of Chemical Engineering, Dong-A University, Hadan 840, SahaGu, Busan 604-701, South Korea
Roman.Petrov@novsu.ru

Abstract. The paper is devoted to the study of mechanism of internal magnetic fields occurrence in the magnetostrictive-piezoelectric gradient structure $CoFe_2O_4/Ni/BaTiO_3$. The insertions of $CoFe_2O_4$ and Ni in this gradient structure in the matrix $BaTiO_3$ is located. The connectivity type of this composite is 0-3 and a modified model of the cell in the form of a parallelepiped is used. This parallelepiped consists of two cubic cells of $BaTiO_3$ matrix. One cubic cell contains of $CoFe_2O_4$ insertion and another cell contains of Ni one. The corresponding magnetic circuit by using the magnetostatics equations was considered. Internal magnetic fields which arise in sublattices of $CoFe_2O_4$ and Ni, taking into account the known parameters of their magnetization curves were found and equal of 860 Oe and of 50 Oe for $CoFe_2O_4$ and Ni, respectively. Further, the pseudo-piezomagnetic constants that correspond to the resulting internal magnetic fields were found by using of the existing dependencies of the pseudo-piezomagnetic coefficients of $CoFe_2O_4$ and Ni on the bias magnetic field. The effective parameters of the structure in the form of a parallelepiped was found.

1. Introduction

The presence of dc magnetic field is required in the practical application of the ME effect. The patent RU2363074C1 shows that in a two-phase magnetostrictive-piezoelectric material with gradients of magnetization and polarization, internal magnetic and electric fields can be created [1]. The use of such gradient materials will make it possible to exclude from the design of magnetoelectric devices the source of an external magnetic field. Hence, the developed devices will have the weight and dimensions are smaller than counterparts.

Earlier, cubic cell model was used at considering the composite material of 0-3 connectivity [2]. A modified model of the cell in the form of a parallelepiped consisting of two cubic cells of the BTO matrix, one of which contains of the CFO cubic inclusion and the other of the Ni cubic inclusion was used in this paper, figure 1.

The appearance of the internal magnetic field in gradient materials $BaCo_xTi_xFe_{12-2x}O_{19}$ and NZFO as theoretically and experimentally was investigated [1,3]. The gradient of magnetization was related with the creation with the help of special technology almost linear concentration gradient of Zn and Co, Ti, Fe, respectively.

ME effect in the bending mode for two-layer composite consisting of a functionally graded NZFO and PZT in the low frequency region and in the electromechanical resonance was investigated [4]. Also important conclusions about the influence of the positivity or negativity of the gradient on the value of ME coefficient was made.

The layered magnetostrictive-piezoelectric composites with gradient material NZFO were investigated [5]. Stepped gradient pseudo-piezomagnetic properties were created through stepwise changes in the concentration of zinc in NZFO. It was studied the ME effect at zero bias DC magnetic field through the use of such a gradient magnetostrictive material. Also, it was obtained a higher value of the maximum of the ME voltage coefficient dependence on the bias field with compared to the respective no-gradient magnetostrictive materials.

The layered magnetostrictive-piezoelectric composites, where as magnetostrictive phase was used stepped gradient material formed from layers of Ni and Metglas theoretically and experimentally was investigated [6,7]. ME effect for bending mode, in low-frequency region and EMR region was discussed. Nonzero ME voltage coefficient without applying an external constant magnetic field was found.

ME effect at low frequencies for bending mode for three-layer composite PZT/$Ni_{0.7}Zn_{0.3}Fe_2O_4$/Ni was theoretically and experimentally investigated [8]. Hysteretic dependence of the magnetoelectric voltage coefficient on the external magnetic field was obtained.

The aim of this work is the study of mechanism of bias magnetic fields occurrence in the magnetostrictive-piezoelectric gradient structure $CoFe_2O_4$/Ni/$BaTiO_3$.

2. Theoretical modeling

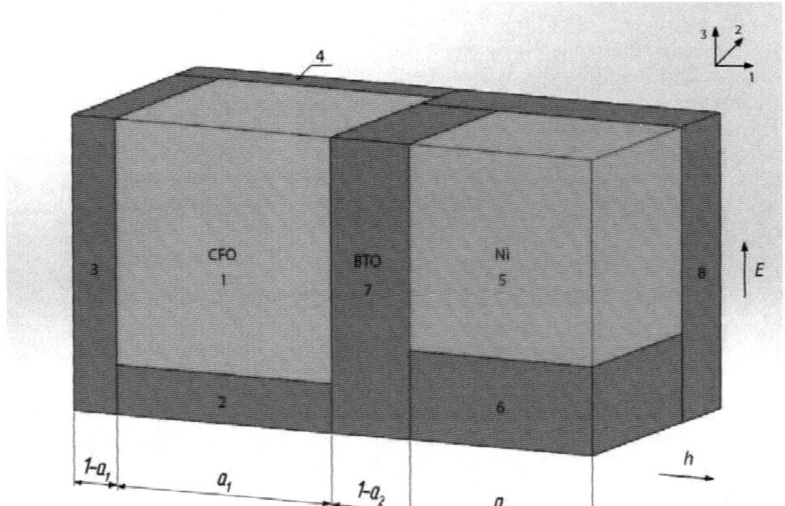

Figure 1. Model of the cell appears in the form of a parallelepiped, which consists of two cubic cells of the BTO matrix, one of which contains of the CFO cubic insertion (insertion 1) and the other contains of Ni cubic insertion (insertion 5).

It is supposed that a sample of the composite was magnetized along the axis 1 a constant magnetic field up to saturation of Ni and of cobalt ferrite, and then the magnetizing field was reduced slowly to zero, as it shown in figure 1. The residual magnetism will be present as the result of it influence in Ni and CFO. In this case, an internal magnetic field directed at the original magnetizing field in the CFO will be present, in addition the internal magnetic field in Ni will be the opposite directed, because the cobalt ferrite is much more hard magnetic material than nickel.

The appropriate models of the magnetic circuit will calculate using the equations of magnetostatics based on the known parameters of the curves of demagnetization of the CFO and Ni. The internal magnetic fields $\tilde{H}_1$, $\tilde{H}_2$ which appearing in them were found for the case $a_1 \geq a_2$:

$$\begin{aligned}
&\tilde{H}_1 = \frac{P_1}{Q_1} \\
&P_1 = H_c\left[B_{r1}\left(\left(a_1^2-1\right)\left(\mu_2-{}^{P}\mu\right)a_2^3+\left(a_1^3-a_1+2\right)\left(\mu_2-{}^{P}\mu\right)a_2^2+\left(a_1^3-a_1+2\right){}^{P}\mu a_1^2\right)+B_{r2}\left(a_1^2-1\right)a_2^3\,{}^{P}\mu\right] \\
&Q_1 = B_{r1}\left(\left(a_1^2-1\right)\left(\mu_2-{}^{P}\mu\right)a_2^3+\left(a_1^3-a_1+2\right)\left(\mu_2-{}^{P}\mu\right)a_2^2+\left(a_1^3-a_1+2\right){}^{P}\mu a_1^2\right)+ \\
&\qquad \mu_0 H_c a_1\,{}^{P}\mu\left(1-a_1^2\right)\left({}^{P}\mu a_1^2+\left(\mu_2-{}^{P}\mu\right)a_2^2\right) \\
&\tilde{H}_2 = \frac{P_2}{\mu_0 Q_1} \\
&P_2 = \mu_0 H_c B_{r1}\,{}^{P}\mu a_1^3\left(1-a_1^2\right)+B_{r2}\left(\mu_0 H_c\,{}^{P}\mu-B_{r1}\right)a_1 a_2^2\left(a_1^2-1\right)-B_{r1}B_{r2}a_2^2\left(a_1^2 a_2-a_2+2\right)
\end{aligned} \quad , (1)$$

for the case $a_1 < a_2$:

$$\begin{aligned}
&\tilde{H}_1 = \frac{P_1}{Q_1} \\
&P_1 = H_c\left[B_{r1}\left(a_1^2 a_2\left(\mu_2-{}^{P}\mu\right)\left(a_2^2-1\right)+\left(a_1 a_2^2-a_1+2\right)\mu_2 a_1^2\right)+B_{r2}\left(a_2^2-1\right)a_2^3\,{}^{P}\mu\right] \\
&Q_1 = B_{r1}\left(\left(a_2^2-1\right)\left(\mu_2-{}^{P}\mu\right)a_1^2 a_2+\left(a_1 a_2^2-a_1+2\right)\mu_2 a_1^2\right)+ \\
&\qquad \mu_0 H_c\,{}^{P}\mu\left(\mu_2\left(a_1^3-2a_1^2+2a_2^2-a_1^3 a_2^2\right)+\left(\mu_2-{}^{P}\mu\right)a_2\left(a_2^2-1\right)\left(a_2^2-a_1^2\right)\right) \\
&\tilde{H}_2 = \frac{P_2}{\mu_0 Q_1} \\
&P_2 = \mu_0\,{}^{P}\mu H_c B_{r1} a_1^3\left(1-a_2^2\right)+\mu_0\,{}^{P}\mu H_c B_{r2}\left[a_1^3\left(a_2^2-1\right)+\left(a_1^2-a_2^2\right)\left(a_2^3-a_2+2\right)\right]+ \\
&\qquad B_{r1}B_{r2}a_1^2\left(a_1-a_1 a_2^2-a_2^3+a_2-2\right)
\end{aligned} \quad , (2)$$

where a_1, a_2 is the relative edge length of CFO and Ni cubic inclusions, μ_0 is the magnetic permeability, B_{r1}, H_c is residual magnetic induction and coercive force of CFO, B_{r2}, μ_2 is the residual magnetic induction and the relative reversible magnetic permeability Ni, ${}^{P}\mu$ is the relative magnetic permeability of BTO.

The volume fraction of the CFO and Ni is equal to 0.15, and for BTO is 0.7. Then $a_1=a_2=0.67$, and internal fields calculated by the formula (1): $\tilde{H}_1 = 68562\ A/m$, $\tilde{H}_2 = 4257\ A/m$.

Further, the pseudo-piezomagnetic coefficients that correspond to the resulting internal magnetic fields were found using the dependence of the magnetostriction coefficients of CFO and Ni on the biasing DC magnetic field, which was given in [9] and [10], respectively.

If an alternating magnetic field directed along the axis 1, the electric field in the direction of the axis 3 occurs in this case. The elastostatic, magnetostatics and electrostatics equations allow to find effective parameters of magnetoelectric composites using the modified model.

Longitudinal components of strain tensor the piezoelectric phase are expressed as

$$\begin{aligned} {}^{n}S_1 &= {}^{\mathrm{p}}s_{11}\,{}^{n}T_1 + {}^{\mathrm{p}}s_{12}\,{}^{n}T_2 + {}^{\mathrm{p}}s_{13}\,{}^{n}T_3 + d_{31}\,{}^{n}E_3 \\ {}^{n}S_2 &= {}^{\mathrm{p}}s_{12}\,{}^{n}T_1 + {}^{\mathrm{p}}s_{11}\,{}^{n}T_2 + {}^{\mathrm{p}}s_{13}\,{}^{n}T_3 + d_{31}\,{}^{n}E_3 \\ {}^{n}S_3 &= {}^{\mathrm{p}}s_{13}\,{}^{n}T_1 + {}^{\mathrm{p}}s_{13}\,{}^{n}T_2 + {}^{\mathrm{p}}s_{33}\,{}^{n}T_3 + d_{33}\,{}^{n}E_3 \\ n &= 2,3,4,6,7,8 \end{aligned} \quad . \tag{3}$$

The third components of the electric displacement vector and the first vector components of the magnetic induction piezoelectric phase are expressed as

$$\begin{aligned} {}^{n}D_3 &= d_{31}\left({}^{n}T_1 + {}^{n}T_2\right) + d_{33}\,{}^{n}T_3 + {}^{\mathrm{p}}\varepsilon_{33}\varepsilon_0\,{}^{n}E_3 \\ {}^{n}B_1 &= \mu_0\,{}^{\mathrm{p}}\mu\,{}^{n}H_1 \end{aligned} \quad . \tag{4}$$

The longitudinal components of the strain tensor magnetostrictive phases are expressed as

$$\begin{aligned} {}^{1}S_1 &= {}^{\mathrm{m1}}s_{11}\,{}^{1}T_1 + {}^{\mathrm{m1}}s_{12}\,{}^{1}T_2 + {}^{\mathrm{m1}}s_{12}\,{}^{1}T_3 + {}^{1}q_{11}\,{}^{1}H_1 \\ {}^{1}S_2 &= {}^{\mathrm{m1}}s_{12}\,{}^{1}T_1 + {}^{\mathrm{m1}}s_{11}\,{}^{1}T_2 + {}^{\mathrm{m1}}s_{12}\,{}^{1}T_3 + {}^{1}q_{12}\,{}^{1}H_1 \\ {}^{1}S_3 &= {}^{\mathrm{m1}}s_{12}\,{}^{1}T_1 + {}^{\mathrm{m1}}s_{12}\,{}^{1}T_2 + {}^{\mathrm{m1}}s_{11}\,{}^{1}T_3 + {}^{1}q_{12}\,{}^{1}H_1 \\ {}^{5}S_1 &= {}^{\mathrm{m2}}s_{11}\,{}^{5}T_1 + {}^{\mathrm{m2}}s_{12}\,{}^{5}T_2 + {}^{\mathrm{m2}}s_{12}\,{}^{5}T_3 + {}^{2}q_{11}\,{}^{5}H_1 \\ {}^{5}S_2 &= {}^{\mathrm{m2}}s_{12}\,{}^{5}T_1 + {}^{\mathrm{m2}}s_{11}\,{}^{5}T_2 + {}^{\mathrm{m2}}s_{12}\,{}^{5}T_3 + {}^{2}q_{12}\,{}^{5}H_1 \\ {}^{5}S_3 &= {}^{\mathrm{m2}}s_{12}\,{}^{5}T_1 + {}^{\mathrm{m2}}s_{12}\,{}^{5}T_2 + {}^{\mathrm{m2}}s_{11}\,{}^{5}T_3 + {}^{2}q_{12}\,{}^{5}H_1 \end{aligned} \quad . \tag{5}$$

The third components of the electric displacement vectors and the first components of magnetic induction vectors of the magnetostrictive phases are expressed as

$$\begin{aligned} {}^{1}D_3 &= {}^{\mathrm{m1}}\varepsilon_{33}\varepsilon_0\,{}^{1}E_3 \\ {}^{1}B_1 &= {}^{1}q_{12}\left({}^{1}T_2 + {}^{1}T_3\right) + {}^{1}q_{11}\,{}^{1}T_1 + {}^{1}\mu_{11}\mu_0\,{}^{1}H_1 \\ {}^{5}D_3 &= {}^{\mathrm{m2}}\varepsilon_{33}\varepsilon_0\,{}^{5}E_3 \\ {}^{5}B_1 &= {}^{2}q_{12}\left({}^{5}T_2 + {}^{5}T_3\right) + {}^{2}q_{11}\,{}^{5}T_1 + {}^{2}\mu_{11}\mu_0\,{}^{5}H_1 \end{aligned} \quad . \tag{6}$$

Taking into account the conditions of compatibility for deformations and boundary conditions for the field vectors one gets a system of 40 equations with 40 unknowns components obtained by substituting the expressions for the longitudinal components of the tensor of deformations, the third component of the vectors of the electric displacement and the first component of the vectors of magnetic induction in the rest of the equation. All unknown components are found by numerical-analytical method of expressing through T_1, T_2, T_3, E_3, H_1.

As the first component of effective deformation tensor is equal to

$$S_1 = \frac{{}^{4}S_1 + {}^{8}S_1}{2} \tag{7}$$

will be obtained the expression

$$S_1 = s_{11}T_1 + s_{12}T_2 + s_{13}T_3 + \tilde{d}_{31}E_3 + q_{11}H_1 . \tag{8}$$

The effective parameters such as the compliance coefficient, piezoelectric coefficient and pseudo-piezomagnetic coefficient are $s_{11} = 5.8 \cdot 10^{-12}\ \frac{\text{m}^2}{\text{N}}$, $\tilde{d}_{31} = -3.88 \cdot 10^{-11}\ \frac{\text{m}}{\text{V}}$, $q_{11} = -2.16 \cdot 10^{-10}\ \frac{\text{m}}{\text{A}}$.

The third component of effective electric displacement vector is

$$D_3 = \left[a_1^2 \cdot {}^1D_3 + (1-a_1)\,{}^4D_3 + a_1(1-a_1)\,{}^3D_3 + a_2^2 \cdot {}^5D_3 + (1-a_2)\,{}^8D_3 + a_2(1-a_2)\,{}^7D_3\right]/2. \quad (9)$$

Unknown components that have previously been found, we substitute in the formula (9) and thus we find

$$D_3 = \tilde{d}_{31}T_1 + \tilde{d}_{32}T_2 + \tilde{d}_{33}T_3 + \varepsilon_{33}\varepsilon_0 E_3 + m_{31}H_1. \quad (10)$$

The effective relative permittivity and effective magnetoelectric susceptibility are $\varepsilon_{33} = 232.8$, $m_{31} = -1.61 \cdot 10^{-9}\ \frac{\text{s}}{\text{m}}$.

The longitudinal oscillations in the magnetoelectric sample investigated above structure by the action of an alternating magnetic field directed along the axis 1 will be discussed below. Composite length is equal $l=10^{-2}$ m and directed along the axis 1, the width and thickness much less than the length. The metal electrodes deposited on the surface of the material. AC voltage is generated on the metal electrodes due to the ME effect. The normal line to the electrodes plane is directed along the axis 3.

In this case, we assume that $T_2=T_3=0$, then

$$S_1 = s_{11}T_1 + \tilde{d}_{31}E_3 + q_{11}h_1, \quad (11)$$

whence

$$T_1 = \frac{1}{s_{11}}S_1 - \frac{\tilde{d}_{31}}{s_{11}}E_3 - \frac{q_{11}}{s_{11}}h_1. \quad (12)$$

The equation of longitudinal oscillations of the membrane

$$\rho\frac{\partial^2 u_1}{\partial \tau^2} = \frac{\partial T_1}{\partial x_1}, \quad (13)$$

where $\rho = {}^{\text{p}}v^{\text{p}}\rho + {}^{\text{m1}}v^{\text{m1}}\rho + {}^{\text{m2}}v^{\text{m2}}\rho = 0.7 \cdot 6017 + 0.15 \cdot 5200 + 0.15 \cdot 8902 = 6327.2\ \text{kg}/\text{m}^3$ is the effective density of the composite.

The first component of strain tensor

$$S_1 = \frac{\partial u_1}{\partial x_1}. \quad (14)$$

The solution to the equation of longitudinal oscillations

$$\begin{gathered} u_1 = A\cos(kx_1) + B\sin(kx_1) \\ k = \omega\sqrt{\rho s_{11}} \end{gathered}. \quad (15)$$

From the boundary conditions for the free sample

$$\begin{gathered} T_1\left(-\frac{l}{2}\right) = 0 \\ T_1\left(\frac{l}{2}\right) = 0 \end{gathered} \quad (16)$$

follows

$$A = 0$$
$$B = \frac{\tilde{d}_{31}E_3 + q_{11}h_1}{k\cos\left(\frac{kl}{2}\right)}. \quad (17)$$

Using the open circuit condition

$$\int_{-\frac{l}{2}}^{\frac{l}{2}} D_3 \mathrm{d}x_1 = 0, \quad (18)$$

ME voltage coefficient is

$$\alpha_E = \frac{E_3}{h_1} = \frac{\tilde{d}_{31}q_{11}(1-w) - m_{31}s_{11}}{\varepsilon_{33}\varepsilon_0 s_{11} + \tilde{d}_{31}^2(w-1)}$$
$$w = \frac{2}{kl}\tan\left(\frac{kl}{2}\right). \quad (19)$$

The dependence of ME voltage coefficient on the frequency of the alternating magnetic field is shown in figure 2. To account for losses in the calculation, the following formula is used

$$\omega = 2\pi\left(1 + \frac{1}{2Q}\mathrm{i}\right)f, \quad (20)$$

where the quality factor of the resonance Q=100.

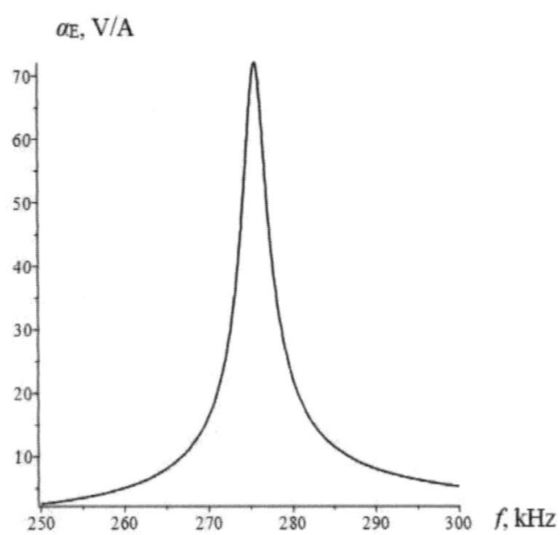

Figure2. The dependence of ME voltage coefficient on the alternating magnetic field frequency. The dimension of ME structure 10×5×0.5 mm. The effective relative permittivity ε_{33}=232.8, and effective magnetoelectric susceptibility m_{31}=-1.61·10^{-9} s/m. The compliance coefficient, piezoelectric coefficient and pseudo-piezomagnetic coefficient are s_{11}=5.8·10^{-12} m^2/N, $\tilde{d}_{31}$ =-3.88·10^{-11} m/V, q_{11}=-2.16·10^{-10} m/A. The effective density of the composite is ρ=6327.2 kg/m^3.

3. Conclusion

In this paper effective parameters of magnetostrictive-piezoelectric composite with the connectivity of 0-3 was found using a modified model of the cell in the form of a parallelepiped based on the equations of elastostatic, magnetostatics and electrostatics. The longitudinal mode of magnetoelectric effect taking into account of gradient structure on the basis of the obtained effective parameters was considered. The use of three-component magnetostrictive-piezoelectric composite gradient structures, similar studied, allows to reduce mass-dimensional characteristics of the ME devices. This is possible due to the fact that there is no need to use a dc magnetic field for ME composite in ME devices.

Acknowledgments

Theresearchwassupported by Ministry of Education and Science of the Russian Federation, the project №8. 11100. 2018/11. 12 and by Basic Science Research Program through the National Research Foundation of Korea (NRF) funded by the Ministry of Education (2016R1C1B1010884).

References

[1] Bichurin M I, Petrov V M, Semenov G A 2009 Magnetoelectric material for components of radioelectronic devices *Patent RU2363074C1*
[2] Bichurin M I, Petrov V M 2003 et al. Modeling of magnetoelectric effects in ferromagnetic/piezoelectric bulk composites, *Proceedings of Int. Conf. «Magnetoelectric Interaction Phenomena in Crystals»* P 65-70.
[3] Mantese J V, Micheli A L Schubring N W 2005 Magnetization-graded ferromagnets: The magnetic analogs of semiconductor junction elements *Applied Physics Letters*, **87**, 082503.
[4] Sudakar C, Naik R, Lawes G, Mantese J V, Micheli A L, Srinivasan G Alpay S P 2007 Internal magnetostatic potentials of magnetization-graded ferromagnetic materials *Applied Physics Letters*, **90**, 062502
[5] Petrov V M and Srinivasan G 2008 Enhancement of magnetoelectric coupling in functionally graded ferroelectric and ferromagnetic bilayers *Physical Review B*, **78**, 184421
[6] Mandal S K, Sreenivasulu G, Petrov V M and Srinivasan G 2010 Flexural deformation in a compositionally stepped ferrite and magnetoelectric effects in a composite with piezoelectrics *Applied Physics Letters*, **96**, 192502
[7] Mandal S K, Sreenivasulu G, Petrov V M and Srinivasan G 2011 Magnetization-graded multiferroic composite and magnetoelectric effects at zero bias *Physical Review B*, **84**, 014432
[8] Sreenivasulu G, Mandal S K, Bandekar S, Petrov V M and Srinivasan G 2011 Low-frequency and resonance magnetoelectric effects in piezoelectric and functionally stepped ferromagnetic layered composites *Physical Review B* **84**, 144426
[9] Laletin U, Sreenivasulu G, Petrov V M, Garg T, Kulkarni A R, Venkataramani N and Srinivasan G 2012 Hysteresis and remanence in magnetoelectric effects in functionally graded magnetostrictive-piezoelectric layered composites *Physical Review B* **85**, 104404
[10] Srinivasan G, Rasmussen E T and Hayes R 2003Magnetoelectric effects in ferrite-lead zirconate titanate layered composites: The influence of zinc substitution in ferrites *Physical Review B* **67**, 014418

Magnetoelectric effects at electromechanical resonance in laminates of lead-free piezoelectric bimorph and magnetostrictive component

V.M. Petrov, M I Bichurin, D.V. Kovalenko and O.I. Malyshev

Novgorod State University, ul. B. St. Petersburgskaya, 41
173003 Veliky Novgorod, Russia
vladimir.petrov@novsu.ru

Abstract. The paper presents a theoretical model for magnetoelectric (ME) effect in multilayers of piezoelectric and magnetostrictive components. We assume the piezoelectric layer to be compositionally-stepped. Taking into account the toxicity of lead-based ceramic, piezoelectric component is made of a lead-free piezoelectric. Analytical expressions for ME voltage coefficients is obtained by substituting the expression for stress at bending mode into open electric circuit condition. Using the layered structure based on the compositionally-stepped piezoelectric components leads to an enhanced magnetoelectric coupling for specific component thickness ratio due to an increase in flexural deformation as compared with the bilayer based on homogeneous piezoelectric layer. For the laminate of lead-free piezoelectric langatate bimorph and magnetostrictive Permendur, ME coupling is stronger due to higher ratio of the piezoelectric coupling coefficient to the dielectric constant. In addition, an 80 % increase in magnetoelectric coupling strength due using the compositionally-stepped piezoelectric is obtained.

1. Introduction

The coupling between the magnetic and ferroelectric ordering in a multiferroic gives rise to magnetoelectric (ME) effect [1-4] due to mechanical deformations. Applied magnetic field H produces the strains of magnetostrictive phase the strains are transferred to the piezoelectric layers. Electric polarization P results from the piezoelectric coupling. ME coupling is described by expression: $P = \alpha H$, with α being the ME susceptibility. In contrast to single-phase materials, the interaction between piezoelectric and magnetic phases in ME composites leads to large values of ME coefficients. In particular, the obtained values of ME susceptibility are several orders of magnitude higher than in the known single-phase materials at room temperature. Thus ME composites are attractive for use in multifunctional devices, such as magnetoelectric converters, attenuators and sensors. Since the ME effect in composite materials is due to the mechanical coupling of the components, a significant enhancement in the ME coupling strength is observed in the electromechanical resonance (EMR) region. [1, 2]. For nominal dimensions of the sample, flexural vibrations occur at significantly lower frequencies than radial and thick oscillations, which makes bending modes preferable for practical applications. A bilayer of functionally graded piezoelectric and magnetostrictive components shows an increased ME coupling at bending mode [5].

The model of low-frequency ME coupling in laminates of piezoelectric bimorph and magnetostrictive layer was discussed recently [6]. Direct ME effect is characterized by the ME voltage coefficient which is equal to the ratio of the induced electric field to the applied magnetic field. It should be emphasized that the magnitude of the ME voltage is proportional to the ratio d/ε_r with d and ε_r denoting the piezoelectric coefficient and dielectric constant for piezoelectric phase, respectively. As a result, the use of appropriate lead-free single-crystal piezoelectrics makes it possible to obtain a value of ME voltage coefficient which exceeds that for piezoceramics based composites containing lead oxide.

This study presents a theoretical model for direct ME effect in the EMR region in multilayers of lead-free single-crystal piezoelectric bimorph and magnetostrictive components. Lead-free composites and single crystal piezoelectrics are the most appropriate piezoelectric materials for environmental protection [7].

2. Theoretical modeling

In this article, we turn to the case of a layered structure with magnetostrictive layer of Permendur and two lead-free single crystal piezoelectric layers of x-cut langatate with opposite directions of x-axis. To reduce demagnetising effects, the ac magnetic field is applied parallel to the length direction, y-axis. In addition, the dc magnetic field is applied to the sample for optimizing the piezomagnetic properties of magnetic layer. The sample is assumed to be a thin plate with thickness t that is small compared to width b. In turn, b is assumed to be small compared to length L. For this case, only the y-components of strain and stress can be taken into consideration.

In our model, the strain ${}^{i}S_2$ and electric displacement ${}^{i}D_1$ for two piezoelectric layers (with superscripts $i=p1$ and $i=p2$) are expressed in terms of stress components ${}^{i}T_2$ and electric field ${}^{i}E_1$ in two piezoelectric layers:

$$
\begin{aligned}
{}^{p1}S_2 &= {}^{p}s_{11}\ {}^{p1}T_2 + {}^{p}d_{11}\ {}^{p1}E_1; \\
{}^{p1}D_1 &= {}^{p}d_{11}\ {}^{p1}T_2 + {}^{p}\varepsilon_{11}\ {}^{p1}E_1; \\
{}^{p2}S_2 &= {}^{p}s_{11}\ {}^{p2}T_2 - {}^{p}d_{11}\ {}^{p2}E_1; \\
{}^{p2}D_1 &= -\,{}^{p}d_{11}\ {}^{p2}T_2 + {}^{p}\varepsilon_{11}\ {}^{p2}E_1;
\end{aligned}
\qquad (1)
$$

where ${}^{p}d_{11}$, ${}^{p}s_{11}$, and ${}^{p}\varepsilon_{11}$ are the piezoelectric coefficient, compliance coefficient at constant electric field, and permittivity of piezoelectric layers, correspondingly.

Similarly, the strain ${}^{m}S_2$ and magnetic field ${}^{m}H_2$ in magnetic layer are expressed in terms of stress components ${}^{m}T_2$ and magnetic induction ${}^{m}B_2$ for magnetic layer:

$$
\begin{aligned}
{}^{m}S_2 &= {}^{m}s_{11}\ {}^{m}T_2 + {}^{m}g_{11}\ {}^{m}B_2; \\
{}^{m}H_1 &= -\,{}^{m}g_{11}\ {}^{m}T_2 + \frac{1}{{}^{m}\mu_{11}}\ {}^{m}B_2;
\end{aligned}
\qquad (2)
$$

where ${}^{m}g_{11}$, ${}^{m}s_{11}$, and ${}^{m}\mu_{11}$ are the piezomagnetic coefficient, compliance coefficient at constant magnetic induction, and permeability of magnetic layer, respectively.

Bending vibrations of a thin plate are governed by following equation for deflection w [8] (displacement in X-direction):

$$\nabla^2\nabla^2 w + \frac{\rho\, t}{D}\frac{\partial^2 w}{\partial \tau^2} = 0 \tag{3}$$

with ρ, t, D, and τ denoting the average density, sample thickness, cylindrical stiffness of the plate, and time, correspondingly. The sample thickness t equals $t_m+t_{p1}+t_{p2}$ with t_m, t_{p1}, and t_{p2}, and being thickness of magnetostrictive layer and two piezoelectric layers, correspondingly.

For solving (1), we use the boundary conditions for the sample with one end rigidly clamped (at y = 0) and one end free (at y = L). Thus the deflection and derivative of deflection $\partial w/\partial x$ equal zero at the rigidly clamped end and the turning moment M and transverse force V equal zero at free end. It should be noted that he lowest frequency is obtained for the sample fixed at one end and free at the other end. The turning moment relative to z-axis can be expressed in terms of stresses as follows:

$$M_z = \int_{A_{p1}} {}^{p1}T_2 dA + \int_{A_{p2}} {}^{p2}T_2 dA + \int_{A_m} {}^{m}T_2 dA \tag{4}$$

where A_{p1}, A_{p2}, and A_m are cross-sections of piezoelectric and magnetostrictive layers.

The transverse force is known to be defined by

$$V_x = \frac{\partial M_z}{\partial y} \tag{5}$$

The following equations can be used to express the strains of both piezoelectric layers and magnetic layer in terms of deflection:

$$\begin{aligned} {}^{p2}S_2 &= -x\frac{\partial^2 w}{\partial y^2}, \quad x_0 - t_{p1} - t_{p2} < x < x_0 - t_{p1}, \\ {}^{p1}S_2 &= -x\frac{\partial^2 w}{\partial y^2}, \quad x_0 - t_{p1} < x < x_0, \\ {}^{m}S_2 &= -x\frac{\partial^2 w}{\partial y^2}, \quad x_0 < x < x_0 + t_m, \end{aligned} \tag{6}$$

where x_0 is the distance of sample middle plane from the interface between piezoelectric and magnetostrictive phases.

We are interested in the ME voltage coefficient $\alpha_E = E_1/H_2$. The magnetic field induced voltage U can be estimated in the following manner:

$$U = \int_{z_0 - {}^{p1}t - {}^{p2}t}^{z_0 - {}^{p1}t} {}^{p2}E_1 dx + \int_{z_0 - {}^{p1}t}^{z_0} {}^{p1}E_1 dx. \tag{7}$$

Finding the electric field induced across the two piezoelectric layer is enabled by using the open circuit condition $\int_G {}^{p1}D_2 dG = 0$ for first piezoelectric layer and $\int_G {}^{p1}2D_2 dG = 0$ for second piezoelectric layer where G is the cross-sectional area of piezoelectric layers normal to the y-axis.

For harmonic vibrations, the deflection as a function of x and τ is determined by expression $w(x,\tau) = w(x)\cdot\cos(\omega\cdot\tau)$ with ω being the angular frequency. Assuming the electromechanical and magnetomechanical coupling coefficients to be small compared to unity, we obtain the following expression for ME voltage coefficient:

$$\alpha_{E31} = \frac{{}^{p}d_{31}}{{}^{p}\varepsilon_{11}}{}^{m}q_{11}t_m(2y_0+t_m)\frac{[2y_0(t_{p2}-t_{p1})+t_{p1}(t_{p1}-2t_{p2})-t_{p2}^2][\cosh(kL)\sin(kL)+\sinh(kL)\cos(kL)]}{4\,{}^{p}s_{11}\,{}^{m}s_{11}DkL(t_{p1}+t_{p2}+t_m)[1+\cosh(kL)\cos(kL)]} \quad (8)$$

where k is the wave number, $k = \omega^{1/2}\left(\frac{\rho t}{D}\right)^{1/4}$.

3. Results and Discussion

Equation (8) shows that ME voltage coefficient is proportional to piezomagnetic coefficient of magnetostrictive phase and ratio of piezoelectric coefficient to dielectric constant of piezoelectric phase. First we consider the ME coupling in a bilayer of magnetostrictive and lead-free single crystal piezoelectric components assuming the thickness of second piezoelectric layer equal to zero, t_{p2}=0 in equation (8). We apply the model developed in the previous section to the bilayer of piezoelectric layer and Permendur, a ferromagnetic alloy (49% Fe, 49 % Co and 2% V) with strong piezomagnetic coupling. Figure 1 shows the dependence of ME voltage coefficient on ratio of the piezoelectric coupling coefficient d to the relative dielectric constant ε_r for layered structure based on Permendur 0.2 mm thick and piezoelectric layer 0.5 mm thick. The following material parameters are used for theoretical estimates: ${}^{m}s_{11}=7.8\cdot10^{-12}$ m²/N, ${}^{p}s_{11}=9.8\cdot10^{-12}$ m²/N, ${}^{m}q_{11}=0.8\cdot10^{-8}$ m/A. The sample length equals 45 mm.

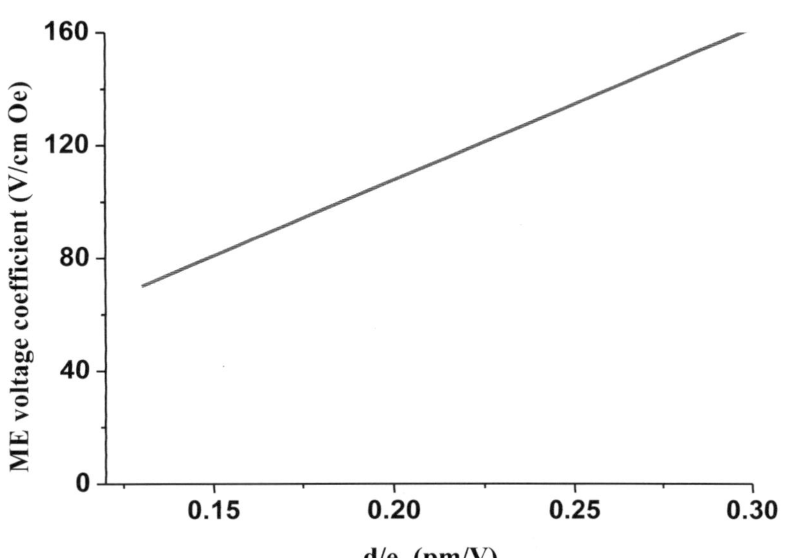

Figure 1. Peak ME voltage coefficient at bending mode as a function of the ratio of the piezoelectric coupling coefficient d to the relative dielectric constant ε_r for bilayer of permendur and piezoelectric components

Note that the losses are taken into account by assuming the angular frequency ω to be complex :

$$\omega = 2\pi(1+0.005i)f \tag{9}$$

where f is the appllied ac magnetic field frequency .

Estimates in Figure 1 shows that the peak ME voltage coefficient at bending mode increases with increasing ratio of the piezoelectric coupling coefficient d to the relative dielectric constant ε_r for permendur-piezoelectric bilayer. Thus the layered structure based on Permendur and lead-free single crystal langatate (LGT) with d/ε_r=0.25 enables one to obtain the approximately 2 times higher ME voltage coefficient compared to PZT based laminate containing lead oxide (d/ε_r=0.13).

A further increase in ME coupling strength can be obtained for layered structure based on Permendur and piezoelectric bimorph. Next we assume that $t_{p1} = t_{p2}$ in equation (8). Figure 2 presents the frequency dependence of ME voltage coefficient for layered structure based on Permendur 0.2 mm thick and LGT single crystal piezoelectric bimorph (two layers 0.5 mm thick). We used the following material parameters for LGT: ${}^pd_{11}$=-6.5·10^{-12} m/V, ${}^p\varepsilon_{11}/\varepsilon_0$=20.

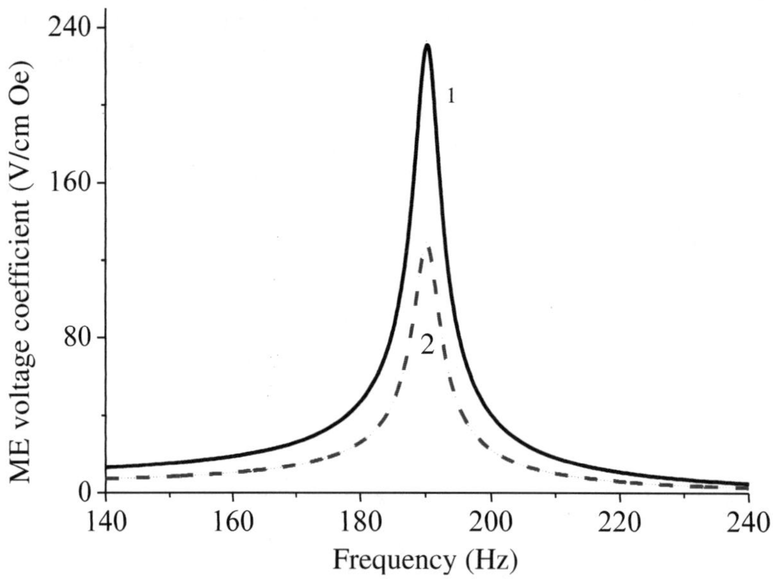

Figure 2. Frequency dependence of ME voltage coefficient for layered structure based on Permendur and LGT single crystal piezoelectric bimorph (1) and for bilayer of Permendur and LGT single crystal (2)

Data in figure 2 reveal an increase in ME voltage coefficient by 80 percents for the structure of langatate bimorph and permendur compared to the bilayer of permendur and homogeneous langatate. It is seen that the peak ME voltage coefficient reaches 230 V/(cm·Oe). The obtained enhancement of ME coupling strength arises due to an increase in the rotational moment and transverse force for piezoelectric bimorph based structure according to equations (4) and (5). The increase in ME voltage coefficient makes environmentally friendly samples of ME laminates potentially suitable for use in the design of magnetic field sensors and converters.

4. Conclusion

In summary, the explicit expression was derived for the ME coefficient in laminates of piezoelectric bimorph and magnetostrictive phase in the bending mode region. Direct ME effect for langatate-

Permendur laminates is shown to be stronger compared to composites with lead-based ceramic ferroelectrics such as PZT.

Direct ME effect was estimated for lead-free piezoelectric langatate bimorph and magnetostrictive Permendur. ME voltage coefficient is approximately 1.8 times higher for the piezoelectric langatate bimorph based structure compared to the homogeneous langatate/Permendur bilayer. The obtained increase in ME coupling strength is due to redistribution of axial strains for compositionally stepped structure due to variation of flexural deformation of the sample. As a result, one obtains the like-sign voltage induced in both langatate layers with opposite x-axis direction. This investigation is of importance for the sensor and technological applications. Distinctive features of lead-free piezoelectric single crystals are the absence of ferroelectric hysteresis and pyroelectric losses. The lead-free composites and single crystal piezoelectrics are the most appropriate piezoelectric materials for environmental protection. The laminates investigated in the present work could potentially be used in ME effect-based devices such as magnetic sensors, energy harvesters etc.

5. Acknowledgments

The research was supported by Grant No: 16-12-10158 from the Russian Science Foundation.

References

[1] Bichurin M I and Petrov V M 2014 *Modeling of Magnetoelectric Effects in Composites* (Dordrecht: Springer)
[2] Nan C W, Bichurin M I, Dong C, Viehland D, and Srinivasan G 2008 *J. Appl. Phys.* 103 031101
[3] Fiebig M 2005 *J. Phys.* D 38 R123
[4] Srinivasan G, Priya Sh, and Sun N X 2015 *Composite Magnetoelectrics* (Cambridge: Woodhead Publishing)
[5] Petrov V and Srinivasan G 2008 *Phys. Rev. B* 78 184421
[6] Petrov V M, Bichurin M I, Kovalenko D V 2017 *Proc. of Progress In Electromagnetics Research Symp.-Spring (Saint Petersburg)* (New York: IEEE) pp 39-41.
[7] Bichurin M, Petrov V, Zakharov A, Kovalenko D, Yang S. C., D. Maurya, V. Bedekar, and S. Priya 2011 *Materials* 4 651
[8] Weaver W, Timoshenko S P, and Young D H 1990 *Vibration problems in engineering* (New York: John Wiley & Sons)

IOP Conf. Series: Materials Science and Engineering **441** (2018) 012039 doi:10.1088/1757-899X/441/1/012039

Method of indirect measurements on the basis of vector-observation

S A Popov

Yaroslav-the-Wise Novgorod State University, ul. B. St. Peterburgskaya, 41, 173004, Veliky Novgorod, Russia

E-mail: Stanislav.Popov@novsu.ru

Abstract. The method of indirect measurements is considered, which allows to increase the accuracy of estimation of the measured sought value by building a multiresponse model of measurements by means of vector-observation. The expressions for calculation of the model coefficient estimates and covariance matrices of measurement errors are given. It is shown, how by means of indirect measurements and the multiresponse model the refined estimation of the sought value and its confidence interval are calculated.

1. Indirect measurements using vector-observation

Indirect measurements refer to such measurements, in which the numerical value of the sought measured value is determined by the known functional dependence through other values that can be directly measured.

$$x = \sum_{i=1}^{m} \{a_i y_i\},$$

where x is the sought measured value, y_i is the value of the indirect measurement, a_i the correcting coefficient, m is the number of indirect measurements of the sought value x .

The order of calculations of the sought value by the measured values in the indirect measurements is set by the Recommendations [1] which set the guidelines for the estimation of the indirect measured sought value and the errors of the measurement result at linear dependence and lack of correlation between errors of measurements of arguments; in case of non-linear dependence and lack of correlation between measurements errors of arguments; for correlated error measurements of arguments when there are rows of individual values of the measured arguments.

To improve the accuracy of the measured sought value in this work it is proposed to use indirect measurements, in which to assess the desired value nonlinear dependence of the sought value and its indirect measurements (calibrating function) is used in the form of vector-observation and covariance matrix of the errors of observations of indirect measurements.

2. Calculation of non-linear calibration function coefficient estimates

Indirect multiple (group) measurements can be used to increase the accuracy of the sought value by increasing the number of measured indirect values and increasing the accuracy of their estimations. For example, a group of sensors can be used to increase accuracy by using an indirect method to measure the same sought value [2]. This article offers the method of group indirect measurement consisting of two stages.

At the stage of the function construction, the calibration function for each sensor is constructed based on the model of dependence of the sought value on sensors responses, adequately connecting the values of indirect measurements with the measured sought value.

In the second stage by means of the calibration function according to the results of measurement of sensors responses, the sought value and its confidence interval are calculated. Here it is shows how on the basis of indirect measurements the estimate of the sought value and its variance are calculated by means of indirect measurements.

The first stage defines the type of multiresponse calibration function and evaluates its coefficients; the multiresponse model of the dependence of the sought value on the sensor responses can be represented as a multiresponse function, known accurate within number coefficients:

$$\mathbf{Y} = \mathbf{F}(\mathbf{B}, x) + \mathbf{E} \quad (1)$$

where $\mathbf{Y} = \{y_1, \ldots, y_m\}^{\mathrm{T}}$ is a vector-observation of the responses of each sensor y_i at the given sought value x, $\mathbf{B} = \{b_1, \ldots, b_k\}^{\mathrm{T}}$ is a vector of the coefficients of the model, $\mathbf{E} = \{e_1, \ldots, e_m\}^{\mathrm{T}}$ is a the vector of additive errors of observation of sensor readings, $\mathbf{F}(\mathbf{B}, x) = \{\mathrm{f}_1(\mathbf{B}, x), \ldots, \mathrm{f}_m(\mathbf{B}, x)\}^{\mathrm{T}}$ is a vector of functions simulating the dependence of sensor responses from the sought value for each sensor.

The vector of estimates of the model coefficients (1) in the general view is calculated at the stage of construction of the calibration function on the results of the experiment in the form of n indirect measurements of the sought value x by means of m sensors on the basis of the following iteration procedure [2]:

$$\mathbf{B}^{s+1} = \mathbf{B}^{s} + \left[\sum_{j=1}^{n} \mathbf{P}(\mathbf{B}^{s}, x_j) \mathbf{P}(\mathbf{B}^{s}, x_j)^{\mathrm{T}} \right]^{-1} \sum_{j=1}^{n} \mathbf{P}(\mathbf{B}^{s}, x_j) [\mathbf{Y}_j - \mathbf{F}(\mathbf{B}^{s}, x_j)], \quad (2)$$

where x_j is the magnitude of the sought value, n is he number of sought values in the experiment,

The matrix of derivative functions $\mathbf{F}(\mathbf{B}, x)$ by coefficients is defined as

$$\mathbf{P}(\mathbf{B}, x) = \frac{\partial \mathbf{F}(\mathbf{B}, x)}{\partial \mathbf{B}} = \left\{ \frac{\partial \mathrm{f}_1(\mathbf{B}, x)}{\partial \mathbf{B}}, \ldots, \frac{\partial \mathrm{f}_m(\mathbf{B}, x)}{\partial \mathbf{B}} \right\}.$$

Covariance matrix of coefficient estimates (2) is calculated by the formula

$$\mathbf{V}_{\mathbf{B}} = \left\{ \sum_{j=1}^{n} \mathbf{P}(\hat{\mathbf{B}}, x_j) \mathbf{V}_{\mathbf{E}}^{-1} \mathbf{P}(\hat{\mathbf{B}}, x_j)^{\mathrm{T}} \right\}^{-1}, \quad (3)$$

where $\mathbf{V}_{\mathbf{E}}$ is the covariance matrix of observational errors.

3. Calculation of non-linear calibration function coefficient estimates

The type of the model in the expression (1), nonlinear within coefficients **B**, is selected on the basis of the study of physical processes that connect indirect and direct measurements. In most cases, it is possible to use polynomial models with a polynomial degree not higher than four. This model can be represented as the following multiresponse, linear within coefficients [3]:

$$\mathbf{Y} = \mathbf{P}^{\mathrm{T}}(x)\mathbf{B} + \mathbf{E}, \quad (4)$$

where x is the sought value, $\mathbf{Y} = \{y_1, \ldots, y_m\}^{\mathrm{T}}$ are the observed values (responses) of each of the m sensors, $\mathbf{B} = \{b_1, \ldots, b_k\}^{\mathrm{T}}$ are the estimated coefficients of the model (1), $\mathbf{E} = \{e_1, \ldots, e_m\}^{\mathrm{T}}$ are the errors of observations of the sought values, $\mathbf{P}(\mathbf{X})$ is function describing the type of dependence of the sought value x from the indirect measurements **Y** which has a dimension equals to the number of model coefficients.

By comparing multiple models as a working model (1), a third-degree polynomial is accepted, for which the number of coefficients is equal $k = 4m$ upon condition that each response is described by the vector of independent coefficients. Then the matrix $\mathbf{P}(x)$ having the dimension $(m \times km)$ takes the following view{4}:

$$\mathbf{P}^{\mathrm{T}}(x)=\begin{Bmatrix} 1 & x_1 & \dots & x_1^3 & 0 & 0 & 0 & 0 & \dots & 0 \\ 0 & 0 & 0 & 0 & 1 & \dots & x_2^3 & 0 & \dots & 0 \\ & & & & & \dots & & & & \\ 0 & 0 & 0 & 0 & 0 & & 0 & 1 & \dots & x_m^3 \end{Bmatrix}. \tag{5}$$

For linear within coefficients model (4) the estimations of the coefficients are calculated by the results of the experiment in the form of indirect measurements of n values of the sought value x for m sensors according to the following formula [2]:

$$\mathbf{B}=\left[\sum_{j=1}^{n}\mathbf{P}(x_j)\mathbf{P}(x_j)^{\mathrm{T}}\right]^{-1}\sum_{j=1}^{n}\mathbf{P}(x_j)\mathbf{Y}_j, \tag{6}$$

where $x_j, j=\overline{1,n}$, are the sought values, set in the experiment, $\mathbf{V}_{\mathbf{E}}$ is the covariance matrix of observational errors, $\mathbf{Y}_j$ is the vector of indirect measurements, n is the number of indirect measurements,

As an estimation of covariance matrix $\mathbf{V}_{\mathbf{E}}$ the matrix $\mathbf{S}_{\mathbf{E}}$ which is calculated by the following formula is used:

$$\mathbf{S}_{\mathbf{E}}=\frac{1}{v-1}\sum_{i=1}^{v}(\mathbf{Y}_i-\overline{\mathbf{Y}})(\mathbf{Y}_i-\overline{\mathbf{Y}})^{\mathrm{T}}, \tag{7}$$

where v is the number of response observations for the constant value of the sought x value.

4. Check the adequacy of the model

As the method of checking the adequacy of the multiresponse model residual analysis is used. To test the hypothesis of equality to zero the mathematical expectation of the residual was used, T^2 is the Hotelling statistics [3] in the form

$$T^2=\frac{1}{nm-k}\sum_{j=1}^{n}\mathbf{R}_j^{\mathrm{T}}\mathbf{S}_{\mathbf{E}}^{-1}\mathbf{R}_j, \tag{8}$$

where $\mathbf{R}_j=\mathbf{Y}_j-\mathbf{F}(\mathbf{B},x_j)$ is the vector of the residues of the mode in j-th observation, and the matrix $\mathbf{S}_{\mathbf{E}}$ was calculated by the formula (7).

If the residues have zero mathematical expectation, then the statistics (8) approximately has the Fischer distribution with degrees of freedom $f_1=n-k/m$ and $f_2=v-1$.

If inequality holds

$$T^2<F_{f_1,f_2,1-\alpha}, \tag{6}$$

where $F_{f_1,f_2,1-\alpha}$ is the quantile of the Fisher distribution with degrees of freedom f_1 and f_2 at the level of confidence $\alpha=0.9$.

Then the hypothesis for equality to zero of mathematical expectation of residues is accepted with confidence probability $1-\alpha$.

5. Design of experiment

The values $x_j, j=\overline{1,n}$, (design of experiment) set in the experiment are used during calculating the calibration coefficient estimates (6). The accuracy of the coefficient evaluation depends on these values (3). The optimal design of experiment is calculated by a numerical method using this iterative procedure.

The design of experiment in this case is a set of control points, $\mathbf{O} = \{x_1, x_2, \ldots, x_n\}$ where n is the number of points in experiment. The *D*-optimality criterion requires such arrangement of points in the design area Σ_x, in which the determinant of the matrix $\mathbf{V_B}$ is minimal. Thus, *D*-optimal design is defined as follows:

$$\det \mathbf{V_B}(\mathbf{O}^o) = \min_{x\in\Sigma_x} \det \mathbf{V_B}(x) = \min_{x\in\Sigma_x} \det \left[\sum_{j=1}^{n} \mathbf{P}(x_j)\mathbf{V_E}^{-1}\mathbf{P}^{\mathrm{T}}(x_j) \right]^{-1}, \tag{7}$$

where $\mathbf{O}^o = \{x_1^o, x_2^o, \ldots, x_n^o\}$ is the optimal design of experiment in terms of the criterion of *D*-optimality.

The covariance matrix of the target function $\mathbf{Y}$ calculated by the model for the vector of independent variables x equals

$$\mathrm{Var}\{\hat{\mathbf{Y}}(\mathbf{O}^o)\} = \mathbf{V_Y}(\mathbf{O}^o) = \mathbf{P}^{\mathrm{T}}(x)\mathbf{V_B}\mathbf{P}(x).$$

Extreme task (7) has a dimension ml, where l is the number of model coefficients , m is the number of observed parameters, is complicated in the sense of convergence computational procedure and the cost of machine time when using standard procedures for the extremum search. In this case the problem is complicated by the condition that the design $\mathbf{O}$ should be discrete, i.e. the control points should belong to some discrete set of permissible points area, and precise, since the number of control points should be precisely defined before starting the experiment. In view of the above limitations the following modified algorithm of the optimal design of experiment search is proposed [4].

1. Specify the number of control points n and determine the arbitrary initial design of the experiment $\mathbf{O}^1 = \{x_1^1, x_2^1, \ldots, x_n^1\}$.
2. The first point x_i is selected to search the local extreme.
3. Local extreme task is solved for one changeable control point x_i and fixed remaining points $x_1, x_2, \ldots, x_i, \ldots, x_n$ and the design of experiment $\mathbf{O}^2$ is determined in the form (7)

$$\mathbf{O}^2 = (x_1^1, x_2^1, \ldots, x_i^2, \ldots, x_n^1) = \max_{\mathbf{X}\in\Sigma_\mathbf{X}} \det \sum_{j=1}^{n} \mathbf{P}(x_j)\mathbf{V_E}^{-1}\mathbf{P}^{\mathrm{T}}(x_j)$$

   For the model linear within coefficients type of three response polynomial of the third degree.
4. The procedure in item 3 is repeated for all points in the design.
5. Items 3-4 are repeated until the specified accuracy of calculations, specified number of iterations or before the expiration of the specified time of the account has been reached.

6. Evaluation of the sought value and its variance

After the first stage of indirect measurements based on the results of the experiment determined the kind of multi-response calibration function (4) and calculated estimates of its coefficients, one can proceed to the measurement of the sought value x by indirect method. In the second stage by means of the calibration function according to the results of measurement of sensors responses, $\mathbf{Y} = \{y_1, \ldots, y_m\}^{\mathrm{T}}$ the sought value x and its confidence interval are calculated.

The iterative procedure of calculating the vector of estimates of the sought value for each sensor can be presented as follows [1]:

$$\mathbf{X}^{s+1} = \mathbf{X}^s - \left[\frac{\partial \Phi(\mathbf{B}, \mathbf{X}^s, \mathbf{Y})}{\partial \mathbf{X}} \right]^{-1} \Phi(\mathbf{B}, \mathbf{X}^s, \mathbf{Y}), \tag{8}$$

where $\Phi(\mathbf{B}, \mathbf{Y}, \mathbf{X}) = \mathbf{Y} - \mathbf{F}(\mathbf{B}, \mathbf{X}), \mathbf{X} = \{x_1, \ldots, x_m\}^{\mathrm{T}}$ is the vector of estimates of the sought value for each sensor.

The covariance matrix (8) for all sensors depends on the matrices $\mathbf{V_B}$ (3) and $\mathbf{V_E}$ (defined in the first stage):

$$\mathbf{V_X} = \mathbf{A}\mathbf{V_B}\mathbf{A}^{\mathrm{T}} + \mathbf{G}\mathbf{V_E}\mathbf{G}^{\mathrm{T}}, \qquad (9)$$

where matrices **A** and **G** are calculated as follows:

$$\mathbf{A} = \left\{ \left[\frac{\partial\Phi(\mathbf{B},\mathbf{X},\mathbf{Y})}{\partial\mathbf{B}}\right]^{\mathrm{T}} \left[\frac{\partial\Phi(\mathbf{B},\mathbf{X},\mathbf{Y})}{\partial\mathbf{X}}\right]^{-1} - \Phi(\mathbf{B},\mathbf{X},\mathbf{Y})^{\mathrm{T}} \left[\frac{\partial\Phi(\mathbf{B},\mathbf{X},\mathbf{Y})}{\partial\mathbf{X}}\right]^{-2} \frac{\partial^2\Phi(\mathbf{B},\mathbf{X},\mathbf{Y})}{\partial\mathbf{T}\partial\mathbf{B}} \right\},$$

$$\mathbf{G} = \left\{ \left[\frac{\partial\Phi(\mathbf{B},\mathbf{X},\mathbf{Y})}{\partial\mathbf{Y}}\right]^{\mathrm{T}} \left[\frac{\partial\Phi(\mathbf{B},\mathbf{X},\mathbf{Y})}{\partial\mathbf{X}}\right]^{-1} - \Phi(\mathbf{B},\mathbf{X},\mathbf{Y})^{\mathrm{T}} \left[\frac{\partial\Phi(\mathbf{B},\mathbf{X},\mathbf{Y})}{\partial\mathbf{X}}\right]^{-2} \frac{\partial^2\Phi(\mathbf{B},\mathbf{X},\mathbf{Y})}{\partial\mathbf{T}\partial\mathbf{Y}} \right\}.$$

The best linear estimate of the sought value $\bar{x}$ is calculated according as follows:

$$\bar{x} = \left(\mathbf{J}^{\mathrm{T}}\mathbf{V_X}^{-1}\mathbf{J}\right)^{-1}\mathbf{J}^{\mathrm{T}}\mathbf{V_X}^{-1}\hat{\mathbf{X}} \qquad (10)$$

where $\mathbf{J}$ is a vector with dimension m whose elements are equal to one, $\hat{\mathbf{X}}$ is a vector of local estimates obtained from formula (8).

Variance s_x^2 of estimate (10) is

$$s_x^{\ 2} = \left(\mathbf{J}^{\mathrm{T}}\mathbf{V_X}^{-1}\mathbf{J}\right)^{-1}, \qquad (11)$$

that is less than the least variance of local estimates.

With independent estimates $\hat{\mathbf{X}}$, i.e. when the covariance matrix $\mathbf{V_X}$ is a diagonal matrix, expressions (10) and (11) are converted into a known formula for of varying accuracy observations.

7. Results of group indirect measurements on the basis of multiresponse functions

Large values of correlation coefficients of local estimates significantly increase the variance of the overall estimate, so if one does not take into account the strong correlation, the variance of the estimate of the sought value will be significantly underestimated.

With large positive correlation coefficients, the estimate of average $\bar{x}$ may be less than the least of the local estimates x_i. Let us, for example, for $m = 2$ (two sensors) known local estimates are $x_1 = 50$ and $x_2 = 60$. Let, also, the covariance matrix $\mathbf{V_X}$ equals

$$\mathbf{V_X} = \begin{Bmatrix} 1 & 1{,}2 \\ 1{,}2 & 2 \end{Bmatrix}.$$

Then from (10) it is possible to get the estimate of the sought value $\bar{x} = 46{,}67$ at variance of estimation of average (11) $s_{\bar{x}}^{\ 2} = 0{,}94$. Without taking into account the correlation, i.e. considering the matrix $\mathbf{V_X}$ as diagonal, we get $\bar{x} = 53{,}33$ and $s_{\bar{x}}^{\ 2} = 0{,}67$. Thus, the error of estimating of the sought value and its covariance without correlation can be 14.3% and 28.5%, respectively. This is a fairly significant error, so in calculating the average on varying accuracy observations, the verification of the independence of local estimates is necessary. If these values are not independent, use expressions (10) and (11) to estimate the average.

The method of indirect measurements based on multiresponse function was used to estimate the magnitude of the homogeneous color (sought value) of computer image by the pixel color values of the photoreceiver matrix [5]. As the object of research the CCD matrix of a photoreceiver ISX455AQF of type with image size 1712 × 1200 pixels (2 MP) was used. Images of homogeneous colors of the RAL D2 system containing 1625 color samples were carried out. Images were executed in macromode several times for each color with the rotation of 45^0; then the color coordinates were averaged. As a result, for standard color coordinates in the CIELAB system with known color coordinates were calculated colors for each pixel. As a result of calculations it was established that the coefficients in the model (4) in this case can be accepted as the same for all pixels of the photoreceiver matrix. Thus, the number of coefficients of the overall image model (4) is assumed to be four, so the matrix $\mathbf{P}^{\mathrm{T}}(x)$ had a dimension $(4\times m)$and took the following form.

$$\mathbf{P}^{\mathrm{T}}(x)=\begin{Bmatrix} 1 & x_1 & \dots & x_1^3 \\ & \dots & & \\ 1 & x_m & \dots & x_m^3 \end{Bmatrix}.$$

Calculation of average values of color coordinates on homogeneous colors of RAL system showed that the error of color reproduction by the photoreceiver matrix is (6-8)%. Calculation of color coordinates according to the formula (8) on the basis of the built model (4) and obtained covariance matrix of these estimates $\mathbf{V_X}$ (8) allowed to adjust this error to the values of (0.8-2)%.

Conclusion

Indirect measurements in the engineering allow to estimate the values for which direct measurements are difficult or impossible. For this purpose the set of sensors is used, according to which indications it is possible to calculate the sought value with minimal error. The more such sensors are, the more accurate the sought value is measured. To ensure the most accurate evaluation of the measured value, it is necessary to take into account the type of dependence of sensor readings from the measured sought value and their statistical relations. It is possible to increase the accuracy of evaluation by using multiresponse model that describes the dependence of sensor readings on the measured sought value. This paper describes the method of estimating indirect measurements based on the building of a multiresponse model. This method involves design and carrying out an experiments to select a type of model, calculate its coefficients, and covariance matrices. Then, on the basis of the received information calculation of point and interval estimation of the measured value is carried out. The carried out experiment on estimation of homogeneous color of digital image has shown, that this method allows in this case to increase accuracy of estimation up to (5-7)%.

References

[1] *Recommendation. State system for ensuring the uniformity of measurements. Indirect measurements. Evaluation of measurement results and estimation of their errors. MI 2083-90* 1991. Standards Publishing House 1-4

[2] Rencher, Alvin C. 2002 *Methods of multivariate Analysis* A John Wiley & Sons, Inc. Publication. Brigham Young University pp339-345

[3] Popov S. A. and Klykov N. N. 2013 *Vestn. Novgorod. Gos. Univ.* **2 (75)** pp 96-98

[4] Popov S. A. 2016 *Vestn.estn. Novgorod. Gos. Univ.* **4 (95)** pp 53-57

[5] Popov S.A., Emelyanov G.M. 2005 *Pattern Recognition and Image Analysis* **15(2)** pp 63-65

IOP Conf. Series: Materials Science and Engineering **441** (2018) 012040 doi:10.1088/1757-899X/441/1/012040

Construction of separating function for objects classification and its use for classification of spherical segments from material with shape memory

S A Popov

Yaroslav-the-Wise Novgorod State University, 41, B.St.Peterburgskaya st., Veliky Novgorod, 173004, Russia
E-mail: Stanislav.Popov@novsu.ru

Abstract. An engineering method for constructing a separating surface for discriminating classes of multiparameter objects is described. The method is based on the proposed approximation of the membership function, which makes it possible to classify with a complex form of the separating surface. An algorithm is given for calculating the estimates of the coefficients of the nonlinear model with respect to the coefficients of the separating surface model using an iterative procedure based on the method of least squares. The results of experiments and calculations on the classification of spherical segments from materials with shape memory, proving the effectiveness of the proposed method are presented.

1. Approach to classification

There is a wide class of problems in which it is necessary to determine regions belonging to different classes by a certain criterion for multiparameter objects in the space of their parameters $\mathbf{X} = \{x_1, x_2, \ldots, x_m\}^{\mathrm{T}}$. Selection of such areas allows you to determine the properties of the objects that provide a given class. Such a problem arises, for example, when determining the optimal geometric parameters of segments from materials with shape memory for different classes by the presence of a rebound [1]. Usually in practice, the number of parameters is large enough and the classes are intermixed, so mathematical-statistical methods are used to distinguish the classes [2].

However, the formal use of such methods often does not allow us to separate classes with sufficient accuracy for practice, since the interfaces between classes are usually quite complex, and their form is unknown. For example, the use of Fisher's discriminant for constructing a class separating classes leads to large class definition errors due to insufficient information about the distribution of components for each class. In these conditions, it is proposed to use the method of least squares to construct the separating surface [3, 4].

2. Classification problems and methods to solve them

In accordance with this method of estimating the coefficients **B**, separating classes surface $\mathrm{g}(\mathbf{X},\mathbf{B})=0$ are constructed so that they provide a minimum of the sum of the squares of the deviations in the form [5]

$$\sum_{i=1}^{n}\left[g(\mathbf{X},\mathbf{B})-y_i\right]^2=\min_{\mathbf{B}}\left\{\sum_{i=1}^{n}\left[g(\mathbf{X},\mathbf{B})-y_i\right]^2\right\}, \quad (1)$$

where $g(\mathbf{X},\mathbf{B})$ is the separating function, $\mathbf{X}=\{x_1,x_2,\ldots,x_m\}^T$ is the vector of parameters, $\mathbf{B}=\{b_1,b_2,\ldots,b_l\}^T$ is the coefficient estimation vector, y is the attribute of belonging to the given class (for example, y = 1 if the object belongs to the first class by test results, and y = -1, second class), n is the sample size.

However, with this approach, contributions to the sum of deviations (1) will be produced by objects with large values of the separating function $g(\mathbf{X},\mathbf{B})$, which leads to significant errors in class recognition, especially in the complex form of the classifying function, which is typical for technical applications.

3. The method of constructing a separating surface for two classes

For a objects, there is a probability of belonging to the class defined q_1 and the probability of belonging to an alternative class (q_2 ($q_1+q_2=1$)). Experimentally, belonging to this class ($y=1$) or not belonging to this class ($y=-1$) is determined by the results of testing objects that give an estimate of the value of the membership function $q(\mathbf{X})=q_1-q_2$ as a function of the magnitude of the component vector. As a result of testing of n articles, the evaluation of the membership vector is determined $\mathbf{Y}=\{y_1,y_2,\ldots,y_n\}^T$.

Approximation of the dependence of the magnitude q on the values of the parameters is proposed to be presented in the form [6]:

$$q(\mathbf{X},\mathbf{B})=sign[f(\mathbf{X},\mathbf{B})]\{1-\exp[-|f(\mathbf{X},\mathbf{B})|]\}, \qquad (2)$$

where $f(\mathbf{X},\mathbf{B})$ is some function whose form and coefficients are determined from the experimental data.

Function (2) is equal to zero on the interface between two classes and tends to 1 (or -1) when moving away from the separating surface. The separating surface is defined by the equation

$$q(\mathbf{X},\mathbf{B})=0. \qquad (3)$$

In this case, the calculation of the coefficient estimates in expression (3) by the least squares method is performed using the following iterative procedure [7,8]:

$$\mathbf{B}^{s+1}=\mathbf{B}^s+\left[\sum_{i=1}^{n}\mathbf{P}(\mathbf{X}_i,\mathbf{B}^s)\mathbf{P}^T(\mathbf{X}_i,\mathbf{B}^s)\right]^{-1}\sum_{i=1}^{n}\mathbf{P}(\mathbf{X}_i,\mathbf{B}^s)[y_i-\mathbf{q}(\mathbf{X}_i,\mathbf{B})], \qquad (4)$$

where s is the iteration number,

$$\mathbf{P}(\mathbf{X},\mathbf{B})=\frac{\partial q(\mathbf{X},\mathbf{B})}{\partial \mathbf{B}}=sign[f(\mathbf{X},\mathbf{B})]\exp[-|f(\mathbf{X},\mathbf{B})|]\frac{\partial f(\mathbf{X},\mathbf{B})}{\partial \mathbf{B}}, \qquad (5)$$

and $\frac{\partial f(\mathbf{X},\mathbf{B})}{\partial \mathbf{B}}=\{1,x_1,x_2,x_3,x_1x_2,x_1x_3,\ldots,x_3^2\}^T$ for a complete quadratic function.

The covariance matrix for estimating the coefficients (4) is calculated by formula

$$\mathbf{V}_{\mathbf{B}}=\sigma_e^2\left[\sum_{i=1}^{n}\mathbf{P}(\mathbf{X}_i,\mathbf{B})\mathbf{P}^T(\mathbf{X}_i,\mathbf{B})\right]^{-1}, \qquad (6)$$

where σ_e^2 is the variance of the class recognition error.

As an estimate of the variance of class recognition, the variance of the model's adequacy error is calculated, by formula:

$$\sigma_e^2=\frac{1}{n-l}\sum_{i=1}^{n}\left[q(\mathbf{X}_i,\mathbf{B})-y_i\right]^2, \qquad (7)$$

where l is the number of coefficients in the model (2).

The diagonal elements of matrix $\mathbf{V_B}$ (6) represent the variances of the estimates of the corresponding coefficients, which makes it possible to calculate their t-statistics and, therefore, to check the significance of the coefficient estimates.

The lower boundary of the one-way confidence interval for the separating function that ensures the probability of recognizing the first class is not less than *P%* is defined as:

$$\mathrm{q}(\mathbf{X},\mathbf{B})=\mathrm{q}(\mathbf{X},\mathbf{B})-t_{P,f}\sigma_e, \tag{8}$$

where t_P is the Student's distribution quantile for the confidence probability P and the number of degrees of freedom f, $\mathbf{B}$ is the coefficient estimations vector.

4. Construction of the model of the dependence of impact force of spherical segments from material with shape memory from geometric parameters

The described procedure for constructing a separating function was used to classify spherical segments from material with shape memory according to the classes of magnitude of the rebound strength when using different geometric dimensions. Experimentally obtained data on the dependence of the impact force of spherical segments on their geometric dimensions were processed by regression analysis [3]. As a model describing the dependence of the impact force of a segment on the obstacle on the ratio of geometric parameters D/R и h/R, we used the quadratic function in form

$$P_{sp}=b_0+b_1z_1+b_2z_2+b_{12}z_1z_2+b_{11}z_1^2+b_{22}z_2^2, \tag{9}$$

where z_1 and z_2 are independent variables, $z_1=D/R$ and $z_2=h/R$, P_{sp} is the specific impact force with a clap (kg).

The coefficients of the model b_o, b_1, b_2, b_{12}, b_{11} and b_{22} were determined by the method of least squares [9,10]. The significance of the coefficients b_i was checked according to the t-criterion [9], when the t-statistics in form $t_i=|b_i|/s_{bi}$ is compared with the student's Student's distribution quantile for a given number of degrees of freedom $f=n–k$, where n is the number of experiments, k is the number of coefficients in the model, significance level (the significance level was assumed to be 0.05). Estimates of the coefficients of model (9) and the values of t-statistics are given in Table 1.

Table 1. Estimates of the coefficients.

b_i	b_0	b_1	b_2	b_{12}	b_{11}	b_{22}
	30.8	-270.8	7522.8	-32407.5	613.2	432437.1
t_i	1.556	-3.455	4.182	-5.098	5.247	4.312

A coefficient is assumed to be insignificant at the level of significance α if inequality is valid,

$$t\le t_{f,1-\alpha}, \tag{10}$$

where $t_{f,1-\alpha}$ is the Student's distribution quantile for the probability $1-\alpha$ with the number of degrees of freedom $f=n-k$, n is the number of observations, k is the number of coefficients in the model equation.

In this experiment, n = 72 and k = 6 for model (9), then f = 66 and with the Student's quantile distribution $t_{f,1-\alpha}$ = 1.67. From Table 1 it follows that the coefficient b_0 is insignificant, so the corresponding term is excluded from equation (9); the values of the coefficient estimates are recalculated for an equation with five coefficients in form

$$P_{sp}=b_1z_1+b_2z_2+b_{12}z_1z_2+b_{11}z_1^2+b_{22}z_2^2. \tag{11}$$

The estimates of the coefficients b_i calculated for equation (11) and the corresponding values of the t-statistics (10) are given in Table 2.

Table 2. The estimates of the coefficients b_i.

b_i	b_1	b_2	b_{12}	b_{11}	b_{22}
	-174.0	6922.9	-31595.3	538.8	432726
t_i	-3.611	3.899	-4.935	5.000	4.270

In this case, f = 67 and $t_{f,1-\alpha}$ = 1.67. From Table 2 is clear that all the coefficients are significant. Thus, the calculation of the strength of clap is carried out in accordance with the expression (3), the coefficients b_i of which are presented in Table 2.

To check the adequacy of the constructed model (10), the residual dispersion s_r^2 was calculated by formula:

$$s_r^2 = \frac{1}{f}\sum_{j=1}^{n}(P_{y\eth j} - P_{ypj})^2 \tag{12}$$

where $P_{s\eth j}$ is the evaluation of the strength of the impact, obtained from the expression (3) for the j-th observation, P_{spj} is the corresponding observed value of the strength of impact, f is the number of degrees of freedom.

For model (11), the residual dispersion (12) is equal to s_r^2=2.847 kg^2. For the model constructed in [4] for the geometric parameters D, R and h, the residual dispersion has the value s_r^2=9.07 kg^2, i.e. model (11), which uses the D/R and h/R ratios, agrees much better with the experimental data.

After substituting the calculated coefficients into equation (11), expression is obtained that describes the dependence of the impact force of a spherical segment on the impeding body on the ratios of the geometric parameters D/R and h/R, in form

$$P_{sp} = -174\cdot\frac{D}{R} + 6923\frac{h}{R} - 316\cdot 10^2\frac{D}{R}\cdot\frac{h}{R} + 539\cdot\left(\frac{D}{R}\right)^2 + 432\cdot 10^3\cdot\left(\frac{h}{R}\right)^2. \tag{13}$$

Our studies show that not all combinations of the parameters D/R and h/R realize clapping and hitting conditions and on obstructive body. In this connection, it is of interest to construct the separating function in the coordinate system D/R and h/R, which allows us to divide the regions where the impact on the obstructive body takes place and in which it is absent.

5. Construction of separating function

Analysis of the experimental data on the strength of the impact, depending on the ratio of the D/R and h/R segments, showed that the segments clasping with clap and impact against the interfering body (force-measuring device) are concentrated within a certain closed region (Figure 1) restoring the form without impact, are beyond the boundaries of this area.

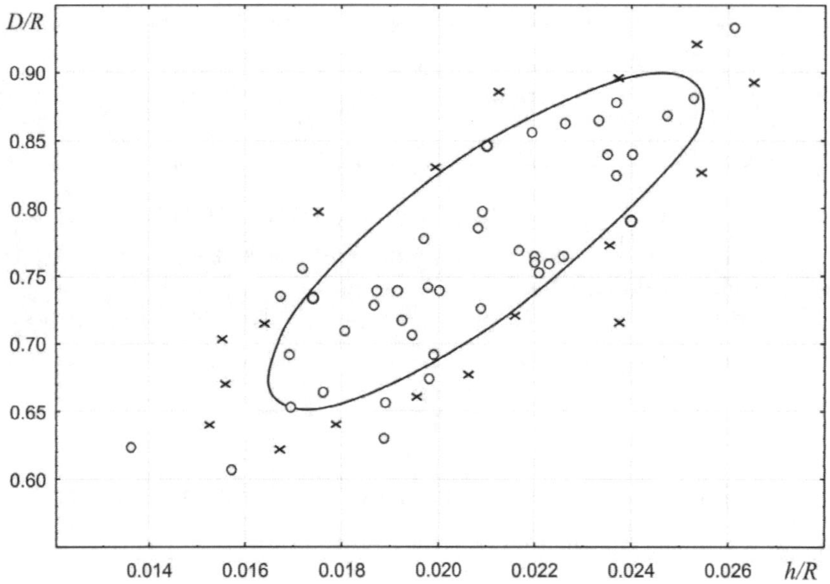

Figure 1. The scattering field for the geometric parameters of segments and the form of the separating function (circles –with impact, crosses - without).

To determine the range of the variables (z_1 and z_2) that provide the real clap of the segment with impact, it is necessary to construct the separating function $g(\mathbf{Z},\mathbf{A})$. As the separating function $g(\mathbf{Z},\mathbf{A})$, a second-order function is chosen in form

$$g(Z,A) = a_0 + a_1 z_1 + a_2 z_2 + a_{12} z_1 z_2 + a_{11} z_1^2 + a_{22} z_2^2 , \tag{14}$$

where $\mathbf{Z} = \left\{ \frac{D}{R}, \frac{h}{R} \right\}^{\mathrm{T}}$ is the vector of geometric parameters z_1, z_2; $\mathbf{A} = \{a_1, a_2, a_{11}, a_{12}, a_{12}\}^{\mathrm{T}}$ is the coefficient vector.

The calculation of the separating function was carried out by the method of least squares [3]. For segments with an impact on the obstacle, the separating function takes positive values, and for segments without impact - negative. To calculate the estimates of the coefficients A, the values of +1 and -1 for the segments without impact were taken as the observed value of the function $g(\mathbf{Z},\mathbf{A})$ for the impacted segments. However, the further the segments in the *D/R* and *h/R* coordinate systems are from the region boundary, the larger the calculated value of the function $g(\mathbf{Z},\mathbf{A})$, which leads to large errors in the construction of the function $g(\mathbf{Z},\mathbf{A})$. To limit the influence of segments that are far from the interface on the value of the separating function $g(\mathbf{Z},\mathbf{A})$ when calculating the coefficients A, equation (14) was transformed by using instead the logarithm function of the module of this function $\mathrm{h}(\mathbf{Z},\mathbf{A})$ with the corresponding sign as follows:

$$\mathrm{h}(\mathbf{Z},\mathbf{A}) = sign[\mathrm{g}(\mathbf{Z},\mathbf{A})]\ln(|\mathrm{g}(\mathbf{Z},\mathbf{A})| + 1), \tag{15}$$

or in the elemental form

$$\mathrm{h}(\mathbf{Z},\mathbf{A}) = sign[\mathrm{g}(\mathbf{Z},\mathbf{A})]\ln\left(\left| a_0 + a_1\frac{D}{R} + a_2\frac{h}{R} + a_{12}\frac{Dh}{R^2} + a_{11}\left(\frac{D}{R}\right)^2 + a_{22}\left(\frac{h}{R}\right)^2 \right| + 1\right),$$

where $sign(x)=1$ for $x \geq 0$ and $sign(x)=-1$ for $x<0$, $|g(\mathbf{Z},\mathbf{A})|$ is the modulus of the function $g(\mathbf{Z},\mathbf{A})$.

The term +1 provides positive values of the function for positive values of the function.

Function $h(\mathbf{Z},\mathbf{A})$ is non-linear with respect to the coefficients, so the coefficients are calculated iteratively. Calculation of estimates of the coefficients A of equation (6) was carried out with the help of the following iterative procedure [5]:

$$\mathbf{A}^{s+1}=\mathbf{A}^{s}+\left[\sum_{j=1}^{n}\mathbf{F}(\mathbf{Z}_j,\mathbf{A}^s)\mathbf{F}(\mathbf{Z}_j,\mathbf{A}^s)^{\mathrm{T}}\right]^{-1}\sum_{j=1}^{n}\mathbf{F}(\mathbf{Z}_j,\mathbf{A}^s)\left[u_j-h(\mathbf{Z}_j,\mathbf{A})\right], \quad (16)$$

where u_j is the observed value of function $h(\mathbf{Z},\mathbf{A})$ (15) for the *j*-th segment (+1 or -1).

Here the derivative of the function $h(\mathbf{Z},\mathbf{A})$ with respect to the coefficients A is equal to

$$\mathbf{F}(\mathbf{Z},\mathbf{A})=\frac{\partial h(\mathbf{Z},\mathbf{A})}{\partial \mathbf{A}}=sign[g(\mathbf{Z},\mathbf{A})]\frac{1}{|g(\mathbf{Z},\mathbf{A})|+1}\frac{\partial g(\mathbf{Z},\mathbf{A})}{\partial \mathbf{A}}, \text{ and } \frac{\partial g(\mathbf{Z},\mathbf{A})}{\partial \mathbf{A}}=\{1,z_1,z_2\}^{\mathrm{T}}.$$

The covariance matrix of the coefficient estimates is calculated by the formula

$$\mathbf{V}_{\mathbf{A}}=s_e^2\left[\sum_{J=1}^{n}\mathbf{F}(\mathbf{Z}_j,\mathbf{A})\mathbf{F}(\mathbf{Z}_j,\mathbf{A})^{\mathrm{T}}\right]^{-1}, \quad (17)$$

where s_e^2 is the variance of the observation error.

As an estimate of variance in the observation of classes, the residual variance calculated using formula:

$$s_r^2=\frac{1}{n-k}\sum_{j=1}^{n}\left[h(\mathbf{Z}_j,\mathbf{A})-u_j\right]^2, \quad (18)$$

where *k* is the number of coefficients in the model (14) (in this case, $k = 6$).

The diagonal elements of the matrix $\mathbf{V}_{\mathbf{A}}$ (17) represent the variances of the estimates of the corresponding coefficients, which makes it possible to calculate their *t*-statistics and, thus, to verify the significance of the coefficient estimates.

The experimental data for calculating the separating function included 72 segments with clap and 65 segments without clap (a total of 137 segments). To calculate the estimates of the coefficients of the separating function from formula (7), a computer program was developed. The results of calculating the estimates of the coefficients of the separating function and *t*-statistics are given in Table 3.

Table 3. The results of calculating the estimates of the coefficients of the separating function.

	a_0	a_1	a_2	a_{12}	a_{11}	a_{22}
a_i	-36.51	163.17	-2268.13	14494.10	-301.87	-215519.28
t_i	-10.46	13.06	-9.94	19.50	-19.40	-18.54

The number of degrees of freedom is $f = 131$. In this case the quantile of the Student's distribution is equal to $t_{f,1-\alpha}= 1.978$ at the significance level $\alpha= 0.05$. Thus, all the coefficients in Table 3 are significant.

When equation (15) is used with the coefficients given in Table. 3, it turned out that seven segments hit the segment class without impact, with impact, i.e. are not recognized correctly.

For practical applications of segments, it is important that the segments are recognized correctly without impact. To reduce the error of segment recognition without impact, the observed values of the function $h(\mathbf{Z},\mathbf{A})$ were assumed with some weight. The higher this weight, the more precisely the segments of the corresponding class are identified.

It was empirically established that with the observed value of the function $h(\mathbf{Z},\mathbf{A}) = -2.5$ for segments without impact, all such segments are identified correctly (it can be shown that this corresponds to the probability of incorrect segment recognition without impact, equal to 0.01). After that, the coefficients of the separating function were recalculated. Table 4 shows the results of calculating the coefficients and *t*-statistics for this case.

Table 4. The results of calculating the coefficients.

	a_0	a_1	a_2	a_{12}	a_{11}	a_{22}
a_i	-99.00	358.35	-3502.02	27422.56	-601.85	-425168.92
t_i	-28.33	27.89	-13.47	33.82	-35.75	-32.69

Figure 2 shows the form of the separating function. The inner area in Figure 4 is the area of acceptable values of the geometric parameters of the segments at which snapping with impact is realized.

6. Calculation of optimal values of geometric parameters

The specific impact strength P_{sp} for the given geometric parameters *D/R* and *h/R* was calculated from the model (3) for segments that were within the allowed range. Optimum values of geometric parameters were determined by the method of statistical tests. The program developed for this calculation performs the following actions:

1. generation of geometrical parameters of the segment *D/R* and *h/R* in the specified range;
2. checking the belonging of a segment to a class with a impact determined by inequality $g(\mathbf{Z},\mathbf{A})>0$; if this inequality is not satisfied, then return to item1;
3. calculation of the impact force P_{sp} for the specified geometric parameters *D/R* and *h/R* according to the model (13);
4. remembering the greatest value of the impact force P_{sp} and the corresponding values of the geometric parameters *D/R* and *h/R*;
5. items 1-4 are repeated predetermined number of times.

Optimization can be performed under given constraints on geometric parameters. To do this, in step 1, for example, you can set the value of *D/R* and the optimization will be performed for this value.

When finding the optimal parameters, 10^6 statistical tests were performed, the accuracy of determination of P_{sp} and parameters was 1.1×10^{-4}. The dependence of the impact force P_{sp} on the parameters *D/R* and *h/R* obtained as a result of this calculation is shown in Figure 2.

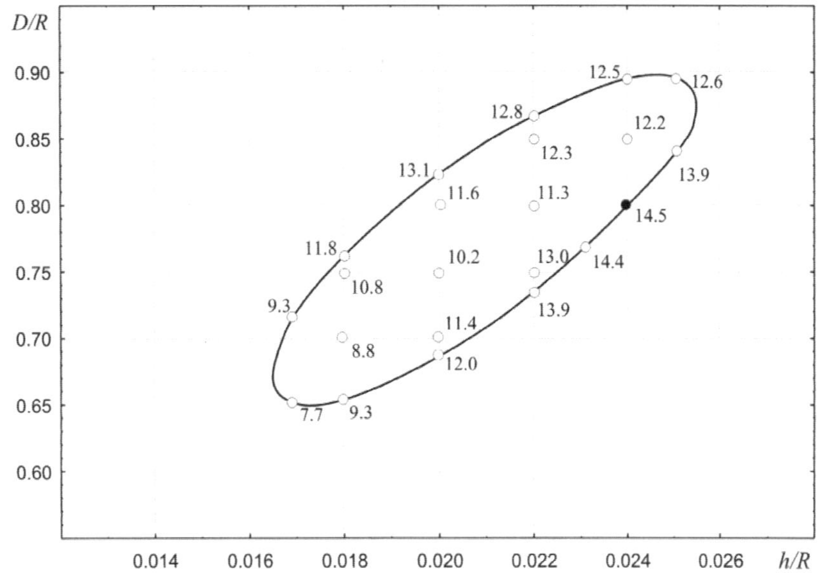

Figure 2. Type of separating function and modeled impact force values (kg) for various parameters *D/R* and *h/R*.

In Figure 2 numbers show the calculated values of the impact force for different values of *D/R* and *h/R*. The black point shows the position of the maximum impact force without any restrictions on geometric parameters. The greatest values of the impact force are in the lower right side of the interface, i.e. at the edge of the allowable area in which clap with impact is performed. The maximum value of the impact force is at the point h/R=0.024 and D/R=0.79; the force being equal to $P_{sp,\max}$ =14.6 kgf.

Thus, using the form of the separating function, it is not difficult to choose the relations of geometrical parameters (*D/R* and *h/R*), at which the impact on the obstacle with the strongest force, or close to it, is performed.

In this model, the diameter of the segment is taken equal to 17 mm. Therefore, if in addition to the diameter *D*, the thickness h of the plate is given, then a constant value of the ratio D/h is provided in the program calculation. For example, given *D* = 16.5 mm and *h* = 0.4 mm, we obtain the values *D/R* = 0.80 mm and *h/R* = 0.019 mm; at these values, the impact force is 12.66 kgf. The resulted calculation results allow to optimize geometrical parameters of spherical segments.

7. Conclusion

The method for constructing separating function for classification of objects with class mixing based on the method of least squares is developed. This method allows one to build a separating function based on experimental data and use it to determine whether objects belong to a particular class.

The model of the dependence of the impact force on the geometrical parameters of the spherical segments (*D/R* and *h/R*) allows us to estimate the impact force of the segment on the obstructing body. The model will allow one to determine the geometrical parameters of segments to obtain the greatest impact force.

The constructed separating function in the space of geometrical parameters *D/R* and *h/R*, allows to determine the regions of geometric parameters of spherical segments from a material with shape memory with impact on obstructing body and without impact. This function enables geometric parameters to identify segments in which the impact is not realized.

The obtained calculations by the model are in good agreement with the experimental results, which makes it possible to determine the optimal sizes of the spherical segments, which ensure the impact of the maximum force.

References

[1] Popov S. A. , Khusainov M.A. Bondarev A.B. Andreev V.A., 2005 *Vestn. Novgorod. Gos. Univ.* **2 (34)** 12–16
[2] Popov S. A. , Andreev V.A., Khusainov M.A. Bondarev A.B. 2006 *Vestn. Novgorod. Gos. Univ.* **2 (39)** 28–30
[3] Khartman K. 1977 *Experiment Design in Investigation of Technological* Prosesses Moscow, "Mir" 552
[4] Bard Y. 1974 Nonlinear Parameter Estimation Academic Press, New York 349
[5] Popov S. A. , Andreev V.A. 2006 *Vestn. Novgorod. Gos. Univ.* **2 (39)** 27–32
[6] Popov S. A. , Zhizhin V.V. 2007 *Vestn. Novgorod. Gos. Univ.* **2 (44)** 11–13
[7] Popov S. A. , Zhizhin V.V. 2009 *Vestn. Novgorod. Gos. Univ.* **2 (50)** 69
[8] Popov S. A. 2010 *Zavodskaya Laboratoriya.* (**4**) 40-42
[9] Khusainiv M.A., Popov S. A. Bondarev A.B. 2012 *Vestn. Novgorod. Gos. Univ.* **2 (67)** 38-40
[10] Khusainiv M.A.Popov S. A. , Malukhina O.A. 2015 *Journal of Applied Physics* 2015. (**8**) 46-52

Investigation of the energy-saving hydraulic drive of a multifunctional automobile with a subsystem of accumulation of compressed air energy

V I Posmetev, V O Nikonov and V V Posmetev

Voronezh State University of Forestry and Technologies named after G. F. Morozov, 112, Lomonosova ave., Voronezh, 394087, Russia

E-mail: posmetyev@mail.ru

Abstract. A mathematical model of the operation of the pneumatic subsystem of the compressed air energy storage device and the computer program developed on its basis are presented, which made it possible to investigate the influence of the geometric and thermophysical parameters of the pneumatic system elements on the efficiency of energy storage and heat losses.

1. Introduction

One of the promising and realizable ways to increase the efficiency of a multifunctional automobile, for example, for the removal of timber or the care for forest shelterbelts, is to equip it with recuperative mechanisms that ensure the repeated use in the work process of energy that is not efficiently dissipated into the environment, and also contribute to improving reliability and simplifying its construction. «Figure 1» presents the scheme of the device developed by the authors for accumulating the energy of compressed air in the hydraulic drive of a multifunctional automobile.

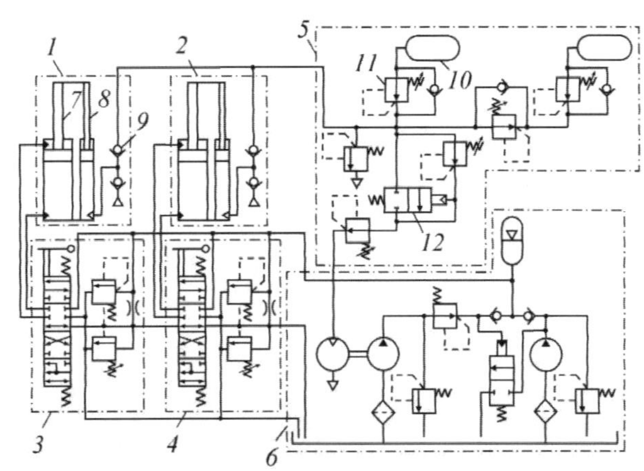

Figure 1. Scheme of the device for accumulating compressed air: *1*, *2* – energy recovery mechanisms; *3*, *4* – hydraulic distributors, *5* – pneumatic accumulator unit; *6* – hydraulic pump unit; *7* – the

hydraulic cylinder, *8* – pneumatic cylinder; *9* – reverse pneumatic valve; *10* – receiver; *11* – reduction pneumatic valves; *12* – pneumatic two-position distributor

One of the serious problems of the practical implementation of the proposed scheme is the importance of increasing the efficiency of accumulation of compressed air energy in the receiver, by reducing the thermal losses of energy stored in it. The main reason for this is the occurrence of free convection, which increases the heat transfer to the walls of the receiver with a significant increase in air temperature during its rapid compression, long storage, and with a sharp decrease in air temperature during its expansion. A negative consequence of this heat and mass transfer is the appearance of moisture, the freezing of the elements of the pneumo-mahogany and the pneumatic engine with the expansion of compressed air [1-6].

2. Materials and methods

To solve this problem, a mathematical model was developed for the operation of the pneumatic subsystem of a multifunctional automobile that encompasses the kinematics of an air cylinder, heat transfer processes in the main pneumatic elements and the processes of transferring compressed air between them. To achieve sufficient universality, the model is based on numerical methods.

From the elements of the pneumatic subsystem, three types of elements that have a significant influence on the process of energy storage and heat losses are chosen for modeling: pneumocylinders, coupled with hydraulic cylinders of the manipulator; pipelines between pneumatic cylinders and receivers; receivers. For each of the listed elements, a separate model has been developed that reproduces both the transfer and change of the state of the gaseous medium and the heat and mass exchange processes. Consider the compilation of the model by the example of the most complex element – pneumatic cylinder.

Modeling of heat propagation is complicated by the need to transfer to the model the elements of the pneumatic system with high detail and spatial resolution, as well as the ability to specify different laws of movement of the rod of the pneumatic cylinder. In this case, it is important to take into account the peculiarities of both single movement of the piston rod during the operation of the multifunction machine, as well as multiple, including periodic motion. Therefore, a simplified analytical model of heat transfer is used, but a grid finite-difference numerical method. The model is based on the basic equations of classical thermodynamics. The complexity of the problem described above is overcome by using the discretization of space with a cubic grid and discretization of time. In the three-dimensional case, the propagation of heat is described by «equation (1)» of thermal conductivity:

$$\frac{\partial}{\partial t}T(\vec{r},t)=\left(\nabla,\ \chi(\vec{r},t)\nabla T(\vec{r},t)\right)+Q(\vec{r},t), \tag{1}$$

Where $T(\vec{r},t)$ – the distribution of temperature in space and its variation over time; $\vec{r}$ – radius of the point of space; t – time; $\nabla=\frac{\partial}{\partial x}\vec{i}+\frac{\partial}{\partial y}\vec{j}+\frac{\partial}{\partial z}\vec{k}$ – differential operator; x, y, z – cartesian coordinates of the space point under study; $\vec{i},\vec{j},\vec{k}$ – unit vectors in the cartesian space; (,) – scalar product; $\chi(\vec{r},t)$ – coefficient of thermal conductivity of matter; $Q(\vec{r},t)$ – the field of heating and cooling sources that changes over time.

To transmit in the model the shape and structure of the pneumatic system elements with sufficient detail, the space in which the simulation is performed is discretized by a rectangular grid with a pitch of h = 5 mm. The grid dimensions in the three spatial directions X, Y, Z are 150 × 60 × 60 knots; the total number of nodes is 540000, which makes the model sensitive to the performance of computer technology. Each grid node for solving the heat conduction equation has six adjacent nodes from which heat can be received, or which transfers heat (Figure 2).

In the finite-difference formulation of the problem «equation (1)» can be rewritten as follows:

$$\frac{\Delta T_{i,j,k}}{\Delta t} = \chi_{i,j,k}\left(\frac{\Delta T_{i,j,k}}{\Delta x} + \frac{\Delta T_{i,j,k}}{\Delta y} + \frac{\Delta T_{i,j,k}}{\Delta z}\right) + Q_{i,j,k}, \tag{2}$$

where (i, j, k) – host indexes; $T_{i,j,k}$ – knot temperature; Δt – step of time sampling (0.1 s); $\Delta x = \Delta y = \Delta z = h$ – step of discretization of space; χ – coefficient of thermal diffusivity; $Q_{i,j,k}$ – supply of heat from the external environment to this cell.

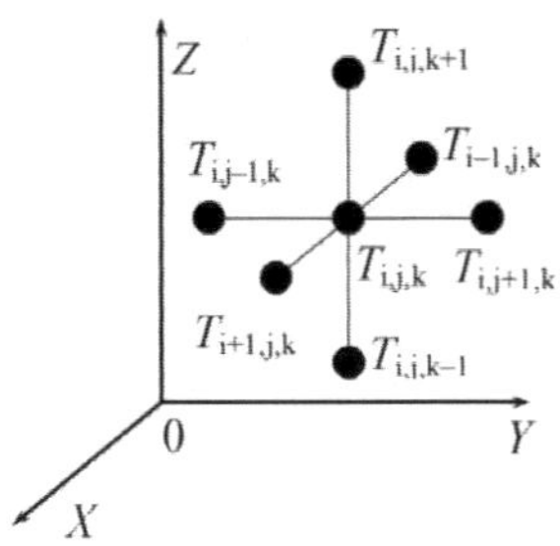

Figure 2. Cross-like scheme of accounting for neighboring nodes in the grid solution of the heat equation.

Describing in detail the finite differences in «equation (2)» and transforming, we obtain the following final «equation 3» for modeling the propagation of heat:

$$\frac{T^{\tau+1}_{i,j,k} - T^{\tau}_{i,j,k}}{\Delta t} = \frac{\chi_{i,j,k}}{h}\left(T^{\tau}_{i+1,j,k} + T^{\tau}_{i-1,j,k} + T^{\tau}_{i,j+1,k} + T^{\tau}_{i,j-1,k} + T^{\tau}_{i,j,k+1} + T^{\tau}_{i,j,k-1} - 6\cdot T^{\tau}_{i,j,k}\right) + Q_{i,j,k}. \tag{3}$$

The last «equation 3» allows us to calculate the temperature $T^{\tau+1}_{i,j,k}$ of each cell (i, j, k) for the next step of time integration $\tau + 1$, based on the current temperature $T^{\tau+1}_{i,j,k}$ of the current integration step τ.

The problem of heat propagation is solved for media of two types: the metal from which the element of the pneumatic system and the heat-insulating material is made, which is surrounded by the element of the pneumatic system. The working gas of the pneumatic system is modeled separately: in the homogeneous gas approximation it is considered that the convective processes in the gas in each pneumatic element proceed much faster than the processes of heat propagation. The heat exchange between the pneumatic system element and the surrounding air is reproduced in the model by setting the boundary conditions on the external surface of the pneumatic system elements. Thus, the grid thermal problem is not solved for the surrounding gaseous medium and the working gas medium, for which the change in the state as a whole and the mass transfer between the elements of the pneumatic system are modeled.

A particular difficulty is the consideration in the thermal problem of the movement of the rod and the piston of the air cylinder. On the basis of the law of motion of the piston $x_c(t)$ at each step of integration τ, the real number x is bound to the grid by rounding to an integer, and shifting left or right in the grid model of the set of nodes belonging to the movable part of the air cylinder (piston with rod). In this case, the displacement of nodes means the assignment of temperatures and types of nodes with a shift to the left or to the right.

Parallel to the solution of the thermal problem, a numerical simulation of the change in the state of the working gas and its transfer between the elements of the pneumatic system are carried out. Simulation is performed by numerical methods on the basis of the Euler method, with the same integration step Δt as for the thermal problem. Consider the example of a pneumatic cylinder algorithm for modeling the

change in the state of the working gas and the displacement between the elements of the pneumatic system (in the finite-difference formulation of the problem).
The movement of the piston of the pneumatic cylinder is a source of disturbance in the system under consideration. In the model, trapezoidal and sinusoidal laws of piston motion are verified. The latter is given by «equation 4»:

$$x_c(t) = x_{c0} + A_x \sin(\frac{2\pi}{T_n} t), \tag{4}$$

Where x_c – distance from piston to the end of the working cavity of the air cylinder; x_{c0} – piston position at the initial time; A_x – piston displacement amplitude; T_p – period of harmonic displacement of the piston.
Based on the current position of the piston $x_ц^\tau$, the current volume of the working cavity of the air cylinder V_c^τ is calculated on the basis of the «equation 5» at this integration step τ:

$$V_c^\tau = x_c^\tau \pi \frac{D_c}{4}, \tag{5}$$

Where D_c – internal diameter of air cylinder.
Then, according to «equation 6», a new working gas pressure is calculated using the ideal gas approximation and the adiabatic process equation. The adiabatic approximation is accepted, since for a short time interval Δt the change in the gas state is more pronounced than the heat removal.

$$P_c^\tau = P_c^{\tau-1}\left(\frac{V_c^{\tau-1}}{V_c^\tau}\right)^{\frac{i+2}{i}}, \tag{6}$$

Where P_c^τ и $P_c^{\tau-1}$ – working gas pressure in the air cylinder at the current and previous integration steps; V_c^τ и $V_c^{\tau-1}$ – the volume of the working cavity of the air cylinder at the current and previous steps of integration; i – number of degrees of freedom of gas molecules (for air with a high degree of accuracy $i = 5$)
Then the gas temperature T_c^τ is calculated from «equation 7» of the ideal gas state:

$$T_c^\tau = \frac{P_c^\tau V_c^\tau}{\nu_c^\tau R}, \tag{7}$$

Where ν_c^τ – the amount of substance in the working cavity of the pneumatic cylinder, which varies as the air enters and leaves the working cavity; R – universal gas constant.
Further, depending on the pressure P_c^τ, the amount of the gas flowing from the atmosphere (at P_c^τ less than atmospheric pressure P_a) or transferred to the receiver (at P_c^τ more than a certain pressure P_p^τ passing gas into the recuperative pneumatic system) is calculated:

$$Q_c^\tau = \begin{cases} k_a\sqrt{P_a - P_c^\tau}, & P_c^\tau < P_a; \\ 0, & P_a \le P_c^\tau \le P_p^\tau; \\ -k_p\sqrt{P_c^\tau - P_p^\tau}, & P_c^\tau > P_p^\tau, \end{cases} \tag{8}$$

Where k_a – coefficient of throttling in the inlet of atmospheric air; k_p – reduced throttling factor for the air path from the pneumatic cylinder to the receiver (in the accumulation mode), or to the pneumatic motor (in use mode). Using the value of the flow Q_c^τ, the change in the current step τ of the amount of

IOP Conf. Series: Materials Science and Engineering **441** (2018) 012041 doi:10.1088/1757-899X/441/1/012041

substance v_c in the working cavity is calculated.

In the same way, heat and mass transfer processes are simulated in other main elements of the pneumatic system: the receiver and the conditional pipeline connecting the air cylinder and the receiver.

3. Results

For the convenience of the simulation, a computer program "Program for modeling the energy-saving hydraulic drive of a multifunctional car with compressed air storage subsystem" in Object Pascal language in the Borland Delphi 7 integrated programming environment was developed. The program allows to set the geometric and thermal parameters of the pneumatic system elements and to investigate their effect on the energy storage efficiency and magnitude of heat losses. In the process of work, the program regularly displays on the screen two projections of the investigated element of the pneumatic system, temperature distribution charts in sections, temperature profiles in specified directions. The program is designed to use a computer with a processor not lower than the Pentium 2.3 GHz, and the amount of RAM is not less than 512 MB. The source code for the program is 12 KB.

Computer experiments with the developed model consisted in repeated movement of the piston of the pneumatic cylinder by a sinusoidal law simulating the work of a combined hydraulic and pneumatic cylinder in the manipulator in the technological cycle. In the course of the experiment, the spatial distribution of the temperature changed and gradually came to the steady-state cycle «Figure 3».

Figure 3. Cartograms of the spatial distribution of temperature in the pneumatic cylinder in different phases *I* .... *IV* of the steady-state duty cycle: *a* – phase *I* (the piston moves to the right); *b* – phase *II* (maximum compression); *c* – phase *III* (the piston moves to the left); *d* – phase *IV* (maximum discharge).

When the piston moves to the right (Figure 3, *a*) gas in the working cavity is compressed (Figure 3, *b*), which leads to its heating and spreading of high temperature into the walls of the pneumatic cylinder and a layer of thermal insulation around the working cavity. The movement of the piston to the left «Figure 3, *c*» leads to the expansion of the working gas, because of which the gas cools down below the ambient temperature (Figure 3, *d*). The reduction in temperature covers the walls of the pneumatic cylinder, piston and rod.

Despite the fact that the cooling area covers a larger volume, the maximum temperature decrease does not exceed 30 K (Figure 4). At the same time, in the region of heating, smaller in volume, the heating of the wall of the pneumatic cylinder can reach 75 K, cyclically varying from 25 to 75 K.

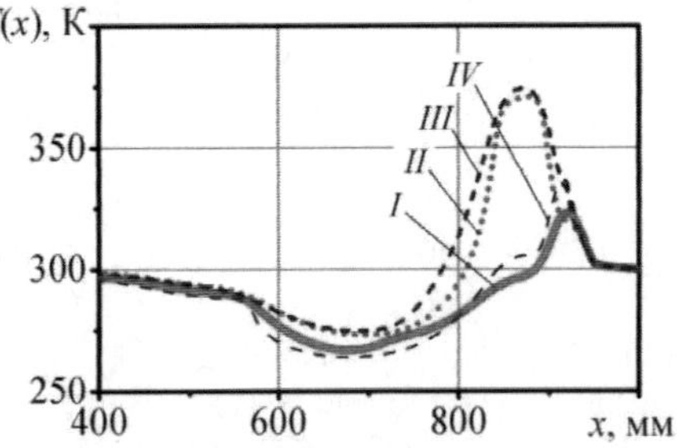

Figure 4. Temperature profiles along the metal wall of the pneumatic cylinder along the length x in different phases I ... IV of the steady-state duty cycle.

Because of the loss of heat, the gas in the working cavity reduces its internal energy, which does not allow the accumulation of a certain part of the mechanical energy of the piston motion by the pneumatic system. To evaluate the energy losses of the compressed air recuperation system in a pneumatic cylinder, a series of computer experiments was performed in which the thickness of the heat-insulating layer b around the air cylinder from 0 to 40 mm in increments of 5 mm was changed. As a material for a heat insulator, a group of inorganic materials with a thermal conductivity of 0.040 W / (m · K) is considered. As a measure of energy loss, the difference $\Delta E_p = E_м - E_g$ is accepted between the mechanical work $E_м$ on the displacement of the piston in a pneumatic cylinder during the inflation of air and the energy of compressed air E_g, available for use in the future. For convenience of perception, the relative energy losses $\Delta E_p/E_м$ are analyzed, that is, the share of energy losses in the total work on the accumulation of compressed air (Figure 5).

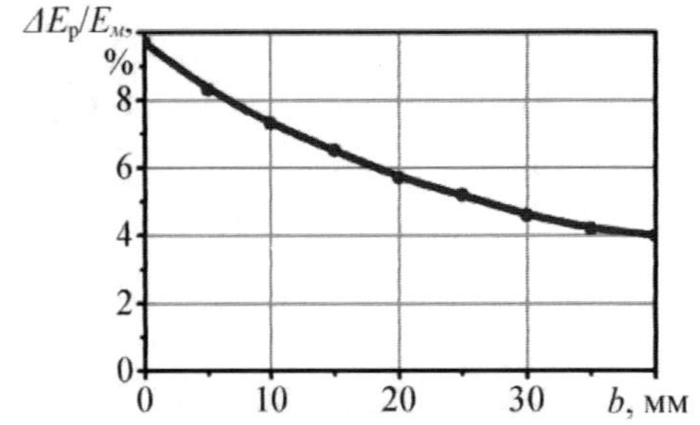

Figure 5. Influence of the thickness b of the thermal insulation layer of the air cylinder on the thermal energy loss $\Delta E_p/E_м$ in it with the accumulation of compressed air

Improvement of thermal insulation leads to a significant reduction in energy losses «Figure 5». The greatest decrease in losses – from 9.7 to 5 % – occurs at a thickness of the heat-insulating layer 20 ... 30 mm. Further increase in the thickness b is not advisable, since it almost does not lead to a reduction in heat losses (about 1 % with an increase in b from 25 to 40 mm), but it requires a considerable increase in the dimensions of the air cylinder.

4. Conclusion

Thus, a mathematical model is developed and implemented in the form of a computer program that allows to investigate the influence of the geometric and thermophysical parameters of the elements of the pneumatic system, as well as the operating conditions of the multifunction machine, on the efficiency of energy storage and heat losses.

References

[1] He W, Wang J 2018 *Renewable and Sustainable Energy Reviews* **87** 77-95
[2] Salyga S, Szablowski L and Badyda K 2016 *Transactions of the institute of fluid-flow machinery* **131** 151-160
[3] Chong S-H 2017 *Energies* **10** 1620
[4] Milewski J, Badyda K and Szablowski L 2016 *Journal of Power Technologies* **96** 245-260
[5] SedighNejad, Igbal T and Quaicoe J 2014 *Electronics* **3** 1-21
[6] Xia C, Zhou Y, Zhou S, Zhang P and Wang F 2015 *Renewable Energy* **74** 718-726

IOP Conf. Series: Materials Science and Engineering **441** (2018) 012042 doi:10.1088/1757-899X/441/1/012042

Method for calculating the minimum thickness of a layer of a heat-resistant coating for protecting the surface of the welded products from drops of molten metal when welding in CO_2

S B Sapozhkov[1], E M Burakova[2]

[1]Novgorod State University, ul. B. St. Petersburgskaya, 41 173003 Veliky Novgorod, Russia
[2]Yurga Institute of Technology TPU Affiliate, Leningradskaya 26, 652055 Yurga, Russia

E-mail: wh13@bk.ru

Abstract. In the industry, welding in shielding gases occupies a leading position, both in Russia and abroad. The main disadvantage of welding in carbon dioxide is the increased spattering of the metal. Reduce the adhesion of molten metal to the product can be, covering the metal with a protective layer in the form of a solution of substances. The thickness of this layer will depend on the metal.

1. Introduction

The work calculates the minimum permissible thickness of the heat-resistant coating to prevent welding of molten metal droplets with the surface of the welded products outside the welding zone. The dependence of the thickness of the protective coating layer on the specific heat consumption (q_n) and its combustion (evaporation), size (r_k) and the initial metal temperature (T_0) drops, as well as on the time of its flight (t) before contact with the surface of the welded product and the details of the welding equipment.

2. Main part

In the welding industry, various heat-resistant coatings are widely used in welding in CO_2 to protect the surface of welded products and welding equipment from drops of molten metal [1-7]. The paper presents a methodology for calculating the minimum thickness of the protective coating layer of welded products and elements of welding equipment.

Solving the task, we take into account that:

1) at the moment of contact of the drop with the surface of the weld metal and the parts of the welding torch, the amount of heat

$$Q_k > Q_{\min}; \quad (1)$$

2) The droplet adheres to the surface of the weld metal after the existing coating burns (evaporates) from the contact surface;

3) in determining Q_k the heat losses of the droplet were taken into account for radiation, convection, and combustion (evaporation) of the coating

$$Q_k = Q_0 - Q_{rad} - Q_{kon} - Q_{sp}, \tag{2}$$

where Q_0 – the amount of heat of the drop metal at the initial instant of time (the moment of detachment from the electrode);

$Q_{rad,kon}$ – loss of heat drops at the time of its flight due to radiation and convection;

Q_{sp} – loss of heat drop on the combustion coating.

The amount of heat of the metal of the electrode drop at the initial instant is determined by the known formula

$$Q_0 = V_k \cdot \gamma_g \cdot T_0 \cdot C_k, \tag{3}$$

where V_k – volume of a drop;

$C_k \gamma_g$ – specific heat and density of liquid metal at temperature T_0;

T_0 – the temperature of the metal droplets at the time of its detachment from the electrode. The loss of heat during the flight of a drop due to radiation is determined by the equation

$$Q_{rad} = m_k \cdot C_k \cdot T_0 \cdot \left(1 - \frac{1}{\sqrt[3]{1 + \frac{3\varepsilon\sigma_0 F_k}{C_k \cdot m_k} T_0^3 t}} \right) \tag{4}$$

and due to convection - by the equation

$$Q_{kon} = at\left(T_0 - 290\right) \cdot F_k. \tag{5}$$

Then the total heat loss, taking into account radiation and convection, is

$$Q_{\text{rad,kon}} = C_k m_k \left[T_0 \left(1 - \frac{1}{\sqrt[3]{1 + \frac{3\varepsilon\sigma_0 F_k}{C_k m_k} T_0^3 t}} \right) + F_k \frac{at(T_0 - 290)}{C_k m_k} \right]. \tag{6}$$

Substituting $F_k = 4\pi \cdot r_k^2$ и $m_k = \frac{4\pi \cdot r_k^3}{3} \gamma_g$, we get

$$Q_{\text{rad,kon}} = C_k m_k \left[T_0 \left(1 - \frac{1}{\sqrt[3]{1 + \frac{9\varepsilon\sigma_0}{C_k \gamma_g r_k} T_0^3 t}} \right) + 3\frac{at(T_0 - 290)}{C_k \gamma_g r_k} \right], \tag{7}$$

Where ε – black ratio;

σ_0 – coefficient of proportionality;

F_k – radiation surface of a drop;

m_k – mass of a metal drop;

t – flight time drops;

r_k – radius of a drop;

a – heat transfer coefficient;

γ_g – density of liquid metal;

$\gamma_g \cdot C_k$ – respectively, the density and heat capacity of the gas at 20^0 C;

V_r – gas flow rate.

Getting to the part, the drop gives some of its heat to the burning (evaporation) of the coating Q_{sp}. This part of the heat can be determined from equation

$$Q_{sp} = q_p V_p, \tag{8}$$

where q_p – specific heat consumption per unit of combustion volume combustion;

V_p – volume of combustible coating.

The volume of the burnable coating depends on the thickness of its layer, the shape and size of the drop, for the determination of which special experiments were performed on surfacing on steel specimens of St3 rollers measuring 400x400x6 mm. Welding was carried out in carbon dioxide gas with a Sv-08Mn2Cr wire with a diameter of 1.6 mm. Welding modes: $I_{св}$=300…320A; $U_{св}$=30…32 B. After welding, the maximum width of the droplet drop zone was determined on the surface of the article and the size of the drops welded with the cleaned surface of the sample at a different distance from the weld. The size of the drops (r_k) was determined by their mass after removal from the surface of the sample. In the control experiments along the welded seam, an asbestos cloth was laid, from which droplets were then removed, weighed and their dimensions determined. It should be noted that the drops collected from the asbestos cloth had a shape close to the ball, and the drops collected from the surface of the metallic sample had significant deviations from this shape, approaching the shape of the hemisphere. That is why the radius of the droplet was determined by calculation and experiment, depending on its mass. Simultaneously, the temperature of the metal droplets was determined by the calculation method at the moment of contact with the surface of the sample, depending on their size and flight distance:

$$T_k = T_0 - \Delta T_k, \tag{9}$$

where T_k – the temperature of the drop metal at the moment of contact with the metal surface;

ΔT_k – drop in the temperature of the drop metal during its flight:

$$\Delta T_k = \frac{Q_{\text{rad,kon}}}{C_k \cdot m_k}. \tag{10}$$

Substituting the value $Q_{\text{рад.кон}}$ from equation (7), we obtain

$$\Delta T_k = T_k \left[1 - \frac{1}{\sqrt[3]{1 + \dfrac{9\varepsilon\sigma_0}{C_k \cdot \gamma_g \cdot r_k} T_k^3 t}} + 3\frac{at(T_k - 290)}{C_k \cdot \gamma_g \cdot r_k} \right]. \tag{11}$$

Based on the results of the calculated data, a graphic dependence of the temperatures on the droplet size was constructed. From Fig. 1 that the temperature of a drop of molten metal depends on its size. It was shown that droplets with a radius of 0.1 10^{-2} m have a temperature of 2600 ... 2790 K, and a radius of 0.2 10^{-2} m - 2850 ... 2920 K.

Knowing the size and shape of the droplet, the speed of its flight and the maximum distance on which it still adheres to the metal surface, using equations (3) - (11), it is possible to determine the minimum values of the heat quantity and the temperature of the droplet metal at the moment of their contact with the surface of the part, sufficient for a strong grip.

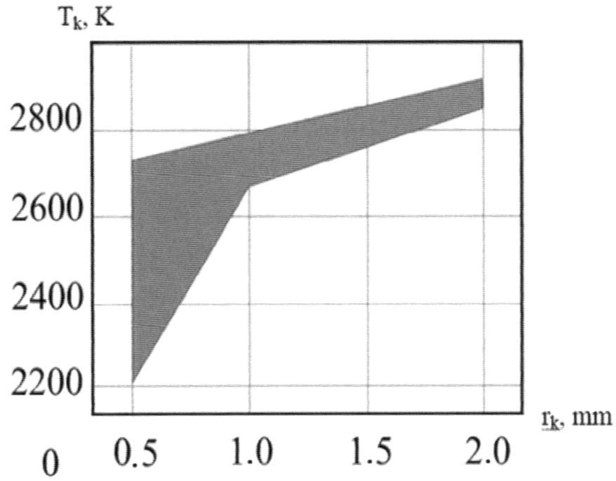

Figure 1. Dependence of the temperature of the metal droplets on their size at the moment of contact with the sample surface

Thus, according to experiments [1-4], droplets of a radius are welded to the surface of the part $r_k \geq 0,5$ mm at a distance of not more than 120 mm from the seam. The calculation shows that at the moment of contact with the surface the drop has about 7 J of heat, and the metal temperature reaches 2253 K.

Therefore, to prevent welding of the molten metal droplets to the surface of the part, it is necessary to apply a protective coating layer, upon combustion of which the droplet consumes all the excess heat and is cooled to a temperature less than 2253 K, i.e. its amount of heat will be:

$$Q_{\min} = 2253 \cdot C_k \cdot \gamma_{2253} \cdot V_k . \qquad (12)$$

To calculate the thickness of such a coating (h_p), it is assumed that the volume of the coating that is burned by a drop equals the volume of the cylinder whose cross-sectional radius is equal to the radius of the drop, and the height to the thickness of the coating layer;

$$V_p = \pi \cdot r_p^2 \cdot h_p . \qquad (13)$$

Then from (13), taking into account (2) - (7), (8), (10), (11) it is possible to obtain after corresponding transformations:

$$h_p \geq \frac{4r_k \cdot \gamma_g \cdot C_k}{3q_p}\left[T_0 - 2253\frac{C_k \cdot \gamma_{2253}}{C_k \cdot \gamma_g} - \right.$$
$$\left. -T_0\left(1 - \frac{1}{\sqrt[3]{1 + \frac{9\varepsilon\sigma_0}{C_k \cdot \gamma_g \cdot r_k}T_0^3 t}}\right) + 3\frac{at\left(T_0 - 290\right)}{C_k \cdot \gamma_g \cdot r_k}\right] \qquad (14)$$

3. Conclusion

Analysis of equation (14) shows that the thickness of the protective coating layer depends on the specific heat consumption (q_p) and its combustion (evaporation), size (r_k) and the initial metal temperature (T_0) drops, as well as on the time of its flight (t) before contact with the surface of the welded product and the details of the welding equipment. Consequently, the minimum required

thickness of the protective coating should be determined based on the maximum possible size of the molten metal droplets and the minimum distance of their flight. Then, knowing the specific heat consumption (q_p) on combustion of a unit of the volume of the coating, it is possible to determine by calculation the minimum necessary thickness for the purpose of preventing the welding of the largest drops with the surface of the products. The consumption of the metal necessary for the combustion (evaporation) of the coating depends on the physical-chemical properties of the coating.
An experimental test showed that the discrepancy between the calculated and experimental values (h_p) is from 1 to 14%. Consequently, the proposed equation allows us to calculate the optimal thickness of the coating layer by calculation, at which the droplets are not welded to the surface of the part.

4. References

[1] Fedko V T, Sapozhkov S B, Sokolov P D and Yastrebov A P. 2004 Elements of theory and technology of anti-spatter surface shielding when welding in carbon dioxide *Monograph. Tomsk: TSU Publishing* 2004 138 p.
[2] Solonenko O P, Kudinov V V, Smirnov A V, Cherepanov A N, Popov V N, Mikhalchenko A A and Kartaev E V 2005 Micro-Metallurgy of Splats: Theory, Computer Simulation and Experiment *JSME International Journal* Series B, Vol.48, No. 3, 2005
[3] Knotek O and Elsing R. Monte Carlo Simulation of Lamellar Structure of Thermally Sprayed Coatings *Surf. Coat. Technol* 1987 № 32 P 261 – 271.
[4] Ghafouri-Azar R, Mostaghimi J, Chandra S and Chamchi M A Stochastic Model to Simulation of a Thermal Spray Coating *Journal of Thermal Spray Technology* 2003 V 12 № 1 P 53-69.
[5] Steinke T, B ker M. Monte Carlo Simulation of Thermal Sprayed Coatings *Proc. ITSC'06* Seattle USA May 15 – 18, 2006 P 329 – 334.
[6] Kuznetsov M A and Kolmogorov D E 2015 Corrosion and Mechanical Properties of Austenic Steel Weld Joints *IOP Conference Series: Materials Science and Engineering* 2015 Vol 91Yurga, Russia May 2015 8 p.
[7] Filonov A V, Kryukov A V and Gusarov D E 2015 Systems to Control Molten Metal Transfer in Arc Welding *IOP Conference Series: Materials Science and Engineering* Vol 91 Yurga Russia May 2015 6 p.

IOP Conf. Series: Materials Science and Engineering **441** (2018) 012043 doi:10.1088/1757-899X/441/1/012043

The mechanism of heat distribution in welded joints made with the use of protective coatings

S B Sapozhkov[1] and E M Burakova[2]

[1]Novgorod State University, ul. B. St. Petersburgskaya, 41 173003 Veliky Novgorod, Russia
[2]Yurga Institute of Technology, Affiliate of National Research Tomsk Polytechnic University, Leningradskaya 26, 652055 Yurga, Kemerovo Region, Russia
E-mail: ssb@novsu.ru
E-mail: burek2@mail.ru

Abstract. In the industry, welding in carbon dioxide takes a leading place, both in Russia and abroad. The main disadvantage of welding in CO_2 is the increased spattering of the metal. Reduce the spraying can be done by coating the metal by protective layer in the form of a solution of substances.

Introduction

Welding in CO_2 found its application in many areas of industry. This method has many advantages, but also there are drawbacks. One of the drawbacks that should be paid attention to is the increased spattering of metal, as well as the spraying of the surface of the welded products, welding equipment parts and assembly and welding devices. Because of this shortage, a greater consumption of the vibrating tool takes place, energy and an increase in surface cleaning operations from splashes of molten metal become more labor-intensive. The deterioration of the protection of the welding zone, leading to the formation of pores in the weld metal, as well as the additional heating of the parts of the welding torch, which leads to premature failure of the equipment, is due to the splashing of the gas supply nozzle of the burner. It was revealed that the incidence of vibration by workers working on cleaning surfaces from splashes of molten metal begins to develop after 7 ... 8 years of operation.

Various foreign and Russian scientists are engaged in identifying methods of struggle and the nature of the spray. To minimize spattering, it is necessary to find the optimal welding regimes in CO_2, providing control over the transfer of electrode metal; to develop welding materials that stabilize arc combustion and affect the transfer of electrode metal [1].

The consequences of spraying has another big problem – it is spraying. To reduce the spraying, a protective coating (PC) can be applied. PC is a protective layer in the form of a solution, which is applied to the surface of the metal to be welded; it dries before welding and anti-adherence spatter to the base metal. To reduce the spatter of welding equipment, PC is applied to the nozzle of the burner or the design of the apparatus is changed.

PC has a number of requirements; the main is the heat resistance of the coating components. Its sufficiency depends on the nature of the interaction of splashes of molten metal with the surface of the welded article. The theory of the interaction of splashes with the surface develops and is constantly supplemented with new information allowing to increase the protective properties of coatings [1].

2 The main part

By applying a protective coating to the surface of the product, a surface-coating system is conditionally formed; the coefficients of thermal conductivity λ and thickness δ are different.

For example, let us take a semi-infinite body with a protective coating of a certain thickness δ_{zp}. applied to it. The contact point (temperature) will be denoted by T. The temperature of the outer layer T_{zp} is known. Thermal contact between surfaces is ideal [2].

We select in the semi-infinite body (Figure 1) an element equal to the thickness of the plate with the initial temperature of the layer T.

The temperature of the outer layer is found from the theory of thermal conductivity. The density of the heat flux in the stationary regime is constant and in all layers has the same values. We write:

$$q = \frac{\lambda_{zp}}{\delta_{zp}}\left(T - T_{zp}\right) \quad (1)$$

The temperature head in the layer through the coating is determined from equation (1):

$$T - T_{zp} = q\frac{\delta_{zp}}{\lambda_{zp}} \quad (2)$$

from here we find the temperature T_{zp}:

$$T_{zp} = T - q\frac{\delta_{zp}}{\lambda_{zp}} \quad (3)$$

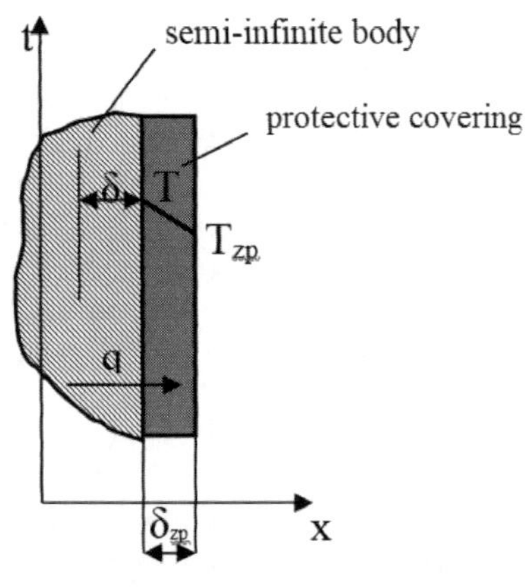

Figure 1. Scheme of temperature distribution using a protective coating.

Thus, the temperature of the outer layer depends on the values of the thermal resistances and the thickness of the protective coating.

Protective coatings have different coefficients of thermal conductivity. Consequently, the heat flux passing through the coating can have different values. The lower the heat flux density, the greater the thermal conductivity of the protective coating. It also affects the distribution of residual stresses. Due to the presence of coating layers, after welding, there is no rapid removal of heat from the plate, which promotes stress relaxation (local tempering).

Another factor that affects the distribution of residual stresses is temperature. About the time of welding, the source of heat is in constant movement. Temperatures of body points increase, reaching a maximum value, and subsequently decrease. At points more distant, the rise and fall of temperature is slower. Due to uneven heating of the metal and its expansion, welding stresses arise. The cooling temperature depends on many factors: the remoteness of the point in question from the seam, the parameters of the welding regime and the thermophysical properties of the material.
The coatings used to protect the welded articles from droplets (splashes) of molten metal when welding in CO_2 are complex. Thermophysical properties are different. Protective coatings are solutions and melts of high molecular weight polymers, colloidal suspensions, concentrated suspensions of various hard or ductile materials, etc. Since the protective coating affects differently the cooling rate, it will also affect the distribution of residual stresses.
The first time derivative of the temperature is the instantaneous cooling rate ω:

$$\omega = \frac{dT}{dt} \tag{4}$$

Determination of the cooling rate for butt-welding of the plates by equation (5):

$$\omega = -2\pi\lambda c\gamma \frac{(T - T_n)^3}{\left[q/(V\delta)\right]^2} \tag{5}$$

where λ – coefficient of thermal conductivity; c – specific heat; γ – density; T_n – heating temperature; T – the temperature to which the metal is heated; q/V – effective linear energy; δ – plate thickness. Welding was done without heating
We will replace $T\text{-}T_n = T\text{-}T_{zp}$ – the temperature flux through the coating, we put this value in equation (5) and after the transformations we obtain:

$$\omega = -2\pi\lambda c\gamma \frac{(T - T_{zp})^3}{\left[q/(V\delta)\right]^2} \tag{6}$$

Cooling rate through coating:

$$\omega_{zp} = -2\pi\lambda_{zp} c_{zp} \gamma_{zp} \frac{\left(T - T_{zp}\right)^3}{\left[q/\left(v\delta_{zp}\right)\right]^2} \tag{7}$$

In equation (7) we substitute the welding energy formula q=ηUI. Calculations showed that the cooling rate through the protective coating layer decreased by 30-40% in comparison with uncoated samples. From equation (7), the dependence of the cooling rate on the welding energy and thermal properties of the coating is seen.
The theoretical data presented in Fig. 2 were obtained by substituting the values of the thermophysical characteristics of the protective coatings in equations (3) and (7).

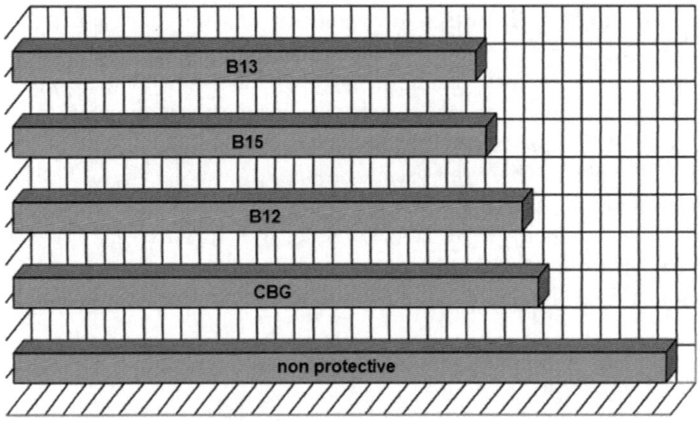

Figure 2. Calculated values of the plate-cooling rate for welding with a protective coating and without it.

With the help of equation (8), we can determine the length of stay of the weld metal above a given temperature:

$$t = \frac{1}{4\pi a\left(T - T_{zp}\right)^2}\left(\frac{q}{v_c \delta_{zp} c_{zp} \gamma_{zp}}\right)^2 \tag{8}$$

The residence time of the weld metal above a predetermined temperature depends on the thermophysical properties of the protective coatings and on their thickness.
The substances have a coefficient of thermal conductivity λ is different and depends on density, pressure, structure, humidity and temperature [3-7]. First of all, it is important to know the dependence of the thermal conductivity coefficient on temperature, since the temperature at different points differs in the distribution of heat. For most materials, the dependence of thermal conductivity on temperature has a linear character:

$$\lambda = \lambda_0 (1 + bt) \tag{9}$$

where λ_0 – coefficient of thermal conductivity at temperature t_0; b – a constant determined experimentally.
In Table 1, the results of studies for determining the thermophysical properties of protective coatings are presented.

Table 1. Thermophysical properties of protective coatings.

№	Protective	γ, 10^{-3} kg/10^{-6} m^3	cγ, $\frac{J}{10^{-3} m^3 \, ^0C}$	λ, $\frac{W}{10^{-3} m^3 \, ^0C}$	a·10^{-6}, 10^{-3} m^2/s
1	MB	2.016	2065.6	0.002604	1.26
2	CBG	1.256	595.4	0.006757	11.35
3	B12	1.521	928.1	0.0074975	8.08
4	B13	1.878	2254.5	0.0085868	3.81
5	ALD	2.758	390.7	0.004307	11.02
6	MGS	1.256	292.1	0.0059701	20.43

7	AGS	2.616	437.4	0.0024218	5.53
8	CPR	1.989	1101.1	0.0064275	5.83

Based on the above data, it is possible to make an assumption about the distribution of residual stresses when welding along the protective coating layer and without it. The rate of plate cooling depends on the composition of the protective coating. The greater the thermal properties and the thickness of the coating, the lower the cooling rate and the residual stresses. However, it is possible, if it was evenly applied.

Equation (10) describes the distribution of temperatures when welding plates without a protective butt-end covering, considering heat transfer:

$$\Delta T = \frac{q/\delta}{4\pi c\gamma at} e^{-r^2/4at}\beta(t) \quad (10)$$

where β(t) – heat transfer function, $\beta(t) = e^{-bt}$; $b = \frac{2\alpha_T}{c\gamma\delta}$.

When welding plates with a protective coating, equation (10) will look like this:

$$\Delta T = \frac{q/\delta}{4\pi c\gamma at} e^{-\left(\frac{r^2}{4at} + \frac{2\alpha_T t}{c_{zp}\gamma_{zp}\delta_{zp}}\right)} \quad (11)$$

By the equations (10) and (11), the temperature fields in the plates with protective coating and without it were built after welding in the period of temperature equalization (Fig. 3).

The graphs show that in the plates with different protective coatings at the same time a diverse temperature distribution.

According to theoretical data, an experiment was conducted. Samples for welding were made of steel sheet St 3 5 mm thick. Figure 4 shows the installation diagram. On the surface of each plate was applied a rectangular grid, the dimensions of the cells 10 × 10mm. In a certain sequence, chromel-alumel thermocouples were located at the grid nodes.

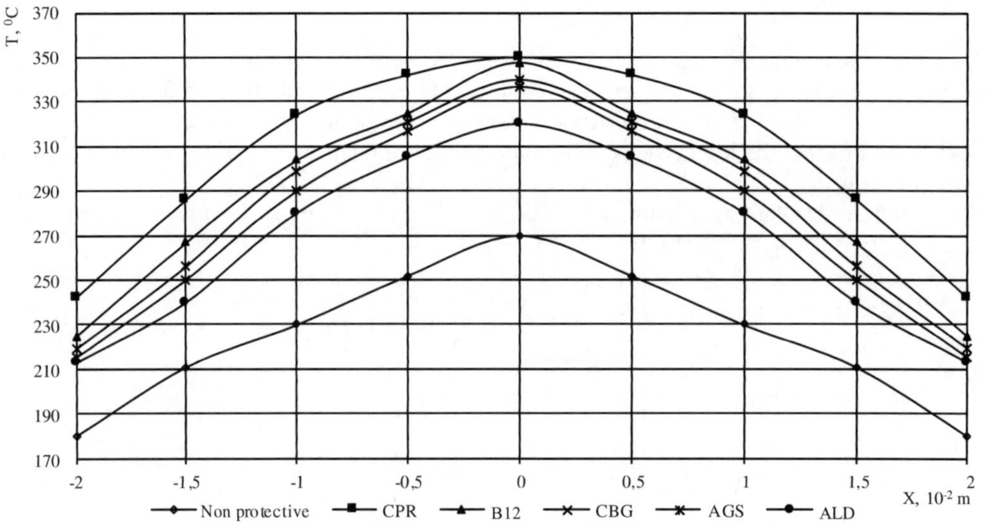

Figure 3. Distribution of temperature increments in sections parallel to the X axis in plates with different protective coatings and without it.

Welding was carried out in carbon dioxide, wire Sv-08Mn2Cr with a diameter of 1.6 mm in one pass (welding current 300 A, arc voltage 28 V).

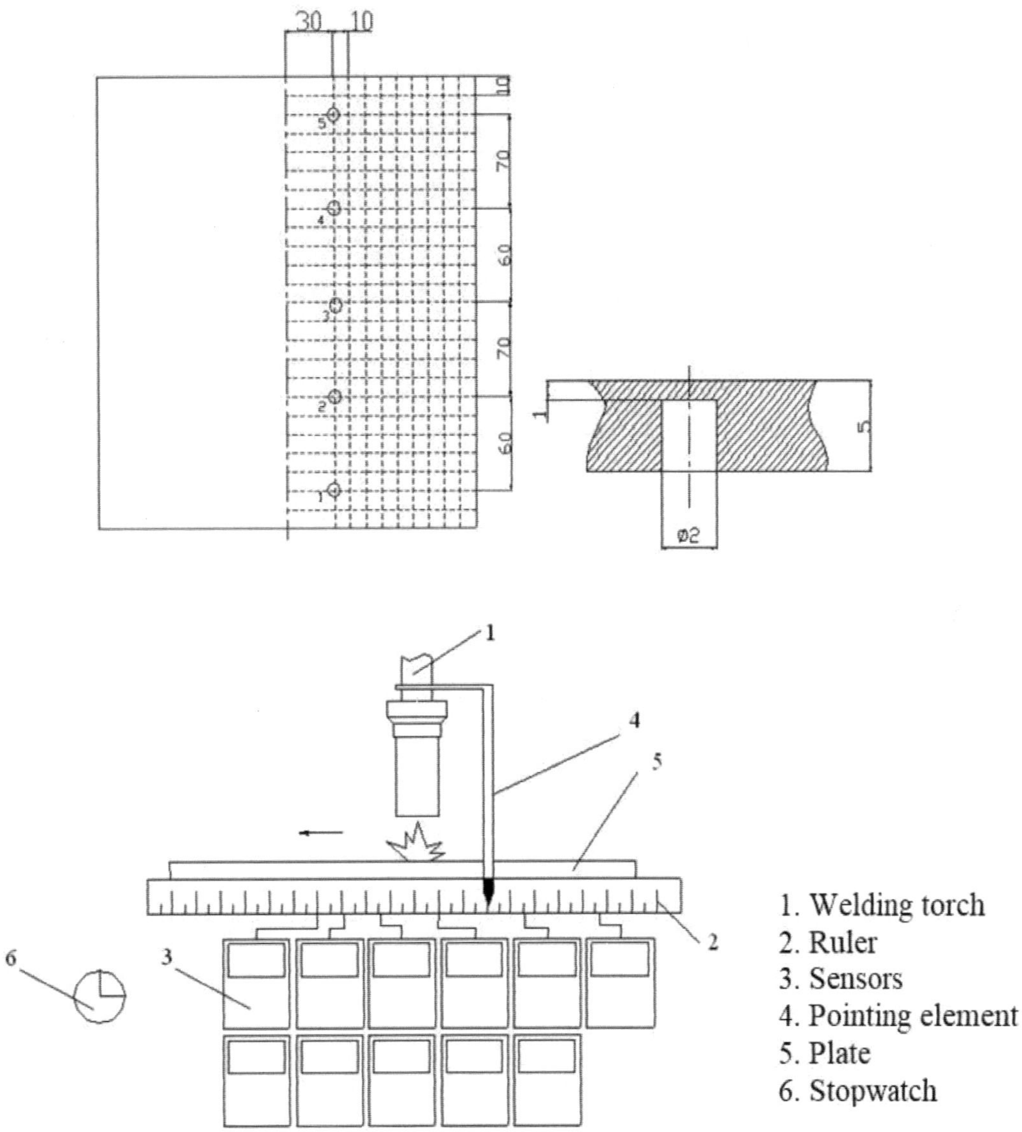

Figure 4. The scheme of the experimental setup.

According to the previously reported data, experimental dependences of the product temperature on the welding time with and without protective coatings for welding in carbon dioxide were constructed.

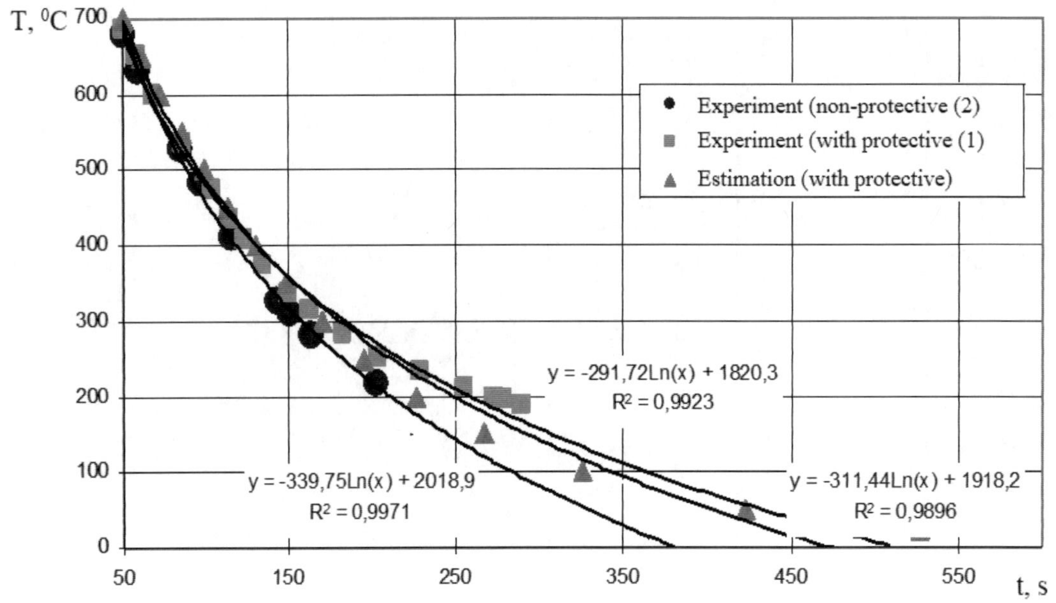

Figure 5. Experimental dependence of the product temperature on the welding time with and without protective coatings for welding in CO_2.

Analyzing the graph (Figure 5), we observe a decrease in the cooling rate of the welded article during welding with the use of protective coatings. There is a decrease in the cooling rate because the protective coating prevents the free flow of heat from the plate. The larger the volume heat capacity, the slower the plate cools down.

Conclusion

A protective coating is the main method of controlling the spraying of the metal surface during welding.

All coatings have different coefficients of thermal conductivity. Passing through the coating heat flow, also has different values. The lower the heat flux density, the greater the thermal conductivity of the protective coating. This affects the distribution of residual stresses.

Analyzing the plotted temperature fields in plates with a protective coating and without it after welding in the period of temperature equalization, it is evident that at the same time in the plates with different protective coatings the temperature distribution is diverse. Protective coating prevents free flow of heat from the plate.

References

[1] Votava J 2013 Protection of welded joints against corrosion degradation. *Acta Universitatis Agriculturae et Silviculturae Mendelianae Brunensis* number 6 pp 1897–1904

[2] Kowalski J, Mazur M and Wojciechowski W The use of welding techniques in protection of welding structures against corrosion *Archives of foundry engineering* Volume 10, Special Issue 3/2010pp 293-298

[3] Bezborodov V and SaraevYuEnhancing the Lifetime of Welded Joints Using the Protective Coating and Causes the Formation of Defects in Them*VII International Scientific Practical Conference "Innovative Technologies in Engineering"*Series: Materials Science and Engineering pp142 (2016) 012023 doi:10.1088/1757-899X/142/1/012023

[4] Solodsky S A, Sarychev V D, Borisov I S 2016 Technology of Ultrasonic Control of Gas-

Shielded Welding Process *IOP Conference Series: Materials Science and Engineering* **142(1)**
[5] Chinakhov D A., Solodsky S A, Grigorieva E G, Mayorova E I 2017 Influence of gas dynamic processes on chemical composition of hardfaced layer when restoring machine parts manufactured from 40H steel *Materials Science Forum* **906** 142-146
[6] Chinakhov D A, Grigorieva, E G, Mayorova E I, Solodsky S A, Torosyan V F 2017 Dependence of Manganese Content in the Weld Metal on the Velocity of Active Shielding Gas Flow *IOP Conference Series: Materials Science and Engineering* **253(1)**
[7] Kuznetsova O V, Begunova L A, Romanenko S V, Solodsky S A 2018 The Determination of Metals in Welding Fume by X-RaySpectrometry *IOP Conference Series: Earth and Environmental Science* **115(1)**

Analysis and prediction of copper surface roughness obtained by selective laser melting

A A Saprykin[1,3], N A Saprykina[1] and A A Saprykin[2]

[1]Tomsk Politechnic University, Tomsk, Russian Federation
[2]Novosibirsk State Technical University, Novosibirsk, Russian Federation
[3]e-mail: sapraa@mail.ru

Abstract. The paper presents the results of experimental studies of the effect of mechanoactivation of the powder, argon shielding gas and the effect of technological modes of melting: the speed of the laser beam, the power of laser radiation, the scanning step, the preliminary temperature of heating the powder material on the surface roughness of the copper powder material obtained by selective laser melting. Experiments on the melting of copper powder are implemented on the installation of layer-by-layer laser melting of the original design, which allows to adjust all technological modes of melting. The surface roughness is determined on the non-contact digital microscope Olympus LEXT OLS4100. The mathematical dependence of the surface layer roughness of copper powder on the technological modes of melting is obtained on the basis of the theory of experiment planning and static processing of the results. The significant parameters of the regime-the power of laser radiation, the speed of the laser beam, the scanning step affecting the roughness of the layer. The positive effect of mechanical activation of powder material and protective atmosphere on the quality of the surface layer is shown.

Introduction

The first commercially successful additive production method was developed in 1987 by Carl Descartes and used to make models and prototypes [1]. Currently, additive technologies are the fastest growing advanced technologies in the world. Unlike traditional methods of material removal, rapid prototyping technologies are aimed at creating complex products by sequentially adding material (materials). By now, a large number of rapid prototyping methods are known which differ in the material used and in the way the product is formed. Innovative is the method of selective laser melting of a physical copy of various objects from metals and alloys based on the 3D CAD-model. The volumetric product is created by layer-by-layer melting of the powdered material laser beam in accordance with the contour of each section. In recent years, the microstructure and mechanical properties of the parts have improved significantly due to improvements in laser processing conditions (increase in laser power, reduction in the diameter of the laser beam, reduction in the thickness of the resulting layer, etc.). Despite significant progress, the technology of selective laser melting still faces surface quality problems [2, 3]. Obtaining a good quality surface is the main issue for preventing delaminating, residual stresses, premature failure.

A review of the literature has shown that the laser power and scanning speed are often presented as the main parameters affecting the surface quality of the product obtained by selective laser sintering [4]. Brecher et al [5] suggested that the scanning speed is preferred in the appointment of modes of melting. The relatively low speed allows for uniform melting of the powder material, but as a result there is a problem of low productivity [6]. Van and co-authors [7] analyzed the effect of scanning speed and pitch to achieve good surface quality. Great importance is given to the impact of scanning strategy. In [8], studies were conducted to assess residual stresses and strains under various scanning strategies for the Nickel-chromium alloy Inconel 718. It is written that the maximum stresses along the x and Y axes were observed in samples with the contour scanning strategy when they were shifted to the center (Fig. 1, h).

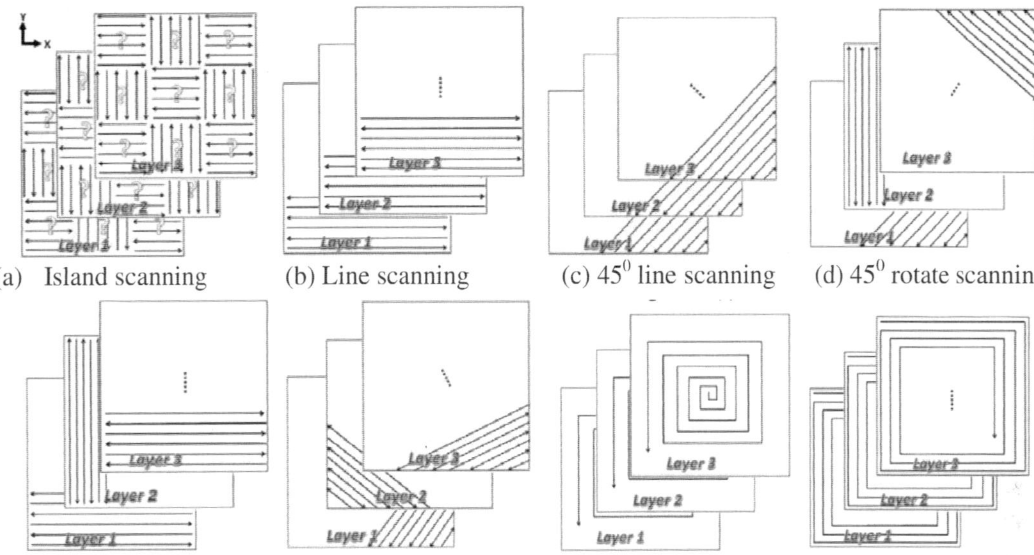

(a) Island scanning (b) Line scanning (c) 45^0 line scanning (d) 45^0 rotate scanning

(e) 90^0 rotate scanning (f) 67^0 rotate scanning (g) In-out scanning (h) Out-In scanning

Figure 1. The scan strategy [8].

Significant stresses were formed with a horizontal scanning strategy (Figure 1, h). All samples have a voltage along the X and Y axes between the single layer and the substrate. It was noted that with oblique scanning at an angle of 45^0, the samples showed slight residual stresses and deformations, in comparison with other scanning strategies (Figure 1, c) [8].

In [9], studies were made of the effect of laser power, the scanning speed, the layer thickness, and the scanning strategy on the surface roughness of the nickel-chromium-iron-molybdenum alloy HASTELLOY X obtained by selective laser melting. It is noted that cohesion has a great influence on the roughness of the surface layer. Coagulation is the fusion of fine powder particles into larger ones under the influence of laser radiation. There was a decrease in coagulation with a decrease in the scanning rate, which is due to an increase in the melting time of the powder and a decrease in the viscosity of the melt. Increasing the scanning step leads to an increase in the surface roughness. The surface roughness has the smallest value of 15 μm at a laser power of 200 W and scanning speed of 3000 mm/s. The authors proposed to improve the roughness of the inclined surface by contour scanning of each layer with increased energy density.

Analysis of the literature showed that the regimes and melting conditions for different materials are determined experimentally and depend on a significant number of parameters [10, 11, 12]. The aim of this paper is to establish the limits in which the surface roughness of a copper powder obtained by selective laser melting can be varied by changing the technological modes of laser processing.

Materials and methods

Experiments on the melting of copper powder are implemented on the installation of layer-by-layer laser melting of the original design, which allows to adjust all technological modes of melting. The device is a technological laser complex for the formation of surfaces of products of complex spatial shape, consisting of ytterbium fiber laser LK – 100 – V, three-coordinate table, personal computer, CNC system and original software. Ytterbium fiber laser with a wavelength of 1.07 microns allows you to adjust the power from 10 to 100 watts. The quality and accuracy of the manufactured products provide constant output power and focus accuracy of the fiber laser. Control of the laser beam with a special program in the working area of 100x100x100 mm allows scanning on any given contour. On the body of the installation, for applying a layer of powder in the melting zone, the carriage and leveling rollers are fixed. Inside the carriage there is a powder hopper that allows you to adjust the density of the applied layer. The powder layer is scanned by the laser beam along the required trajectory at the specified modes. After receiving a single layer, the table is lowered by a stepper motor to the layer thickness [13]. With the help of a carriage, the platform is filled with powder material to form the next layer. The cycle is repeated until the complete creation of the product, then the table is moved to its original position and the finished part is removed. To eliminate the interaction of the powder with oxygen and nitrogen, it is possible to supply a protective gas to the laser melting zone.

Copper stabilized powder is widely used in various fields of industry. The powder particles have a spherical shape with a nominal size of 0.007 mm, bulk density of 1.25-1.9 g/cm3. The roughness of the layer was determined by a microscope OLYMPUS LEXT OLS 4100.

During the experiment, the technological modes of melting were changed: the power of laser radiation, scanning speed, scanning step, the temperature of heating the powder composition. Studies have been conducted on the influence of argon protective gas and mechanical activation of powder on the roughness of the surface layer of the copper powder obtained by the method of selective laser melting. The protective environment allows to exclude interaction of powder products with oxygen and nitrogen, and also to strengthen a surface of a product. Mechanical activation is used to increase the dispersion, the presence of defects in the crystal lattice of the powder [14]. Mechanical processing of the powder was carried out in the centrifugal-planetary mill AGO-2.

Results and discussions

At the first stage, preliminary exploratory experiments were carried out to determine the acceptable melting modes. Under them meant modes when a single layer did not crumble, i.e. had some mechanical strength without significant deformation. Experiments have shown that the laser power is less than 14 W and the laser beam speed is more than 3000 mm/min are insufficient for melting the powder material. Power more than 30 W and speed less than 200 mm/min leads to intensive oxidation and ignition of the powder. Scanning step of more than 0.3 mm does not allow single tracks to be sintered together, which prevents the formation of a single layer. Increasing the temperature of heating the powder material has a positive effect on the strength of the single layer.

To establish the empirical dependence of the layer roughness of copper powder on the technological modes of melting, a four-factor experiment was carried out on the program of the second order central composite planning. For this purpose, sixteen experiments of the four-factor experiment were implemented; seven experiments were conducted in the center of the plan and supplemented with eight experiments in "star" points. Copper powder was determined by the following ranges of conditions of melting are: the radiation power P=(14-30) W, the moving speed of the laser beam V=(200-3000) mm/min, scanning step s=(0.1-0.3) mm; the heating temperature of the powder composition varied in the range t=(26-200) ^{0}C. Surface roughness was determined using a digital contactless microscope Olympus LEXT OLS4100, figure 2.

Figures 3-8 show the effect of technological regimes and conditions of melting on the appearance of the surface of a single layer of copper powder obtained by selective laser melting.

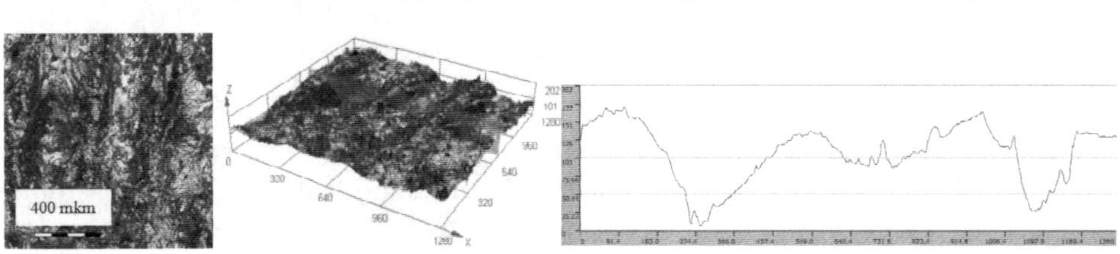

а *б* *в*

Figure 2. Determination of surface roughness on the microscope Olympus LEXT OLS4100:

a-photo of a single layer of copper powder obtained by selective melting in the following modes: P=30 W, V=2000 mm / min, s=0.2 mm, t=300^{0}C; b-cross section of the surface profile; b-profilogram of the surface layer Rz, μm.

The effect of laser power on the surface roughness of a single layer is shown in figure 3. The power increase from 14 to 30 W at V=1600 mm/min, s=0.2 mm leads to a change in Rz from 480 to 725 μm.

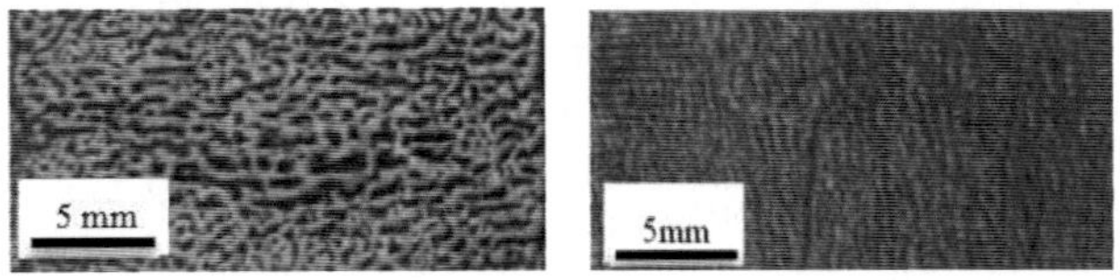

a b

Figure 3. Photos of the surface of a single layer of copper powder obtained by selective melting at the modes:
a-P=30 W, V=1600 mm/min, t=114 ^{0}C, s=0.2 mm, Rz=725 μm,
b-P=14 W, V=1600 mm/min, t=114 ^{0}C, s=0.2 mm, Rz=480 μm.

The effect of the laser beam velocity on the surface roughness of a single layer is shown in figure 4. Increasing the speed from 200 to 3000 mm/min at P=22 W, s=0.2 mm leads to a change in Rz from 750 to 480 microns. In figure 4, a there are areas of uniform powder melting.

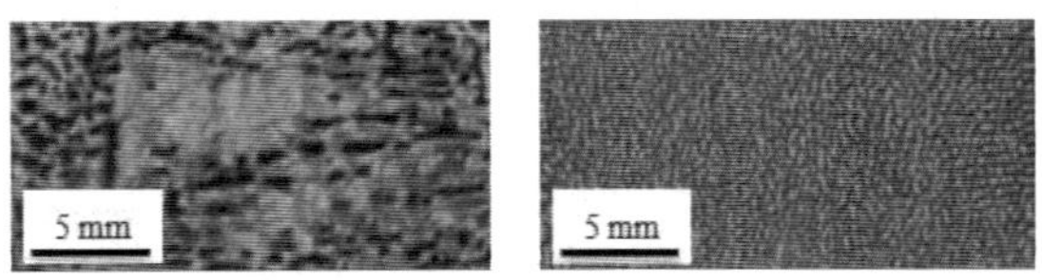

a b

Figure 4. Photos of the surface of a single layer of copper powder obtained by selective melting at the modes:
a- P=22 W, V=200 mm/min, t=114 ^{0}C, s=0.2 mm, Rz=750 μm,
b- P=22 W, V=3000 mm/min, t=114 ^{0}C, s=0.2 mm, Rz=480 μm.

The effect of the heating temperature of the powder material on the surface roughness of a single layer is shown in figure 5. The temperature increase from 20 to 200 ^{0}C at P=30 W, V=3000 mm / min, s=0.3 mm leads to a change in Rz from 540 to 525 microns.

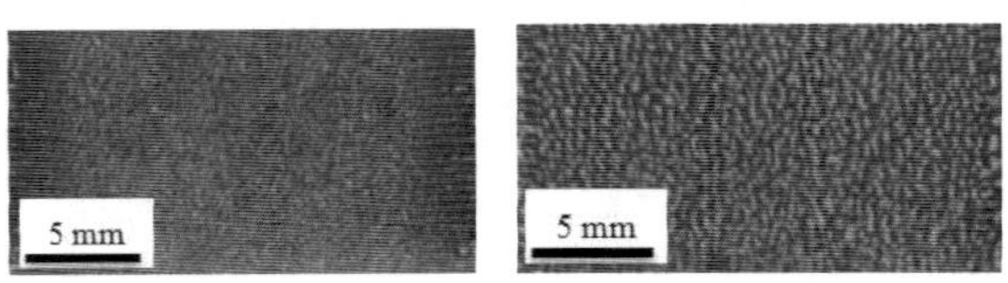

a b

Figure 5. Photos of the surface of a single layer of copper powder obtained by selective laser melting at the modes:
a- P=30 W, V=3000 mm/min, t=200 ^{0}C, s=0.3 mm, Rz=525 μm,
b- P=30 W, V=3000 mm/min, t=20 ^{0}C, s=0.3 mm, Rz=540 μm.

The effect of the scanning step on the surface roughness of a single layer is shown in figure 6. Increasing the scan step from 0.1 to 0.3 mm at P=30 W, V=3000 mm / min, t=200 ^{0}C leads to a change in Rz from 740 to 525 microns.

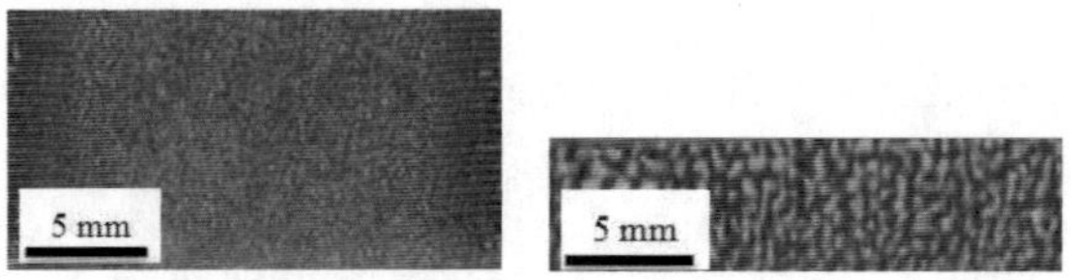

a b

Figure 6. Photos of the surface of a single layer of copper powder obtained by selective laser melting at the modes:
a- P=30 W, V=3000 mm / min, t=200 ^{0}C, s=0.3 mm, Rz=525 μm,
b- P=30 W, V=3000 mm / min, t=200 ^{0}C, s=0,1 mm, Rz=740 μm.

Figure 7 shows the effect of shielding gas on the surface roughness of a single layer. During the sintering of copper powder in argon, a change in the color of the sintered surface was observed, it acquired a Golden color and the surface roughness of Rz decreased from 650 to 500 μm.

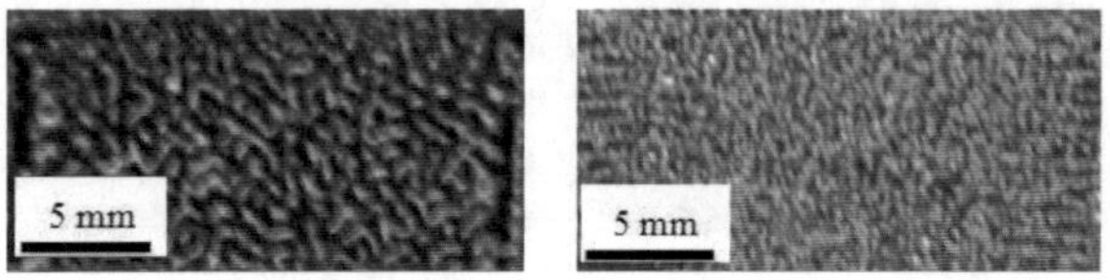

a b

Figure 7. Photos of the surface of a single layer of copper powder obtained by selective melting in different modes:
a- P=22 V, V=1600 mm / min, t=114 ^{0}C, s=0.3 mm, Rz=650 μm, air sintering;
b- P=22 V, V=1600 mm / min, t=114 ^{0}C, s=0.3 mm, Rz=500 μm, sintering in argon.

Influence of mechanical activation of powders on the surface roughness of a single layer is shown in figure 8. The effect of laser radiation on the powder subjected to three-minute activation leads to a decrease in the diameter of the coagulated particles and their uniform distribution over the surface. Surface roughness is significantly reduced from 600 to 125 microns, at P=14 W, V=200 m/ min, t=20 ^{0}C, s= 0.1 mm.

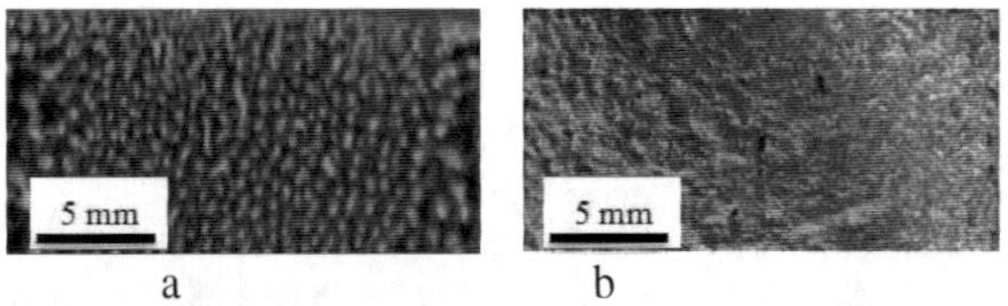

a b

Figure 8. Photos of the surface of a single layer of copper powder obtained by selective melting: a-P=14 W, V=200 mm/min, t=20 ^{0}C, s=0,1 mm, Rz=600 μm; b – P=14 W, V=200 mm / min, t=20 ^{0}C, s=0,1 mm, Rz=125 μm, after three minute activation.

As a result of planning and conducting experiments and static processing [15] the results obtained, the mathematical dependence of the surface roughness of the copper powder on the melting regimes is obtained, which allows identifying significant parameters:

$$R_Z = 356 + 15 \cdot P - 0.1 \cdot V - 0.057 \cdot t + 425 \cdot S \quad (1)$$

The dependence of the roughness of a single layer on the melting modes is shown in figure 9 [16, 17].

The roughness of the surface layer of copper powder is significantly affected by the speed of the laser beam. The change in V from 200 to 3000 mm / min leads to a decrease in Rz by 56.25%, at P=22 W, s=0.2 mm, t=114 ^{0}C, in accordance with figure 4. Power also has a big impact on Rz. Increasing P from 14 to 30 W R_Z increases by 62%, at V=1600 mm / min, t=114 ^{0}C, S=0.2 mm, figure 3. Change S from 0.1 to 0.3 mm reduces Rz by 40%, at P=30 W, t=200^{0}C, V=3000 mm / min, according to figure 6. The heating temperature of the powder material on the Rz affects slightly.

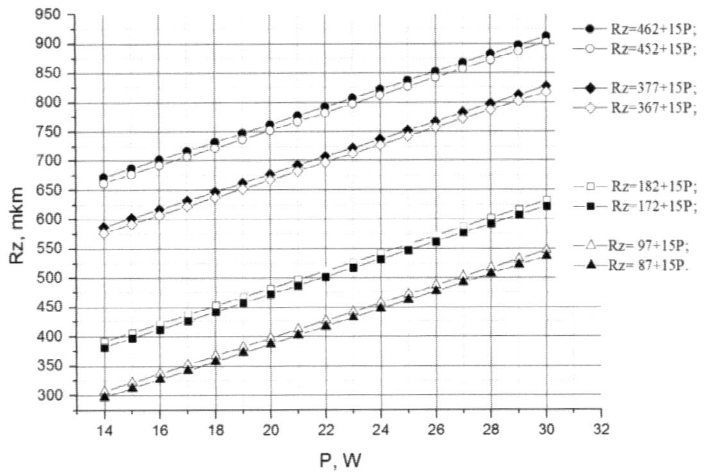

Figure 9. Diagram of the influence of melting modes on the roughness of a single layer *Rz* :
-•- V=200 mm/min, t=26 ^{0}C, S=0.3 mm; -○- V=200 mm/min, t=200 ^{0}C, S=0.3 mm;
-◆-V=200 mm/min, t=26 ^{0}C, S=0.1 mm; -◇- V=200 mm/min, t=200 ^{0}C, S=0.1 mm;
-□- V=3000 mm/min, t=26 ^{0}C, S=0.3 mm; -▪- V=3000 mm/min, t=200 ^{0}C, S=0.3 mm;
-△-V=3000 mm/min, t=26 ^{0}C, S=0.1 mm; -▲- V=3000 mm/min, t=200 ^{0}C, S=0.1 mm.

Thus, the speed of the laser beam, the laser radiation power and the scanning step are the main parameters that affect the roughness of the sintered surface layer of copper powder.

Conclusion

The conducted research allowed to determine the rational areas of technological modes of selective laser melting of copper powder: P=(14-30) W, V=(200-3000) mm/min, s=(0.1-0.3) mm; t=(26-200) ^{0}C, to determine the nature of the effect of selective laser melting modes and to obtain an empirical dependence of the roughness of a single surface layer of copper powder, which allows to control the melting process in order to obtain a quality product. The positive effect of protective atmosphere and mechanical activation of metal powder materials on the quality of the surface layer is noted. To reduce roughness, improve internal structure and strength properties, melting is recommended to be carried out in argon using metal powder materials subjected to mechanical activation. The studies show that the roughness of the surface layer Rz can be changed within significant limits, changing the technological modes of laser processing. The roughness of the sintered surface layer of PMS-1 is most influenced by the speed of the laser beam. The change in V from 200 to 3000 mm / min leads to a decrease in Rz by 56.25%, at P=22 W, s=0.2 mm, t=114 ^{0}C. Power also has a big impact on Rz. Increasing P from 14 to 30 W R_Z increases by 62%, at V=1600 mm/min, t=114 ^{0}C, S=0.2 mm. Change S from 0.1 to 0.3 mm reduces Rz by 40%, at P=30 W, t=200^{0}C, V=3000 mm / min. the temperature of heating of the powder material on Rz affects slightly. Thus, the speed of the laser beam, the laser radiation power and the scanning step are the main parameters that affect the roughness of the sintered surface layer of copper powder.

References

[1] Beaman J J and Deckard C R 1990 Selective laser sintering with assisted powder handling US patent no 4938816

[2] Shishkovsky I B 2006 Selective laser sintering and synthesis of functional structures Diss Dr Phys-Mat Sciences 01.04.17 Samara RGB p 390
[3] Agarwala M, Bourell D, Beaman J, Marcus H and Barlow J 1995 Direct selective laser sintering of metals Rapid Prototyping Journal Vol 1 **1** pp 26–36
[4] Mumtaz K and Hopkinson N 2009 Top surface and side roughness of Inconel 625 parts processed using selective laser melting Rapid Prototyping Journal 15 (**2**) pp 96–103
[5] Brecher S, Jeschke G, Schuh S, Aghassi J, Arnoscht F and Bauhoff F 2012 Welter, Integrative Production Technology for High-wage Countries Springer Berlin Heidelberg pp 17–76
[6] Gibson I, Rosen D and Stucker B 2009 Additive manufacturing technologies Rapid Prototyping to direct digital manufacturing New York USA Springer p 459
[7] Wang D, Liu Y, Yang Y and Xiao D 2016 Theoretical and experimental study on surface roughness of 316L stainless steel metal parts obtained through selective laser melting Rapid Prototyping Journal 22 (**4**)
[8] Cheng B, Shrestha S and Chou K 2016 Stress and deformation evaluations of scanning strategy effect in selective laser melting Additive Manufacturing **12** pp 240–251
[9] Tian Y, Tomus D, Rometsch P and Wu X 2017 Influences of processing parameters on surface roughness of Hastelloy X produced by selective laser melting Additive Manufacturing **13** pp 103–112
[10] Kruth J, Wang X, Laoui T and Froyen L 2003 Lasers and materials in selective laser sintering. Assembly Automation **23** pp 357–371
[11] Jhabvala J, Boillat E and Glardon R 2013 Study of the interparticle necks in selective laser sintering Rapid Prototyping Journal Vol 19 **2** pp 111–117
[12] [12] Kruth O et al 2004 Selective laser melting of iron-based powder Journal of Materials Processing Technology **149** pp 616–622
[13] Saprykin A and Saprykina N 2012 7th International Forum on Strategic Technology IFOST 2012 6357719
[14] Saprykina N, Saprykin A, Arkhipova D and Borovikov I 2016 Improvement of the Sintered Surface and Bulk of the Product Via Differentiating Laser Sintering (Melting) Modes Source of the Document IOP Conference Series: Materials Science and Engineering **142**(1) 012089
[15] Spiridonov A A 1981 Planning of experiment in the study of technological processes Moscow Machine Building p 184
[16] [16] Saprykin A A, Saprykina N A and Arkhipova D A 2017 The effect of layer-by-layer laser sintering on the quality of copper powder sintered surface layer Proceedings - 2016 11th International Forum on Strategic Technology IFOST 2016 16760286
[17] Saprykin A A, Saprykina N A and Arkhipova D A 2016 Influence of Shielding Gas and Mechanical Activation of Metal Powders on the Quality of Surface Sintered Layers IOP Conf Ser: Mater Sci Eng **125** 012016

Non-alloyed ohmic Cr/Pt/Au contacts to GaN based structures

A V Zhelannov[1], B I Seleznev[2]

[1] OKB-Planeta PLC, Fedorovskiy Ruchey 2/13, 173004, V. Novgorod, Russia
e-mail: ZhelannovAV@okbplaneta.ru
[2] Yaroslav-the-Wise Novgorod State University, B. Sankt-Petersburg str., 41, Novgorod-the-Great, 173003, Russia

Abstract. The study of non-alloyed ohmic contacts based on Cr/Pt/Au is presented. Metallization was made on the epitaxial (Sample A) and ion-doped (Sample B) GaN layers. The deposited Au/Pt/Cr/n-GaN contacts exhibited an ohmic behavior with specific contact resistance 2.8×10^{-6} $\Omega\cdot cm^2$ and 3.5×10^{-7} $\Omega\cdot cm^2$ for the samples A and B respectively. The diode structures were made using Cr/Pt/Au system.

1.Introduction

The direct-gap semiconductor gallium nitride and solid solutions based on it are promising materials for creating devices in various areas of electronics. The use of nitride structures is currently not limited to optoelectronic applications. Such structures are used to develop the component base of power and microwave electronics, such as Schottky diodes and microwave elements. However, in order to realize the advantages of nitrides for obtaining highly efficient devices, an advanced technology is needed to obtain stable ohmic contacts to the semiconductor. Low specific resistivity, high reliability at elevated temperatures and good reproducibility of ohmic contacts are the most important factors that limit the output characteristics of devices. To date, a large number of metallization systems have been proposed and studied for the formation of ohmic contacts of instrumental structures based on gallium nitride. Among them, the most widespread is the Ti/Al-based system. Such metallization can be described as Ti/Al/X(Ni, Ti, Mo, Pt, Pd)/Au [1, 2]. To obtain low-resistance contacts based on this metallization system, one needs to select the thickness of individual layers and the temperature-time processing regimes [3]. As a rule, annealing temperatures for such systems are in the range 800-900° C.

The use of high-temperature treatment for multilayer metallization systems (above 800 ° C) leads to a deterioration in the morphology of the metallization surface, which makes it difficult to carry out further operations to form the gate region. The presence of an uneven edge in ohmic contacts ultimately affects the parameters of the instrument structures. At the same time, a decrease of annealing temperature leads to an increase in contact resistance. To solve this problem, it is necessary to develop a technology for creating ohmic contacts without high-temperature treatments.

2. Experimental method

For the experimental work, epitaxial structures of gallium nitride grown on 2 inch sapphire substrates by the method of chemical vapor deposition from the gas phase using organometallic compounds were used. In the framework of the work, the parameters of ohmic contacts formed on epitaxial and ion-doped GaN layers are compared (Figure 1).

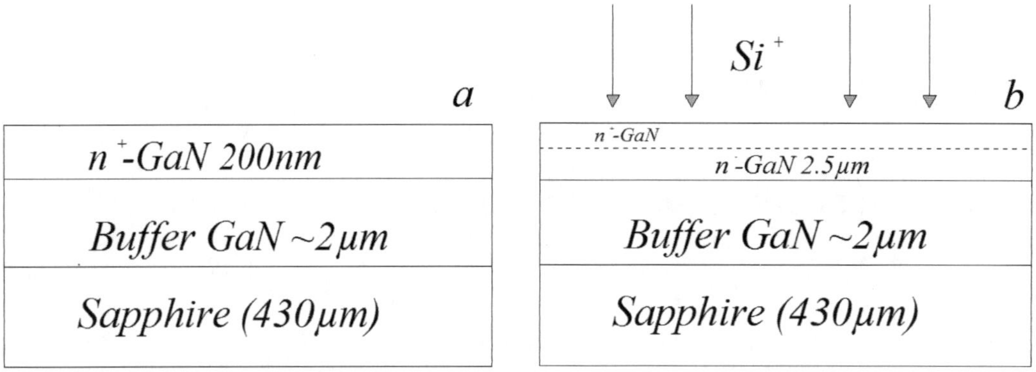

Figure 1. Initial GaN structures: *a* - structure with epitaxial n^+ layer; b - structure with ion-doped n^+ - layer.

In the structure shown in Fig. 1a, an epitaxial layer of gallium nitride with a charge carrier concentration of 4 10^{18} cm^{-3} is used. In the sample in Fig. 1b, the n^+ GaN layer was made by ion doping. To form the layer, the technology of ion implantation of Si^+ in GaN was used [4]. Embedding of the impurity was carried out with an energy of 50 keV and a dose of 10^{15} cm^{-2}. To prevent the channeling effect, the wafers were placed at an angle of 7° to the normal of the incident beam. To activate the embedded impurity, photon annealing was used. The annealing was carried out at 1250° C during 1 minute in nitrogen atmosphere [5]. As a protective coating, before the photon annealing, a SiO_2 film was deposited on the GaN surface. After activation, the concentration of charge carriers in the n^+ GaN layer was $2 \cdot 10^{19} cm^{-3}$.
As a metallization of ohmic contacts for the structures under consideration, a Cr/Pt/Au-based system was used. The contacts were formed by electron-beam deposition of the metallization system using liftoff photolithography. The quality of the ohmic contacts was estimated from the specific contact resistance determined by the TLM method [6].

3. Measurement results

The resistance of the contacts was measured using test elements at the semi-automatic station Cascade Microtech. As a measuring part for the control of parameters on a direct current, we used a meter for the characteristics of semiconductor devices Agilent B1505A.
After deposition, the contacts showed ohmic characteristics with a resistance of 2.8 10^{-6} Ω cm^{-2} and 3.5 10^{-7} Ω cm^{-2} for samples with epitaxial and ion-doped n^+ GaN layers, respectively. To evaluate the parameters of contacts, a study was made of the effect of annealing on the characteristics of the contacts. The annealing was carried out in the temperature range 400 - 800° C during 1 minute.
The characteristics of the tested samples are shown in Figure 2.

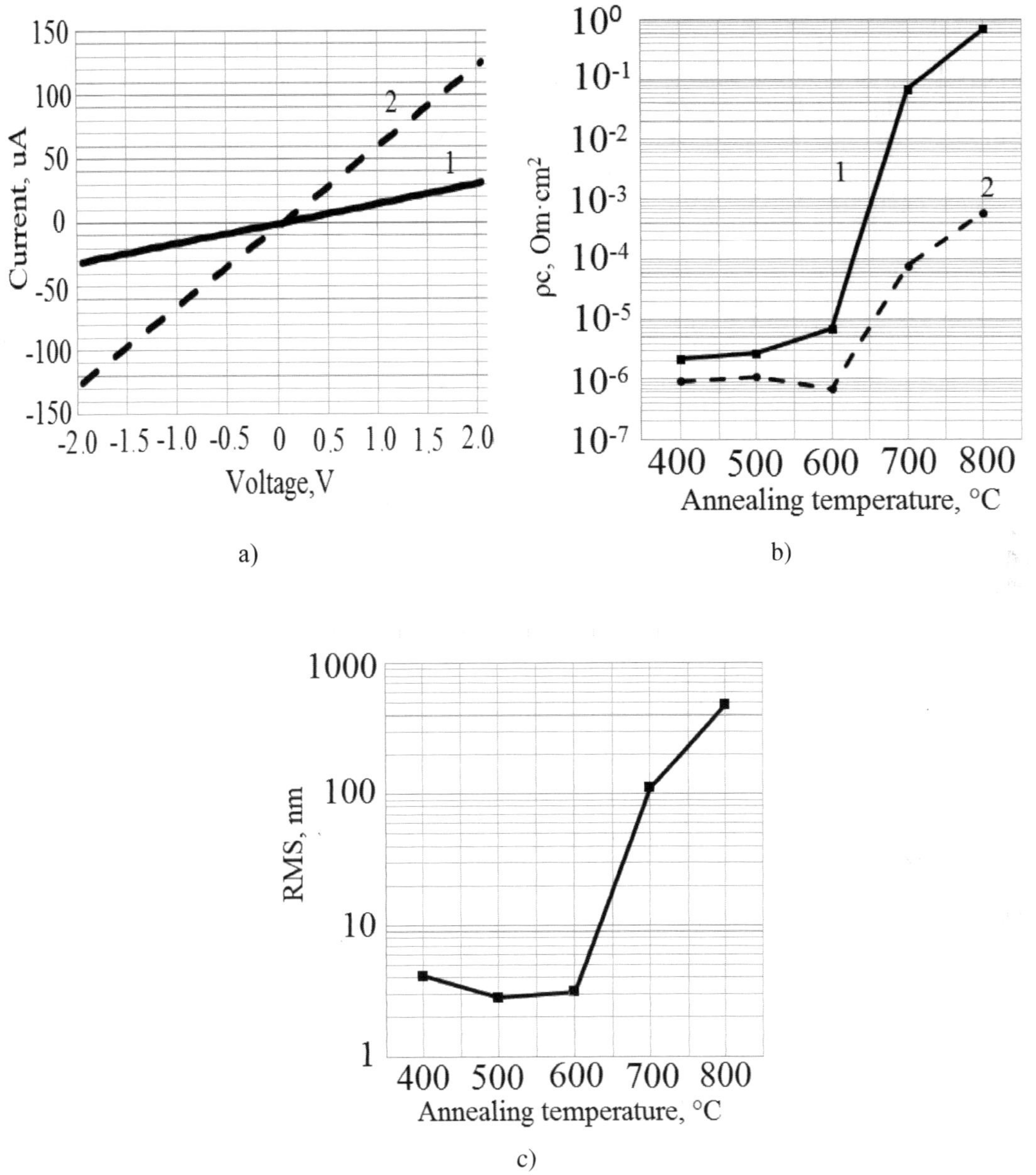

Figure 2. Effect of the heat treatment on the characteristics of the contacts for GaN samples
1 - epitaxial layer of GaN; 2 - ion-doped GaN layer:
a - current-voltage characteristics of contacts to the structures;
b - the effect of the annealing temperature on the contact resistance of the contacts;
c - the influence of the annealing temperature on the value of the root-mean-square roughness of the surface.

The contacts obtained show ohmic characteristics up to an annealing temperature of 600° C, while the specific contact resistance changes comparatively little. The contacts become nonohmic at annealing temperatures from 700° C. Deterioration of characteristics is caused by the processes of diffusion of metals into a semiconductor under high-temperature treatment. As shown in Fig. 2c, there is a sharp deterioration in the surface morphology, determined by the rms surface roughness. We also compare the system based on Cr/Pt/Au with the Ti/Al/Ni/Au-based system of the contacts to epitaxial layers of gallium nitride. In this case, Ti/Al/Ni/Au contacts show nonohmic characteristics after deposition and heat treatment up to 600° C. At high temperatures, there is a noticeable interaction of metals in the metallization system, leading to a decrease in the contact resistance, as shown in Fig. 3. To obtain ohmic contacts in the case of Ti/Al/Ni/Au system, heat treatment at a temperature above 600°C is needed.

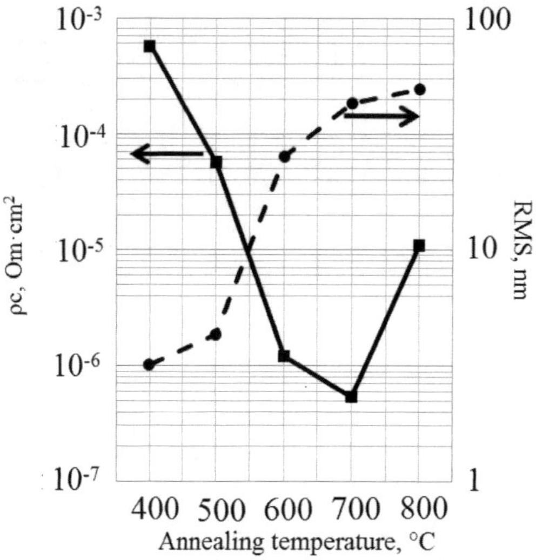

Figure 3. Effect of heat treatment on the characteristics of contacts for the system Ti/Al/Ni/Au.

Using the obtained data, instrument structures of Schottky diodes were formed. The Ni/Au (0.07/0.3 μm) system was used as the metallization of the barrier contacts. The diameter of the Schottky contact was 500 μm. Passivation of the structure was carried out by a film of silicon nitride 150 nm thick, obtained by plasma-chemical deposition using a gas mixture of monosilane, ammonia and nitrogen. The study of the current-voltage characteristics of the structures obtained (Fig. 4) shows the perspectives of using the ion doping technology for the formation of subcontact regions. From the current-voltage characteristic, the value of the direct voltage drop at the direct current level of 100 A/cm^2 was estimated. For structures with ohmic contacts based on Cr/Pt/Au, it is 0.7 V, and for ones based on Ti/Al/Ni/Au it is 1.35 V. From the current-voltage characteristics, the Schottky contact parameters are determined. The Schottky barrier value for structures with ohmic contacts based on Cr/Pt/Au was 0.63 eV, and the nonideality coefficient was -1.15. These parameters are close to theoretical values.

4. Conclusion

Thus, in the course of the work, a Cr/Pt/Au metallization system was investigated to form ohmic contacts to epitaxial and ion-doped gallium nitride layers. It is shown that the use of this system makes it possible to obtain ohmic contacts to GaN without the use of high-temperature annealing. This system makes it possible to obtain contacts with a specific contact resistance of the order of 2.2 10^{-6} $\Omega \cdot cm^2$ for the epitaxial n^+ GaN layer and 9.1 10^{-7} Ω cm^2 for the ion-doped n^+ GaN layer. Diode structures with the use of non-alloyed Cr/Pt/Au metallization were fabricated, their characteristics were measured and a comparison was made with the characteristics of structures with a Ti/Al/Ni/Au system. The advantage of the Cr/Pt/Au system is demonstrated.

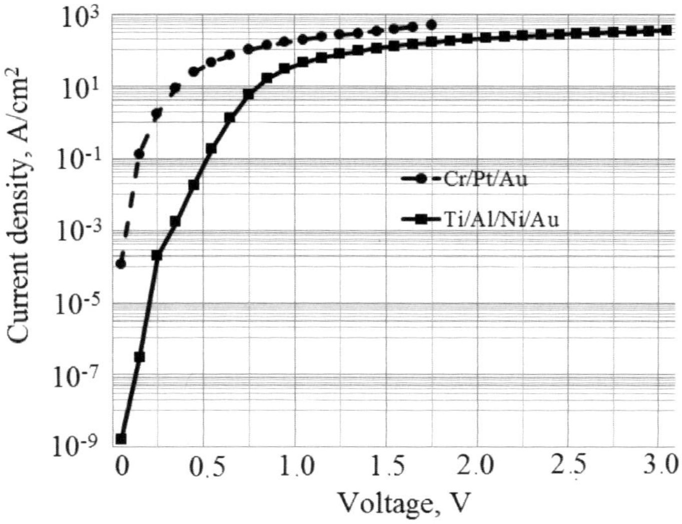

Figure 4. Current-voltage characteristics of diode structures with ion-doped subcontact layers for different ohmic contact systems.

The publication was prepared with the financial support of the Ministry of Education and Science of Russia in the framework of the project part of the State Assignment, project No. 3.33572.2017.

References

[1] Daele B. Van at al 2005 *Appl.Phys. Lett* **87** 061905
[2] Davydov AV at al. 2005 *Phys. Status Solidi C* **2**.2551
[3] Jacobs B at al. 2002 *Journal of Crystal Growth* **241** 15-8
[4] Lee M-L, Sheu J K, Hu C C 2007 *Appl. Phys. Lett.* **91** 182106
[5] Seleznev B I,. Fedorov D G. Moskalev G Y 2016 *Semiconductors* **50** 832-38
[6] Reeves G K, Harrison H B 1982 *IEEE Electr. Dev. Leter.* **EDL-3** 111-13

IOP Conf. Series: Materials Science and Engineering **441** (2018) 012046 doi:10.1088/1757-899X/441/1/012046

Electromagnets friction surfaces contamination diagnostics method based on weber-ampere characteristics analysis

D V Shaykhutdinov, K M Shirokov, N D Narakidze, E V Kirievskiy, N I Gorbatenko

Platov South Russia State Polytechnic University (NPI), 132, Prosveschenija St., Novocherkassk, 346400, Russia

E-mail: d.v.shaykhutdinov@gmail.com

Abstract. In the production of electrical products, an important element is conducting exit and intermediate control operations to reduce risks. A promising approach to the diagnosis of electromagnetic devices is the analysis of their weber-ampere characteristics. Control of the magnetic properties of products allows identifying defects without the need for long-term operations of disassembly / assembly mechanisms. The introduction of sensors of magnetic magnitudes into the design of electromagnets is impractical in terms of cost and reliability. With the help of the "working" elements of electromagnets already available in the design, it is possible to measure the weber-ampere characteristics. The article solves the problem of analysis and identification of the relationship between the critical defect of armored-type electromagnets with the weber-ampere characteristic of the product and the development of a method for the technical diagnosis of a malfunction.

1. Introduction

Output and intermediate monitoring of electromagnetic devices are mandatory steps in the manufacture of products. To solve this problem, a number of methods [1-4] and devices [5-10] were developed earlier. In the course of the work, new possibilities were identified for diagnosing faults in electromagnetic devices. A promising approach to the diagnosis of electromagnetic devices is the analysis of their weber-ampere characteristics [11]. Control of the magnetic properties of products allows identifying defects without the need for long-term operations of disassembly / assembly mechanisms. However, the most informative characteristics of electromagnets, which include web-ampere characteristics, are practically impossible to obtain with the help of known magnetic-value sensors [12, 13], since their "introduction" into the finished device without violating its integrity is in most cases impossible. Thus, the actual task is the development of methods for technical diagnostics of the most common faults of electromagnetic devices.

The most common faults of electromagnetic mechanisms include contamination of contacting working surfaces, increasing the coefficient of friction and interfering with the work of the product for the given parameters of electrical signals.

2. Materials and Methods

When developing elements of the theory of the magnetic field, methods of the theory of experimental design, measurement theory, mathematical modeling using licensed packages of GMSH, Octave application software were used.

Consider the magnetic system (Fig. 1), the main elements of which are: the magnetizing winding of the product with the number of turns w connected to the current source i(t), two ferromagnetic elements (for example, the core and the armature of the electromagnet) with an air gap between these elements.

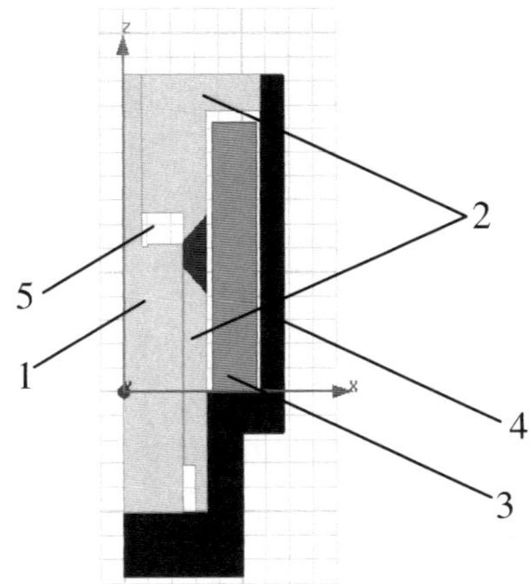

Figure 1. Electromagnet. 2D axisymmetric model: 1 - armature, 2 - core, 3 - magnetizing coils, 4 - case made of non-magnetic material, 5 - air gap

To analyze the influence of contamination of the contacting working surfaces of the core and the armature of the electromagnet increasing the friction coefficient between them, let us consider an equation describing the distribution of forces acting on the armature of an electromagnet and describing its motion:

$$F_{em-z} + m\frac{d^2z}{dt^2} - P + F_{em-x} \cdot k_{tr} \cdot sign\frac{dz}{dt} = 0, \qquad (1)$$

where F_{em-z} - functional dependence of the electromagnetic force acting on the armature along the z-axis, i - instantaneous current in the magnetizing windings of the product, z - the relative position of the armature along the axis z; m - armature weight; P - gravity at anchor, $P=mg$, g – acceleration of gravity; F_{em-x} - functional dependence of the electromagnetic force acting on the armature across the axis z, along the axis x; k_{tr} - coefficient of friction between the contacting surfaces of the armature and the core; $sign\frac{dz}{dt}$ - parameter that defines the direction of motion of the armature.

The mechanical force in a magnetic field is expressed as a derivative of the energy of the magnetic field along the coordinate [14] is:

$$F = \left|\partial W_M / \partial x\right| = \left|\partial(Li^2)/(2\partial x)\right|, \qquad (2)$$

where W_M – magnetic field energy; L – coil electromagnet inductance; i – magnetizing current in the coil of an electromagnet.

The energy of the magnetic field in a gap of length $l_в$ is:

$$W_M = BHSl_B/2, \qquad (3)$$

where B – magnetic induction in the gap, H – magnetic field strength in the gap; S – cross-sectional area of the magnetic circuit, $l_в$ – air gap length.

So the force tending to reduce the air gap, determined by the derivative in the direction perpendicular to the plane of the air gap (along the z axis), according to (2) taking into account (3):

$$F = \frac{\partial W_M}{\partial l_B} = \frac{BH}{2} S = \frac{B^2}{2\mu_0} S, \tag{4}$$

where $\mu_0 = 4\pi 10^{-7}$ Gn/m – magnetic constant.

Let's represent formula (4), expressing the induction B in the air gap through the parameters of the electromagnet. We take the cross section of the magnetic circuit constant. Using the law of total current for a magnetic core with an air gap in the approximation of the absence of scattering fields and homogeneity of the induction B of the magnetic field in the cross section S of the magnetic circuit, we write:

$$Hl + H_B l_B = iw,$$

where $H = B/\mu\mu_0$ and $H_B = B/\mu_0$ – the magnetic field strength in the magnetic circuit and in the air gap; w – number of turns of the coil of electromagnet; l – ength of the centerline of the magnetic circuit; μ - magnetic permeability of magnetic material.

Consequently, we will have:

$$B = \frac{\mu_0 iw}{(l/\mu + l_B)}. \tag{5}$$

Substituting the induction expression B (5) into formula (4), we obtain the expression for the force acting along the z axis and tending to decrease the air gap l_B, in the form:

$$F_{\text{em-}z} = \frac{\mu_0 (\mu i W)^2 S}{2(l + \mu l_B)^2}. \tag{6}$$

If the anchor is placed concentrically with respect to the hole of the yoke, then because of the symmetry of this arrangement, the total radial force acting on the anchor across the z axis along the x axis (Fig. 1) is zero [15]. However, in most cases the condition of concentric arrangement is not fulfilled and the anchor is located eccentrically relative to the yoke. In this case, the radial force of unilateral attraction acts on the anchor. This force, in accordance with [15], has the form:

$$F_{\text{em-}x} = 4{,}06 \cdot 10^{-8} \cdot B^2 \frac{\pi d^2}{4} \frac{d\varepsilon}{4h_\varepsilon \sqrt{e^2 - \varepsilon^2}}, \tag{7}$$

where d – anchor diameter, ε – the amount of armature displacement from the center, e – half the air gap between the friction surfaces of the anchor and the yoke, h_ε – height of anchoring in the yoke.

Substituting (5) into (7), taking into account the fact that in this case it is necessary to consider the value of e as the air gap, we obtain:

$$F_{\text{em-}x} = 4{,}06 \cdot 10^{-8} \cdot \left(\frac{\mu_0 iw}{(l/\mu + e)} \right)^2 \frac{\pi d^2}{4} \frac{d\varepsilon}{4h_\varepsilon \sqrt{e^2 - \varepsilon^2}}. \tag{8}$$

Taking into account (6) and (8), expression (1) takes the form:

$$\frac{\mu_0 (\mu i W)^2 S}{2(l + \mu l_B)^2} + m\frac{d^2 z}{dt^2} - mg + 4{,}06 \cdot 10^{-8} \cdot \left(\frac{\mu_0 iw}{(l/\mu + e)} \right)^2 \frac{\pi d^2}{4} \frac{d\varepsilon}{4h_\varepsilon \sqrt{e^2 - \varepsilon^2}} \cdot k_{\text{тр}} \cdot sign\frac{dz}{dt} = 0. \tag{9}$$

Let us consider the stationary state of the electromagnetic system of an electromagnet in which: e=const, ε=const, d=const, w=const, h_ε=const, μ=const, S=const, $l_в$ =const. Then, valid for further calculations, will be the notation:

$$b_1 = \frac{\mu_0 (\mu W)^2 S}{2(l + \mu l_B)^2} \text{ and } b_2 = 4{,}06 \cdot 10^{-8} \cdot \left(\frac{\mu_0 w}{(l/\mu + e)} \right)^2 \frac{\pi d^2}{4} \frac{d\varepsilon}{4h_\varepsilon \sqrt{e^2 - \varepsilon^2}}.$$

Then, expression (9) takes the form:

$$b_1 i^2 + m\frac{d^2 z}{dt^2} - mg + b_2 \cdot i^2 \cdot k_{\text{тр}} \cdot sign\frac{dz}{dt} = 0\,. \tag{10}$$

For the motion of the armature of the electromagnet, the following condition must

$$m\frac{d^2 z}{dt^2} \ge 0\,.$$

In addition, in this case, $sign\frac{dz}{dt} = -1$. Consequently, we will have:

$$0 \le m\frac{d^2 z}{dt^2} = -b_1 i^2 + mg + b_2 \cdot i^2 \cdot k_{tr} \text{ or } b_1 i^2 - mg - b_2 \cdot i^2 \cdot k_{tr} \le 0\,.$$

If we will solve this inequality with respect to the friction coefficient k_{tr}:

$$k_{\text{тр}} \ge \frac{b_1}{b_2} - \frac{mg}{b_2 i^2}\,. \tag{11}$$

If we will solve this inequality with respect to the current in the magnetizing windings i:

$$i \ge \sqrt{\frac{mg}{b_1 - b_2 k_{tr}}}\,. \tag{12}$$

3. Results and Discussion

Weber-ampere characteristic of the product is a dependence of the form Φ(i), where Φ – magnetic flux in the magnetic series of the core and the armature of the electromagnet. In this case, the magnetic flux is determined both by the properties of the materials of the ferromagnetic elements and by the dimensions of the air gap (Fig. 1). On the basis of the obtained dependence (12), we conclude that with an increase in the coefficient of friction k_{tr}, for example, with clogging or oxidation of the contacting surfaces of the electromagnetic system, the amount of current necessary to ensure the movement of the armature should increase. And this dependence is nonlinear. On the basis of this conclusion, and based on the expression (11), we obtain: an increase in the current i_k, it is necessary to achieve a certain value of the magnetic flux $\Phi_k(i_k)$, with all other conditions being equal, is a sign of the presence of additional opposing forces, in particular frictional forces. On the basis of this conclusion, an algorithm is proposed for conducting technical diagnostics of the presence of clogging in electromagnetic devices on the basis of an analysis of their web-ampere characteristics, which consists of the following:

1. The measurement of the weber-ampere characteristic is performed in the form of $\psi_j(i_{sj})$ of suitable electromagnetic device and stored in memory for later comparison (j=1,2..m, where m – number of points on the weber-ampere characteristic).
2. The measurement of the web-ampere characteristic $\psi_j(i_{dj})$ of a diagnosed electromagnetic device is performed.
3. If the condition is $i_{sj}(\psi_j) << i_{dj}(\psi_j)$ – a conclusion about the clogging of an electromagnet is made.

4. Conclusion

Analysis of the dependencies between the change in the configuration of the magnetic system of a typical electromagnetic mechanism and its magnetic characteristics showed that the determination of the presence and type of a product malfunction can be carried out by monitoring its weber-ampere characteristic. Derived analytical dependencies can be used to diagnose not only electromagnets, but also other electrical products..

Acknowledgments

This article is based on results of the project No. SP-359.2018.1, implemented under the program "Scholarship provided by the President of the Russian Federation for young scientists and graduate students engaged in advanced research and development in priority areas of modernization of the

Russian economy" using equipment of shared facility "Diagnosis and energy-efficient electrical equipment" (NPI).

References

[1] Leukhin R I, Shaykhutdinov D V and Leukhin V I 2016 *Proc. Eng.* **150** 347-353
[2] Shaykhutdinov D, Lankin A, Narakidze N, Grechikhin V, Shirokov K and Gorbatenko N 2015 *Research J. of Appl. Sci.* **10(10)** 555-557
[3] Zohoori A, Vahedi A, Meo S and Sorrentino V 2016 *J. of Intellig. and Fuzzy Syst.* **30(1)** 159-169
[4] Kiran Kumar B, Siva Reddy Y V, Vijaya Kumar M 2015 *Int. J. of Control Theory and Appl.* **8(1)** 71-76
[5] Zolfaghari M, Taher S A and Munuz D V 2016 *Ain Shams Eng. J.* **7(2)** 729-740
[6] Choi J-S, Ko J-S, Chung D-H 2006 *2006 SICE-ICASE Int. Joint Conf.* 690-695
[7] Meng F, Zhang H, Cao D and Chen H 2016 *IEEE/ASME Transactions on Mechatronics* **21(3)** 1742-1753
[8] Grechihin V V, Lankin M V, Lankin A M 2015 *Proc. Eng.* **129** 793-798
[9] Gorbatenko N I, Lankin A M and Lankin M V 2016 *Proc. Eng.* **150** 1027-1031
[10] Jiang J and Tian J 2015 *J. of Chongqing University* **38(4)** 75-79
[11] Shaikhutdinov D V, Gorbatenko N I, Naraqidze N D, Leukhin R I, Shirokov K M, Dubrov V I, Stetsenko I A, Ahmedov Sh V 2014 *Mod. prob. of sci & education* **5** 119
[12] Sakhovova A A, Shirokov K M, Janvarev S G 2013 *Mod. prob. of sci & education* **5** 111

Modification of wear-resistant coatings of Fe-Cr-C system based on the Cr3C2 obtained with help of SHS method

A V Shchegolev[1], V F Aulov[2], A V Ishkov[1], V V Ivanayskiy[1], N T Krivochurov[1]

[1]Altai State Agricultural University, 98, Krasnoarmeyskiy ave., Barnaul, Russia
[2]Federal Scientific Agroengineering Center «VIM», 5, first Institutskiy travel, Moskow, Russia
E-mail: qqq681@mail.ru

Abstract. The article deals with new wear-resistant materials of the Fe-Cr-C system, designed to strengthen the working bodies of machines by high-performance inductive surfacing. Materials are presented by a mixture of powders (burden) based on high-alloy chromium-plated cast iron of type PG-USC25 (in Russian) and flux P-0.66, modified by cermet based on Cr3C2, obtained by the SHS method. The peculiarity of the modifier proposed by the authors is that the open pores of this ceramic are filled with the main, matrix material of the future wear-resistant coating - cast iron PG-CS-25 (in Russian), which is used in the SHS as an inert. X-ray diffraction studies of cermet and metallographic studies of surfaced coatings have been performed.It was shown that the introduction into the prepared burden for inductive surfacing up to 25% by weight specified cermet allows to increase hardness (up to 1320 HV) and wear resistance of the welded material. The effect is explained by formation in the hardening coating of up to 20% by vol. of new phase - the needle-shaped carbide Cr7C3 due to recrystallization and change in the composition of carbides formed from artificially introduced Cr3C2 and already available in matrix eutectic cast iron, chromium and carbon.

1. Introduction

Wear-resistant coatings of the Fe-Cr-C system are widely used in mechanical engineering for strengthening wear of workpieces and tools surfaces, as well as razor edges of working mechanisms of agricultural machines [1]. Various methods and technologies are used for application hardenable surfaces [2], and in particular, inductive overlaying [3]. However, in some cases, the wear resistance of such surfaces is insufficient [4].

In order to improve the wear resistance without significant changing the material composition and strengthening technology, the finished burden is modified by adding various materials (powders of refractory and wear-resistant compounds, nanoparticles, carbide particles, etc.) [5-7], which can be obtained separately. The processes of recrystallization, growth or crushing of chromium carbides are initiated in the modified materials and leads to an improvement of its mechanical properties [8, 9].

The promising technique of obtaining modificators is self-spreading high-temperature synthesis (SHS), based on exothermicity interaction reactions of majority kinds of metals with nonmetals and formation of target products (carbides, borides, silicides and nitrides) [10-12].

The purpose of the article is modification of the surfacing material PG-US25 (in Russian) in oder to improve its wear resistance, powder cermet based on Cr_3C_2, obtained by the SHS method.

2. Experimental procedure

As initial reagents was used: aluminum powder of PAP-1 mark, chromium oxide (III) reactive purity and technological graphite of GLS mark. In oder to secure from oxidation of reagents and products, the synthesis was carried out in argon in a special SHS reactor, initiating combustion of burden with a nichrome coil.

Cylindrical samples with dimensions of 20 × 60 mm were prepared by extrusion of burden from the fractionated powders of the starting reagents and wetted in the binder- of 5% alcohol solution of boric acid.

The finished products (powders, cermets, alloys) after complete cooling in a reactor in argon flow, were crushed, lixiviated with 10% aqueous alkali solution, washed and dried on a filter.

Hardening coatings were obtained by inductive overlaying of a modified burden, by weight. %: PG-UC25 (in Russian) material - 85; Flux P-0.66 (in Russian) - 15 [13] on 65G steel samples. The overlaying was carried out by the industrial HFC hardening installation ELSYT-100/75 (in Russian).

X-ray phase analysis of products was performed on a DRON-6 (in Russian) diffractometer (Cu Ka-radiation, l = 1.5418 Å), phase identification was made with the help of PDWin software package using card indexes: PDF-2, ICDD [14]. The metallographic analysis of microstructures of hardening coatings was performed on a CARL ZEISS AXIO OBSERVER-Z1 M microscope and MH-6 microhardness tester according to ASTM E standard methods: 3, 407, 1558, 883, 384 [15].

3. Results and discussion

Cr_3C_2 chromium carbide used as a modificator and formed according to the known SHS reaction [16]:

$$3Cr_2O_3 + 6Al + 4C \rightarrow 2Cr_3C_2 + 3Al_2O_3,\ T_{ad.} = 2500K; \quad (1)$$

It emerged that after lixiviation, product Al2O3 has high porosity and low density and when you add it to the known burden for inductive overlaying (mixture of PG-UC25 (in Russian) powders and flux P-0.66 (in Russian)), it appears in the metal melt and removes with slag. In oder to get the effective introduction into the hardening coating a ceramet based on Cr_3C_2 was obtained, pores of which are filled with surfacing material. In this case, PG-CS25 (in Russian) was used as an inert [10, 16], and injected in amounts of 5, 10, 15 and 25%, from the mass of the metal material.

Using it in SHS reactiona high-alloy nickel cast iron of the ChN mark as an inert and without further introduction into the graphite mixture, the resulting ceramic products appeared to be a mixture of two higher carbides Cr_3C_2 and Fe_3C with a small fraction of reduced iron (Figure1).

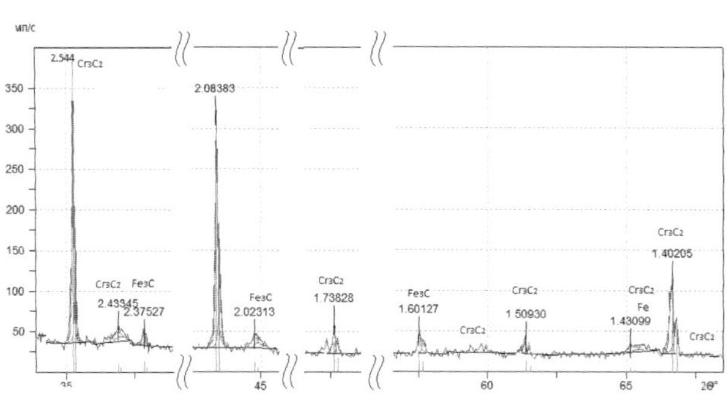

Figure 1. X-ray diffraction pattern and results of identification phases at SHSin the system Cr_2O_3 + Fe_3O_4 + Al, with the addition of 35% by weight. Inert - cast iron of ChN mark.

While using as the inertia of the main surfacing material - highly alloy cast iron of PG-USC25 (in Russian) brand, the SHS product is a ceramet based on a single carbide Cr_3C_2. The introduction of the powder of this modifier in the quantity of 25% in the ready-to-fill burden was optimal and provided the maximum hardness and wear resistance of the hardening coating.

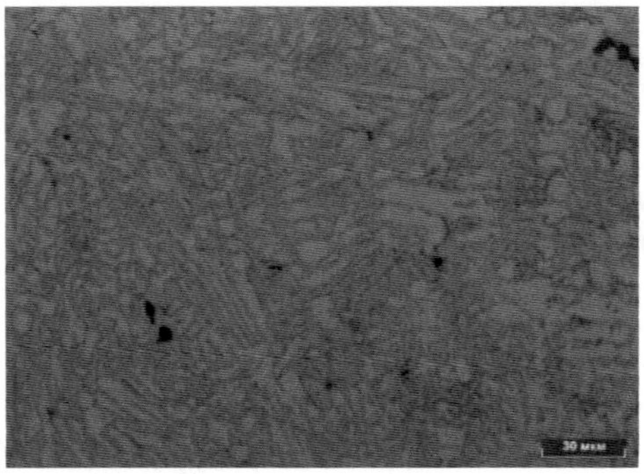

Figure 2. Microstructure of the Fe-Cr-C system deposited metalwith a cermet content based on Cr_3C_2 - 25% by weight. (500×)

Deposited metal of this composition (Figure2) has a complex structure of the composite, in the matrix, which consists of ledeburite and finely-divided carbides, the needle-like particles of the chromium carbide Cr_7C_3 are scattered. This carbide is formed during the inductive surfacing of the modified PG-US25 (in Russian) material from the introduced Cr_3C_2 carbide, as well as the chromium and carbon present in the carbide eutectics, the basic phase in the Fe-Cr-C system (Table 1).

Table 1. Characteristics of the structural components of the deposited material of the Fe-C-Cr system modified by the Cr_3C_2 based cermet

Structural composition	Share, %	Hardness HV (load of 100 g)
Chromium carbide	15-20	1324
Matrix (eutectics)	75-80	838
Metal average hardness	-	1081

The high wear resistance of the deposited metal is provided precisely by chromium carbide Cr_7C_3 (Figure3), which possesses one of the highest hardness among Cr-C carbides [17] and it is the main coating phase. It is interesting to note that this carbide is formed in a deposited layer at a much lower temperature than the equilibrium one - 1766 °C [18].

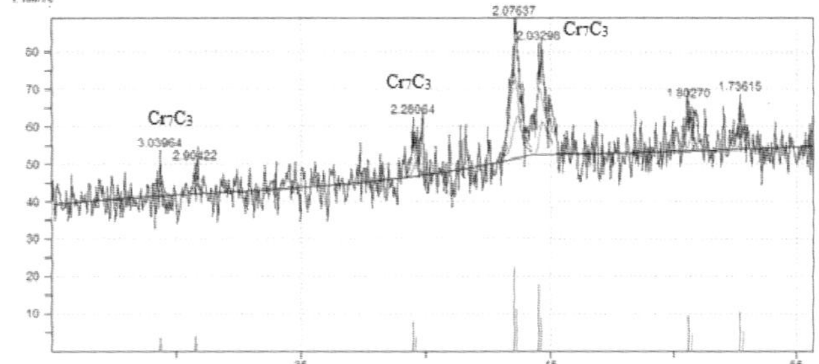

Figure3. X-ray diffraction pattern and results of the identification phases of the deposited metal with the content of Cr_3C_2 cermet - 25% by weight

The hardness of the wear-resistant coatings of the Fe-Cr-C system is determined by the carbide formation, the hardness of the deposited layer will practically not depend on the technological factors: the temperature of the metal heating during surfacing and its cooling rate while cooling down [8, 9].

This circumstance allows us to recommend the modification of the surfacing material of PG-US25 (in Russian) cermet based on Cr_3C_2 as an effective way to increase its wear resistance.

4. Conclusion

On the basis of carried out experiments and study results of composition, structure and characteristics of deposited wear-resistant material, the following conclusions can be made:
1. It is proposed to increase the durability of surfacing material of Fe-Cr-C system (PG-UC25) (in Russian) by modifying them with cermet based on Cr_3C_2.
2. Cermet is obtained by the SHS method, and open pores of the ceramics are filled with matrix material of the coating used as an inert for synthesis.
3. Introduction 25% modificator in the ready burden for inductive surfacing provided an increase in the hardness and wear resistance of the deposited layer due to the formation of Cr_7C_3 chromium carbide in it.

References

[1] Laird G, Gundlach R, Röhrig K 2000 Abrasion-resistant Cast Iron Handbook. *Ed. American Foundry Society Schaumburg USA* 1-222.
[2] Hasui A., Morigaki O. Surfacing and Sputtering 1985 *Mechanical Engineering.* 1-240 (in Russian)
[3] Tkachev V N 1970. *Induction hardfacing of hard alloys* (M. Mechanical Engineering) 1-183.
[4] Zum K and Eldis G 1980 *Abrasive Wear of White Cast Irons* Wear. **Vol. 64** 175-194.
[5] Hou Y, Wang Y, Pan Z and Yu L 2012 Journal of Rare Earths **Vol. 30** 283-288.
[6] Ryabtsev I A, Panfilov A I, Babinets A A, Ryabtsev I I, Gordan G N, Babiichuk I L 2015. *Automatic welding* **5-6** (**742**) 84-88.
[7] Albertin E and Sinatora A 2001. Effect of carbide fraction and matrix microstructure on the wear of cast iron balls tested in a laboratory ball mill. Wear **Vol. 250** 492-501.
[8] Bedolla A, Arias L and Hernández B 2003 Journal of Materials Engineering and Performance. - **Vol. 12**. 371-382.
[9] Liu F, Jiang Y, Xiao H and Tan J 2015 *Journal of Alloys and Compounds* **Vol. 618** 380-385.
[10] Amosov A P, Borovinskaya I P and Merzhanov A G 2007 Powder technology of self-propagating high-temperature synthesis of materials. M. Mechanical engineering 1 1-567.
[11] Gorshkov V A, Miloserdov P A, Luginina M A, Sachkova N V and Belikova A F 2017. *Inorganic Materials* **Vol. 53 3** 271-277.
[12] Ilyushchenko A, Vitias P, Beliaev A and Talako T 2001 *Proceedings of the International Thermal Spray Conference* «Thermal Spray: New Surfaces for a new millennium» Ed. Berndt C C, Khor K A, Lugscheider E F. Publ by ACM International, Materials Park Ohio, USA 1299-1302.
[13] Surviving materials of the member states of the CMEA 1979 catalog Kiev M. VINITI 1-619.
[14] Brady J B, NewtonRM and BoardmanSJ 1995 New Uses for Powder X-ray Diffraction Experiments in the Undergraduate Curriculum. *Jour. Geol. Education* **Vol. 43**. **№ 5** 466-470.
[15] ASTM International - Standards Search: E -https: www.astm.org Standard/alpha-lists/E.html
[16] Koizumi M 1998 Chemistry of Synthesis of Combustion, Ed. M. Koizumi (M. World) 1-247.
[17] Mehdi Mazar Atabaki, Sajjad Jafari and Hassan Abdollah 2012 Abrasive Wear Behavior of High Chromium Cast Iron and Hadfield Steel A Comparison. *Journal of Iron and Steel Research, International* **Vol. 19** Issue **4** 43-50.
[18] Giessen B and Wang S Alloy Phase Diagrams Ed. Bennett L H, Massalski T B and Giessen B C 1983 North-Holland (New York USA) 289-294.

IOP Conf. Series: Materials Science and Engineering **441** (2018) 012048 doi:10.1088/1757-899X/441/1/012048

Deformation Criteria for Calculation of Composite Constructions

A I Shein[1], O G Zemtsova[1]

[1] Penza State University of Architecture and Construction, 28, Titova str., Penza, 440028, Russia

E-mail: shein-ai@yandex.ru

Abstract. In this paper we consider the calculation of reinforcement and cross-sections of reinforced concrete structures: beams, columns, diaphragms of rigidity, etc. The article is divided into two parts. In the first part, the criterion for limiting deformations in given zones of the most stressed sections of composite systems is proposed as the basis for calculating reinforced concrete structures. The introduction of the appropriate limitations makes it possible to determine, from the equilibrium conditions of forces in a given section, one of the main geometric parameters of this section: height, width, cross-section of the working armature or the maximum possible load. An algorithm for solving the problem of selecting sections of a statically indeterminate system is shown. This approach is called the method of limited deformations (MLD). The method is illustrated by a number of examples. In the second part, the definition of points of nucleation of cracks and trajectories of their development is established by solving the problem of determining the deformed state of a composite under conditions of a plane stress state. Determination of the main deformations is performed by a modernized finite element method. The finite element calculation is based on specially developed composite elements describing the joint work of concrete and reinforcement. Two new (planar and spatial) composite finite elements are proposed. We show the scheme of formation of the general matrix of rigidity of a composite, as superposition of global matrixes of rigidity of a binder and armature.

1. Introduction

Mechanical systems, consisting of composite materials, are becoming more and more popular. They are widely used in construction, aircraft building, shipbuilding. Such materials usually consist of two or more components. The basis of the composites is binder. Strength in given directions is provided by reinforcing material - steel or carbon fiber reinforcement, fibers, threadlike crystals. In this regard, the work of construction from composite highly depends on the location of the reinforcing material and the compound of the composite.

Reinforced concrete can be qualified as a composite, in which a relatively compliant under stretch concrete is reinforced with high-strength fibers in the form of steel (reinforced concrete), carbon plastic or fiberglass reinforcement. The diagrams of steel reinforcement can have an area of fluidity (so-called building steel) or have only a limited descending branch (high-strength reinforcing steel). Composite reinforcement, as a rule, has an elastic-linear relationship between stresses and deformations up to destruction.

Calculation of reinforced concrete constructions, based on the limit equilibrium theory, is extremely easy to use, but the complete disregard for the deformations of the mechanical system makes it difficult to design a practical application. Furthermore, actual structures should not form a single plastic hinge even in the static loading conditions, and especially when we have the alternating dynamic system [1-5]. The author for a long time had to deal with the examination of buildings and structures. The presence of cracks in reinforced concrete elements has always been regarded as a circumstance requiring, usually, to conduct additional researches (aggressive environment analysis, evaluation of fire resistance, evaluation of opportunities and the value of dynamic effects, etc.) and to carry out the development of strengthening project. In this regard, there was the intention to make a strain sign a basis for design of reinforced concrete structures. This article provides a simple and reliable method of composite systems calculation, based on the specified limitation of cracks development – limited deformation method (MLD).

Most often, the calculation of composite systems by the finite element method (FEM) contains a procedure of "smearing" the characteristics of reinforcing structures along the body of a finite element [6]. In practice, this consists of selecting orthotropic constants for solving plane problems or anisotropy constants for volume projects. In the ANSYS computational complex [7], it is possible to create combined finite elements from cubic binder elements with hinge assemblies and hinged-rod reinforcing elements.

In addition, in ANSYS it is possible to represent a composite in the form of rectangular parallelepipeds consisting of sets of layers of materials with different orientations and properties of orthotropic material in each layer. A similar approach is proposed in the monograph [8]. In this paper we offer an alternative simple, reliable and physically clear method of finite element analysis of a composite. At the same time, this work continues the development of the method of bounded deformations [9] for the problems of plane stressed state.

2. Methodological basis of calculation MLD and algorithm of the method

The limit equilibrium method of reinforced concrete structures is based on the idea of complete destruction. To prevent destruction of the projected structures, a system of coefficients of reserve is introduced. These coefficients ensure the strength of the structure, but its economy and rationality become extremely doubtful. In contrast to the limit equilibrium method, the basis of the method of limited deformations is based on the opposite idea: the construction should not collapse under any possible loads [4, 5]. With an unfavorable combination of loads in the structure only limited, predefined, surface fractures can occur and that can appear only on one, the most stress-strained section of the element.

The limit equilibrium method could be called a "method of unlimited deformations". In fact: in order for plastic deformations to develop in a middle (in height) zone of a symmetric curved beam, many times greater than maximum possible elastic deformations tens must occur. As a result of this approach, a serious contradiction arises between elastic calculation and the design task.

Brittle fracture of the concrete in the stretched areas of reinforced concrete elements is associated with the formation and development of cracks. Adopting approximating dependencies, we will use the concepts of the limit of compressibility $[\varepsilon^-]$ and ultimate tensile of concrete $[\varepsilon^+]$, i.e., concrete deformation at the fracture point. Limited compressibility of concrete while bending and eccentric compression can be taken as $[\varepsilon^-]\approx(2{,}0\div3{,}5)\cdot10^{-3}$; and the ultimate elongation $[\varepsilon^+]\approx(1{,}5\div3)\cdot10^{-4}$, i.e., an order of magnitude less.

Before reaching certain limits in the extreme fiber tensile stress (or strain), we will disclose the internal static indeterminacy basing on the usual hypothesis of plane sections and the equilibrium conditions of the solid composite nonlinear stress. When cracking in cross section *j* we will assume hypothesis of plane sections as fair, outside of the crack boundary (Figure 1).

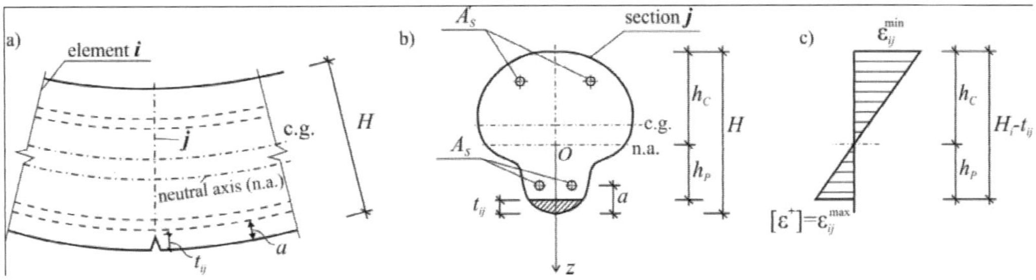

Figure 1. A fragment of a reinforced concrete element

Suppose in this concrete element *i* we know the most intense cross-section *j*. We put the condition for this section:

$$t_{ij} \le [t_{ij}], \tag{1}$$

where t_{ij} – depth of the crack in the cross section *j* of element *i*; $[t_{ij}]$ – allowable depth of a crack in a cross section *j* of element *i*. In the design tasks, it is advisable to take:

$$0 \le [t_{ij}] \le a, \tag{2}$$

where *a* – protective element layer thickness of *i*. Then the strain of the stretched composite zone at the crack border will be equal to $[\varepsilon^+]$. In the same cross section, outside of the boundary fracture, deformations are distributed linearly.

Similarly, the restriction to a depth of crushing zone is introduced. However, here it is expedient to introduce the following restrictions:

$$\left.\begin{aligned}[t_{ij}^-]=0\\ t_{ij}^- \ge 0\end{aligned}\right\}. \tag{3}$$

For bending elements fiber deformation at a distance from the neutral axis *z* (Figure 1) is defined by:

$$\varepsilon = z \cdot w''. \tag{4}$$

Here w'' – the curvature of the axis of the element. On the border of the crack the strain is:

$$[\varepsilon^+]_{ij} = (h_p)_{ij} \cdot (w'')_{ij}, \tag{5}$$

where h_P – the height of the tension zone. At any other point of the cross section, the deformation can be calculated by the formula:

$$\varepsilon_{ij} = \left(\frac{z}{h_p}\right)_{ij} \cdot [\varepsilon^+]_{ij}. \tag{6}$$

Note that the maximum compressibility can be achieved only at the height of the compressed zone

$$z = h_c > 10 h_p. \tag{7}$$

What is more usually realized that in the columns than in the girders.

Let longitudinal force N_{ij} and bending moment M_{ij} of external forces in the *j* cross section correspond to a given strain and generalized load *P*. We impose the following conditions to effective dimensions (not affected by the crack) of the section:

$$\left.\begin{aligned}\int_{A_э} \sigma \cdot dA_э = N_{ij}\\ \int_{A_э} \sigma z \cdot dA_э = M_{ij}\end{aligned}\right\}, \tag{8}$$

where the unknown value is or one of the dimensions that define the area of the cross-section of the concrete *A*, or a cross-section of tensile reinforcement or generalized load *P*.

The most rational dimensions of cross sections of reinforced elements are obtained by performing the pre-tensioning of reinforcement. At the same time, the effort in the prestressing rod in the design section is equal to the algebraic sum of the efforts without pre-stress N_1, corresponding deformation $\varepsilon_a(z)$, and tension reinforcement efforts, taking into account losses N_0 (N_0 corresponds to the non-deformed concrete in view of losses).

Algorithm of Limited deformations method of calculation of statically indeterminate systems consists of the following:

1. Preliminary appointing the dimensions of the elements.
2. FEM calculations are performed with the construction of envelope diagrams of moments.
3. Using $M_{ij\,max}$ from equations (8), parameters of the cross-section of *i*-th elements are selected.
4. Introducing improved values of stiffness. The FEM is recalculated with the calculating of new efforts parameters.
5. Refinement the $M_{ij\,max}$ and parameters of the cross-section of *i*-th elements.
6. Introducing improved values of stiffness. Final calculation are performing.

Here is an example of statically indeterminate system (Figure 2a), calculated by this algorithm. The unknown variable is considered to be the width of the cross-section b_1 or b_2.

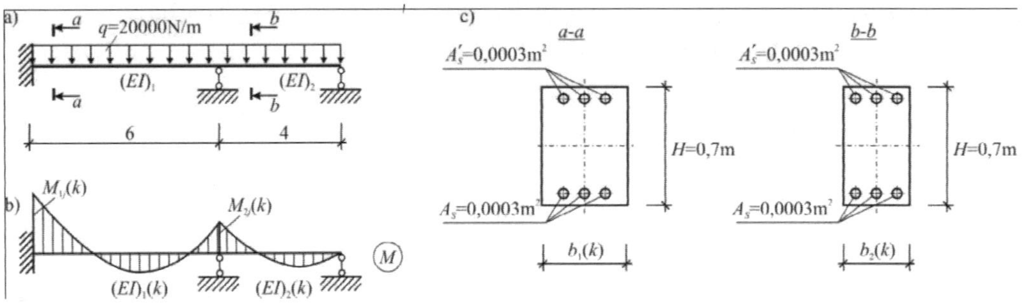

Figure 2. Selection of a statically indeterminate system's cross section

Characteristics of the concrete and reinforcement are the same as in the previous example. The calculations are summarized in Table 1.

Table 1. Results of calculations

The number of calculate ***k***	Stiffness ratio		M_{ij}, N·m		Dimensions of cross-section	
	$(EI)_1$	$(EI)_2$	M_{1j}	M_{2j}	b_1	b_2
1	1	1	64707	50588	155	109
2	1.289	1	65339	49323	157	104
3	1.343	1	65444	49117	157	104

3. Calculation of reinforced concrete systems by FEM

Consider the calculation and selection of reinforcement and cross-sections near the support of reinforced concrete beams or diaphragms of rigidity.

The inclined cracks in the area near the support of the beams originate from the combined effect of cross force and bending moment. Inclined cracks trajectory on the sections of reinforced concrete beams near the supports, which lines are defined by trajectory of the main deformations, present some theoretical interest. Determination of the points of nucleation of cracks and trajectories of their development can be established by solving problems of determination of the deformed state of the composite under conditions of plane stressed state. The initiation of a crack will occur at that point of the support zone of the beam, where the deformations have reached one of the limiting values

$$\varepsilon^+_{max} = [\varepsilon^+], \quad (9)$$

$$\varepsilon^-_{min} = [\varepsilon^-], \quad (10)$$

where ε^+_{max}, ε^-_{min} – the main strains of tension and compression at a point, $[\varepsilon^+]$, $[\varepsilon^-]$ – deformation, corresponding to the breaking and crushing of concrete fibers.
As we know, the main deformations in the plane problem can be determined from the equations:

$$\left.\begin{aligned}\varepsilon^+_{max} &= \frac{1}{2}(\varepsilon_x+\varepsilon_y)+\frac{1}{2}\sqrt{(\varepsilon_x-\varepsilon_y)^2+\gamma^2_{xy}}\\ \varepsilon^-_{min} &= \frac{1}{2}(\varepsilon_x+\varepsilon_y)-\frac{1}{2}\sqrt{(\varepsilon_x-\varepsilon_y)^2+\gamma^2_{xy}}\end{aligned}\right\}. \quad (11)$$

The definition ε_x, ε_y and γ_{xy} can be performed by a modernized finite element method (FEM). Modernization of FEM will be done in a way that it is possible to solve the problems of calculating composite systems. In this case, finite elements will be formed not by choosing orthotropic constants for the description of reinforcement (which is very approximate and conditional), but by developing special composite elements, describing the joint work of concrete and reinforcement. For a plane problem of the theory of elasticity, such a finite element can be obtained by imposing a rod finite element on one side of the triangular finite element of the plane problem (Figure 3). Here the rod element will have the characteristics of the reinforcement, and the flat triangular element will describe the concrete object.

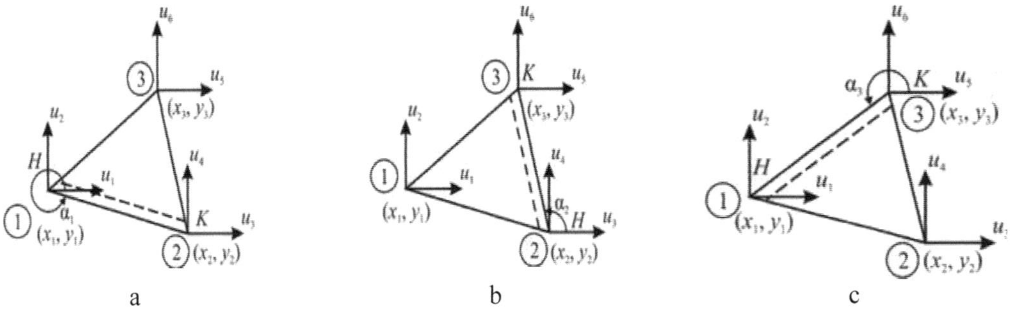

Figure 3. Composite finite element for solving a flat problem for a composite – a triangle describes a concrete object, a rod (dashed line) describes an armature.

The stiffness matrix of a triangular element [10] has the form:

$$K_{pl}=\frac{E_1h}{4A(1-v_1v_2)}\begin{bmatrix} y^2_{23}+a_1x^2_{23} & (a_1+v_2)x_{32}y_{23} & y_{23}y_{31}+a_1x_{23}x_{31} & a_1x_{32}y_{31}+v_2x_{13}y_{23} & a_1x_{12}x_{23}+y_{12}y_{23} & a_1x_{32}y_{12}+v_2x_{21}y_{23}\\ (a_1+v_2)x_{32}y_{23} & \frac{v_2}{v_1}x^2_{23}+a_1y^2_{23} & a_1x_{13}y_{23}+v_2x_{32}y_{31} & \frac{v_2}{v_1}x_{23}x_{31}+a_1y_{23}y_{31} & a_1x_{21}y_{23}+v_2x_{32}y_{12} & \frac{v_2}{v_1}x_{12}x_{23}+a_1y_{12}y_{23}\\ y_{23}y_{31}+a_1x_{23}x_{31} & a_1x_{13}y_{23}+v_2x_{32}y_{31} & y^2_{31}+a_1y^2_{31} & (a_1+v_2)x_{13}y_{31} & a_1x_{12}x_{31}+y_{12}y_{31} & a_1x_{13}y_{12}+v_2x_{21}y_{31}\\ a_1x_{32}y_{31}+v_2x_{13}y_{23} & \frac{v_2}{v_1}x_{23}x_{31}+a_1y_{23}y_{31} & (a_1+v_2)x_{13}y_{31} & \frac{v_2}{v_1}x^2_{31}+a_1y^2_{31} & a_1x_{21}y_{31}+v_2x_{13}y_{12} & \frac{v_2}{v_1}x_{12}x_{31}+a_1y_{12}y_{31}\\ a_1x_{12}x_{23}+y_{12}y_{23} & a_1x_{21}y_{23}+v_2x_{32}y_{12} & a_1x_{12}x_{31}+y_{12}y_{31} & a_1x_{21}y_{31}+v_2x_{13}y_{12} & y^2_{12}+a_1x^2_{12} & (a_1+v_2)x_{21}y_{12}\\ a_1x_{32}y_{12}+v_2x_{21}y_{23} & \frac{v_2}{v_1}x_{12}x_{23}+a_1y_{12}y_{23} & a_1x_{13}y_{12}+v_2x_{21}y_{31} & \frac{v_2}{v_1}x_{12}x_{31}+a_1y_{12}y_{31} & (a_1+v_2)x_{21}y_{12} & \frac{v_2}{v_1}x^2_{12}+a_1y^2_{12} \end{bmatrix},$$

where $x_{ij}=x_i-x_j$; $y_{ij}=y_i-y_j$, h – beam or diaphragm thickness, A – area of the final element, $a_1=\frac{G}{E_1}(1-v_1v_2)$, G – shear modulus

The stiffness matrix of the reinforcing rod has the form:

$$K_a=\frac{E_aA_a}{l}\begin{bmatrix} cc & cs & -cc & -cs\\ sc & ss & -sc & -ss\\ -cc & -cs & cc & cs\\ -sc & -ss & sc & ss \end{bmatrix}, \quad (12)$$

where E_a, A_a – the modulus of elasticity and the cross-sectional area of the reinforcement, l –distance between nodes, c, s – cosine and sine of the angle of inclination of the face of the triangle with the armature to the axis$x$.

The composite stiffness matrix is found by superimposing:

$$K_{COM} = K_{pl} + K_a. \tag{13}$$

To solve the spatial problem, the composite finite element can be obtained in an similar manner – by imposing a rod element on one of the faces of a concrete tetrahedron (Figure 4).

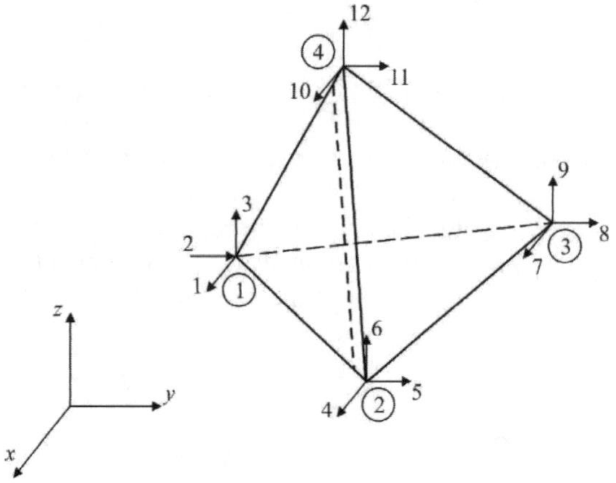

Figure 4. Composite finite element for solving a spatial problem for a composite - a tetrahedron describes a concrete, a rod (dashed line) describes an armature

The stiffness matrix of the composite finite element for solving a spatial problem

$$K = K_T + K_a, \tag{14}$$

where K_T for isotropic material is determined by the expression (17.186) [10, page 267], and the stiffness matrix of the reinforcing bar has the form:

$$K_{ai} = \frac{E_a A_{ai}}{l_i} \begin{bmatrix} (c_1c_1)_i & (c_1c_2)_i & (c_1c_3)_i & -(c_1c_1)_i & -(c_1c_2)_i & -(c_1c_3)_i \\ (c_2c_1)_i & (c_2c_2)_i & (c_2c_3)_i & -(c_2c_1)_i & -(c_2c_2)_i & -(c_2c_3)_i \\ (c_3c_1)_i & (c_3c_2)_i & (c_3c_3)_i & -(c_3c_1)_i & -(c_3c_2)_i & -(c_3c_3)_i \\ -(c_1c_1)_i & -(c_1c_2)_i & -(c_1c_3)_i & (c_1c_1)_i & (c_1c_2)_i & (c_1c_3)_i \\ -(c_2c_1)_i & -(c_2c_2)_i & -(c_2c_3)_i & (c_2c_1)_i & (c_2c_2)_i & (c_2c_3)_i \\ -(c_3c_1)_i & -(c_3c_2)_i & -(c_3c_3)_i & (c_3c_1)_i & (c_3c_2)_i & (c_3c_3)_i \end{bmatrix}. \tag{15}$$

where c_j –cosine of the slope angle of the face ***i*** with the ***j*** axis of coordinates.
If the bonding material in the plane problem is represented as an isotropic body, then the approximate solution of the plane problem for the finite element takes the following form:

$$\begin{cases} \varepsilon_x = \dfrac{\partial u}{\partial x} = ((y_3 - y_2)u_1 + (y_1 - y_3)u_3 + (y_2 - y_1)u_5)/(2A); \\ \varepsilon_y = \dfrac{\partial v}{\partial y} = ((x_2 - x_3)u_2 + (x_3 - x_1)u_4 + (x_1 - x_2)u_6)/(2A); \\ \gamma_{xy} = \dfrac{\partial u}{\partial y} + \dfrac{\partial v}{\partial x} = ((x_2 - x_3)u_1 + (x_3 - x_1)u_3 + (x_1 - x_2)u_5 + (y_3 - y_2)u_2 + (y_1 - y_3)u_4 + (y_2 - y_1)u_6)/(2A). \end{cases} \tag{16}$$

4. The general model of a composite

In many cases, it is more convenient to formulate a stiffness matrix not from composite elements, but using the superposition principle (at least in increments) of the global matrixes of stiffness of the binder and reinforcement. Let us imagine separately the kinematic scheme of the reinforcement, which approximates the reinforcing material in the best way, and the kinematic finite element scheme of the body of the binder, which is most suitable for its calculation. We put only one condition: **all** nodes of the kinematic diagram of the reinforcement must coincide with a part of the nodes of the kinematic

IOP Conf. Series: Materials Science and Engineering **441** (2018) 012048 doi:10.1088/1757-899X/441/1/012048

scheme of the binder body.

We show this method using the example of a finite element model of a reinforced concrete beam symmetrical relatively to the central vertical axis. Figure 5 shows the kinematic finite element scheme of the concrete beam body. In Figures 6 and 7 shows the kinematic diagram of the reinforcement.

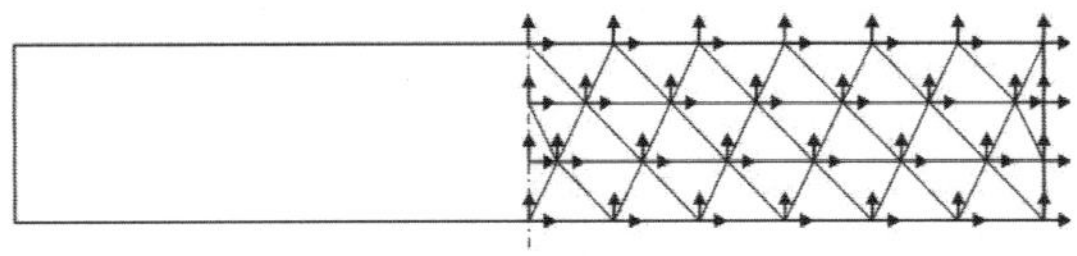

Figure 5. Kinematic model of the concrete body (to the left of the axis of symmetry)

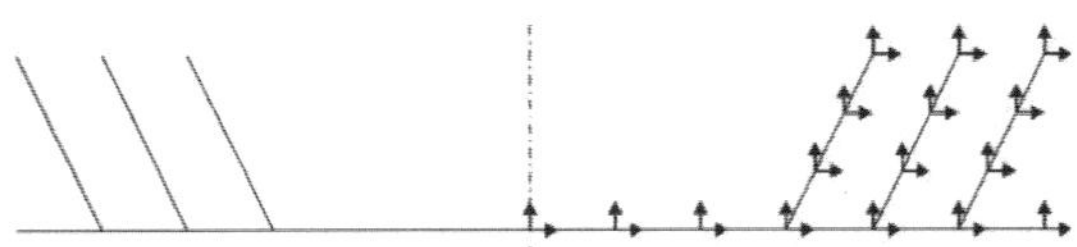

Figure 6. Kinematic model of the reinforcement (to the left of the axis of symmetry) with hinge-rod assemblies

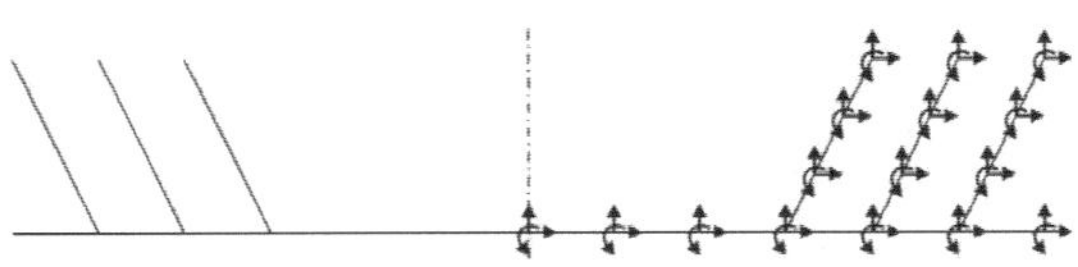

Figure 7. Kinematic model of the reinforcement (to the left of the axis of symmetry) with frame-rod assemblies

The separate global stiffness matrices of a concrete body and reinforcement must have the identical dimension. Then both matrices are summed, thus forming a global stiffness matrix of the composite system - of the reinforce concrete construction. In some cases, it can have null rows and columns. Physically, this means that the reinforcing rods can be nonconnected in one carcass.

When the kinematic schemes are superimposed on each other, the stiffness matrices fold together.

Calculation under cracking conditions is conveniently carried out in the form of a series of numerical experiments with increasing load. If, in performing any calculation, the deformations of the finite element exceed the limitations, then the number of displacements of one of the nodes of this final element of the concrete body should be doubled. In Figure 8a we show the kinematic diagram of the concrete body after breaking of the shaded element. In addition, the process of occurrence and movement of cracks can be modeled by introducing additional two nodes into the body of that concrete element which deformations exceed the limitations (Figure 8b). Trajectories of cracks will be the lines along which we are doubling the number of displacements.

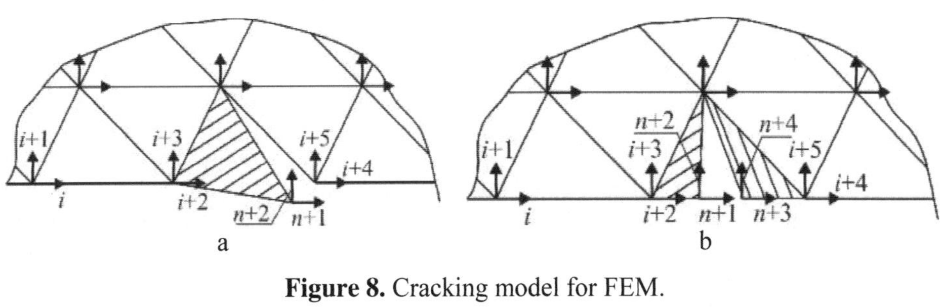

Figure 8. Cracking model for FEM.

Determination of the parameters of beam or diaphragm reinforcement can be performed using the following algorithm. We accept some redundant reinforcement scheme. After the finite-element calculation of the beam (diaphragm), we determine the deformations of the elements ε_x, ε_y, γ_{xy} and the main deformations of the elements $\varepsilon^+_{max}, \varepsilon^-_{min}$. If everywhere $\varepsilon^+_{max} < [\varepsilon^+]$, $\varepsilon^-_{min} > [\varepsilon^-]$ then reduce the amount of reinforcement. Otherwise, we finish the calculation.

5. Conclusion

Thus, the work deals with the calculation and selection of reinforcement and cross-sections of reinforced concrete beams, columns or diaphragms of rigidity. The main provisions of bounded deformations method are given. The basis for the calculation of this method is the criterion for limiting deformations in given zones of the most stressed normal sections of composite structures. On the basis of the method, the solutions of the problems of selecting the parameters of sections of reinforced concrete beams with a pre-stressed and non-stressed reinforcement are performed. The algorithm of the method for calculating statically indeterminate systems is presented and numerically realized.

When calculating the beams and the stiffening diaphragms, it is proposed to determine the points of nucleation of cracks and the trajectory of their development by solving the problem of determining the deformed state of a composite under conditions of a plane stressed state. Determination of the main deformations is performed by a modernized finite element method. The finite element calculation is based on specially developed composite elements describing the joint work of concrete and reinforcement. Two new (planar and spatial) composite finite elements are proposed. The scheme of formation of the general model of a composite is given. The algorithm for selecting the reinforcement of a composite system is shown.

Using the proposed model, the following design actions can be performed:

- perform a verification calculation with the definition of stresses and deformations in concrete and reinforcement;
- determine the trajectories of the development of cracks;
- optimize the mass and location of the reinforcement;
- determine the deflections of the reinforced concrete mechanical system.

References

[1] Shein A I 2005 Method of Grid Approximation of Elements in Problems of Structural Mechanics of Nonlinear Rod Systems, *Publishing house of Penza State University of Architecture and Construction*

[2] Shein A I 2003 Calculation of rod systems on the basis of the refined theory and the method of grid approximation of elements *Building materials, equipment, technologies of the XXI century*. **1** 38-39

[3] Shein A I 2002 Solution of the multiparameter problem of the dynamics of rod systems by the grid approximation method of elements, *Industrial and Civil Engineering* 2 27-31

[4] Shein A I, Zavyalova O B 2012 Calculation of monolithic reinforced concrete skeletons taking into account the sequence of erection, physical nonlinearity and creep of concrete *Construction mechanics and design of structures* **5** 64-69.

[5] Zavyalova O B, Shein A I 2015 Application of Grid Approximation Method for the Calculation of Monolithic Reinforced Concrete Frame Taking into Account Construction Sequence and Concrete Creep, *ICAET 2014, Applied Mechanics and Materials*. **752-753** 617-623.

[6] Karpenko N I 1996 General Models of Reinforced Concrete Mechanics (Stroyizdat, Moscow)

[7] Ibrahim Amer M, Mahmood Sh 2009 Mohammed Finite element modeling of reinforced concrete beams strengthened with FRP laminates *European Journal of Scientific Research* **4** 526-541

[8] Klovanich S F, Mironenko I N 2007 Method of Finite Elements in Ferro-Concrete Mechanics, (Publishing house of Odessa National Maritime University)

[9] Shein A I 2017 Method of limited deformation in solving design problems for reinforced

concrete structures *Naukovedenie* **2** 104-112.
[10] O M Paliy 1982 *Handbook of Ship Mechanics*. Vol **2** (Sudostroyeniye, Leningrad)

Study of the temperature processes at sensorless determination of electrical product magnetic characteristics

K M Shirokov, D V Shaykhutdinov, N D Narakidze, E V Kirievskiy, N I Gorbatenko

Platov South Russia State Polytechnic University (NPI), 132, Prosveschenija St., Novocherkassk, 346400, Russia

E-mail: d.v.shaykhutdinov@gmail.com

Abstract. Actual is the use of senseless devices for determining the webere-ampere characteristics of electrical products. However, these devices are characterized by the error in determining the flux linkage, caused by inaccurate determination of the active component of the winding resistance during the test. In classical senseless devices, the active component of the winding resistance is determined immediately before measuring the webere-ampere characteristic and is then applied as a constant value. Thus, the error in determining the active component of the winding resistance is due to the error in determining its initial value and the temperature change in the active component of the winding resistance during the determination of the magnetic characteristic. The analysis of the temperature error of the sensorless devices is carried out. A functional relationship between the error in determining the flux linkage and the electrical parameters of the winding was revealed. This makes it possible to evaluate both the absolute and relative error in the determination of flux linkage and conclude that it is necessary to apply methods for compensating for the error in determining the active component of the winding resistance. The obtained analytical results are confirmed by modeling in the program of circuit analysis. The developed procedure for estimating the error of the flux-linkage determination is also verified on the experimental data.

1. Introduction

The complexity of the designs of electrical products, the trend of miniaturization and the reduction of material consumption make ever higher demands on the quality of the component parts. The quality of the parts is determined primarily by their mechanical, electrical and magnetic properties [1, 2]. And it is from the magnetic properties that the operational characteristics of electrical products depend, and the use of parts with high magnetic properties, for example as magnetic circuits, reduces the dimensions of electrical products. Information on the magnetic properties of parts allows you to detect defects at the assembly stages, and to identify shortcomings of other parts of the product by testing electrical assemblies. Using the results of tests of blanks and products in the process control subsystem for the production of electrical products, it is possible to increase the yield of suitable products and, thereby, significantly reduce economic costs [3,4].

Equally important is the diagnostics and monitoring of the technical condition of the machines in operation, as well as the forecasting of their technical condition, the sorting of the output, depending on the provision of the required electrical and mechanical parameters, as well as, for example, acoustic ones [5, 6], which are provided at a high level, in some cases very critical (the noise of submarine

engines).
The complexity of the task of monitoring magnetic properties is that the most informative characteristics of electrical products (EP) can not be accurately determined without the use of special measuring transducers, which greatly increases the test time and makes it difficult to conduct continuous monitoring [7, 8]. A promising direction in solving this problem is the use of sensorless control methods. In this case of magnetic monitoring as a source of magnetic field it is used the working winding of electrical products.

2. Materials and Methods

When developing elements of the theory of the magnetic field, methods of the theory of experimental design, measurement theory, mathematical modeling using licensed packages of GMSH, Octave application software were used [9].
During the test, heating occurs and, as a consequence, changes in the active component of the winding resistance, which causes an inadmissible error in determining the magnetic flux and, as a consequence, the webere-ampere characteristics of the EP (Fig. 1). In Fig. 1 we introduce next notation: ψ'_r, ψ''_r – flux linkages corresponding to the remanent magnetization of the sample; I'_c, I''_c – currents corresponding to the coercive force; ψ'_m, ψ''_m – saturation flux linkages; ψ_r, I_c ψ_m – their actual values; $\Delta\psi_r$, ΔI_c, $\Delta\psi_m$ and $\delta\psi_r$, δI_c, $\delta\psi_m$ – their absolute and relative errors, respectively. In this cases, it is important to develop methods and devices for sensorless magnetic control with compensation for the temperature error [10].
Consider the magnetic system (Figure 1), the main elements of which are: the magnetizing winding of the product with the number of turns w connected to the current source i(t), two ferromagnetic elements (for example, the core and the armature of the electromagnet) with an air gap between these elements [11].

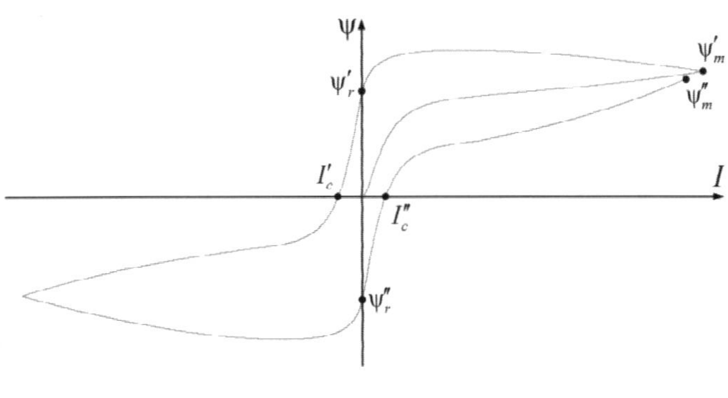

Figure 1. Magnetic characteristic with characteristic points

In classical senseless devices, the active resistance component of the working winding EP is preliminarily determined and then used in calculating the magnetic flux as a constant value [12]. Therefore, classical sensorless devices determine the flux linkage by expression:

$$\psi = \int_0^t (u - iR_1)dt + C,$$

where u - winding voltage, i - current flowing through the winding, R_1 - the active component of the winding resistance, determined immediately before measuring the magnetic characteristic; C - integration constant.
The actual value of flux linkage is determined by the formula:

IOP Conf. Series: Materials Science and Engineering **441** (2018) 012049 doi:10.1088/1757-899X/441/1/012049

$$\psi_d = \int_0^t \left(u - i \cdot \left(R_I + \Delta R^{\Sigma}\right)\right) dt + C,$$

where ΔR^{Σ} – total accumulated absolute error in determining the current value of the active component of the winding resistance.

Due to this, the absolute error in the determination of the flux linkage is determined by the expression:

$$\Delta\psi = \psi - \psi_d = \int_0^t \left(i \Delta R^{\Sigma}\right) dt. \quad (1)$$

In case of accurate initial determination of the value of the active component of the winding resistance – $\Delta R^{\Sigma} \approx 0$. Then, during the test, a current change occurs (increase, decrease, change of direction), leading to heating of the EP and, as a consequence, to a change ΔR^{Σ}. After integrating the sign of $\Delta\psi$ corresponds to the sign of the expression $i\Delta R^{\Sigma}$. And when expression $i\Delta R^{\Sigma}$ changes its sign, the direction of $\Delta\psi$ change is changes too. Thus, the change in the error in determining the flux linkage is nonmonotonic in nature.

Let's analyze the thermal processes of the EP winding. Two thermal processes lead to a change in the temperature of the winding: an increase in temperature due to heat generation during the flow of electric current along it and a change in the temperature of the winding, due to heat exchange with the environment (air, other EP-details). The resulting change in winding temperature is determined by the amount of heat received or given, in accordance with expression:

$$dQ_1 = cm\Delta T,$$

where dQ_1 – amount of heat received or returned; c – specific heat of winding; m – winding mass; ΔT - temperature variation. According to the Joule-Lenz law, the amount of heat received by the winding when an electric current flows through it is given by

$$dQ_2 = i^2 R(t°\mathrm{C}) dt.$$

Let us consider a special case in which the subject EP has a steady room temperature and the process of measuring the magnetic characteristic is fast enough, so that the heat exchange between the winding EP and the surrounding medium can be neglected. In this case, all the amount of heat released in the winding, due to the flow of electric current through it, will go to increase its temperature – $dQ_1 = dQ_2$ and consequently:

$$cm\Delta T = i^2 R(t°\mathrm{C}) dt.$$

We express from this relation the change in temperature:

$$\Delta T = \frac{i^2 R(t°\mathrm{C}) dt}{cm}.$$

The current value of the active component of the winding resistance from temperature is given by:

$$R(t°\mathrm{C}) = R_I \cdot (1 + \alpha\Delta T),$$

where α – copper temperature coefficient.

Substitute the value of ΔT in this expression:

$$R(t°\mathrm{C}) = R_I \cdot \left(1 + \alpha \cdot \frac{i^2 R(t°\mathrm{C}) dt}{cm}\right).$$

Expressing the current value of the active component of the winding resistance from this expression, we get:

$$R(t°\mathrm{C}) = \frac{cmR_I}{cm - \alpha R_I i^2 dt}. \quad (2)$$

Using the expression (2), it is possible to calculate the value of the active component of the winding resistance at any time during the determination of the EP magnetic characteristic, using i as the value of the effective current, while not taking into account the heat transfer between the winding and the environment, the change in specific heat due to temperature change, and The change in the active component of the winding resistance between its initial change and the current calculated value. Iterative calculation avoids the last drawback and uses instantaneous current values in the calculation:

$$R(t°C)_k = \frac{cmR(t°C)_{k-1}}{cm - \alpha R(t°C)_{k-1} i_k^2 \cdot \Delta t}, \quad (3)$$

where i_k - the current value for each k measurement (for k = 1..N, where N is a the number of measured points, when determining the magnetic characteristic); Δt - step of digitization; wherein $R(t°C)_0 = R_I$.

Let's analyze the expressions (2), (3) taking into account the world tendency to reduce the weight and size parameters, when $m \to 0$:

$$R(t°C) = \lim_{m\to 0} \frac{cmR_I}{cm - \alpha R_I i^2 dt} = \lim_{m\to 0} \frac{\frac{cmR_I}{cm}}{\frac{cm}{cm} - \frac{\alpha R_I i^2 dt}{cm}} = \lim_{m\to 0} \frac{R_I}{1 - \frac{\alpha R_I i^2 dt}{c} \cdot \frac{1}{m}} = \lim_{m\to 0} \frac{R_I}{1 - \frac{k}{m}},$$

coefficient $k = \frac{\alpha R_I i^2 dt}{c} > 0$ is a positive constant, because tabular coefficients of copper α, c and the active component of the winding resistance is always positive; the square of the current value of the current is positive $i^2 > 0$; when the instantaneous current value assumes a zero value, the active component of the winding resistance does not change; $dt > 0$ - positive period of time. In this way $\lim_{m\to 0} \frac{k}{m} \to +\infty$. In case $R(t°C) > 0$ it is true that $R(t°C) = \lim_{m\to 0} \frac{R_I}{1 - \frac{k}{m}} > 0$; and in case of $R_I > 0$ it is true that $1 - \frac{k}{m} \geq 0$, or $\frac{k}{m} \leq 1$. Thus we obtain a system under which the expressions (2) and (3) have meaning:

$$\begin{cases} \lim_{m\to 0} \frac{k}{m} \to +\infty; \\ \frac{k}{m} \leq 1. \end{cases}$$

So $\frac{k}{m} = 1$ - is the maximum value for which the expressions (2) and (3) have a physical meaning. So we have:

$$R(t°C) = \lim_{m\to 0} \frac{R_I}{1 - \frac{k}{m}} = \lim_{m\to 0} \frac{R_I}{1 - 1} = +\infty.$$

The analysis of expressions (2) and (3) shows that the trends of miniaturization lead both to a greater absolute value of the active component of the winding resistance, and to its greater increment in the process of sensorless determination of the magnetic characteristic of EP.

3. Results and Discussion

By using of KEK-03-MMM-150 a proportional electromagnet was tested with a sinusoidal current and the necessary electrical parameters for $R(t°C)$ calculation corresponding the equation (3). When the expression (3) was realized, it was also taken into account that the winding had a mass $m = 0{,}027\ kg$ and it is made of copper, the specific heat of which at room temperature is

$c = 385$ $J/(kg{\cdot}K^{-1})$. Fig. 2 shows the calculated active component of the winding resistance, the relative change in the active component of the winding resistance was $\delta R = 0{,}2\,\%$.

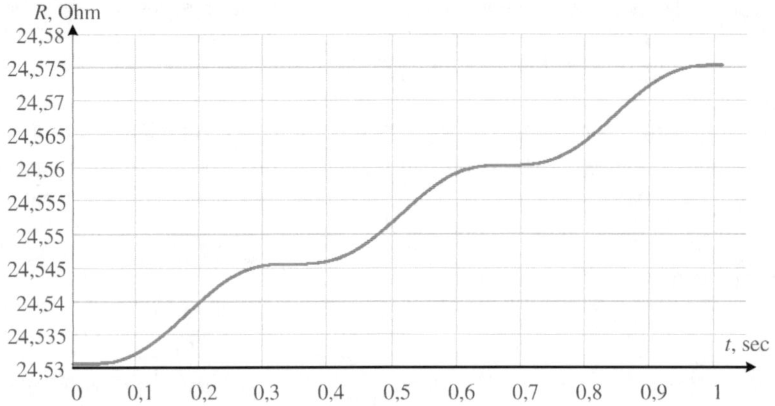

Figure 2. Time diagram of the calculated active component of the EP-winding resistance

Modeling of the sensorless device in the Micro-Cap circuit analysis program was carried out. The results of simulation of a sensorless device are given in Table 1.

Table 1. Results of a sensorless device simulation

Parameters	Experiments No.							
	1	2	3	4	5	6	7	8
R_I, Ohm	12.25	24.5	24.5	49	98	12.25	49	98
R_{II}, Ohm	12.2745	24.5	24.549	49.098	98.196	12.2745	49.098	98.196
δR, %	0.2	0	0.2	0.2	0.2	0.05	0.8	3.2
cm	14.6	-	29	58.5	117	58	14.5	7.25
I_c, A	-0.055398							
ψ_r, Wb	0.034006							
ψ_m, Wb	0.164811							
I'_c, A	-0.056444	-0.055479	-0.057262	-0.058124	-0.063877	-0.055398	-0.06898	-0.187729
$\Delta I'_c$, A	-0.001046	-8.1*E*-05	-0.001864	-0.002726	-0.008479	0.034006	-0.013582	-0.132331
$\delta I'_c$, %	1.9	0.15	3.4	4.9	15.3	0.16	24.5	239
I''_c, A	0.057167	0.055394	0.058968	0.0625	0.069669	-0.055782	0.084681	0.350093
$\Delta I''_c$, A	0.001769	4*E*-06	0.00357	0.007102	0.014271	-0.000384	0.029283	0.294695
$\delta I''_c$, %	-3.2	-0.01	-6.4	-12.8	-25.8	0.7	-53	-532
ψ'_r, Wb	0.03462	0.034026	0.037321	0.035653	0.039203	0.05584	0.042263	0.105603
$\Delta\psi'_r$, Wb	0.000614	2E-05	0.003315	0.001647	0.005197	0.000442	0.008257	0.071597
$\delta\psi'_r$, %	1.8	0.06	9.7	4.8	15.3	-0.8	24.3	210.6
ψ''_r, Wb	-0.035064	-0.033974	-0.036169	-0.038327	-0.042721	0.034486	-0.051608	-0.17753
$\Delta\psi''_r$, Wb	0.001058	3.2E-05	0.002163	0.004321	0.008715	0.00048	0.017602	0.143524
$\delta\psi''_r$, %	3.1	0.1	6.4	12.7	25.6	1.4	51.8	422
ψ'_m, Wb	0.164817	0.164681	0.16487	0.164868	0.16615	-0.034248	0.166256	0.180226
$\Delta\psi'_m$, Wb	6*E*-06	-0.00013	5.9E-05	5.7E-05	0.001339	0.000242	0.001445	0.015415
$\delta\psi'_m$, %	0.01	-0.08	0.04	0.04	0.8	0.7	0.9	9.4
ψ''_m, Wb	-0.164706	-0.164628	-0.164554	-0.165327	-0.16529	0.164629	-0.16675	-0.179639
$\Delta\psi''_m$, Wb	0.000105	0.000183	0.000257	0.000516	0.000479	-0.000182	0.001939	0.014828

Parameters	Experiments No.							
	1	2	3	4	5	6	7	8
$\delta\psi''_m$, %	0.06	0.1	0.16	0.3	0.3	-0,11	1.2	9.0

In Table 1 the following notation is used: R_{II} - the final value of the active component of the winding resistance; experiments No. 1, 3-5 were carried out at $\delta R = 0{,}2\,\%$, to ensure that the coefficient *cm* in manual mode was selected; experiments No. 2 $\delta R = 0\,\%$ demonstrates a high accuracy of the determination of the characteristic points in the absence of an error in determining the active component of the winding resistance; experiments No. 6-8 take into account that the decrease in the mass-dimension parameters of the EP-winding leads to both an increase in the active component of the winding resistance and a decrease in the *cm* coefficient, so increases the rate of change in the active component of the winding resistance during the determination of the magnetic characteristic.

4. Conclusion

The obtained results confirm the theoretical conclusions made and prove the high increasing relevance of the development of sensorless methods and devices for determining the electrical products magnetic characteristics.

Acknowledgments

The reported study was funded by RFBR according to the research project № 18-38-00682 using equipment of shared facility “Diagnosis and energy-efficient electrical equipment” (NPI).

References

[1] Jiang J and Tian J 2015 *J. of Chongqing University* **38(4)** 75-79

[2] Shaykhutdinov D, Lankin A, Narakidze N, Grechikhin V, Shirokov K and Gorbatenko N 2015 *Research J. of Appl. Sci.* **10(10)** 555-557

[3] Sakhovova A A, Shirokov K M, Janvarev S G 2013 *Mod. prob. of sci & education* **5** 111

[4] Kiran Kumar B, Siva Reddy Y V, Vijaya Kumar M 2015 *Int. J. of Control Theory and Appl.* **8(1)** 71-76

[5] Zolfaghari M, Taher S A and Munuz D V 2016 *Ain Shams Eng. J.* **7(2)** 729-740

[6] Leukhin R I, Shaykhutdinov D V and Leukhin V I 2016 *Proc. Eng.* **150** 347-353

[7] Choi J-S, Ko J-S, Chung D-H 2006 *2006 SICE-ICASE Int. Joint Conf.* 690-695

[8] Meng F, Zhang H, Cao D and Chen H 2016 *IEEE/ASME Transactions on Mechatronics* **21(3)** 1742-1753

[9] Gorbatenko N I, Lankin A M and Lankin M V 2016 *Proc. Eng.* **150** 1027-1031

[10] Zohoori A, Vahedi A, Meo S and Sorrentino V 2016 *J. of Intellig. and Fuzzy Syst.* **30(1)** 159-169

[11] Shaikhutdinov D V, Gorbatenko N I, Naraqidze N D, Leukhin R I, Shirokov K M, Dubrov V I, Stetsenko I A, Ahmedov Sh V 2014 *Mod. prob. of sci & education* **5** 119

[12] Grechihin V V, Lankin M V, Lankin A M 2015 *Proc. Eng.* **129** 793-798

Technology of hard alloy tool working capacity increase by applying composite coatings

I V Shvetsov[1], **O A Malukhina**[1], **D V Zhelnov**[2], **B Y Mokritsky**[3]

[1] Novgorod State University, ul. B. St. Petersburgskaya, 41, 173003 Veliky Novgorod, Russia
[2] Department of Expertise and Certification of the Union "St. Petersburg Chamber of Commerce and Industry", 20, pr. Lesnoy, St. Petersburg, 194044, Russia
[3] Komsomolsk-on-Amur State Technical University, 27, pr. Lenin, Komsomolsk-on-Amur, 681013, Russia

E-mail: Igor.Shvetsov@novsu.ru

Abstract. Technological methods of increasing the working capacity of metal-cutting tools from hard alloy are reviewed. Data on the tool hardening efficiency is given. Comparative tests of the tool subjected to various types of strengthening thermoplastic processing show a relative increase in its performance. Paper presents a comparative evaluation of hard alloys and their coatings. Attention is paid to the technological control of the working capacity of the carbide tool using the coating. The structural features of working capacity control are considered in the example of composite end-to-end carbide cutters and tangential plates for restoring wheel-rolling processing.

The current choice of the instrument design and its material is carried out by the technologist in accordance with generally accepted recommendations. All the informational material is, in fact, made up experimentally for certain processing conditions. If the machining requirements differ from the standard ones, the technologist does not have the option to select the tool, and he must perform a set of trial studies or carry out a selection for risk.

This led to the following:

- tool-making enterprises produce tools for typical processing conditions,and making tools for special (non-standard) conditions is problematic because of the need to perform research and development work and small production volumes;

- enterprise-consumers of the instrument are obliged to organize production sites where the tool purchased for standard machining conditions is adapted to meet their non-standard machining requirements.

First of all, untypical processing conditions here are described as the processing of specific materials that are not covered by general engineering recommendations, for example, a number of alloyed steels from a group of shipbuilding, low-magnetic, stainless materials, high-strength cast-iron, titanium alloys and others; machining under shock-cycling loading conditions, for example, turning a cylinder with a discontinuous surface or with variable cut geometry, for example, unbalanced milling or gear milling.

In such processing conditions, the tool life period is much shorter than that predicted by the recommendations for typical machining methods. This entails significant costs for the tool component and forces to work on improvement of the performance of a typical tool. The possibilities to carry out the improvements of the working capacity of the tool for a consumer enterprise are limited by many reasons. One of which is the lack of technological equipment on the market for tool hardening.

In some cases, especially when cutting hard-to-work materials, such requirements can be brought to the coating [1, 2], which can not be satisfied with the same coating composition. Selecting the composition of the coating for strength characteristics can reduce other indicators, which include adhesion interaction with the processed metal, thermal conductivity, or incompatibility with the coefficient of thermal expansion [3, 4, 5]. For conditions where a mediated solution of the coating features is required, it can be achieved by creating a coating using composition materials. This should be understood as a structure which consists of an alloy of more than one metal, for example, when the alloy contains titanium nitride and zirconium nitride, i.e. (TiZr)N.

The spectrum of applied coatings is very wide, they differ in composition, in the number of layers, in thickness, in the method of coating, etc. Figure 1 presents the illustrations of coatings made via electron HITACHI S-3400N microscope with a magnification of x3000.

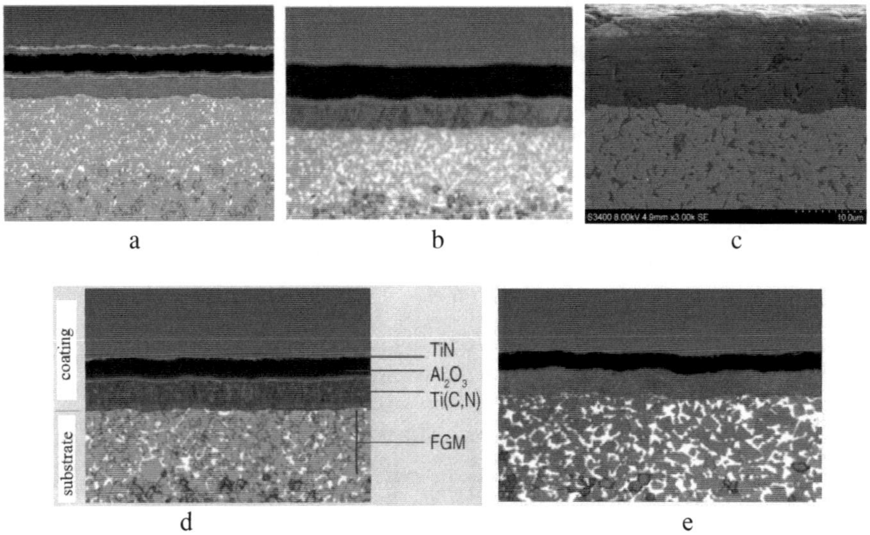

Figure 1. Catalog illustration of Pramet coatings: a - alloy 9210; b - alloy 9315; c) coating layers of alloy 9215 of Pramet LNUX 301940SN-DM; d - alloy 9230; e - alloy 9235.

The preparation of a composite-based coating is possible in several ways. It is simpler to obtain a composite coating on the basis of the CIB method. Let's consider several ways. The first is as follows: the construction of the cathode used as a source of metal ions to form a refractory compound in the plant chamber is prefabricated, and the ratio of the metal working areas corresponds to the ratio of the nitrides of these metals in the coating, Fig. 2.

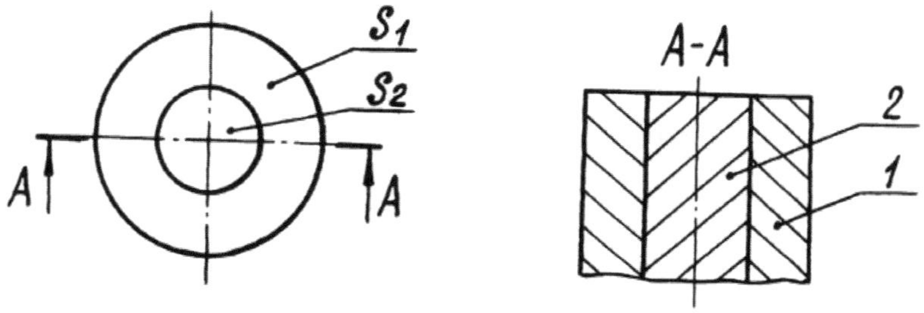

Figure 2. Assembled cathode.

Part 1 of the cathode is made of titanium, part 2 is made of zirconium. At the ratio of the areas of the burning surface of the arc (spark) at the cathode S_2: S_1 = 1: 3, the coating composition will have approximately the same ratio of the nitrides of these elements, i.e. 60-70% TiN and 40-30% ZrN.

Technologically this method is simple, but it is significantly more complicated if a composition of more metals is needed. Moreover, far from always the required refractory metals are present in the parameters of the billet in terms of geometry.

The next way of obtaining composite coatings is the technology on the basis of which the given coating composition is made by evaporation of a cathode, the composition of which is selected from powder components and sintered by the powder metallurgy method.

The data on the use of sintered cathodes for obtaining complex composite coatings is presented below.

A) The sintered cathode is made by the method of powder metallurgy from a mixture of Mo molybdenum powders and silicon nitride Si3N4. The ratio of the components was varied so that when the cathode was evaporated by the CIB method, a wear-resistant coating (MoSi)N with a ratio of Mo_2N and Si_3N_4 from 4: 1 to 7: 3 was obtained in the Bulat equipment chamber. When comparing the performance of such a coating with Mo_2N coating, a significant increase in efficiency is shown, figure 3. Variant 1 in Figure 3 is taken as a traditional basic one. The performance was assessed when drilling blind holes 45 mm deep, 32 mm in diameter in specialized shipbuilding steel AK-29 with a cutting speed of 60 m / min and 0,12 mm/ rev feed to wear of 0,4 mm.

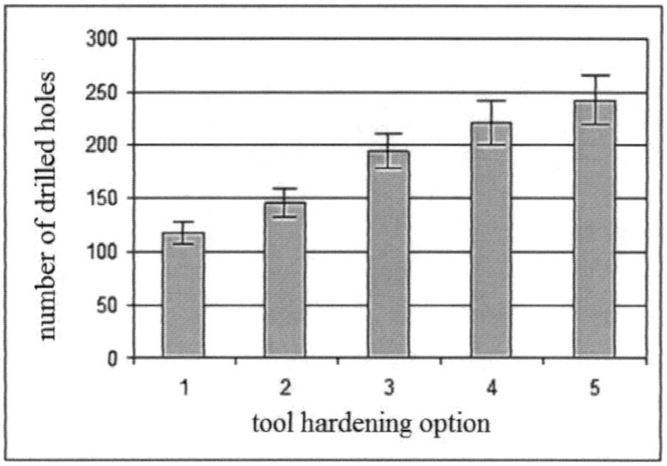

Figure 3. Illustration of the dependence of the performance of carbide drills on the way they are hardened:
1-BK6 + TiN,
2-BK6 + MoN,
3-BK6 + TiC + TiN + TiCN + Al2O3,
4-BK6 + (MoSi) N with the ratio Mo2N and Si3N4 4: 1.
5-BK6 + (MoSi) N with a ratio of 7: 1 for Mo_2N and Si_3N_4.

Option 2 is taken as more efficient in relation to the base one due to a better thermal conductivity of the MoN coating. Option 3 is taken as representative of modern multi-layer coatings. Variants 4 and 5 are taken as examples of developed composite coatings with a different ratio of components. From these data it follows that the performance of composite coatings is somewhat higher.

B) The cathode is made from a mixture of powders of molybdenum, zirconium, boron, silicon and titanium in a predetermined ratio. The structure of this coating is close-packed with a substantial content of silicides and borides on the surface. The results of testing such a tool with an intermittent turning of specialized steel AK-32PK at a cutting speed of 80 m / min, feed of 0.2 mm / rev and a depth of 1.5 mm are shown in Fig. 4. Here, for comparison, the tests results of another developed composite coating and two typical (TiN and TiC) coatings are presented.

The performance of the instrument with composite coatings is much higher. Position 4 in Fig. 4 shows the wear rate of a tool with a composite coating obtained by evaporation of a cathode, the manufacture of which uses powders of not only pure chemical elements but also their compounds, namely titanium, molybdenum, niobium diboride, tantalum diboride, zirconium diboride. The performance of composite coatings commensurates with the solutions, but judging by the intensity of wear, the tool is advisable to use in finishing stages, where significant wear values are unacceptable.

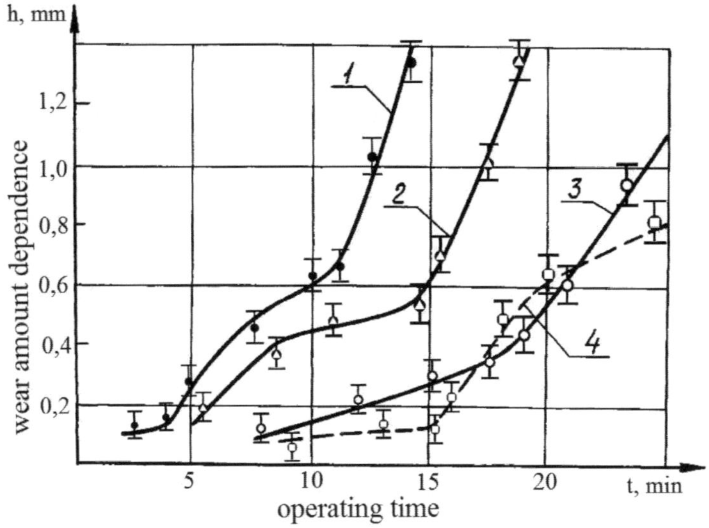

Figure 4. Dependence of the amount of wear on the back edge on the working time of carbide tool VK8 with:

1 - TiN coating (CIB);
2 - coating with TiC (HT);
3 - composite coating obtained by evaporating a sintered cathode containing molybdenum, zirconium, titanium, boron, silicon;
4 - by a composite coating obtained by evaporating a sintered cathode containing titanium, molybdenum, niobium diboride, tantalum diboride, zirconium diboride

The tool performance with a simpler composite coating of Cr + (Hf, Zr) N depending on the ratio of the components of the composite coating is illustrated in Figure 5 through the wear value for end milling of the titanium alloy VT-20 at a cutting speed of 80 m / min, feed 0.16 mm / tooth and a depth of 2 mm. It follows from Fig. 5 that the problem of optimizing the ratio of components takes place, but for any of the relations considered, the performance of the composite coating is higher than that of a simple nitride coating.

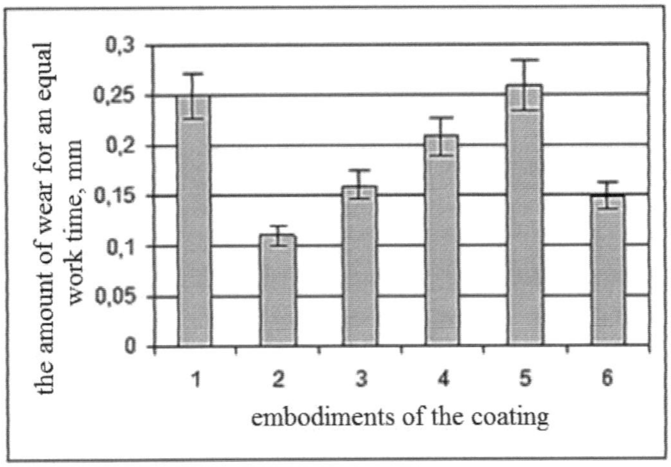

Figure 5. The amount of wear for an equal work time of the VK8 tool with a different ratio of components of the composite coating: 1 – Cr+HfN,
2 – Cr+ HfN(80%) & ZrN(20%),
3 – Cr+ HfN(60%) & ZrN(40%),
4 – Cr+ HfN(40%) & ZrN(60%),
5 – Cr+ HfN(80%) & ZrN(20%),
6 – Cr+ HfN(70%) & ZrN(30%).

This direction of increasing the efficiency of the tool using composite coatings is in agreement with the data of researches [6, 7, 8, 9-11].

Figure 6 illustrates an example of tool performance management under the same cutting conditions due to optimization of a coating composition in comparison with a modern coating.

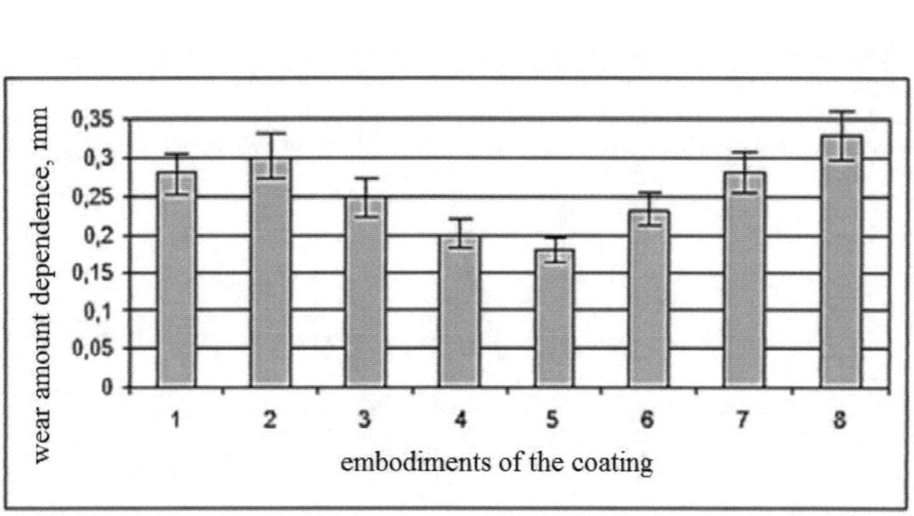

Figure 6. Wear value, mm, for equal working time of BK8 coated tool: 1 - modern industrial coating TiC + TiCN + Al_2O_3; 2, 3, 4, 5, 6, 7, 8 - the developed composite coating of Mo + (Ti, Zr) N + Mo + ZrN with the amount (volume%) of titanium nitrides, respectively: 50, 40, 30, 25, 20, 10

It follows from the figure that at a ratio (volume%) of titanium nitrides and zirconium nitrides 50:50, the efficiency of modern and developed coatings is commensurable. With a decrease in the proportion of titanium nitrides, i.e. with an increase in the share of zirconium nitrides, the effectiveness of coatings varies, the extremum of the efficiency of the developed coating with respect to the current coating is observed at (25-30)% of the content of titanium nitrides. In the absence of titanium nitrides, the coating efficiency is not maximal.

The solution presents the results of the development of a composite coating for turning (speed 160 m / min, feed 0.1 mm / rev, depth 1.5 mm) with a hard alloy T15K6 coated with Mo + (Ti, W) N + (TiW) + ZrN hard- (stainless) steel grade 31H19N9MVBT. The first layer is molybdenum, the second layer is titanium and tungsten nitrides, the third layer is titanium and tungsten, the fourth layer is zirconium nitride. They are illustrated in Figure 7 for an example of optimizing the composition of a composite (second) nitride layer with the corresponding same optimization of the third (metallic composite) layer.

Figure 7 shows an extreme dependence with an optimum in the region of 70%.

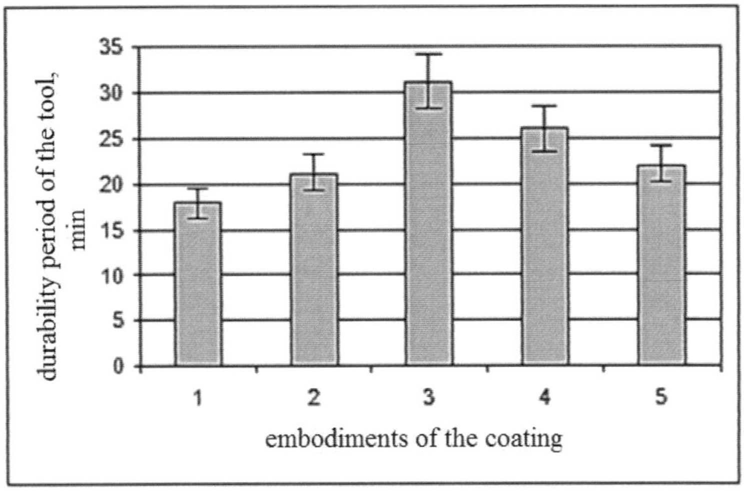

Figure 7. Durability period of the tool, min, with a percentage of titanium nitrides in the second layer and the same titanium content in the third layer: 1 - 50%; 2 - 60%; 3 - 70%; 4 - 80%; 5 - 90%.

Conclusion

In general, on the use of composite coatings it can be concluded that, with an unimportant complication of the technology of their application, it is possible to obtain a noticeable increase in the performance of the tool in difficult processing conditions.

It is most simple to realize the application of a composite coating by the method of cathode-ion bombardment due to the use of compound composite cathodes obtained using the powder metallurgy method.

References

[1] Grigoriev S. N., Vereschaka A. A., Vereschaka A. S. and, Kutin A. A., 2012 *Procedia CIRP* **1** 318–323

[2] Volosova M. A., Grigoriev S. N., 2010 *J. Uprochnyayushchie tekhnologii i pokrytiya [J. Hardening technologies and coatings]* **6** 37-42 [in Russian]

[3] Zotov E. S., Vereshchaka A. A., Vereshchaka A. S., Keramicheskie rezhushchie instrumenty [Ceramic cutting tools] *publishing house FGBOU VPO "MGTU Stankin» izd-vo FGBOU VPO «MGTU Stankin»* 2013 148 [in Russian]

[4] Mokritsky B. Y., Sablin P. A., Vereshchaka A. A., Instrumental'noe obespechenie sovremennykh mashinostroitel'nykh proizvodstv [Instrumental support of modern machine-building productions: monograph] *Komsomolsk-on-Amur: FGBOU VO "KnAGTU»* 2017 200 [in Russian]

[5] Vereschaka A. A., 2014 *Key Eng. Materials* **581** 62–67 (DOI: 10.4028/www.scientific.net / KEM.581.62)

[6] Vereschaka A. A., Vereschaka A. S., Batako A. D., Mokritskii B. J., Aksenenko A. Y., and Sitnikov N. N., 2017 *J. Nanomaterials and Nanotechnology* **7** 1–13 (DOI: 10.1177/1847980416680805 journals.sagepub.com/home/nax. SAGE)

[7] Vereschaka A. A., Mokritskiy B. Y., Sitnikov N. N., Oganyan V. G., and Aksenenko A. Y., 2017 *J. of Nano Research* **45** 110-123

[8] Zhao W, Wang S, Han Z and He N 2015 *Procedia CIRP* **35** 1–7

[9] Volkhonskii A. O., Vereshchaka A. A., Blinkov I. V., Vereshchaka A. S., and Batako A. D. 2016 *Int. J. Adv. Manuf. Technol* **84** 1647–60 (DOIi:10.1007/s00170-015-7821-8)

[10] Malushin1 N N, Valuev D V, Il'yaschenko D P, Trifonov V A, Nekrasov R Yu Technological Improvement of Surfacing of Parts of Hammer Crushers Used in Coke-Chemical Industry *J. Materials Science Forum* **927** 168-175

[11] Sarychev V D, Chinakhov D A, Solodsky S A, Granovskii A Y, Nevskii S A 2017 Impact of the

Chemical Elements Upon the Convective Flows in the Molten Metal of the Weld Pool *IOP Conference Series: Earth and Environmental Science* **66(1)**

IOP Conf. Series: Materials Science and Engineering **441** (2018) 012051 doi:10.1088/1757-899X/441/1/012051

Control of rotary equipment unbalance with using statistical criteria comparison of vibration spectrum

I V Shvetsov[1], Yu V Vankov[2], A R Zagretdinov[2]

[1] Novgorod State University, ul. B. St. Petersburgskaya, 41, 173003 Veliky Novgorod, Russia
[2] Department of Expertise and Certification of the Union "St. Petersburg Chamber of Commerce and Industry", 20, pr. Lesnoy, St. Petersburg, 194044, Russia
[3] Kazan State Power Engineering University, Kazan

E-mail: Igor.Shvetsov@novsu.ru

Abstract. The article offers statistical criteria for comparison of vibration spectra for detecting the unbalance of rotor equipment. The method of rejection is described. Experimental studies were carried out on an axial fan.

Introduction

Rotary equipment (electric motors, pumps, fans, compressors, etc.) is widely used in almost all industries. One of the most common defects of such equipment is the unbalance of the rotor (unbalance).

An unbalance arises from the action of a centrifugal force, which can cause the rotor to vibrate in the radial and axial directions. The causes of the unbalance can be: heterogeneity of the rotor material, manufacturing and assembly errors, wear of the assemblies, sedimentation of polluting particles, etc.

The method of spectral analysis of vibroacoustic signals has been obtained a wide application for the control of rotary equipment [1-3]. By changing the frequency components of the vibration spectrum, it is possible to identify the main equipment defects, including the unbalance of the rotor.

Due to the presence of a large number of harmonic components in the spectra, such analysis is quite often difficult.

Due to the presence of a large number of harmonic components in the spectra, such analysis is quite often difficult.

The control method and statistical comparison criteria

The proposed methodology is implemented in the application package [4-8].

The rejection of equipment is proposed to be performed depending on the results of comparison of the reference and current spectra from the values of three statistical comparison criteria:

– correlation coefficient;
– Spearman's nonparametric rank estimate;
– statistics of Fisher signs.

To calculate the correlation coefficient, use the following relationship:

$$r = \frac{\sum a_i a_{si} - \left(\sum a_i \sum a_{si}\right)/n}{\sqrt{\left(\sum a_i^2 - \left(\sum a_i\right)^2 / n\right)\left(\sum a_{si}^2 - \left(\sum a_{si}\right)^2 / n\right)}},$$

where a_i is the amplitude at the *i-th* frequency of the current spectrum; a_{si} is the amplitude at the *i-th* frequency of the reference spectrum, *n* is the number of spectra compared to the standard.

In addition to the correlation evaluation between the tested and reference spectra, a nonparametric rank estimate of the Spearman is calculated:

$$r = 1 - \frac{6}{n\left(n^2 - 1\right)} \sum \left(rank\, a_i - rank\, a_{si}\right)^2,$$

where rank a_i is the rank of the amplitude a_i in the variational series of amplitudes of the tested spectrum; rank a_{si} - also for the reference spectrum.

Statistics of Fisher signs *S* is calculated as the number of frequencies of the positive amplitude of the difference spectrum (half of the number of frequencies of zero amplitude is added to this sum). The Iman approximation is calculated:

$$S^* = \left(S - n/2\right)/\sqrt{n/4},$$

To classify the equipment as suitable or defective, the used approach is typical for the procedures for rejecting anomalies: the program interprets the set of computed comparison criteria ($p_1, p_2, \ldots, p_m$) as a set of measured values of an abstract parameter and applies the following procedure to this set of values [9,10]:

1) calculate the position estimate $\bar{p}$;

2) calculate the spread estimate *S;*

3) construct a confidence interval for a given level of significance α

$$\bar{p} \pm St\left(1 - \frac{a}{2}, m - 2\right),$$

where $t\ (\alpha,\ m)$ is the α-quantile of the Student's distribution with *m* degrees of freedom.

If the parameters of the monitored product go beyond the confidence interval, the program automatically issues an alarm and the control object is recognized as defective.

Information-measuring complex

The block diagram of the developed information-measuring complex is shown in figure. 1.

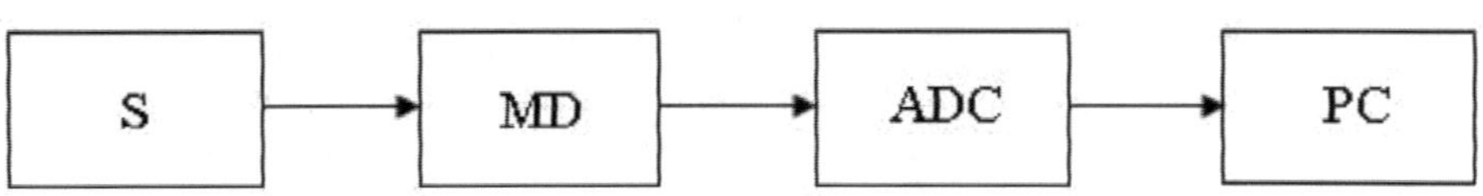

Figure 1. Structural diagram of the information-measuring complex:
S - vibration acceleration sensor AP2038P, MD - matching device AG01-3;
ADC - analog-to-digital converter NI USB-6229, PC - personal computer with installed package of applications.

Vibration transducer AP2038P allows simultaneous measurement of vibration acceleration along three orthogonal axes: *x, y, z.* Processing and analysis of signals is performed with the help of a set of programs implementing the proposed principle of rejection of equipment.

Experimental research

The 8-blade axial fan with a rotation speed of 2830 rpm was chosen as the monitoring object. The three-component vibration transducer was mounted on the bearing support of the fan shaft. The vibration signal was registered simultaneously in three directions (figure 2): x - axial, y - transverse horizontal, z – vertical.

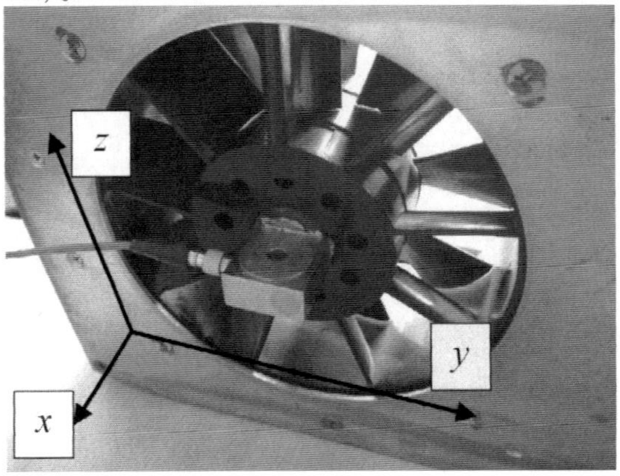

Figure 2. The control object with the notation of the direction of the vibration detection axes.

The unbalance of the fan shaft was created by alternately fastening to its blade cargoes in masses of 0.56 and 1.12 grams. The reference spectrum was formed by averaging the spectra of signals taken from a defect-free fan. The sampling rate of the ADC was 44100 Hz. The length of the signal for the formation of the spectra is 22050 counts. The limits of the confidence intervals for all comparison criteria are formed in accordance with the significance level of 0.05.

The obtained spectra of vibro-acoustic signals are shown in figures 3-5. The spectra were compared with the standard in the frequency range from 0 to 2 kHz.

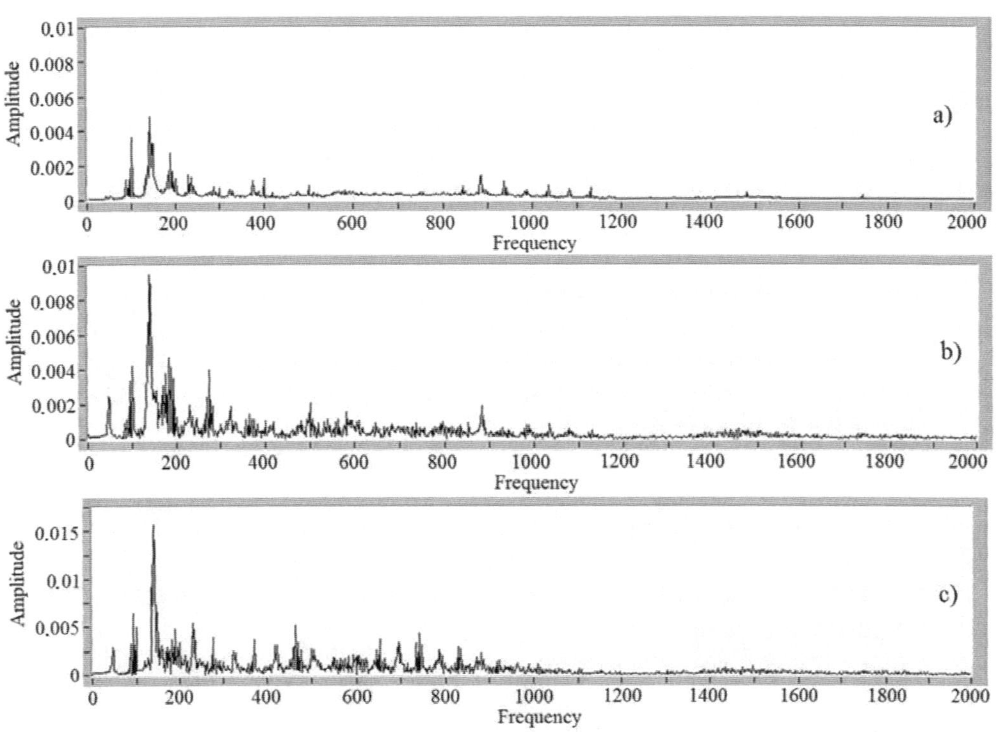

Figure 3. Spectra of vibro-acoustic signals along the x-axis: a) reference spectrum; b) a fan spectrum with a load of 0.56 g; c) fan spectrum with a load of 1.12 g.

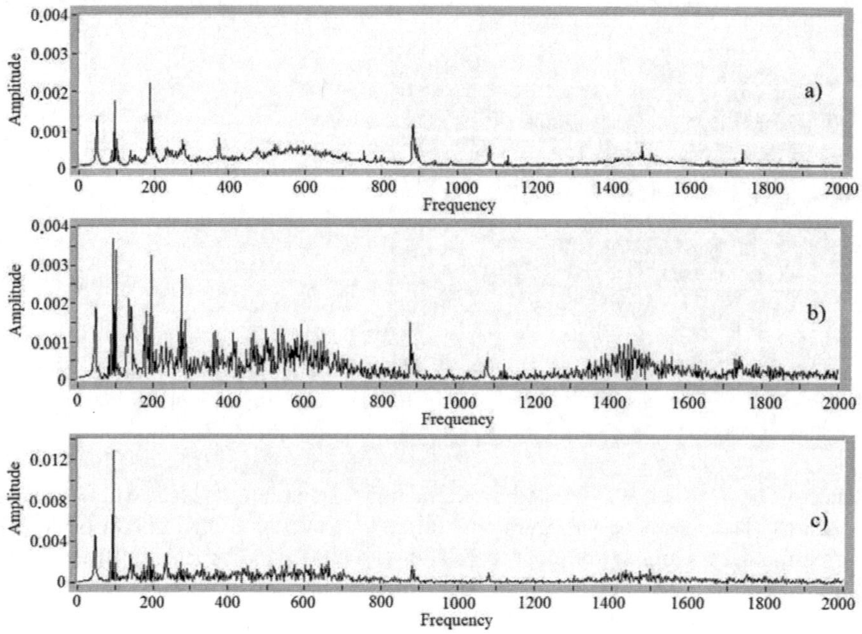

Figure 4. Spectra of vibro-acoustic signals along the y axis: a) reference spectrum; b) a fan spectrum with a load of 0.56 g; c) fan spectrum with a load of 1.12 g.

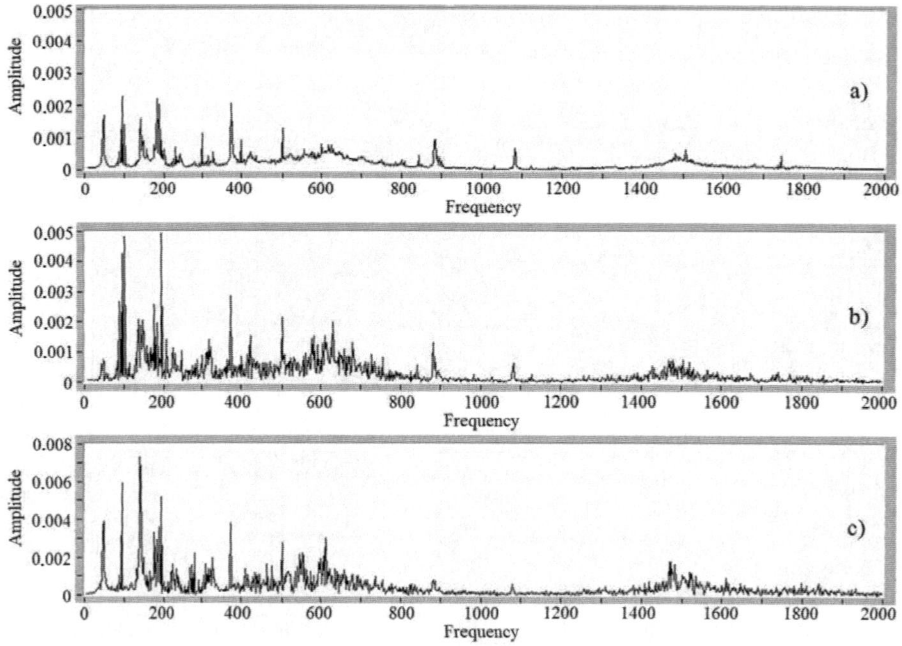

Figure 5. Spectra of vibro-acoustic signals along the z axis: a) reference spectrum; b) a fan spectrum with a load of 0.56 g; c) fan spectrum with a load of 1.12 g.

The proposed statistical criteria for comparing the spectra showed a stable determination of the unbalance of the fan rotor in all three measurement axes. In figures 6-8 the results of a comparison of the vibration spectra along the *y* axis are shown. It can be seen from them that the values of the comparison criteria for signals taken from a fan with a load go beyond the limits of confidence intervals.

The greatest sensitivity to determining the degree of development of unbalance was shown with the statistics of Fisher signs. The larger the unbalance of the fan is, the farther away its values are from the limits of the confidence intervals. To make a reliable decision about the defectiveness of the control object, the results obtained for a series of experiments are recommended to be subjected to median averaging.

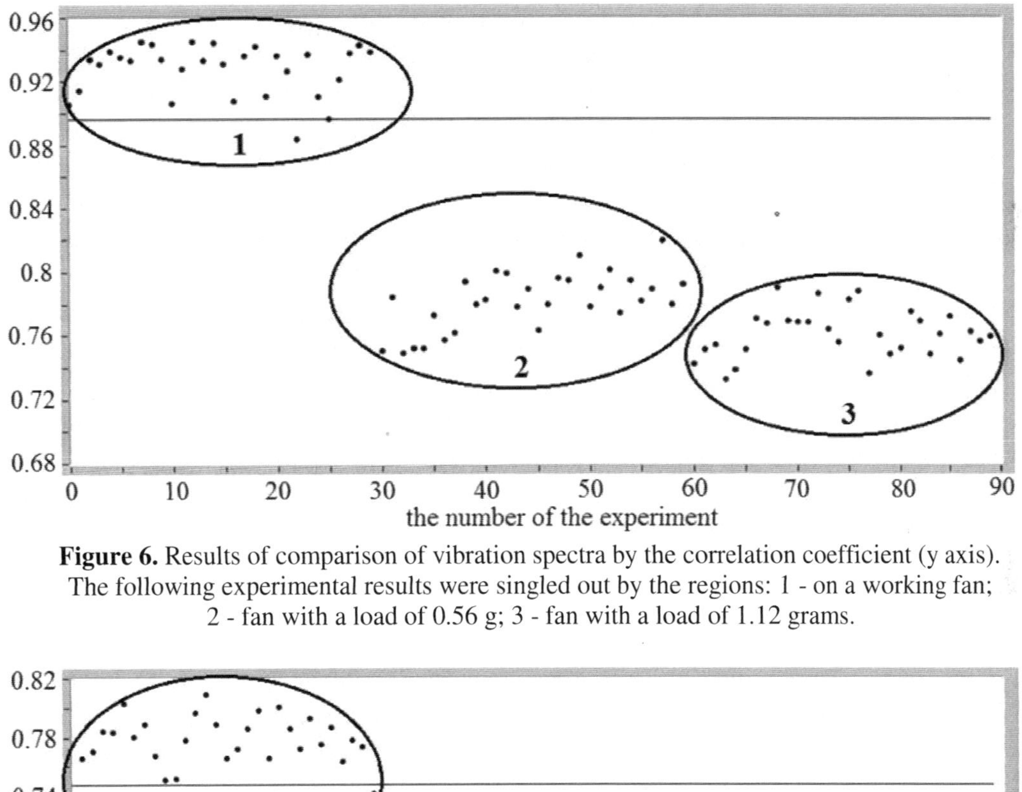

Figure 6. Results of comparison of vibration spectra by the correlation coefficient (y axis). The following experimental results were singled out by the regions: 1 - on a working fan; 2 - fan with a load of 0.56 g; 3 - fan with a load of 1.12 grams.

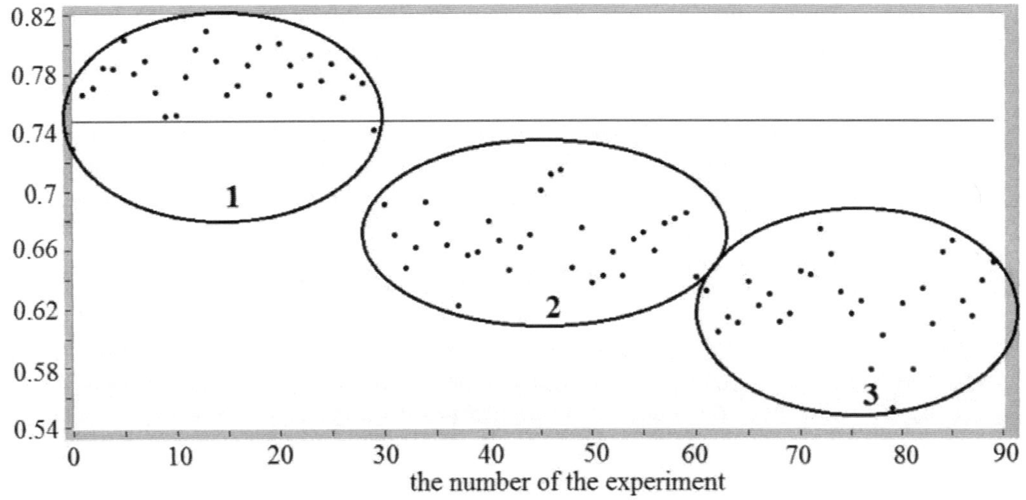

Figure 7. Results of comparison of vibration spectra according to Spearman's rank estimate (y-axis). The following experimental results were singled out by the regions: 1 - on a working fan; 2 - fan with a load of 0.56 g; 3 - fan with a load of 1.12 grams.

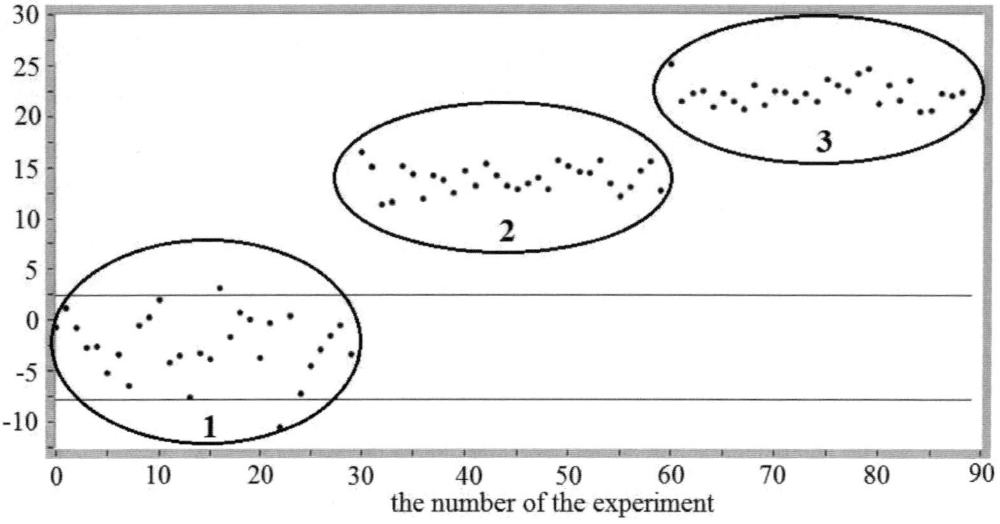

Figure 8. Results of comparison of vibration spectra according to statistics of Fisher signs (y axis).

The following experimental results were singled out by the regions: 1 - on a working fan; 2 - fan with a load of 0.56 g; 3 - fan with a load of 1.12 grams.

Conclusion

The proposed statistical comparison criteria allow analyzing the spectra of oscillations over a wide frequency range. In the course of experimental studies, the possibility of their application to control the unbalance of rotor equipment has been revealed.

The developed information-measuring complex with a package of applied programs makes it possible to carry out express control of the unbalance of rotor equipment in the process of its production and operation.

References

[1] Kluev V V 2005 *Non-destructive testing and diagnostics. Mechanical engineering* (Moscow) 656

[2] Genkin M D and Sokolova A G 1987 *Vibro-acoustic diagnostics of machines and mechanisms* (Moscow: Engineering) 282

[3] Goldin A S 1999 *Vibration of rotating machines. Mechanical engineering* 344

[4] Vankov Yu V, Ivshin I V, Zagretdinov A R, Izmaylova E V and Khalilov R G 2014 *Certificate of state registration of the computer program 2014613685.* (Software for the analysis of vibro-acoustic signals. Registered on 02.04.2014).

[5] Ivshin I V, Vankov Yu V, Izmaylova E V, Zagretdinov A R and Nizamiev M F 2014 *Certificate of state registration of computer program 2014613692* (Software package for express control of engine parts KAMAZ. Registered 02.04. 2014)

[6] Zagretdinov A R, Izmailov E V and Nizamiev M F 2014 The software package for Express control of the turbo compressor body of the KAMAZ gas engine *X international scientific*

and technical conference of young specialists "Study, design and manufacturing technology of compressor machinery", dedicated to the 90th anniversary from the birthday of B. N. Snapp (Kazan)

[7] Vankov Yu V, Ivshin I V, Zagretdinov A R and Nizamiev M F 2016 Software and algorithmic support of Express control of the turbocharger body of the KAMAZ engine *Bulletin of the University of Technology* 141-143

[8] Vladimirov O V, Zagretdinov A R, Ivshin I V and, Nizamiev M F *Investigation of the effect of defects on the natural oscillation frequencies of power plant parts*

[9] Vankov Yu V 2003 *Low-Frequency control methods. Method of free oscillations* (Publishing house of KSPEU, Kazan) 140

[10] Ivshin I V 2009 *Development of new low-frequency vibro-acoustic methods for monitoring the technical condition of parts, components and mechanisms of weapons and military equipment* (Publisher KHMCS, Kazan) 170

Comparative tests of the metal cutting tools performance in the processing of stainless steels

I V Shvetsov[1], B Ya Mokritskij[2], O A Malukhina[1], A H Rahmonov[2], V N Belyakov[1]

[1]Yaroslav-the-Wise Novgorod State University, ul. B. St. Petersburgskaya, 41, 173003 Veliky Novgorod, Russia
[2]Komsomoljskij-on-Amur state technical university, 27, Lenin pr., Komsomoljsk-on-Amur, 681013, Russia

E-mail: Igor.Shvetsov@novsu.ru

Abstract. Corrosion-resistant stainless steels like 09X17N7U do not fit into the generally accepted system of recommendations for the selection of a rational metal cutting tool and parameters of the cutting regime for them. An individual approach is required. A comparative study of the period of wear resistance of the developed domestic instrumental material has been carried out, and recommendations for their use have been given.

Introduction

Demand in specialized (developed for the needs of shipbuilding) stainless steel grade 09H17N7U grows both in shipbuilding and in other industries (oil and gas engineering, chemical, etc.). But the specificity of its composition, structure and physical and mechanical characteristics hinders its use because of the complexity of the choice for its processing of a rational metal-cutting tool and the processing regime for it. There is no foreign analogue of this, and accordingly there are no clear recommendations of foreign instrumental firms. There are general recommendations; they are typical for the processing of simpler engineering stainless steels of general purpose. The recommended cutting speed is more than 120 m/min. For steel 09H17N7U it is impracticable because of its specific heat conductivity and heat capacity, which leads to thermoplastic destruction of the carbide cutting tool.

Known recommendations on the use of domestic tools are obsolete for various reasons, including due to the bankruptcy of their producers.

There was a need to create a bank of recommendations on the use of tool materials for such steel. This article is aimed at solving this problem.

Discussion of the results

The uniqueness of the physical and mechanical characteristics and performance properties, Table 1 and Table 2, of specialized stainless steel grade 09H17N7U* creates difficulties [1,2] when machined with a blade metal cutting tool. High strength and toughness of this steel do not allow using high cutting speeds, the wear tool life of the cutting tool is low, which is not acceptable in cases when changing the cutting edge of the cutting insert or changing the cutting tool is unacceptable.

Table 1. Chemical compositions of steel grade 09H17N7U.

C	Si	Mn	Ni	S	P	Cr	Al
до 0.09	до 0.8	до 0.08	7 - 8	до 0.02	до 0.03	16 - 17.5	0.5 - 0.8

*GOST 5632-72 Steel highly alloyed and alloys corrosion-resistant, heat-resistant and heat-resistant. Grades.

Table 2. Physical and mechanical compositions of steel grade 09H17N7U.

Strength limit $\sigma_в$, MPa	Yield limit σ_T, MPa	Relative elongation at break δ_5, %	Impact strength KCU, kJ/m^2	Heat treatment
830	735	12	490	Hardening and high release

We carried out comparative studies of various instrumental materials [3, 4, 5, 6] to identify the rational ones for these operating conditions.

The external turning is considered by passing cutters with mechanical fastening of cutting inserts. Only the four-sided square plates (diameter of the circumscribed circle 17.5 mm) with and without a central fixing hole are shown here, with and without a flute groove. Information on the used tool materials is given below.

The diameter of the workpiece changed during processing from 280 to 60 mm. The cutting speed was maintained in the range of 50 - 55 m/min due to the variation in the number of spindle turns. Screw-cutting lathe of 16K25 model was used. The feed of the tool from the roughness requirements of the machined surface is chosen to be 0.21 mm/turn blanks. The cutting depth was taken to be 0.5 mm for the finishing conditions and 1 mm for the roughing conditions. In both cases, the maximum allowable amount of wear on the back face was 0.5 mm. The cutting tool was compared for the period of wear resistance, i.e., by the time of operation of the cutting inserts with a proper roughness, until a wear of 0.5 mm on the back face. To measure the attained wear value, the treatment was interrupted every 15 minutes. Wear measurements were made on a multi-sensor measuring center (video measuring machine) of the Micro Vu Sol 161. The results of the tests were duplicated and documented. Below is the information only on the wear data h_3 on the back surface.

Figure 1 shows the experimental wear amount dependence h_3 on time t of the hard alloy work (grade VK8) of the tool.

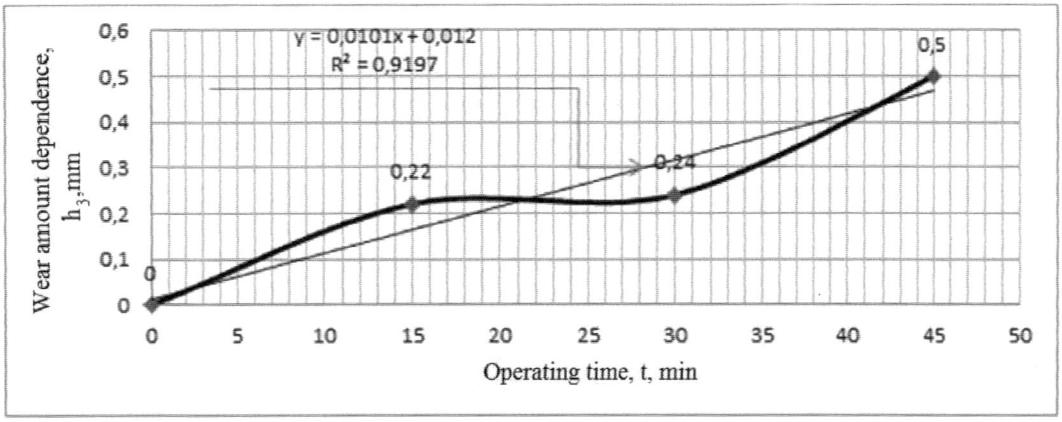

Figure 1. An example of wear amount dependence $h_з$ on time t hard alloy work (grade VK8) of the tool. (cutting mode: v = 55 m/min, n = 160 turns/min, s = 0.21 mm/turn, t = 1 min).

Similar graphs are constructed for different tool materials under identical operating conditions. An example of a comparative assessment of their wear resistance is shown in Figure 2.

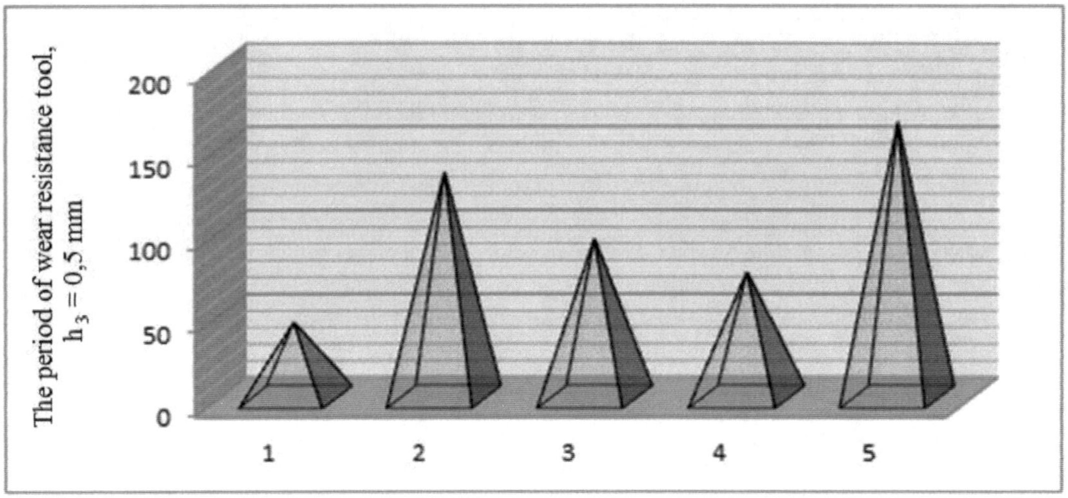

Figure 2. An example of comparing the period of wear resistance of various tool materials: 1 – BK8; 2 – VK8 + покрытие**№1; 3 – VK8 + покрытие** №2; 4 - VK8 + coating** №3; 5 – VK8 + are subjected to a diamond thermojet [2] along the front surface and subsequent polishing on the back surface of the plate.

A comparative analysis of the results shows that the tool materials are significantly unequal in terms of wear resistance in the processing of this steel. The use of coatings [7, 8, 9, 10-14] on the domestic solid alloy grade VK8 significantly increases the efficiency of cutting plates. Technological acceptance of modification [2] of the cutting edges from the front and back surfaces also significantly improves the working capacity of the tool.

In Figure 3, several replaceable cutting plates of different shapes are conditionally shown in the working position in order to be able to visually assess their angle in plan.

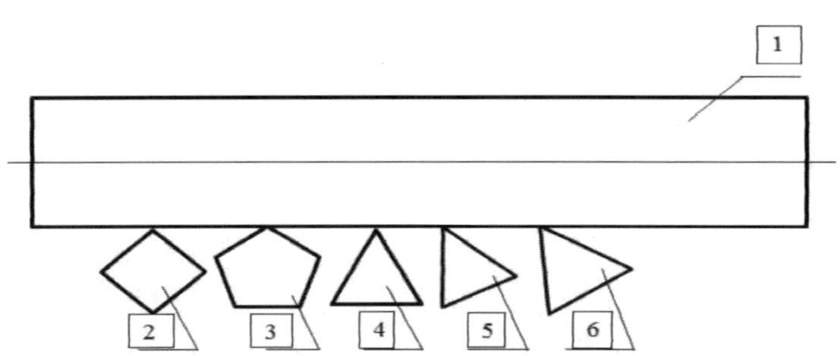

a

b

c

Figure 3. Examples of cutting plates of different shapes. The conditional scheme of the arrangement of the plates for longitudinal turning external processing (a): 1 - the workpiece to be processed; 2 - square plate; 3 - pentahedral plate; 4, 5 and 6 - plates of triangular shape, installed in different positions in relation to the surface of the workpiece to be treated 1; examples of actual plates (b); examples of inserting plates into the tool body (c).

Comparative tests of wear resistance of such plates under comparatively identical operating conditions are carried out. Under wear resistance is meant here the operating time of the plate until a certain amount of wear on the rear surface is reached, this value is taken equal to 0.5 mm. The identity of the operating conditions could not be fully ensured, but the parameters of the cutting regime were the same in all cases (cutting speed 50-55 m/min, feed 0.21 mm/turn of workpiece, cutting depth 1 mm). The front corner at cutting at different plates differed not essentially. In all cases, the brand of the used instrumental material was VK8. The cutting forces are not analyzed here.

Table 3 shows the results obtained for individual plates.

Table 3. Information on separate used plates.

N	Information on the cutting plate, her photo	Period of wear resistance, min.	Character of wear on the back surface	Example of shavings formed under specified operating conditions
1	A square plate (the diameter of the circumscribing circle is 17.5 mm) with a central hole 	45	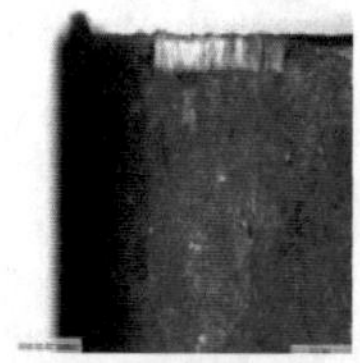	
2	Pentagonal plate without chip shaving groove 	100	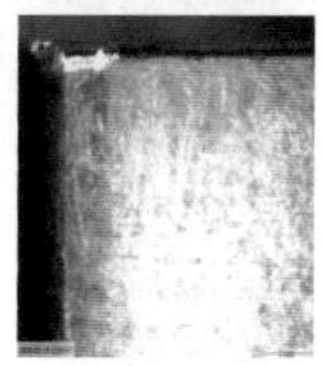	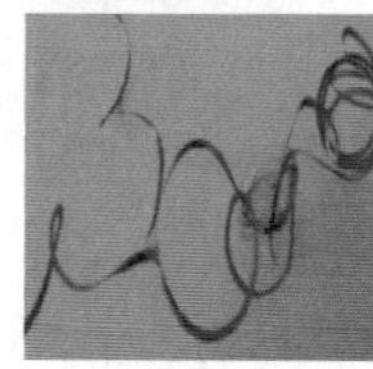
3	Rhombic plate with two cutting edges CCMT 120404 – HMP Korea KORLOY. Inc 	90		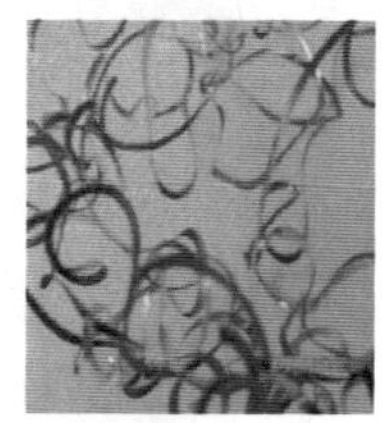
4	Triangular (with broken lines of the sides) plate (the diameter of the circumscribing circle is 21.5 mm) 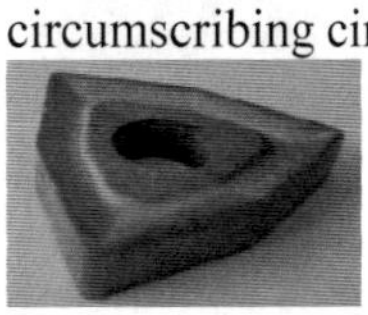	17	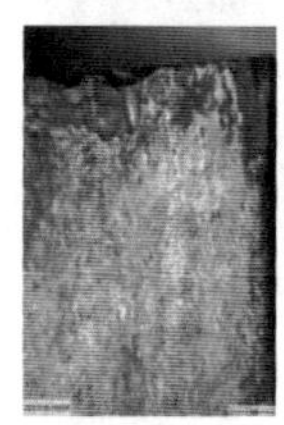	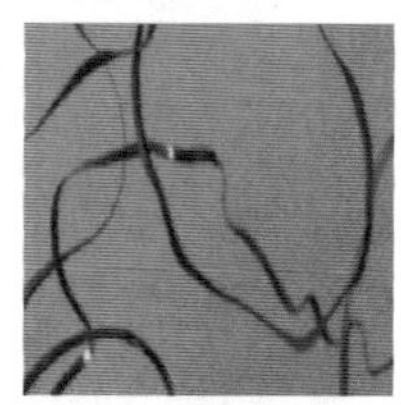
5	The square plate is double-sided (the diameter of the circumscribing circle is 21.5 mm) 	65		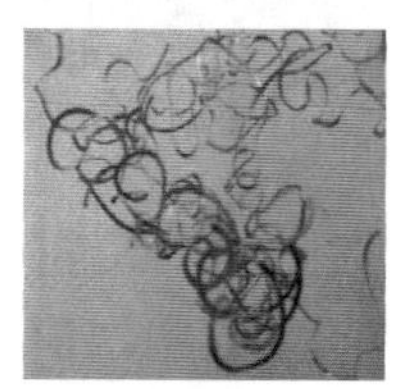

Conclusion

1. The insignificant difference (rows 1, 2 and 3 of the table) in the dimensions and shape of plates of medium dimension (diameter of the circumscribing circle from 16.5 to 18 mm) entails a significant difference in the period of wear resistance, the nature of wear of the cutting plate. In this case, the pentagonal shape of the plate is preferred, not significantly, but inferior to the rhombic shape (line 3). The type of formed chips is similar.

2. For plates of higher dimension (diameter of the circumscribing circle 21.5 mm), the shape of the plate essentially determines its wear-resistance period (lines 4 and 5), with the quadrangular square shape being preferred.
3. At this stage of the work, it is not possible to qualitatively explain the advantage of a cutting plate of one form or another, but the information given can be used as recommendations. The obtained conclusions do not contradict the existing ideas about the cutting processes presented in various information sources and they are based on the synergetic mechanisms of the process of blade processing of viscous materials and offer recommendations for the selection and strengthening of domestic carbide materials for processing of this stainless steel.

References

[1] Mokrickij B Y, Vereshchaka A A, Belyh S V and Mokrickaya E B 2016 *J. Uprochnyayushchie tekhnologii i pokrytiya [J. Hardening technologies and coatings]* **5** 3-6 [in Russian]
[2] Mokrickij B Y, Povyshenie rabotosposobnosti metallorezhushchego instrumenta putyom upravleniya svojstvami instrumental'nogo materiala [Hardening of composite coatings of inserts for machining stainless steel 09KH17N7JU]. – *Vladivostoki izd-vo Dal'nauka* 2010 232 [in Russian]
[3] Grigor'ev S N 2004 *J. Tekhnologiya mashinostroeniya* [*J. mechanical engineering*] **3** 20 [in Russian]
[4] Andreev A A and Grigor'ev S N Iznosostojkie vakuumno-dugovye pokrytiya na osnove titana v instrumental'nom proizvodstve [Wear-resistant titanium-based vacuum arc coatings in tool production] STIN, 2006, no. 2, pp. 19-24
[5] Tabakov V P Formirovanie iznosostojkih ionno-plazmennyh pokrytij rezhushchego instrumenta [Formation of wear-resistant ion-plasma coatings of the cutting tool]. Moscow, "Mashinostroenie", 2008. 311 p.
[6] Moiseev V F and Grigor'ev S N Instrumental'nye materialy. Izdanie vtoroe. Monografiya [Instrumental materials. Second edition. Monograph] Moscow, IC MGTU «Stankin», YAnus-K, 2005. 248 p.
[7] Grigor'ev S N, Tabakov V P and Volosova M A Tekhnologicheskie metody povysheniya iznosostojkosti kontaktnyh ploshchadok rezhushchego instrumenta: monografiya [Technological methods of increasing wear resistance of contact pads of cutting tools: monograph] Staryj Oskol: TNT, 2011, 380 p.
[8] Uhlmann E, Huehns T, Richarz S et al. 2009 *J. Industrial ceramics* **2** 113-18 [in Russian]
[9] Grigor'ev S N, Smolencev E V and Volosova M A Tekhnologiya obrabotki koncentrirovannymi potokami ehnergii: uchebnoe posobie [The technology of processing by the concentrated streams of energy: a training manual] izd. Pererab. i dop. Staryj Oskol: TNT, 2010, -280 p.
[10] Elyutin A V, Anikin V N, Blinkov I V and Vereshchaka A S Metodologicheskie principy sozdaniya vysokoprochnoj gradientno-sloistoj keramiki s pokrytiem dlya skorostnoj lezvijnoj obrabotki bez primeneniya SOTS [Methodological principles of creation of high-strength gradient-layered ceramics with coating for high-speed blade processing without the use of coolants] Trudy kongressa "Konstruktorsko-tekhnologicheskaya informatika" 2000. Tom 1. Moscow, MGTU «STANKIN», 2000. pp. 180 – 182
[11] Malushin1 N N, Valuev D V, Il'yaschenko D P, Gizatulin R A, Nekrasov R Yu Improvement of Research Procedures into the Highly Rigid Welded Metal Using Thermal Microscopy Facilities *J. Materials Science Forum* **927** 148-152
[12] Valuev D V, Serikbol A 2014 Stress state and its definition in hardfacing heat-resistant steel cold rolling mill rollers *J. Advanced Materials Research* **1040** 703-707
[13] Valuev D V, Danilov V I 2012 Reasons for negative formation of structures in carbon steel processing of pressure, *J. Proceedings - 2012 7th International Forum on Strategic Technology, IFOST 2012* **2** 151-154
[14] Malushin N N, Valuev D V, Valueva A V, Serikbol A 2014 Kinetic study of the effect of

plasticity and its role in stress relaxation in the weld speed steel during the martensitic transformation *J. Applied Mechanics and Materials* **682** 58-63

IOP Conf. Series: Materials Science and Engineering **441** (2018) 012053 doi:10.1088/1757-899X/441/1/012053

Integro-differential equation of impedance curvilinear dipole antenna.

A V Sochilin[1], S I Eminov[1]

[1]Novgorod State University, ul. B. St. Petersburgskaya, 41 173003 Veliky Novgorod, Russia
E-mail: eminovsi@mail.ru,

Abstract. In this paper, a new one-dimensional integro-differential equation with respect to the current for the impedance curvilinear dipole antenna is obtained. The applicability of the Galerkin method is shown, efficiency is demonstrated. The results of numerical calculations are analyzed.

Introduction

In the work [1], a linear impedance vibrator was investigated, work [2] is devoted to a curvilinear dipole antenna with an ideally conducting surface. In this paper, we study the impedance curvilinear vibrator.

1. The initial equation

Consider a curvilinear vibrator whose generatrix in space is described by the relations

$$x = x(\tau),\ y = y(\tau),\ z = z(\tau),\ -1 \le \tau \le 1. \tag{1}$$

The vibrator is assumed to be thin, its radius a is much smaller than the wavelength and length of the antenna. The Lame coefficient is determined by the formula

$$H_\tau(\tau) = \sqrt{x'^2 + y'^2 + z'^2}\ .$$

On the surface of the vibrator antenna, an impedance boundary condition of the form [3]

$$E_\tau + E_\tau^0 = \xi j_\tau, \tag{2}$$

where E_τ - secondary electric field, E_τ^0 - primary electric field, j_τ - surface current density, ξ - surface impedance. Passing from the density of the longitudinal component to the total current by the formula $I(\tau) = 2\pi a j_\tau(\tau)$, as in [2], we obtain the integro-differential equation

$$\frac{1}{4\pi}\int_{-1}^{1} I(t)\frac{\partial^2}{\partial\tau\partial t}S(\tau,t)dt + iH_\tau\sqrt{\frac{\varepsilon}{\mu}}\frac{\xi}{2\pi a}I(\tau) -$$

$$-\frac{1}{4\pi}\int_{-1}^{1} I(t)(kH_\tau)(kH_t)(\vec{e}_\tau\cdot\vec{e}_t)S(\tau,t)dt = iH_\tau\sqrt{\frac{\varepsilon}{\mu}}E_\tau^0(\tau). \tag{3}$$

Here $\vec{e}_\tau$ and $\vec{e}_t$ - orths, tangent to the generator of the vibrator at points τ and t,

$$S(\tau,t) = \frac{1}{\pi}\int_0^{\pi}\frac{\exp(-ikR)}{kR}d\varphi,\quad R = \sqrt{R_0^2 + 4a^2\sin^2\frac{\varphi}{2}} \tag{4}$$

$$R_0 = \sqrt{[x(\tau)-x(t)]^2 + [y(\tau)-y(t)]^2 + [z(\tau)-z(t)]^2}\ .$$

2. Allocation of the main operator

Let us clarify the nature of the singularity of $S(\tau,t)$, which enters into the kernel of the integral equation. We get

$$S(\tau,t)=\frac{1}{\pi}\int_0^{\pi}\frac{\exp(-ikR)}{kR}d\varphi=$$

$$=\frac{1}{\pi}\int_0^{\pi}\left(\frac{\exp(-ikR)}{kR}-\frac{1}{k\sqrt{R_0^2+a^2\varphi^2}}\right)d\varphi+\frac{1}{\pi}\int_0^{\pi}\frac{1}{k\sqrt{R_0^2+a^2\varphi^2}}d\varphi.$$

The last integral is calculated analytically, as the elementary

$$\frac{1}{\pi k}\int_0^{\pi}\frac{1}{\sqrt{R_0^2+a^2\varphi^2}}d\varphi=\frac{1}{\pi ka}\left[\ln\left(\pi ka+\sqrt{(\pi ka)^2+(kR_0)^2}\right)+\ln\frac{1}{kR_0}\right]=$$

$$=\frac{1}{\pi ka}\left[\ln\left(\pi ka+\sqrt{(\pi ka)^2+(kR_0)^2}\right)+\ln\frac{|\tau-t|}{kR_0}\right]+\frac{1}{\pi ka}\ln\frac{1}{|\tau-t|}. \quad (5)$$

Write the function $S(\tau,t)$ taking into account (5)

$$S(\tau,t)=\frac{1}{\pi ka}\ln\frac{1}{|\tau-t|}+S_1(\tau,t), \quad (6)$$

where

$$S_1(\tau,t)=\frac{1}{\pi}\int_0^{\pi}\left(\frac{\exp(-ikR)}{kR}-\frac{1}{k\sqrt{R_0^2+a^2\varphi^2}}\right)d\varphi+$$

$$=\frac{1}{\pi ka}\left[\ln\left(\pi ka+\sqrt{(\pi ka)^2+(kR_0)^2}\right)+\ln\frac{|\tau-t|}{kR_0}\right].$$

Function $S_1(\tau,t)$ is continuous, when τ coincides with t, function $S(\tau,t)$, as follows from (5), has a logarithmic singularity in the kernel. And now we rewrite equation (3) with allowance for the transformation (6)

$$\frac{1}{4\pi(ka)}(AI)(\tau)+\left(iH_\tau\sqrt{\frac{\varepsilon}{\mu}}\frac{\xi}{2\pi a}\right)I(\tau)+(MI)(\tau)+(NI)(\tau)=iH_\tau\sqrt{\frac{\varepsilon}{\mu}}E_\tau^0(\tau), \quad (7)$$

where

$$(AI)(\tau)=\frac{1}{\pi}\frac{\partial}{\partial\tau}\int_{-1}^{1}I(t)\frac{\partial}{\partial t}\ln\frac{1}{|\tau-t|}dt,$$

$$(MI)(\tau)=\frac{1}{4\pi}\int_{-1}^{1}I(t)\frac{\partial^2}{\partial\tau\partial t}S_1(\tau,t)dt,$$

$$(NI)(\tau)=-\frac{1}{4\pi}\int_{-1}^{1}I(t)(kH_\tau)(kH_t)(\vec{e}_\tau\cdot\vec{e}_t)S(\tau,t)dt.$$

The operator of multiplication by a constant, integral operators M and N are limited. Operator A is a symmetric positive-definite operator in a Hilbert space $L_2[-1,1]$. As it is shown [4], inverse operator A^{-1} is completely continuous in the energy space H_A of operator A. And for any bounded operator B operator $A^{-1}B$ is completely continuous in space H_A. Consequently, equation (7) is equivalent to the Fredholm equation of the second kind in a Hilbert space H_A. An approximate solution, found by Galerkin's method [2], converges to an exact solution.

3. Results of numerical calculations

As an example, we will investigate a helical vibrator antenna with a reactive surface impedance. The generator of a helical antenna in space is described by the relations

$$x=r_0\cos(\varphi_0\tau)\ \ y=r_0\sin(\varphi_0\tau),\ z=b\tau,\ -1\le\tau\le1. \quad (8)$$

Further, the right-hand side of the integro-differential equation (7) is given in the form

$$H_\tau E_\tau^0(\tau)=U_0 f(\tau),\; f(\tau)=\frac{1}{2T}\begin{cases}1,|\tau|\le T,\\ 0,|\tau|>T,\end{cases},$$

voltage $U_0=1V$.

As a result of solving the equation, we first find the current $I(\tau)$, and then the input resistance $Z=\frac{U_0}{I(0)}$. The results of the calculations are shown in figure 1 and figure 2.

Analysis of the results allows us to conclude that the resonance is shifted to a region of smaller antenna lengths with an increase in the reactive surface impedance. Increasing the radius of the vibrator leads to a decrease in the maximum values of the input resistance when the antenna length is changed.

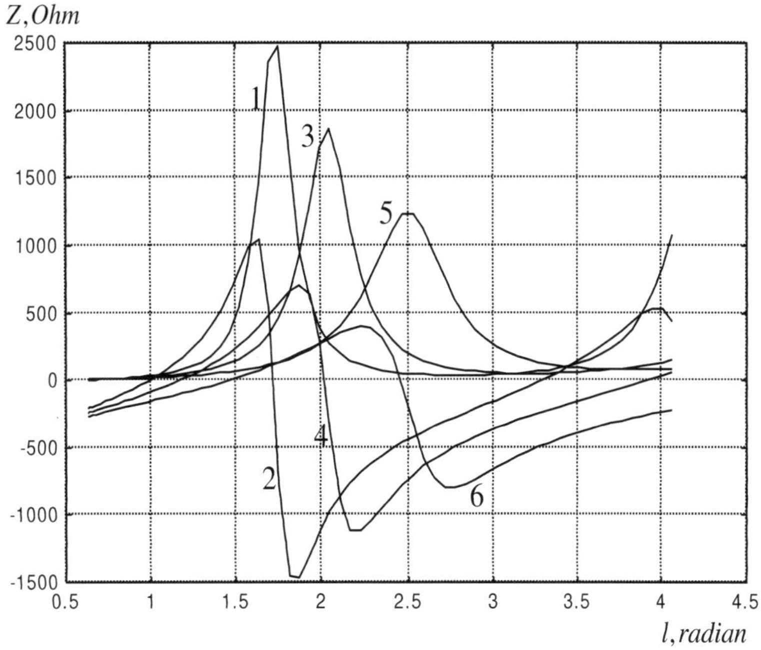

Figure 1. Dependence of the input resistance of the vibrator Z of the electric length of the shoulder l for parameters:

$T=0.01$, $r_0=0.2\lambda$, $b=0.1\lambda$, $a=0.005\lambda$

$\xi\approx i37.7Ohm$ (Re(Z) - 1, Im(Z) - 2)

$\xi\approx i18.85Ohm$ (Re(Z) - 3, Im(Z) - 4)

$\xi=0$ (Re(Z) - 5, Im(Z) - 6)

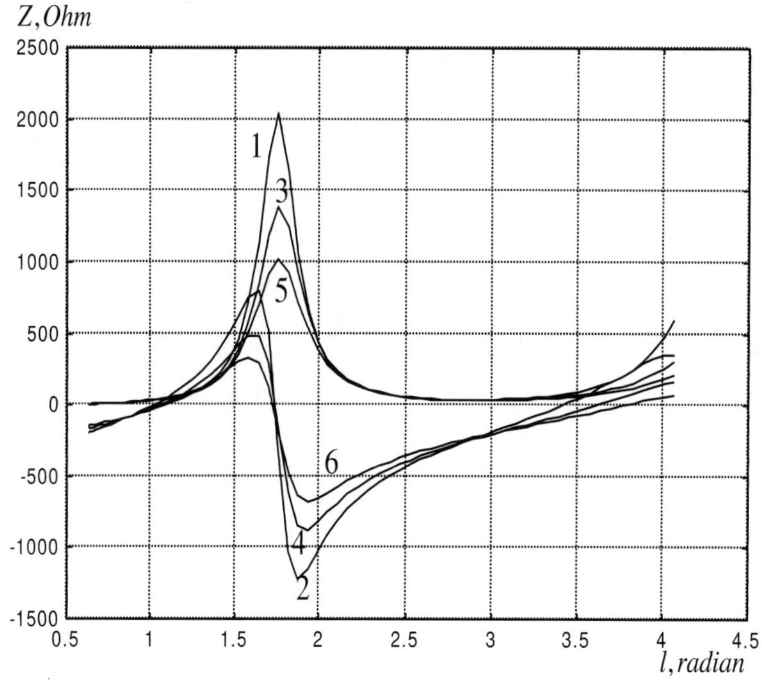

Figure 2. Dependence of the input resistance of the vibrator Z of the electric length of the shoulder l for parameters:

$$\xi \approx i37.7 Ohm,\ T = 0.01,\ r_0 = 0.2\lambda,\ b = 0.1\lambda$$

$a = 0.006\lambda$ (Re(Z) - 1, Im(Z) - 2)

$a = 0.008\lambda$ (Re(Z) - 3, Im(Z) - 4)

$a = 0.01\lambda$ (Re(Z) - 5, Im(Z) - 6)

References

[1] Daniljchuk V.L., Eminov S.I. Theory of the integral equation of an impedance vibrator // *Journal of Technical Physics*. 1993. V. **38**. #**12**. P.201-204.

[2] Sochilin A.V., Eminov S.I. Numerical-analytical method for calculating curvilinear vibrator antennas // *Radio engineering and electronics*. 2017. V. **62**. #1. P.59-64.

[3] Vasiljev E.N. Excitation of bodies of revolution. M: Radio and communication, 1987.

[4] Eminova V.S., Eminov S.I. Substantiation of the Galerkin method for hypersingular equations // *Journal of Computational Mathematics and Mathematical Physics*. 2016. V. **56**. #3. P.432-440.

IOP Conf. Series: Materials Science and Engineering **441** (2018) 012054 doi:10.1088/1757-899X/441/1/012054

Composite coatings of nickel-vanadium-phosphorus-nitride boron

A A Tikhonov, D A Filippov

Novgorod State University 41, B. St. Peterburgskaya, Velikiy Novgorod, 173003, Russia
E-mail: Dmitry.Filippov@novsu.ru

Abstract The effect on the micro hardness of nickel - vanadium coatings from the concentration of sodium hypophosphite in the range 10-30 g / l, the concentration of boron nitride from 0 to 50 g / l, and from the temperature in the range of 50 - 90 o C was studied. The influence of these factors on the rate of coating , the appearance of the coating and the thickness of the coatings was researched. The effect of heat treatment regimes on the micro hardness of coatings is studied. Regressions equations are obtained that allow calculating the micro hardness of coatings when the composition, the temperature of the electrolyte and the heat treatment temperature change. The wear resistance of the obtained composite coating is determined. The composite coating of nickel-vanadium-phosphorus-boron nitride can be recommended for increasing the resistance of deforming tools operating at temperatures up to 600°C.

1. Introduction

In modern engineering, the problem of increasing the micro hardness and wear resistance of coatings applied to parts and tools is very relevant. For this purpose, galvanic chrome plating is mainly used. However, this process has a number of disadvantages, first of all, high toxicity of chromium electrolyte components, low scattering power, small current output, etc. Most often, for the replacement of galvanic chrome plating is using nickel-based composite electrochemical deposition processes are proposed, as well as chemical and galvanic nickel-phosphor coatings [1]. This work is devoted further development of these methods.

2. Materials and methods

Metal samples from carbon tool steel with a diameter of 20 mm and a height of 20 mm were used as the test samples. Before coating, samples were ground, polished, degreased with Viennese lime, decapitated in a 10% solution of sulfuric acid, washed with tap water and distilled water. The electrolyte was prepare from distilled water, chemically pure electrolytic materials and cubic boron nitride (elbor) were used. The grain size of the main fraction (95%) was 1 μm and smaller. The electrolyte was stirred with a propeller stirrer and thermostatic.

3. Investigation of the effect of the electrolyte composition and electrolysis regimes on microhardness and coating quality.

The object of the research were electrolytes for coatings, the basis of which was electrolyte No. 1 of the following composition: nickel sulphate 40 g / l, boric acid 20 g / l, vanadic acid 5 g / l, sodium nitrilotriacetic acid.30 g / l. As the anode, a graphite electrode was used. The cathode current density is 1A / dm^2, the pH of the solution is 4.0. Sodium hypophosphite and cubic boron nitride were added to the basic composition of the electrolyte, the temperature of the electrolyte was varied according to the table 1. The influence of three factors on the micro hardness of coatings (Y) was studied: the concentration of sodium hypophosphite, the concentration of boron nitride and temperature. The upper level for sodium hypophosphite z_1^{max} = 30 g / l, lower z_1^{min} = 10 g / l, z^0 = 20, Δz = 10; the upper level for boron nitride z_2^{max} = 50 g / l, the lower z_2^{min} = 0 g / l, z_2^0 = 25, Δz = 25; the upper level is temperature z_3^{max} = 50 ° C, the lower z_3^{max} = 90 ° C, z_3^0 = 70, and Δz = 20. The planning matrix of experiment and the microhardness measurements in kg / mm2 (yield Y) are shown in Table 1

Table 1 Planning matrix and micro hardness measurements in kg / mm^2

Values of factors on a natural scale				Values of factors in a dimensionless coordinate system			Yield Microhardness kg / mm^2
№	z_1	z_2	z_3	x_1	x_2	x_3	Y
1	10	0	50	–1	–1	–1	350
2	30	0	50	+1	–1	–1	322
3	10	50	50	–1	+1	–1	446
4	30	50	50	+1	+1	–1	464
5	10	0	90	–1	–1	+1	792
6	30	0	90	+1	–1	+1	724
7	10	50	90	–1	+1	+1	824
8	30	50	90	+1	+1	+1	946

The yield (micro hardness) was approximated by the linear equation

$$Y = b_0 + b_1 x_1 + b_2 x_2 + b_3 x_3 \tag{1}$$

The coefficients b_i were determined by the formula

$$b_J = \frac{1}{N}\sum_{i=1}^{N} x_{ji} y_i \tag{2}$$

Using the results presented in Table 1, the regression equation is obtained in the form

$$Y = 608.5 + 5.5x_1 + 61.5x_2 + 213x_3 \tag{3}$$

It follows from Equation 3 that all three factors increase the micro hardness, while the micro hardness increases the most with increasing temperature of the electrolyte. This is due to an increase in the phosphorus content in the coating. Increases the hardness of the coatings and increases the concentration of boron nitride in the electrolyte, as this leads to an increase in the content of this very solid compound in the coating. The concentration of sodium hypophosphite is least affected by the micro hardness of the coatings.
The results of the effect of heat treatment on the micro hardness are presented in Table 2.
From the results obtained, the regression equation was obtained to determine the microhardness of the coatings after heat treatment at 400 ° C in the form

$$Y = 751.25 + 33x_1 + 46.5x_2 + 212.5x_3 \tag{4}$$

Table 2. Planning matrix and micro hardness measurement in kg / mm^2 after heat treatment

Values of factors on a natural scale				400° C	600° C
№	z_1	z_2	z_3		
1	10	0	50	554	894
2	30	0	50	511	995
3	10	50	50	545	439
4	30	50	50	545	322
5	10	0	90	841	995
6	30	0	90	913	1064
7	10	50	90	933	439
8	30	50	90	1168	1354

Thermal treatment of coatings at 400 ° C increases the micro hardness of coatings due to the formation of nickel and vanadium phosphides.

During the second heat treatment, the temperature of the furnace was increased to 600 ° C, and the holding time was 1.5 hours. From the obtained results of micro hardness measurements (Table 2), it follows that there is a further increase in the micro hardness of the coatings of the samples, but a decrease in micro hardness in experiments Nos. 3, 4 and 7 is also observed. This seems to be due to the strengthening effect due to an increase in the concentration nickel and vanadium phosphides and with the softening effect of boron nitride arising at high temperatures.
The regression equation for this regime has the form

$$Y = 812.75 + 121x_1 - 174.25x_2 + 150.23x_3 \quad (5)$$

Further heat treatment at a higher temperature of 800 ° C led to softening of the coating. Table 3 shows the results of the effect of the concentration of sodium hyposite (Z1), the concentration of boron nitride (Z2) and the temperature of the electrolyte (Z3) on the current efficiency, the coating rate, the appearance of the coating and the thickness of the coatings. Coating time in all experiments is 1 hour.

Table 3 Influence of the main factors on the current output, speed and type of coating

Temperature and additives introduced into the electrolyte-base No. 1.	Current yield %	Speed of application, μm/min	The thickness of the coating μm	Appearance of the coating.
Z1= 10, Z2=50, Z3=50	84.35	0.184	11.041	Gray, frosted
Z1=30, Z2=50, Z3=50	73.37	0.160	9.603	Semi-shiny
Z1=30, Z2=0, Z3=90	104.32	0.227	13.633	Shiny
Z1=10, Z2=50, Z3=90	110.48	0.241	14.461	Light gray, frosted
Z1=30, Z2=50, Z3=90	108.41	0.236	14.189	Semi-shiny

Several processes often take place on electrodes during electrolysis, they are divided into basic processes (in our case this is primarily the separation of nickel and vanadium by electrolysis) and by-products (hydrogen evolution). The current yield was determined by the formula

$$CY = m(pr)/m(th)\cdot 100\% \quad (6)$$

where $m(pr)$ is the practical mass of the coating; $m(th)$ is the theoretical mass of the coating, which should have been obtained in accordance with the Faraday law. Since the vanadium content did not exceed 0.22%, it was not taken into account.

As can be seen from Table 3, the current output at low temperatures of the electrolyte is lower than at high temperatures. This is due to the fact that at a temperature of 90°C, the coating is not only increased by electrolysis, but also because of the chemical reactions of the reduction of nickel and vanadium ions with sodium hypophosphite, so the current efficiency in these cases exceeded 100%. The quality of the coating, first of all, improved with an increase in the temperature of the electrolyte and the concentration of sodium hypophosphite, and also with a decrease in the concentration of boron nitride.

To study the wear resistance coating on the sample was applied from electrolyte No. 1 with additives of sodium hypophosphite 30 g / l and boron nitride 50 g / l. The electrolyte temperature was 90 ° C, the cathode current density was 1A / dm2. The heat treatment of the coated sample was carried out at a temperature of 600 ° C for 1.5 hours. Wear of this coating was 3.9 × 10-6 g / m. For comparison, a steel sample was tested synchronously with a heat-treated chemical nickel-phosphor coating deposited from an acetic acid solution with sodium hypophosphite as a reducing agent [3-8]. The wear of nickel-phosphoric coatings was more than 6.7 times.

4. Conclusion

It is shown that, due to the introduction of vanadic acid and boron nitride into the electrolyte, the wear resistance of coatings increases. The resulting regression equations make it possible to calculate the micro hardness of the coatings when the composition, the temperature of the electrolyte and the heat treatment temperature change. The composite coating of nickel-vanadium-phosphorus-boron nitride can be recommended for increasing the resistance of deforming tools, especially molds operating at temperatures up to 600°C.

Reference

[1] Morgunov A V, Skopintsev V D 2008 Galvanotechnika I obrabotka poverchnosti [Electroplating & surface treatment] XVI, 30-34

[2] Vyacheslavov P M, Shmeleva N.M. Kontrol' electrolitov I pokritij [Control of electrolytes and coatings] Spb Mashinostroenie , 1985, 98 p [in Russian]

[3] Vansovskaya K M Metallicheskie pokritiya, nanesennie khimicheskim sposobom [Metallic coatings applied chemically] Spb Mashinostroenie , 1985, 103 p [in Russian]

[4] Fedoseev S N, Valuev D V, Mamadaliev R A, Sokolov P N 2017 Application of integratedmethods to improve the technological properties of steel *Key Engineering Materials* **736** 110-115

[5] Malushin N N, Valuev D V, Valueva A V, Serikbol A 2014 Kinetic study of the effect of plasticity and its role in stress relaxation in the weld speed steel during the martensitic transformation *Applied Mechanics and Materials* **682** 58-63

[6] Malushin N N, Kovalev A P, Valuev D V, Shats E A, Borovikov I F 2016 Development and study of hard-facing materials on the base of heat-resisting high-hardness steels for plasma-jet hard-facing in shielding-doping nitrogen atmosphere *IOP Conference Series: Materials Science and Engineering* **142**

[7] Gizatulin R A, Valuev D V, Valueva A V, Serikbol A, Borovikov I F 2015 Optimizing technological parameters of the reduction processes in treating steels in a ladle furnace *IOP Conference Series: Materials Science and Engineering* **91**

[8] Solodsky S A, Lugovtsova N Y, Borisov I S 2014 The technology of Mig-Mag welding with low-frequency modulation of arc current *Applied Mechanics and Materials* **682** 426-430

On the program implementation of a Markov homogeneous monotonous random search algorithm of an extremum

A S Tikhomirov[1]

[1]Novgorod State University, ul. B. St. Petersburgskaya, 41 173003 Veliky Novgorod, Russia

E-mail: Alexey.Tikhomirov@novsu.ru

Abstract. A program that implements a Markov homogeneous monotonous random search algorithm of an extremum is presented. This program allows to solve a fairly wide class of problems of finding the global extremum of an objective function with a high accuracy.

1. Introduction

Let the objective function $f:\mathbf{R}^d \mapsto \mathbf{R}$ takes a minimum value at a single point x_*. Consider the problem of finding the global minimum point x_* with specified accuracy ε. One way to solve this problem is to apply random search algorithms for the extremum of a function (see [1-17]). Such methods have long been used successfully in solving complex optimization problems. A theoretical study of the rate of convergence of certain Markov search algorithms was carried out in [3, 11-17]. This paper is a continuation of [12-14] and is devoted to a computer program [18] realizing one of the obtained algorithms for a homogeneous Markov monotonic search for an extremum.

2. Statement of the problem

As an optimization space, we consider the space $X = \mathbf{R}^d$ with metric

$$\rho_\infty(x, y) = \max_{1\le n\le d} | x_n - y_n |, \tag{1}$$

where $x = (x_1,\dots,x_d)$ and $y = (y_1,\dots,y_d)$. Closed ball of radius r with the centre in point x we denote by $B_r(x) = \{y \in \mathbf{R}^d : \rho_\infty(x, y) \le r\}$. By μ denote d-dimensional Lebesgue measure. Metrics ρ_∞ is chosen for reasons of simplicity of modeling of the considered random search. The fact is that the simulation of the search is based on modeling of uniform distributions in balls. A ball in the metric (1) is a cube. Simulating a uniform distribution in a multidimensional cube is simple, while it is quite difficult to effectively model a uniform distribution in an "ordinary" ball given by the Euclidean metric.

To search for the minimum point, we use a homogeneous Markov monotonic random search (see [3, 11-14]), described later with the aid of the modeling algorithm. Notation "$\eta \leftarrow P(\cdot)$" reads like this: "get the realization of a random vector η with distribution P". For numbers and points in the optimization space, operations of the form $k \leftarrow 1$ и $\xi \leftarrow x$ denote the usual assignment operations.

Algorithm 1

Step 1. $\xi_0 \leftarrow x$, $k \leftarrow 1$.

Step 2. $\eta_k \leftarrow P(\xi_{k-1},\cdot)$.

Step 3. If $f(\eta_k) \le f(\xi_{k-1})$, then $\xi_k \leftarrow \eta_k$, otherwise $\xi_k \leftarrow \xi_{k-1}$.

Step 4. If $k < N$, then $k \leftarrow k+1$ and go to step 2, otherwise finish the algorithm.

Here x is a starting point of search, N is a number of search steps, $P(x,\cdot)$ is Markov transition function (see [3, 11–14]).

On the first step of Algorithm 1, random search is initialized. The starting point of the search is the point x (operator $\xi_0 \leftarrow x$), and the number of the next step of the search k becomes equal to unity (operator $k \leftarrow 1$).

On the second step of Algorithm 1 we choose a new "test" point η_k in the optimization space. A new "test" point is chosen randomly, using the distribution $P(\xi_{k-1},\cdot)$. Distribution $P(\xi_{k-1},\cdot)$ depends on the position of the "old" search point ξ_{k-1}. This dependence makes it possible to improve the efficiency of random search. The transition function $P(x,\cdot)$ we will call *the trial transition function*. We will consider the search, the trial transition function $P(x,\cdot)$ which does not depend explicitly on the step number k. Such a search is called *homogeneous*. A search whose trial transition functions clearly depend on the step number k, is called *nonhomogeneous*. Of course, due to heterogeneity, you can improve the search efficiency. But such a relationship complicates the choice of search parameters, and the "right" choice of search parameters can be a very difficult task (see, for example, [7]). Therefore, for the sake of ease of use of the search, we confine ourselves to a homogeneous search.

In the third step of Algorithm 1 we compare a new test point η_k with old search point ξ_{k-1}. If a new test point η_k no worse than the old search point ξ_{k-1} (that is, if inequality $f(\eta_k) \le f(\xi_{k-1})$ holds), then the search goes to a new point η_k (the operator $\xi_k \leftarrow \eta_k$), otherwise the search remains at the old point (the operator $\xi_k \leftarrow \xi_{k-1}$).

In the fourth step of Algorithm 1, we check the search stopping condition. In this case, a very simple criterion for stopping the search is selected. The search simply performs a predetermined number of steps N, and stops after that.

Note that the second, third, and fourth steps of Algorithm 1 are repeated N times. The first step of Algorithm 1 is performed only once. We also note that the introduced random search is *monotone*, in the sense that inequalities $f(\xi_k) \le f(\xi_{k-1})$ are performed for all $k \ge 1$.

3. Simulation of random search

The key question of choice for the type of search being investigated is the choice of the type of trial transition function $P(x,\cdot)$. When choosing a transition function, two criteria are usually used. First, the search must be sufficiently effective (requiring not too many steps to solve the task). In addition, the simulation of the distribution $P(x,\cdot)$ should be simple enough.

We use a homogeneous Markov monotonic random search, the transition function $P(x,\cdot)$ which minimizes the estimate of the complexity of random search investigated in [3, 12-14] when optimizing the simplest objective function. This search has several advantages. On the one hand, this search (for nondegenerate objective functions) ensures a good order of dependence of the obtained labor intensity estimates on ε (see [3, 12–14]). On the other hand, this search is easily modeled. And the given examples of the use of the program show an acceptable efficiency in the optimization of not too complicated objective functions.

Consider a random search for algorithm 1, trial transition functions $P(x,\cdot)$ which have symmetric densities of the form $p(x,y) = g(\rho_\infty(x,y))$, where ρ_∞ is metric, and g is nonincreasing nonnegative function defined on the semiaxis $(0,+\infty)$. Function g we call a search form. Suppose that $0 < \nu < \gamma \le \Gamma$ (ν, γ, Γ are search parameters) search form g is given by the following formula:

$$g(r) = \frac{1}{\Lambda}\begin{cases} \nu^{-d}, 0 < r \le \nu, \\ r^{-d}, \nu < r \le \gamma, \\ \gamma^{-d}, \gamma < r \le \Gamma, \\ 0, \quad \Gamma < r, \end{cases} \quad (2)$$

where $\Lambda = 2^d (d\ln(\gamma/\nu) + \Gamma^d/\gamma^d)$ is normalizing constant (providing the necessary normalization condition for the density), d is the dimension of the optimization space.

By $U_r(x,\cdot)$ we denote the uniform distribution in the ball $B_r(x)$, i.e. we put $U_r(x,\cdot) = \mu(\cdot \cap B_r(x))/\mu(B_r(x))$. Let α be uniformly distributed on a segment $[0,1]$ random value. Let $p = d2^d \ln(\gamma/\nu)/\Lambda$ and $q = \Lambda 2^{-d}/d$. Using the results of [12], we obtain the following algorithm for modeling a random vector η, having a distribution $P(x,\cdot)$ with the form (2).

Modeling Algorithm 1

Step 1. Get α.

Step 2. If $\alpha \ge p$, then $r \leftarrow \Gamma$, otherwise $r \leftarrow \nu\exp(\alpha q)$.

Step 3. $\eta \leftarrow U_r(x,\cdot)$, complete the algorithm.

Since in the case of a metric ρ_∞ simulate distribution $U_r(x,\cdot)$ very simple, then, in general the simulation algorithm 1 is very easily programmed.

By virtue of the recommendations of [12], we choose the value $\gamma = \Gamma/2^{1/d}$. Thus, we use the distribution $P(x,\cdot)$ with the form (2) and two parameters ν и Γ for modeling random search. In fact (see [12]), the distribution $P(x,\cdot)$ with form (2) is a mixture of uniform distributions in balls with center at the point x and radii from the segment $[\nu,\Gamma]$. In the modeling algorithm 1, we first simulate the radius of the ball, and then simulate a uniform distribution in a ball with a selected radius. The presented simulation algorithm is only slightly more complicated than the algorithm for simulating the simplest random search (the so-called "blind search" [3, 5]) in which the uniform distribution in the pre-fixed area of the optimization space is used. The modeling algorithm 1 is simpler than most of the random search modeling algorithms used [3, 5-7].

4. Description of the program

The program is written in C# in the integrated development environment of Microsoft Visual Studio Professional 2010. The program has a graphical user interface, written using Windows Forms. You can download the program at www.novsu.ru/doc/study/tas1 from the folder "Random_search". The program is also available as an executable file MarkovMonotonousSearch.exe and as a project containing the source code of the program and allowing the user to edit the program at his discretion.

Microsoft .NET Framework 4 is required to run the executable file of the program. It is usually already installed on the computer, but you can download it from the Microsoft website if necessary. To edit a project, you need to install the Microsoft Visual Studio development environment. This development environment can be used for free, and therefore this development environment can serve as a convenient tool for scientific calculations.

For calculations in the program, a numeric double type is used, providing an accuracy of 15-16 characters. Note that this numerical format limits the possible accuracy of the solution of the problem. Arguing somewhat simplistically, we obtain the following conclusions. If the target function behaves approximately as a quadratic function in the vicinity of the global minimum, then with accuracy of approximation by argument in the order 10^{-8}, we obtain the accuracy of approximation by the value of function in the order 10^{-16}. If the minimum value of the objective function is different from zero, then the numeric double type, providing an accuracy of 15-16 characters, will not allow us to calculate the value of the function with accuracy higher 10^{-16}. Thus, the typical accuracy of the solution of the problem will be of the order 10^{-7} of the approximation with respect to the argument, and of the order 10^{-14} of approximation by the value of the function. Such accuracy, as a rule, is sufficient from a practical point of view. And this accuracy of the solution of the problem can be obtained by using the considered random search program when solving not too complicated optimization problems. Of course, if the minimum value of the objective function is zero, and the minimum point is also at zero, then the problem can be solved with much higher accuracy.

To apply the search, you need to specify the target function, the search form parameters, the starting point of the search, and the number of search steps. The search results will be the end point of the search (approximating the point of the global minimum) and the value of the objective function at the end point of the search.

The submitted search has only four parameters. Parameters of the search form are positive numbers ν and Γ, for which inequalities must hold $0<\nu<\Gamma$. The choice of search parameters depends on the success of the search and its laboriousness. Value ν can be chosen close to the required accuracy of the solution of the problem when approximating with respect to the argument. Value Γ can be chosen close to or the expected accuracy of the initial approximation (the distance from the initial point of search to the minimum point) or to the diameter of the investigated region in the optimization space. When choosing Γ one can use the upper bound.

The third parameter N is number of search steps. It is desirable to take the number of search steps large enough. When solving a single task, you can, for example, execute a billion search steps, even if the task is fairly simple and can be solved much faster. Modern personal computers may well perform similar amounts of computation, at least for not too complex target functions. True, for such amounts of calculation, the code of the target function must be programmed.

In addition, you must select the starting point of the search. It is clear that the starting point is better located closer to the point of global extremum.

Numbers ν, Γ, N and the starting point of the search is easy to set in the main program window.

It is more difficult to specify an objective function. The target function can be defined in two ways. First, you can write the function code directly in the program code (in C#). To do this, open the source project in Microsoft Visual Studio Professional 2010 (or a compatible version). In the project, you need to open the ObjectiveFunction.cs file. In this file, you need to specify the dimension of the optimization space and the code that calculates the value of the objective function. Then you need to recompile and run the program (for example, by pressing the F5 key). C# language is an object-oriented programming language, oriented to programming using classes. The ObjectiveFunction.cs file contains the definition of the ObjectiveFunction class, which contains only the definition of the objective function and the dimension of the optimization space. There is no other complex code in this file. When writing code that calculates the value of an objective function, as a rule, minimal information about a programming language such as C, C++, C#, Java is sufficient.

Second, the target function can be specified in the search program itself (without using Microsoft Visual Studio). To do this, click the button "Set the formula" in the program, select the "Use formula" option in the dialog box, specify the dimension of the optimization space, and write the formula that defines the objective function.

The method of specifying the objective function is shown in the text field with the inscriptions "We use the function code" or "We use the formula".

The size of the optimization space is specified when specifying a function (or a constant in the ObjectiveFunction.cs file or a parameter when specifying a formula).

The rules for writing the formula are as follows. The formula can contain numbers, point coordinates, parentheses, arithmetic signs, function calls and constants. A decimal notation of numbers with a decimal separator is used (you can use both a comma and a period).

The coordinates of the point are written in the form x[k] (or X[k]). The index k must be a nonnegative integer constant, which is smaller than the dimension of the optimization space. For example, you can write x[0] + x[1].

Eight symbols of arithmetic operations are used: +, - (unary and binary minus), *, /, % (% operator calculates the remainder after dividing the first operand by the second), ^ (exponentiation), binary operation | computes the maximum of two numbers, the binary operation & computes a minimum of two numbers.

The operations of calculating the maximum and minimum have the lowest priority, followed by addition and subtraction operations, then the operations of multiplication, division and calculation of the remainder, then there is an exponentiation, then a unary minus.

You can use 18 functions. C# names are used (for example, Math.Exp), abbreviated names (without the Math class), and the Russian variants of writing function names. For example, the natural logarithm can be written in the following three ways: Math.Log, log, ln. In names you can use both uppercase and lowercase letters (they are equivalent). Function arguments are written in parentheses. A complete list of functions is provided in the "Help" menu of the program.

In the formula, you can use two constants, these are the numbers π and e.

As usual, you can include spaces and newline characters in the formula. If you set the value of "Number of search steps N" to zero, the program will calculate the value of the function at the starting point of the search. This can be used to calculate the value of the objective function at a given point.

In the program, you can specify output formats for the value of the objective function and the point coordinates (the C# formatting rules are used). The format E specifies an exponential representation of numbers. The format F specifies the representation of numbers with a fixed comma (period). The format G specifies the shorter of the two E or F formats. Precision specifies the number of digits in the fractional part for E and F, and the number of significant digits for G.

You can write comments to the problem being solved. Comments are written in text format and saved in a file along with parameters and search results.

To perform a search, the program uses a pseudo-random number generator. It can be initialized either by a value depending on the system time of the computer, or by a given value. The program can store data in XML format, and export key search characteristics in text format. Note that using the old Microsoft Visual Studio 2010 development environment when writing the program allows you to work with the project even for users of computers running the operating system Windows XP.

5. The choice of search parameters

It is important to note that the choice of search parameters can have a major impact on the effectiveness of the random search method [3, 5, 7]. In this case, many search algorithms contain a large number of heuristic parameters, and it is very difficult for the user of such an algorithm to find "good" parameter values that are suitable for the optimized function. We cite a quote from [7] relating to the ultrafast annealing method proposed by L. Ingber: "Among the shortcomings of this method, one can call it that, in view of the large number of parameters, it sometimes takes several months to fine-tune it for a specific task." At the same time, with proper selection of parameters, the method of ultrafast annealing can show very good results [6, 7].

The proposed search algorithm is largely free from insurmountable difficulties in the choice of parameters. To apply the search, you need to specify two parameters of the search form, the starting point of the search and the number of search steps. Considerations for choosing the search parameters described in the previous section can be considered quite simple. In particular, in the numerical examples of the next section, the minimum parameter selection was performed, consisting of just a few attempts to start the program with different parameter values.

6. Examples of using the program

Here are some examples of using the presented program to solve optimization problems. For computing, a personal computer with an Intel Core i5-4460 processor was used.

6.1. Example 1

We use an example from [5]. Here $X = \mathbf{R}^2$, $x = (x_1, x_2)$,

$$f(x) = f(x_1, x_2) = x_1^4 + x_1^2 + x_1 x_2 + x_2^2.$$

Function *f* takes a minimum value at a single point $x_* = (0,0)$ and $f(x_*) = 0$. The starting point of the search is selected $x = (1,1)$ and $f(x) = 4$. The number of search steps (the number of calculations of the objective function) *N* here takes the value 10000.

Algorithm B of the book [5] receives the minimum value of the objective function 2.7×10^{-6}. Algorithm B corresponds to the search for algorithm 1 using the normal probability distribution as a transition function.

Algorithm C of the book [5] receives the minimum value of the objective function 2.5×10^{-7}. Algorithm C also uses the normal probability distribution as a transition function, but it is a more complex search option, in which the movement performed in the previous step of the algorithm is taken into account when constructing a new search point.

Algorithm 1 with parameters $\nu = 2 \times 10^{-24}$ and $\Gamma = 0.7$ receives the minimum value of the objective function 9.9×10^{-49}. In this example, the search for algorithm 1 turned out to be much more accurate than algorithms B and C of the book [5].

For the same objective function and search with parameters $N = 10^6$, $\nu = 10^{-165}$, $\Gamma = 0.7$ we obtain the value of the objective function equal to zero (i.e., less than the value 5×10^{-324}, which defines a range of values of the double type of the C# programming language) and a minimum point $(7 \times 10^{-163}, -3 \times 10^{-163})$. The execution time of a million search steps was 0.06 seconds. Note that in this case, the extreme accuracy with which you can perform calculations on C# when using the numeric format double (due to the fact that you can not more accurately calculate the value of the objective function) is achieved.

6.2. Example 2

Here the optimization space $X = [-8,8]^2$, $x = (x_1, x_2)$,

$$f(x) = f(x_1, x_2) = \frac{1}{2}\left((x_1^4 - 16x_1^2 + 5x_1) + (x_2^4 - 16x_2^2 + 5x_2)\right).$$

Function f has four local minima, one of which is global. The starting point of the search is selected $x = (4.0, 6.4)$ and $f(x) = 537.18$. Search for Algorithm 1 with parameters $\nu = 10^{-8}$ and $\Gamma = 10$ when it finds the minimum value of the target function -78.3323314075428 and the minimum point (-2.903534, -2.903534). Note that here we have reached the extreme accuracy with which we can perform calculations on C# using the numerical format double (because we can not calculate the value of the objective function more accurately).

6.3. Example 3

Here the space $X = [-4,4]^{10}$, $x = (x_1, x_2, \ldots, x_{10})$,

$$f(x) = f(x_1, x_2, \ldots, x_{10}) = \sum_{n=1}^{5}\left(100(x_{2n} - x_{2n-1}^2)^2 + (1 - x_{2n-1})^2\right).$$

Function *f* is a known test function of Rosenbrock, used for local optimization methods. Function *f* assumes the minimum value $f(x_*) = 0$ in point $x_* = (1,1,\ldots,1)$. The starting point of the search is selected $x = (-1.2, 1, -1.2, 1, \ldots, 1)$ and $f(x) = 121$. Search for Algorithm 1 with parameters $\nu = 10^{-17}$ and $\Gamma = 4$ with $N = 1000000$ finds the minimum value of the objective function 2.84483×10^{-29}. The search time was 1.8 seconds.

6.4. Example 4
Consider an example with a very simple objective function, but in an optimization space of very large dimension for random search methods. Here the space $X = \mathbf{R}^{1000}$, $x = (x_1, x_2, \ldots, x_{1000})$, $f(x) = \sum_{n=1}^{1000} x_n^2$. Function *f* assumes the minimum value $f(x_*) = 0$ in a single point. The starting point of the search is selected $x = (1,1,\ldots,1)$. Search for Algorithm 1 with parameters $\nu = 10^{-10}$ and $\Gamma = 10$ at $N = 1000000$ finds the minimum value of the objective function 1.55934×10^{-14}. The search time was 13 seconds.

7. Conclusion
The obtained results show that the presented random search program can be successfully used to solve optimization problems. The program itself is easy to use, and selecting search parameters is not a difficult task. In this case, the program allows you to solve problems with the maximum accuracy, which can be obtained by using the numerical format double of the C# programming language.

References
[1] S.M. Ermakov, A.A. Zhiglyavsky, On the random search for a global extremum, *Theory of Probability and its applications*. No. **1** (1983) 129-136.
[2] S.M. Ermakov, A.A. Zhiglyavsky, M.V. Kondratovich, On the comparison of some procedures for the random search for a global extremum, *Journal of Computational Mathematics and Mathematical Physics*. Vol. **29**, No. **2** (1989) 163-170.
[3] A. Zhigljavsky, A. Zilinskas, Stochastic global optimization, Springer-Verlag, Berlin, 2008.
[4] A. Zhigljavsky, A. Zilinskas, Stochastic global optimization: a review on the occasion of 25 years of Informatica, *Informatica*, Vol. **27**, No. **2** (2016) 229-256.
[5] J.C. Spall, Introduction to stochastic search and optimization: estimation, simulation, and control, Wiley, New Jersey, 2003.
[6] L. Ingber, Very fast simulated re-annealing, *Mathl. Comput. Modelling*. V. **12** (1989) 967-973.
[7] A.S. Lopatin, Simulated Annealing Method, *Stochastic optimization in computer science.* Issue **1** (2005) 133-149.
[8] O.N. Granichin, B.T. Polyak, Randomized estimation and optimization algorithms under almost arbitrary noise, Science, Moscow, 2003.
[9] Yu.A. Sushkov, On a method for the organization of random search, Operations research and statistical. modeling. *Leningrad University Publishing House*, Leningrad, Vol. **1** (1972) 180-186.
[10] D. Tarlowski, On the convergence rate issues of general Markov search for global minimum, *Journal of Global Optimization*. Vol. **69**, Issue **4** (2017) 869-888.
[11] V.V. Nekrutkin, A.S. Tikhomirov, Speed of convergence as a function of given accuracy for random search methods, *Acta Applicandae Mathematicae*. **33** (1993) 89-108.
[12] A.S. Tikhomirov, On the Markov homogeneous optimization method, *Computational Mathematics and Mathematical Physics*, Vol. **46**, No. **3** (2006) 361-375.
[13] A.S. Tikhomirov, On the convergence rate of the Markov homogeneous monotone optimization method, *Computational Mathematics and Mathematical Physics*. Vol. **47**, No. **5** (2007) 780-790.
[14] A. Tikhomirov, T. Stojunina, V. Nekrutkin, Monotonous random search on a torus: integral upper bounds for the complexity, *Journal of Statistical Planning and Inference.* Vol **137**, Issue **12** (2007) 4031-4047.
[15] A.S. Tikhomirov, On the convergence rate of the simulated annealing algorithm, *Computational Mathematics and Mathematical Physics*. Vol. **50**, No. **1** (2010) 19-31.
[16] A.S. Tikhomirov, Lower bounds on the convergence rate of the Markov symmetric random search, *Computational Mathematics and Mathematical Physics.* Vol. **51**, No. **9** (2011) 1524-1538.

[17] A.S. Tikhomirov, On the rate of convergence of one inhomogeneous Markov algorithm of search for extremum, *Vestnik St. Petersburg University. Mathematics.* Vol. **44**, No. **4** (2011) 309-316.
[18] A.S. Tikhomirov, Computer program "MarkovMonotonousSearch", Certificate of state registration of the computer program No. 2016663645 (2016).

IOP Conf. Series: Materials Science and Engineering **441** (2018) 012056 doi:10.1088/1757-899X/441/1/012056

Modeling of truncated probability distributions

M S Tokmachev[1]

[1]Novgorod State University, ul. B. St. Petersburgskaya, 41 173003 Veliky Novgorod, Russia

Abstract. A method for modeling truncated probability distributions based on the transformation of some basic distribution is developed. One of the parameters of the simulated distributions is a quantile of arbitrary order. The cases of the initial multiplication of the basic density function by linear and quadratic factors are considered. In the case of the basic normal law with linear multipliers, expressions for the density and the first moments are generally presented. Using different base distributions and varying parameters leads to a wide class of new distributions. The results of the work can be useful in the processing of statistics with a non-standard structure.

1. Introduction

In classical probability distributions, often, the theoretical values of random variables are distributed on infinite intervals. The corresponding sampling values are distributed over finite intervals. And then the so-called "tails of distributions" for the distribution of the general totality are required to be taken into account in some way. One such method is the use of truncated distributions. The truncation of the distribution over a smaller interval in one way or another leads to a rearrangement of the distribution. The variety of truncation methods serves as a source of modeling of new probability distributions.

The truncation essence is the translation of the probabilities, determined by the "distribution tails", to the probability concentration interval after the truncation operation. Geometrically for the density curve is the addition of the figure to the rest of the curvilinear trapezoid by the area equal to the area of the cut off part. Thus, for the density, the normalization condition is guaranteed. And what will be the shape of this added figure - a question at the discretion of the researcher. The shape of a given area can be represented in various ways.

Classical and most simple when truncating the distribution to some interval $[a;b]$ is (see, for example, [1, 2]) the method of multiplying the density function $f(x)$ on the correcting factor, which guarantees the validity of the normalization condition:

$$g(x) = \begin{cases} \dfrac{f(x)}{\int_a^b f(x)\mathrm{d}x} & to\ x \in [a;\ b] \\ 0 \ \ to\ x < a\ \ or\ \ x > b\ . \end{cases}$$

Modifications of the method are possible. In particular, B.F. Kiryanov [3] proposed in the classical truncation to additionally use the shift of the abscissa axis to the intersection with the density curve at one of the closest points to the axis, a or b (for symmetry, the intersection can be in both boundary points).

We note that in this way the distribution obtained on the interval, in general, does not repeat the configuration of the initial distribution. With a different truncation technique, the configuration can be saved.

The paper presents other possible ways of modeling distributions based on truncation, two-sided or one-sided. A more complicated way of truncating is density function transformation $f(x)$ in given intervals by multiplying it by linear or quadratic factors.

With different methods of transformation, the truncation operation has a different degree of computational complexity. For a number of simulated distributions, computer programs have been developed that not only form the models of truncated distributions numerically, but also test and fit the theoretical distribution of the species in question according to the sample data.

2. Simplest methods of modeling

2.1. Models with configuration saving

Let there be given a distribution with a density function $f(x)$. In particular, $f(x)$ can be the density of a normal law $N(m;\sigma)$. We reduce the distribution to the interval $[a;b]$ with the preservation of trends in the density function. In addition, we take a point x_0, $x_0 \in [a;b]$. The density of the truncated $[a;b]$ distribution is assumed equal to

$$g(x) = \begin{cases} f(x) + h_1 \ to\ x \in [a;\ x_0] \\ f(x) + h_2 \ to\ x \in [x_0;\ b] \\ 0 \ to\ x < a \ or\ x > b \ . \end{cases} \quad (1)$$

The shape of the obtained curve $g(x)$ is shown in Figure 1.

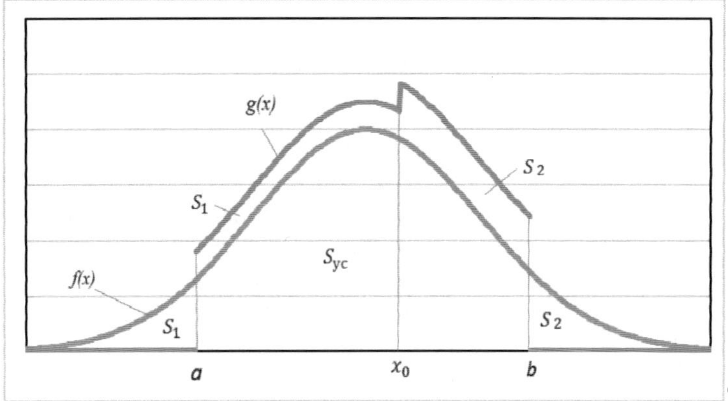

Figure 1. Graph of the density of a truncated distribution $g(x)$ kind (1).

Curve $g(x)$ in Fig. 1, in contrast to the classical truncation, preserves the configuration of the original curve $f(x)$ on intervals $[a;x_0]$ and $(x_0;b]$. The total area of the figures above the curve $f(x)$ is necessarily equal to the sum of the areas of the curvilinear trapezoids of the "distribution tails" $S_1 + S_2$. This condition allows you to select values h_1 and h_2, see. (1), in different ways.

For example, $h_1 = \frac{S_1}{x_0 - a}$, $h_2 = \frac{S_2}{b - x_0}$, where S_1 and S_2 – the area of curvilinear trapezoids of "tails of distribution". Let us check the normalization condition:

IOP Conf. Series: Materials Science and Engineering **441** (2018) 012056 doi:10.1088/1757-899X/441/1/012056

$$\int_{-\infty}^{+\infty} g(x)\,dx = \int_{a}^{x_0} (f(x) + h_1)\,\mathrm{d}x + \int_{x_0}^{b} (f(x) + h_2)\,\mathrm{d}x$$

$$= \int_{a}^{x_0} f(x)\,\mathrm{d}x + \int_{x_0}^{b} f(x)\mathrm{d}x + h_1(x_0 - a) + h_2(b - x_0) = S_{yc} + S_1 + S_2 = 1.$$

Note that the areas S_1 and S_2 are expressed in terms of the distribution function of a non-truncated random variable: $S_1 = F(a)$, $S_2 = 1 - F(b)$

Assuming in (1) $h = h_1 = h_2$, we come at a continuous $[a; b]$ function $g(x)$, wherein, $h = \frac{S_1+S_2}{b-a} = \frac{F(a)+1-F(b)}{b-a}$. Consequently,

$$g(x) = \begin{cases} f(x) + \frac{1-(F(b)-F(a))}{b-a} \; to \; x \in [a;\, b] \\ 0 \;\; to \;\; x \notin [a;\, b] \end{cases}. \tag{2}$$

Thus, unlike the classical truncation in this case, the density function is transformed in an additive manner.

In a model with a probability density of the form (1), it is possible to use other relations. For example,

a) $h_1 = \frac{S_2}{x_0-a}$, $h_2 = \frac{S_1}{b-x_0}$; b) $h_1 = \frac{\alpha(S_1+S_2)}{x_0-a}$, $h_2 = \frac{(1-\alpha)(S_1+S_2)}{b-x_0}$, where $0 < \alpha < 1$.

Distribution function $G(x)$ at a density of $g(x)$ kind (1) on interval $[a; b]$ differs from $F(x)$ on the linear summand:

$$G(x) = \begin{cases} F(x) - F(a) + h_1(x - a) \; to \; x \in [a;\, x_0] \\ F(x) - F(a) + h_1(x_0 - a) + h_2(x - x_0) \; to \; x \in [x_0; b]. \end{cases}$$

Similarly, one-sided truncations can be made.

2.2. The use of parabolic components

Let $f(x)$ be the density function of the initial distribution. We perform its truncation on an interval using additions of parabolic type. A graph of the corresponding truncation of the distribution is shown in Figure 2.

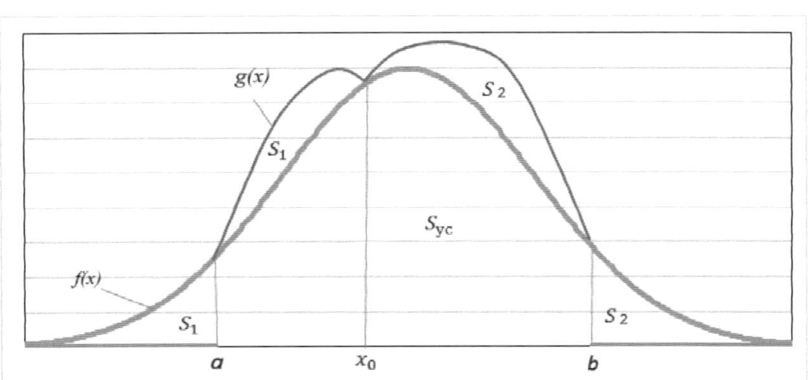

Figure 2. Graph of density $g(x)$ with superstructures of parabolic type.

Add-ons above the curve $f(x)$ figures should have given areas, for example, S_1 and S_2, and each of the curves of a parabolic type must pass through two given points, for example, $f(a)$, $f(x_0)$ and $f(x_0)$, $f(b)$. Wherein, x_0 is some value assigned by the researcher, $x_0 \in (a;b)$.
Then the density function of the truncated distribution $g(x)$ has the form

$$g(x) = \begin{cases} A_1x^2 + B_1x + C_1 \ \ to\ x \in [a; x_0] \\ A_2x^2 + B_2x + C_2 \ \ to\ x \in [x_0; b] \\ 0 \ \ to\ x \notin [a; b]\ . \end{cases} \quad (3)$$

Thus, the density curve in the truncation interval consists of two pieces of parabolas determined by the density $f(x)$ of basic distribution. Coefficients $A_1, B_1, C_1, A_2, B_2, C_2$ are calculated from the values of the function $f(x)$ in the boundary points, as well as the values of the areas of the superimposed figures. For the areas of the corresponding superstructures, in particular, S_1 and S_2 there derived two equations which are linear in regard to the mentioned coefficients:

$$\int_a^{x_0} \left(A_1x^2 + B_1x + C_1 - f(x)\right)dx = S_1$$

$$\int_{x_0}^{b} \left(A_2x^2 + B_2x + C_2 - f(x)\right)dx = S_2$$

Consequently, the coefficients in formula (3) can be found as a solution of two systems consisting of three linear algebraic equations each. In particular, in the case shown in Figure 2 (with areas of superstructures $S_1 = F(a)$ and $S_2 = 1 - F(b)$), we get

$$\begin{cases} A_1a^2 + B_1a + C_1 = f(a) \\ A_1x_0^2 + B_1x_0 + C_1 = f(x_0) \\ A_1\left(\frac{x_0^3 - a^3}{3}\right) + B_1\left(\frac{x_0^2 - a^2}{2}\right) + C_1(x_0 - a) = F(x_0) \end{cases} \quad (4)$$

$$\begin{cases} A_2x_0^2 + B_2x_0 + C_2 = f(x_0) \\ A_2b^2 + B_2b + C_2 = f(b) \\ A_2\left(\frac{b^3 - x_0^3}{3}\right) + B_2\left(\frac{b^2 - x_0^2}{2}\right) + C_2(b - x_0) = 1 - F(x_0) \end{cases} \quad (5)$$

The main determinants of each of the systems (4), (5) are respectively equal $\frac{(x_0-a)^4}{6}$ and $\frac{(b-x_0)^4}{6}$, and hence are different from zero for $x_0 \in (a;b)$. Therefore, the coefficients in the density function (3), as solutions of systems, are uniquely determined. In this case, as can be seen from the systems, they directly depend on the values a, x_0, b and the base distribution with the density function $f(x)$ and the distribution function $F(x)$.

Note that the area of the figures in parabolic superstructures when truncating the base distribution can be related not only as S_1 and S_2, but also differently, for example, S_2 and S_1 or $\alpha(S_1 + S_2)$, $\beta(S_1 + S_2)$, where $\alpha + \beta = 1$. The only restriction that guarantees the fulfillment of the normalization condition for the truncated distribution: the sum of the added areas is equal to $S_1 + S_2$, and $S_1 + S_2 = 1 - (F(b) - F(a))$.

The superstructure in the truncation of the distribution can also consist of one parabolic curve. In this case, only three unknown coefficients. They are calculated as the solution of the corresponding system of three linear algebraic equations constructed according to the above algorithm. We represent the expressions for the density function for such truncations.

1). $g(x) = \begin{cases} Ax^2 + Bx + C \ \ to\ x \in [a; b] \\ 0\ to\ x \notin [a; b] \end{cases}$

The resulting truncated distribution is a modification of the symmetric classical parabolic distribution. In this case, for arbitrary and different basic functions $f(x)$ symmetry, generally speaking, is absent.

$$2).\ g(x) = \begin{cases} Ax^2 + Bx + C \ to\ x \in [a; x_0] \\ f(x)\ \ to\ x \in (x_0; b] \\ 0\ \ to\ x \notin [a; b], \end{cases}$$

where $x_0 \in (a; b)$.

$$3).\ g(x) = \begin{cases} f(x)\ to\ x \in [a; x_0] \\ Ax^2 + Bx + C\ \ to\ x \in (x_0; b] \\ 0\ \ to\ x \notin [a; b], \end{cases}$$

where $x_0 \in (a; b)$.

$$4).\ g(x) = \begin{cases} f(x)\ to\ x \in [a; x_0]\ \cup [x_1; b] \\ Ax^2 + Bx + C\ \ to\ x \in (x_0; x_1) \\ 0\ \ to\ x \notin [a; b], \end{cases}$$

where $a < x_0 < x_1 < b$.

$$5).\ g(x) = \begin{cases} A_1x^2 + B_1x + C_1\ \ to\ x \in [a; x_0] \\ f(x)\ \ to\ \ x \in (x_0; x_1) \\ A_2x^2 + B_2x + C_2\ \ to\ x \in [x_1; b] \\ 0\ \ to\ x \notin [a; b], \end{cases}$$

where $a < x_0 < x_1 < b$.

The computer implementation of this method of truncating different distributions makes it easy to calculate the coefficients, the distribution function, and the numerical characteristics associated with the moments.

Varying by position parameters a, b, x_0, x_1 and the ratio of areas in superstructures above the density function graph $f(x)$. The basic distribution leads to a large variety of simulated truncated distributions.

Similarly, one-sided truncations can be considered, using a larger number of parabolic-type components, as well as other more complex configurations and their combinations.

3. Simulation of truncated distributions with a given quantile and additional factors

Let $f(x)$ is density of the base distribution of a random variable X. We transform this distribution by truncating to a given interval $[a; b]$. We introduce the restriction in the form of quantiles of order α with the value of this quantile x_α. At the same time, the interval $[a; b]$ is divided into two intervals $[a; x_\alpha]$, $(x_\alpha; b]$. On each of the intervals obtained for the density function we introduce additional factors.

For linear factors, the density of the modeled distribution $g(x)$ has the form:

$$g(x) = \begin{cases} (B_1x + C_1)f(x)\ \ to\ x \in [a; x_\alpha] \\ (B_2x + C_2)f(x)\ \ to\ x \in (x_\alpha; b] \\ 0\ \ \ to\ x \notin [a; b]\,. \end{cases} \quad (6)$$

In this case, $g(x)$ is the density of a certain distribution, it is sufficient that it is nonnegative and that the normalization condition $\int_a^b g(x)\mathrm{d}x = 1$.

Coefficients B_1, C_1, B_2, C_2 are found as solutions of the system of four linear equations that guarantee and the fulfillment of the normalization condition:

$$\begin{cases} \int_a^{x_\propto}(B_1 x + C_1)f(x)dx = \propto \\ \int_{x_\propto}^{b}(B_2 x + C_2)f(x)dx = \beta \\ (B_1 x_\propto + C_1)f(x_\propto) = g(x_\propto) \\ (B_2 x_\propto + C_2)f(x_\propto) = g(x_\propto) \end{cases} \quad (7)$$

Numerical values $\propto$, $g(x_\alpha)$ are chosen by the researcher, and the probabilities α and β are related by $\propto + \beta = 1$. Consequently, $a, b, \propto, x_\alpha, g(x_\alpha)$ are parameters of the simulated truncated distribution. The non-negativity of the function $g(x)$ is limited by $g(a) \geq 0,\ g(b) \geq 0,\ g(x_\propto) \geq 0$

We note that the solution of the system (7) essentially depends on the type of the base distribution with the density function $f(x)$, which is clearly present in each of the equations of the system. We give the result under the basic normal distribution.

Theorem, [4]. Let $f(x)$ be the density of a Gaussian random variable $X\!: X \sim N(m;\sigma)$. The following parameters have been fulfilled: $x_\propto \in (a;b)$, $g(x_\propto) \geq 0$, $0 < \alpha < 1$, $\beta = 1 - \alpha$. Then the coefficients of the density function $g(x)$ of a simulated distribution of the form (6) satisfy the relations

$$B_1 = \frac{\propto - \frac{g(x_\propto)}{f(x_\propto)}\left[\Phi\left(\frac{x_\propto - m}{\sigma}\right) - \Phi\left(\frac{a-m}{\sigma}\right)\right]}{\sigma\left[\varphi\left(\frac{a-m}{\sigma}\right) - \varphi\left(\frac{x_\propto - m}{\sigma}\right)\right] + (m - x_\propto)\left[\Phi\left(\frac{x_\propto - m}{\upsilon}\right) - \Phi\left(\frac{a-m}{\sigma}\right)\right]};$$

$$B_2 = \frac{\beta - \frac{g(x_\propto)}{f(x_\propto)}\left[\Phi\left(\frac{b-m}{\sigma}\right) - \Phi\left(\frac{x_\propto - m}{\sigma}\right)\right]}{\sigma\left[\varphi\left(\frac{x_\propto - m}{\sigma}\right) - \varphi\left(\frac{b-m}{\sigma}\right)\right] + (m - x_\propto)\left[\Phi\left(\frac{b-m}{\sigma}\right) - \Phi\left(\frac{x_\propto - m}{\sigma}\right)\right]};$$

$$C_1 = \frac{g(x_\propto)}{f(x_\propto)} - B_1 x_\propto;\ C_2 = \frac{g(x_\propto)}{f(x_\propto)} - B_2 x_\propto,$$

$$B_1 a + C_1 \geq 0,\ B_2 b + C_2 \geq 0,$$

where $\varphi(x) = \frac{1}{\sqrt{2\pi}} e^{-\frac{x^2}{2}}$ is the probability density of the standard normal law, $\mathrm{F}(x) = \frac{1}{\sqrt{2\pi}} \int_{-\infty}^{x} e^{-\frac{t^2}{2}} dt$ is Laplace function.

Theorem, [4]. Let $f(x)$ be the density of a Gaussian random variable $X\!: X \sim N(0;1)$. The density of a truncated distribution $g(x)$ satisfies the relation (6). Then for the mathematical expectation and variance of the truncated distribution, the following relations hold:

$$M(X) = B_1[a\varphi(a) - x_\propto\varphi(x_\propto)] + B_1[\Phi(x_\propto) - \Phi(a)] - C_1[\varphi(x_\propto) - \varphi(a)]$$
$$+B_2[x_\propto\varphi(x_\propto) - b\varphi(b)] + B_2[\Phi(b) - \Phi(x_\propto)] - C_2[\varphi(b) - \varphi(x_\propto)];$$

$$D(X) = B_1[a^2\varphi(a) - {x_\propto}^2\varphi(x_\propto)] + C_1[\Phi(x_\propto) - \Phi(a)] + C_1[\varphi(a) - \varphi(x_\propto)]$$
$$+B_2[{x_\propto}^2\varphi(x_\propto) - b^2\varphi(b)] + C_2[\Phi(b) - \Phi(x_\propto)] + C_2[\varphi(x_\propto) - \varphi(b)] - M^2(X).$$

IOP Conf. Series: Materials Science and Engineering **441** (2018) 012056 doi:10.1088/1757-899X/441/1/012056

Similarly, using functions $\varphi(x)$ and $\Phi(x)$, in distributions modeled on the basis of the basic normal law, we obtain expressions for the distribution function, for the characteristic function, and for higher-order moments.

For quadratic multipliers, the density of the model distribution $g(x)$ has the form:

$$g(x) = \begin{cases} (A_1x^2 + B_1x + C_1)f(x) \ \ to\ x \in [a; x_\alpha] \\ (A_2x^2 + B_2x + C_2)f(x) \ \ to\ x \in (x_\alpha; b] \\ 0 \ \ to\ x \notin [a; b]\,. \end{cases} \tag{8}$$

Coefficients A_i, B_i, C_i, $(i = 1, 2)$ in the relation (8) are both solutions of systems of linear algebraic equations with respect to these coefficients

$$\begin{cases} \int_a^{x_\propto}(A_1x^2 + B_1x + C_1)f(x)dx = \propto \\ (A_1x_\propto{}^2 + B_1x_\propto + C_1)f(x_\propto) = g(x_\propto) \\ (A_1a^2 + B_1a + C_1)f(a) = g(a) \end{cases} \qquad \begin{cases} \int_{x_\propto}^{b}(A_2x^2 + B_2x + C_2)f(x)dx = \beta \\ (A_2x_\propto{}^2 + B_2x_\propto + C_2)f(x_\propto) = g(x_\propto) \\ (A_2b^2 + B_2b + C_2)f(b) = g(b) \end{cases} \tag{9}$$

When $\propto + \beta = 1$, the normalization condition is satisfied. We note that after integration of each of the systems (9) in the first equation, we arrive at precisely the linear algebraic equation with respect to the coefficients A_i, B_i, C_i. When $x_\propto \in (a; b)$, the solution of systems is unique. Again $a, b, \propto$, $x_\alpha, g(x_\alpha)$ are the parameters of a simulated truncated distribution based on the base distribution with density $f(x)$.

Under the conditions formulated, an arbitrarily given set of parameters does not guarantee a probabilistic solution of systems (9). For the density function to be nonnegative $g(x)$ the nonnegativity of its quadratic multipliers is necessary:

$$\begin{cases} (A_1x^2 + B_1x + C_1) \geq 0 \ to\ x \in [a; x_\alpha] \\ (A_2x^2 + B_2x + C_2) \geq 0 \ \ to\ x \in (x_\alpha; b] \end{cases}.$$

For verification, we can use two lemmas.

Lemma 1. If the square trinomial $Q(x) = Ax^2 + Bx + C$ on interval $[r; s]$ satisfies the conditions $Q(r) = 0, Q(s) = 0, A < 0\ A < 0$, then $Q(x)$ is nonnegative on $[r; s]$.

Lemma 2. If the square trinomial $Q(x) = Ax^2 + Bx + C$ has no roots on the interval $(r; s)$ and $Q(r) \geq 0,\ Q(s) > 0$ (or $Q(r) > 0,\ Q(s) \geq 0$), then $Q(x)$ is nonnegative on $[r; s]$.

For the basic function of the standard normal law, the corresponding formulas for computing the coefficients, the distribution function, and the moments of even and odd orders are presented in [5]. In addition, along with the functions $\varphi(x)$ and $\Phi(x)$, In the relations obtained, we use double factorials and polynomials of a certain class. In [6], specific relationships were found for calculating the coefficients, the distribution function, and the moments for the basic exponential distribution.

By the same algorithm for constructing a distribution with density $g(x)$ The uniform distribution on the segment, the distribution with a sinusoidal density and the distribution [7, 8] of the hyperbolic cosine type are investigated as a basis. In all cases, depending on the parameters, non-trivial distributions with a complex structure are obtained. Example of density function graph $g(x)$ of kind (8) the simulated distribution is shown in Figure 3. Note that the configuration of the graph is very sensitive to the distribution parameters, and also to the choice of the basic function.

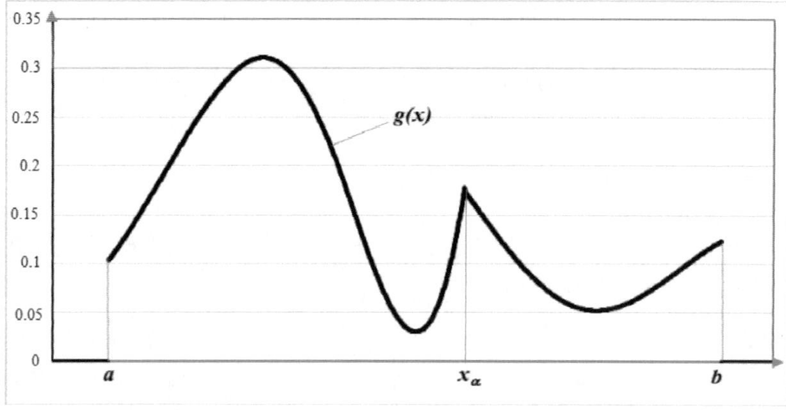

Figure 3. Graph of the density function $g(x)$ at the basic normal distribution.

As a result of the transformations, the final distribution turns out to be a mixture of different distributions, which, as a rule, corresponds to the multifactority and inhomogeneity of the impact in the formation of the indicator under study, based on real data. The input of quantiles makes it possible to take into account possible qualitative transitions corresponding to the data structure.

According to the form of the function (8) for any base function $f(x)$ the mixture turns out to consist of six truncated distributions: three for each of the intervals $[a; x_\alpha]$, $(x_\alpha; b]$. In particular, with the basic normal distribution, the components of the mixture are: initial density $f(x)$, displaced Rayleigh distribution density $\left(\frac{x-m}{\sigma^2}e^{-\frac{(x-m)^2}{2\sigma^2}},\ x > m\right)$ and the displaced Maxwell distribution density $\left(\frac{2(x-m)^2}{\sigma^3\sqrt{2\pi}}e^{-\frac{(x-m)^2}{2\sigma^2}},\ x > m\right)$. At the base density of the exponential distribution, in addition to the initial density, the components of the mixture are the Erlang distribution densities of the second and third orders.

When modeling a particular type of function $g(x)$ in accordance with real data, the computer program checks compliance with Pearson's consent criterion. Distribution parameters are adjusted to minimize the value χ^2_{view}.

4. Conclusion

Statistical indicators are formed as a result of the interaction of many factors, heterogeneous and multidirectional. Therefore, for many real statistical data, as a rule, not only normal, but also other classical distribution laws do not work. Naturally, in statistical data possessing a complex nonstandard structure, the theoretical probability distribution must also be structurally complex. It is necessary to model non-standard distributions. The obtained distributions allow us to consider and analyze the estimated inhomogeneous general populations, and also help in justifying their decomposition into homogeneous components.

The considered method of modeling based on truncation with further transformations defines an extensive class of probability distributions with non-standard structure. Let's formulate the algorithm of actions:

1) basic distribution (selection of the basic function $f(x)$;
2) truncation, two-sided or one-sided (border selection, a and b);
3) truncation, two-sided or one-sided (border selection α and the corresponding quantile x_α);
4) density boundary values (choice of values $g(a), g(x_\alpha), g(b)$);

5) calculation of coefficients $g(x)$ with the control of the non-negativity of the density function.

The results of modeling new probability distributions can be very useful when processing statistical data with nonstandard probability ratios.

References

[1] Cramer G 1975 *Mathematical Methods of Statistics* (Moscow: Mir) p 631
[2] Vadzinsky R N 2001 *Handbook of Probability Distributions* (SanktPetersburg: Science) p 295
[3] Kiryanov B F, Tokmachev M S 2009 *Mathematical Models in Public Health* (Veliky Novgorod: Publishing House of Novgorod State University) p 279
[4] Tokmachev M S, Ryazantsev P P 2010 Simulation of truncated distributions *Bulletin of NovSU.* No. **55** pp 34–36
[5] Tokmachev M S 2011 Simulation of truncated distributions with a given quantile *Bulletin of KSTU A N Tupolev.* No. **2** pp 157–162
[6] Tokmachev M S 2011 Simulation of truncated distributions with a given quantile based on the exponential distribution *Bulletin of NovSU. S.: Techn. science.* No. **65** pp 97–100
[7] Tokmachev M S 1995 Constant regression of quadratic statistics on linear statistics *Bulletin of NovSU. S: Natural. and techn. Science.* No. **1** pp 139–141
[8] Tokmachev M S 2017 Investigation of the probability distribution of hyperbolic cosine type *Bulletin SUSU. S: Mathematics. Mechanics. Physics.* T. 9, No. **3** pp 18–26

Boltzmann principle of superposition in the theory of wood creep for deformations in time

K A Varenik [1], A S Varenik [1], R C Sanzharovskij [1]

[1]Yaroslav-the-Wise Novgorod State University, ul. B. St. Peterburgskaya, 41, 173004, Veliky Novgorod, Russia

[1]e-mail: vkirillv89@mail.ru

Abstract. A number of theoretical orientations can be distinguished in the field of wood creep study. The first orientation includes models based on different combinations of the simplest linear models: Hooke model of elastic body, viscous element of Newton, element of dry friction of Coulomb. The second orientation is based on the parallel connection of the elastic element with the viscous element. In the third orientation the elastic and viscous element are arranged sequentially. In the fourth orientation the Boltzmann theory of elastic inheritance is used. This article considers the feasibility and practicability of the study of wood creep within a framework of the latter of these orientations. The evidence of the erroneous of using this principle is given.

1. introduction

In accordance with the Boltzmann principle [1, 2], the linear creep equation of wood was recorded in 1954 in the following form (in the final notations):

$$\varepsilon(t) = \sigma(\tau_1)\delta(t,\tau_1) + \int_{\tau_1}^{t} \frac{d\sigma(\tau_1)}{d\tau}\delta(t,\tau)d\tau, \tag{1}$$

where $\delta(t,\tau) = \frac{1}{E(\tau)} + \left(C_0 + \frac{A_1}{\tau}\right)\left[1 - e^{-\gamma(t-\tau)}\right]$;

$\frac{1}{E(\tau)}$ is elasto-instantaneous deformation,

$C_0 = 0{,}32 \cdot 10^{-5}\,\mathrm{cm}^2/_{\mathrm{kgf}}$, $A_1 = 5 \cdot 10^{-5}\,\mathrm{cm}^2/_{\mathrm{kgf}}$, $\gamma = 0{,}04$ are wood creep constants derived from extensive experimental studies.

Prokopovich I.E. and Zedgenidze V.A. [3] recorded the measure of wood creep on the basis of experiments in the form:

$$C(t,\tau) = (C_0 + Ae^{-\gamma\tau})\left(1 - B_1 e^{-\gamma_1(t-\tau)}\right),$$

где $C_0 = 2{,}87 \cdot 10^{-5}(\mathrm{MPa})^{-1}$;
$A = 10{,}95 \cdot 10^{-5}(\mathrm{MPa})^{-1}$;
$\gamma_1 = \gamma = 0{,}15\left(^1/_{\mathrm{day}}\right)$;
$B_1 = 1$.

After integrating by parts, the basic equation of wood creep takes the form:

$$\varepsilon(t) = \frac{\sigma(\tau_1)}{E(t)} - \int_{\tau_1}^{t} \sigma(\tau) \frac{d}{d\tau} \delta(t,\tau) d\tau, \tag{2}$$

where $\varepsilon(t)$ is the full deformation.

In the reasoning of equation (2), Prokopovich I.E. and Zedgenidze V.A. consider it possible to observe the influence of wood moisture "'as a result of the peculiar aging associated with the drying of wood". Such variant of the theory is developed by a number of scientists on the basis of experimental investigations of Ivanov U.M. At direct consideration of humidity parameter $\omega(t)$, integral equation (2) is written in the form:

$$\varepsilon(t) = \frac{\sigma(\tau_1)}{E[\omega(t)]} - \int_{\tau_1}^{t} \sigma(\tau) \frac{d}{d\tau} \delta[\omega(\tau), t, \tau] d\tau, \tag{3}$$

where $\delta[\omega(\tau), t, \tau] = \frac{1}{E[\omega(t)]} + \frac{\varphi[\omega(\tau),t,\tau]}{E[\omega(t)]}$,

$\varphi[\omega(\tau), t, \tau]$ is creep characteristic.

For example, Orlovich R.B. [4] gets:

$$\varphi[\omega(\tau), t, \tau] = \left\{\frac{\varphi_\infty(\omega(\tau))}{t-\tau} \int_0^t F[\omega(\tau+\rho)] d\rho \right\} \left[1 - e^{-\beta(\omega)(t-\tau)}\right].$$

In a simpler assumption (Prokopovich I.E., Zedgenidze V.A.) it can be written:

$$\varphi[\omega(\tau), t, \tau] = \varphi_\infty[\omega(\tau)] \left[1 - e^{-\beta(\omega)(t-\tau)}\right]. \tag{4}$$

The equations (1) and (2) are based on two fundamental assumptions:

-the principle of classical linear connection:

$$\varepsilon(t,\tau) = \sigma(\tau) \cdot \delta(t,\tau) \text{ , or } \varepsilon[\omega(\tau), t, \tau] = \sigma(\tau) \cdot \delta[\omega(\tau), t, \tau]; \tag{5}$$

-the Boltzmann principle of linear superposition.

2. Application of Boltzmann principle in wood creep theory

Classical linear connection (5) is usually established by means of thorough experiments (simple loading) of the process of wood deformation with long time constant strain $\sigma_0 = \sigma(\tau_1)$. This is due to the fact that to detect a similar connection at variable strains $\sigma(t)$ the known behaviour of wood deformation at single constant strains is used. This known behaviour is described by a creep measure $C[\omega(\tau), t, \tau] = \frac{\varphi[\omega(\tau),t,\tau]}{E[\omega(t)]}$. We emphasize that there are non-stationary variables, because of the dependence on humidity - the function of wood aging $\varphi_\infty[\omega(\tau)]$ and modulus of elasticity $E[\omega(\tau)]$.

The Boltzmann principle of linear superposition is true only at very strong limitations on the function of aging and modulus of elasticity. In the case of structure (5) these restrictions are not feasible, and the theory of wood creep, built on the basis of this principle, proves to be deeply erroneous.

In non-stationary properties of wood, the Boltzmann principle of superposition incorrectly builds the creep core, incorrectly describes the processes of change of instantaneous deformations and creep deformations due to incorrect determination of the full rate of deformation value corresponding to (5).

According to structure (5), the deformation rate is equal to:

$$v_\varepsilon = \dot{\sigma}(\tau)\delta[\omega(\tau), t, \tau] + \sigma(\tau)\frac{\partial\delta[\omega(\tau),t,\tau]}{\partial\omega(\tau)}\dot{\omega}(\tau) + \sigma(\tau)\frac{\partial\delta[\omega(\tau),t,\tau]}{\partial(\tau)} + \sigma(\tau)\frac{\partial\delta[\omega(\tau),t,\tau]}{\partial t}.$$

It can be seen from this expression that in integral equation (1) the last three components of the deformation rate, caused by the rate of change of the full unit deformation, are lost $\delta[\omega(\tau), t, \tau]$. These losses are as follows:

$$-\sigma(\tau)\frac{E'[\omega(\tau)]}{E^2}\dot{\omega}(\tau)\sigma(\tau)\frac{1}{E[\omega(\tau)]}\frac{\partial\delta[\omega(\tau),t,\tau]}{\partial\omega(\tau)}\dot{\omega}(\tau) + \sigma(\tau)\frac{1}{E[\omega(\tau)]}\frac{\partial\varphi[\omega(\tau),t,\tau]}{\partial(\tau)} +$$

$$\sigma(\tau)\frac{1}{E[\omega(\tau)]}\frac{\partial\varphi[\omega(\tau),t,\tau]}{\partial t} - \sigma(\tau)\varphi[\omega(\tau), t, \tau]\frac{E'[\omega(\tau)]}{E^2[\omega(\tau)]}\dot{\omega}(\tau).$$

They are comparable in order of importance to the remaining (1) component. These losses cause significant discrepancies between the theory and the experiment described in the scientific literature on wood creep.

In application of the Boltzmann principle, errors are made with (and without necessity) opposite to actions under the important result of classical theory (5): first is differentiation and then integration. In the process of differentiation, many components of the rate of full wood deformation are lost, after integration the losses pass to the deformation value and then into practical application.

The Boltzmann principle distorts the classic linear relationship causing three types of errors: I- incorrectly finds the values of short-term deformations; II – incorrectly defines the core of an integral equation; III – erroneously adds instantaneous strains to the notion of creep deformations.

Case I. The rate of elastic deformation equals:

$$\dot{\varepsilon}_y(\tau) = \dot{\sigma}(\tau)\frac{1}{E[\omega(\tau)]} + \sigma(\tau)\frac{\partial}{\partial\tau}\frac{1}{E[\omega(\tau)]}.$$

Integrating, we have:

$$\varepsilon_y(t) - \varepsilon_y(\tau_1) = \int_{\tau_1}^{t}\frac{1}{E[\omega(\tau)]}\dot{\sigma}(\tau)d\tau + \int_{\tau_1}^{t}\sigma(\tau)\frac{\partial}{\partial\tau}\frac{1}{E[\omega(\tau)]}d\tau.$$

Integrating here the first term in parts and reducing, we will find the value of short-term deformation:

$$\varepsilon_y(t) = \frac{\sigma(t)}{E[\omega(\tau)]}. \tag{6}$$

It can be seen that the first term under the integral sign (2) is superfluous and usage in (1) and (2) the Boltzmann principle:

$$\varepsilon_y(t) = \frac{\sigma(\tau_1)}{E[\omega(\tau_1)]} + \int_{\tau_1}^{t}\frac{1}{E[\omega(\tau)]}d\sigma(\tau) = \frac{\sigma(t)}{E[\omega(t)]} - \int_{\tau_1}^{t}\sigma(\tau)\frac{\partial}{\partial\tau}\frac{1}{E[\omega(\tau)]}d\tau. \tag{7}$$

is deeply mistaken.

The Boltzmann principle of superposition erroneously reconstructs the real elastic model of wood with modulus of elasticity $E[\omega(\tau)]$; it adds to it nonexistent model of linear liquid viscous with coefficient of viscosity $k_y(\tau) = \frac{E^2[\omega(\tau)]\,1}{\dot{E}[\omega(\tau)]\dot{\omega}(\tau)}$,forming in such a way the Maxwell scheme.

Consider, the example, putting in (6), (7) $\sigma(t) = \sigma_0 = const$, we will get:

$$\varepsilon_y(t) = \frac{\sigma_0}{E[\omega(\tau)]} \text{ and } \varepsilon_y(\tau_1) = \frac{\sigma_0}{E[\omega(\tau\tau_1)]}.$$

Comparison of these deformations is shown in Figure 1.

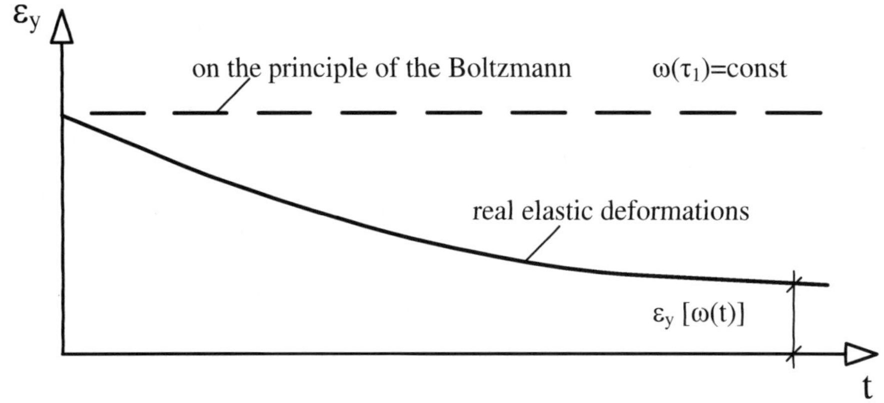

Figure 1. Comparison of deformation graphs

Case II. In the field of linear creep, the number of additional (fictitious) bodies arising due to the wrong scheme of creep core (inherited function of the 1-st kind) increases significantly. It depends on the type of function $\varphi[\omega(\tau), t, \tau]$, describing the non-stationary characteristic of wood creep. Let's write it down, for example, in the form (6).

In this case, the basic law (5) forms the set of redundant (non-existent) bodies connected in series with each other: two bodies of Voigt type and two viscous elements. The creep deformations of these bodies are equal:

$$\varepsilon_1(t) = \int_{\tau_1}^{t} \sigma(\tau) \frac{1}{k_1(\tau)} e^{-\beta(t-\tau)} d\tau, \quad k_1(\tau) = \frac{E[\omega(\tau)]}{\varphi'_\infty[\omega(\tau)]\dot{\omega}(\tau)}; \tag{8}$$

$$\varepsilon_2(t) = \int_{\tau_1}^{t} \sigma(\tau) \frac{1}{k_2(\tau)} d\tau, \quad k_2(\tau) = \frac{E^2[\omega(\tau)]}{E'[\omega(t)]\varphi_\infty[\omega(t)]\dot{\omega}(\tau)}; \tag{9}$$

$$\varepsilon_3(t) = \int_{\tau_1}^{t} \sigma(\tau) \frac{1}{k_3(\tau)} e^{-\beta(t-\tau)} d\tau, \quad k_3(\tau) = -\frac{E^2[\omega(\tau)]}{E'[\omega(t)]\varphi_\infty[\omega(t)]\dot{\omega}(\tau)}; \tag{10}$$

$$\varepsilon_4(t) = \int_{\tau_1}^{t} \sigma(\tau) \frac{1}{k_4(\tau)} d\tau, \quad k_4(\tau) = -\frac{E[\omega(\tau)]}{\varphi'_\infty[\omega(t)]\dot{\omega}(\tau)}; \tag{11}$$

where k_1, k_2, k_3, k_4 are viscosity factors or internal friction factors of fictitious bodies; moreover, Voigt body (10) and body (11) of the viscous element expand at compression.

Creep deformations (8)-(11) caused by the influence of the superposition principle are summed up with a short-term fictitious deformation:

$$\varepsilon_5(t) = \int_{\tau_1}^{t} \sigma(\tau) \frac{\partial}{\partial\tau} \frac{1}{E[\omega(\tau)]} d\tau,$$

and introduce large errors in the value of full deformation determined by the creep law (3).

Essence and the secondariness of the Boltzmann scheme for the theory of wood creep are investigated on the rheological model described by Ugolevym B.N. [5] and Latishenko V.A. [6], in the record of Ishlinsky A.U.:

$$\frac{\partial\sigma}{\partial t} + r\sigma = b\frac{\partial\varepsilon}{\partial t} + bn\varepsilon,$$

what gives:

$$\varepsilon(t) = \frac{\sigma(t)}{b} - \int_{t_0}^{t} \sigma(t) \frac{1}{b} \frac{\partial\varphi(t-\tau)}{\partial(\tau)} d\tau \tag{12}$$

where $\varphi(t-\tau) = b\frac{r-n}{b}\left[1 - e^{-n(t-\tau)}\right]$.

The Boltzmann case is obtained from (12) by its series transformations, mathematically appropriate only in the conditions of stationary properties:

$$\varepsilon(t) = \sigma_0\left[\frac{1}{b} + \frac{1}{b}\varphi(t-t_0)\right] + \int_{t_0}^{t}\left[\frac{1}{b} + \frac{1}{b}\varphi(t-\tau)\right] d\sigma(\tau). \tag{13}$$

In the transformation (13), as opposed to (12), ductility function is used, which attracted attention of scientists. However, transformation (13) is possible only with very strong limitations. With the broad interpretation of ductility, in the conditions of non-stationary wood, these restrictions were not taken into account, and the theory of creep proves to be deeply erroneous.

First, instantaneous deformation with extremely simple physical meaning for arbitrary time t is imposed the property of the process that creates the temptation to expand the theory and turns into the above-mentioned gross error at non-stationary variables $E[\omega(t)]$.

Secondly, it is necessary to integrate (12) in parts, that at broad interpretation of ductility function in the conditions of aging (4) creates another temptation, traditionally leading to another blunder when finding the core of the integral equation; as it is known, at non-stationary properties of wood creep deformation is obtained from another solution of differential equation, solution, written in more complex form:

$$\varepsilon_n(t) = e^{-F(t)}\left[\varepsilon_{n0} + \int_{t_0}^{t} \sigma(t) \frac{1}{\eta(t)} e^{F(t)} dt\right]; \; F(t) = \int_{t_0}^{t} n(t) dt,$$

when parameters $b(t)$ and $n(t)$ are time functions.

In the wood of Ugolev B.A., Latishenko V.A. the deformation rate is degenerated due to the difference core. The Boltzmann principle distorts the essence of non-stationary wood model: one classical wood creep body it replaces by the chain model of sequentially connected bodies with set of erroneous properties.

In the theory of wood creep there is a case when at a difference core of integral equation, the broad interpretation of ductility function is unacceptable. For example, the creep core is represented as (second case):

$$k(t-\tau)=\frac{Ae^{-n(t-\tau)}}{(t-\tau)^{\propto-1}}.$$

This kinematic equation of motion, in relation with the expansion of the reverse problem of mechanics, corresponds to certain forces. Here, from the analysis of the differential equation of creep it is found out that in this core there is a force of resistance with coefficient of viscosity, equal $\eta(t,\tau')=A_1(t-\tau)^{\propto-1}$, that it is impossible within the meaning of the coefficient of viscosity in the frames of resistance forces of Newton mechanics.

Case III. Corresponds to the broad interpretation of the ductility function in the '' chain model ''. It is presented in the theoretical rheology, and as a repetition in the theory of wood creep (Schanzlin J., Toratti T., Becker P., Martensson A., Hanhijarvi A. and others).

Let us pre-write the Boltzmann scheme for the Maxwell body:

$$\varepsilon(t)=\sigma_0\left[\frac{1}{E_0}+\frac{1}{\eta}(t-t_0)\right]+\int_{t_0}^{t}\left[\frac{1}{E_0}+\frac{1}{\eta}(t-\tau)\right]d\sigma(\tau), \quad (14)$$

where η is a stationary viscosity coefficient.

In the '' chain model '', by serial connection of the bodies (12) and (14), we have expansive record of ductility function:

$$\sigma(t-\tau)=\frac{1}{E_0}+\frac{1}{E_0}\varphi(t-\tau)+\frac{1}{\eta}(t-\tau)\ , \quad (15)$$

A pair of integral equations corresponding to the expansive scheme (15) and permitted either in relation to deformation or relative to strains in theoretical rheology are called '' equations of Boltzmann-Volterra ''; It is also noted that this pair '' represents the full mathematical formulation of the principle of linear superposition ''.

However, such '' chain model '' with its broad interpretation of the ductility coefficient, is essentially erroneous. This is demonstrated by bringing it to differential form:

$$\ddot{\varepsilon}(t)\frac{\eta}{n}+\dot{\varepsilon}(t)\eta=\ddot{\sigma}(t)\frac{\eta}{E_0 n}+\dot{\sigma}(t)\left(\frac{\eta}{E_0}+\frac{1}{n}+\frac{\eta}{c_0}\right)+\sigma(t), \quad (16)$$

where $c_0=\frac{b}{r-n}$.

From (16) it can be seen that there is a force of resistance $\ddot{\varepsilon}(t)\frac{\eta}{n}$ proportional to acceleration which is incompatible with classical mechanics, and, in connection with article 5.1.1 (3) P EN 1990, '' chain model '' is inconsistent to calculation model.

Components of force of the calculation model can be function from position $\varepsilon(t)$, rate $\dot{\varepsilon}(t)$, time and other variables. If there is (among others) force proportional to acceleration $\ddot{\varepsilon}(t)$, the fundamental principle of mechanics about independence of actions of forces turns out to be violated. The well-known scientist Pars L. found unacceptability of such forces both in tasks of mechanics, and in applications [7].

3. Conclusion

'' Chain models '' are often used to investigate the wood creep, beginning with the publication of Burgers [8], complemented the classic Kelvin scheme with a viscous element, which led to the occurrence of resistance forces proportional to acceleration. The same defect is found in investigations of wood, which complement classic Voigt scheme with viscous element.

In case of non-linear creep and short-term non-linearity of wood, the application of the Boltzmann scheme is also erroneous.

References

[1] Boltzmann L. Zur Theorie der elastischen Nachwirkung. – *Sitzungsberichte Kaiserliche Akademie wissenhaft Wien mathematische-Naturwissenhaft*, 1874, B. **70**, H. **2**, p. 275-305.
[2] Boltzmann L. Zur Theorie der elastischen Nachwirkung. – *Annalen der Physik und Chemie*, 1876, Bd. **7**, H. **4**, p. 624-655.
[3] Prokopovich I E, Zedgenidze V A. Prikladnaja Teorija polzuchesti. – M., 1980. – 239 p. (In Russian)
[4] Orlovich R B. Dlitel'naja prochnost ' i deformativnost ' konstrukcij iz sovremennyh drevesnyh materialov pri osnovnyh jekspluatacionnyh Vozdejstvijah: Avtoref. Diss. D-Ra. Tehn. Nauk- L., 1991-50 p. (In Russian)
[5] Ugolev B N. Deformativnost ' Drevesiny i naprjazhenija pri sushke. – M.: Lesnaja promyshlennost ', 1971. – 176 p. (In Russian)
[6] Latishenko V A. Diagnostika zhestkosti I prochnosti Materialov. – Riga: Znanie, 1968. – 320 p. (In Russian)
[7] Pars L. Analiticheskaja Dinamika.-M., Nauka, 1971.0-636 p.
[8] Burgers J.M. First report on viscosity and plasticity. – Amsterdam: N. V. Noord-Hollandsche, 1935.-256 p.

Designing of the digital casting process for the gas turbine engine blades with a single-crystal structure

R A Vdovin[1, 3], V G Smelov[1], V S Sufiiarov[2], E V Borisov[2]
[1] Samara University, 34, Moskovskoe sh., Samara, 443086 Russia
[2] National Technology Initiative Center of Excellence in Advanced Manufacturing Technologies at Peter the Great St. Petersburg Polytechnic University, 29, Polytechnicheskaya, St. Petersburg, 195251 Russia

[3] E-mail: Vdovin.ssau@gmail.com

Abstract. The work represents the approaches for designing the digital casting process of turbine blades. The development of the mathematical model for determining the non-stationary temperature distribution in the "furnace-moulding-flask-filler-metal-stopper" process system has appeared to be the core of the digital blade manufacturing process considering particular structural, geometry and thermal-physical features of this system. The approbation of the developed mathematical model as well as the computer designing methods under the pilot production conditions through manufacturing of the GTE turbine blades has demonstrated the decreased amount of defects in the macrostructure of blanks of turbine blades with the longitudinally-oriented and single-crystal structure by 30% in total.

1. Introduction

Presently, one of the most important features of engine building is the enhanced reliability of gas turbine engines. Since the reliability of gas turbine engines is, among other things, ensured by process methods, the principal efforts shall be directed to improving the production processes for manufacturing the main motor parts. First of all, this refers to the manufacturing technologies of working path parts, such as turbine blades, since they are the most critical parts of the engine experiencing the maximum loads and conditioning the engine service life.

Due to a complex spatial shape and the absence of extended surfaces for the accurate location of the blades, the most effective and, in some cases, the only method to manufacture the blade blanks is the airfoil non-tolerated method for manufacturing the blade castings by the single-crystal casting method and a subsequent machining of the root area and shroud platform.

The overview of the existing domestic processes for the manufacture of gas turbine engine (GTE) blades [1-6] allowed establishing that the manufacture of non-tolerance turbine blade blanks is extremely unstable and is characterized by a high percentage of defects (approximately 50%). In the first turn, it is related to the instability of process conditions. A considerable part of the total amount of defects is formed by the inconsistency in microstructure and geometric parameters of the airfoil, whose reasons are the non-stationarity of temperature conditions as well as the imperfection of the local airfoil polishing process.

The existing process solutions aimed at reducing the defects during the manufacture of the GTE turbine blades is currently at a sufficiently high level of elaboration. However, their common deficiency is a weak utilization of the production computing aids. This deficiency is caused by a number of reasons: 1) low grade of convergence of the computer simulation results and the results of full-scale experiments at different options of control of process parameters; 2) the absence of the methods for determining the optimal process parameters at different turbine blade manufacturing stages; 3) the absence of a complex simulation twin-model for multi-optional numerical studies of the manufacturing process of turbine blades.

The work is aimed at improving the quality, performance and reducing the level of defects in manufacturing the blanks of GTE turbine blades obtained by the method of directional crystallization of single-crystal castings due to the elaboration of the finite-element model of the process.

The objectives of this work are:

- development of the mathematical model for determining the non-stationary temperature distribution in the "furnace-moulding-flask-filler-metal-stopper" process system considering particular structural, geometry and thermal-physical features of this system;
- development of the finite-element simulation method for the manufacturing process of GTE turbine blade blanks;
- checking the fitness of the developed mathematical model and method for designing the manufacturing process by comparing the results (casting structure parameters, presence and location of casting defects) with full-scale experiments during pilot production of the turbine blades of the 1st stage GTE.

2. Description of the model

2.1 Geometric model

At present, turbine blades are manufactured at the industrial plants by casting with directional crystallization of heat-resistant alloys at PMP-2 units, UVNK high-speed directional crystallization units and VIP-NK vacuum smelting units. The work considers the manufacturing process for blades at the PMP-2 unit.

The geometric model of the process diagram designed in the NX CAD software package, is given in figure 1 a. The process diagram comprises a graphite moulding flask, a graphite filler for positioning the ceramic mould, the stopper for tapping the molten metal at a definite moment of time, a plug to add rigidity to the process system, blade blanks and seed.

Finite element model [7] of the process system was designed in the MSC Apex Modeler software and in the Visual-Mesh module of the ProCast software product, allowing to obtain the mesh reliably coupled with the blade casting lattice [8].

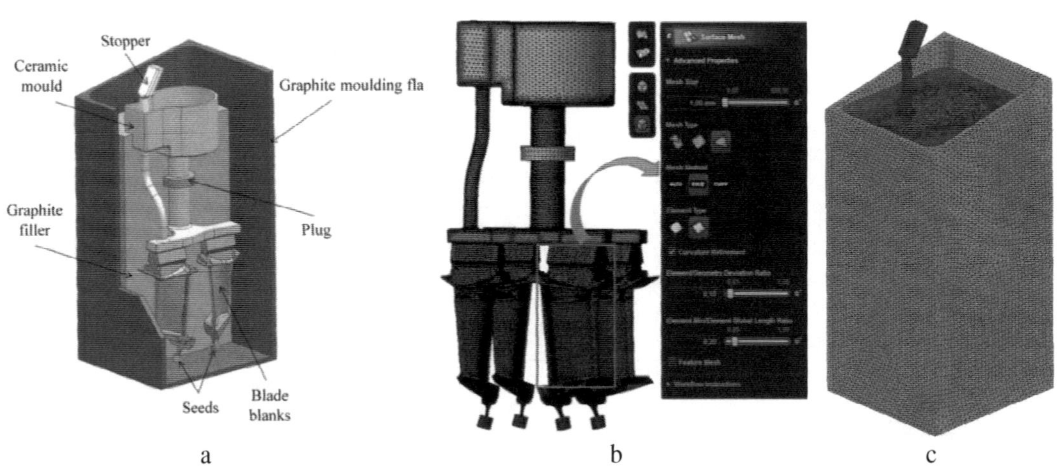

Figure 1. Geometric model of process system: a – 3D-model of turbine blade casting and casting block; b – finite element model of the gating and feeding system (LPS) and turbine blade casting;

c – finite-element model of casting block.

A characteristic dimension of the element in the blade blank is 1 mm; in perforation (cooling channels) - 0.5 mm; at the blade root (edge) - 0.3 mm; in the gating system - 2 mm; in the pouring cap and central feeder - 3 mm. The computational region consisted of 3,471,693 nodes and 17,771,077 elements.

2.2 Mathematical model

This section is devoted to the development of the mathematical model for determining the non-stationary temperature distribution in the process system of "furnace-moulding-flask-filler-metal-stopper" considering particular structural, geometric and thermal-physical features of this system. It was established that the main control effect may be produced on the process through changing: the power of heaters, temperature conditions of each of the zone of PMP-2 unit, the moving speed of the flask train in the unit (pushing cycle). For manufacturing of turbine blade blanks considered in the process, the heat exchange process is the most relevant for defining the blade single-crystal structure quality.

In the volume of the material of solid bodies furnace-moulding flask-filler-metal-stopper, the heat transfer processes along the entire range of temperatures are described by a classic heat-transfer equation (1):

$$c(T)\rho(T)\frac{\partial T}{\partial t} = div(\lambda(T)\, gradT) \tag{1}$$

where $T(t,\vec{r})$ – temperature at the moment of time t in the point with coordinate $\vec{r}$ inside of each body of the system; $c(T)$, $\rho(T)$ and $\lambda(T)$ – specific heat capacity, density and thermal conductivity factor of the materials, different for each system.

For the metal weight, considering that during heating the material should be transferred to the liquid phase, the heat conductivity equation shall be represented in the following form (2) considering the possibility of its smelting in the solidus and liquidus temperature range:

$$c_{эф}(T)\rho(T)\frac{\partial T}{\partial t} = div(\lambda(T)\, gradT) \tag{2}$$

where the parameter $c_{эф}$ is determined from the equation (3):

$$c_{эф} = c(T) - \frac{L}{T_L - T_S}\varphi\left(\frac{T-T_S}{T_L-T_S}\right)\left[\eta\left(\frac{T-T_S}{T_S}\right) - \eta\left(\frac{T-T_L}{T_L}\right)\right] \tag{3}$$

where T_L and T_S – solidus and liquidus temperature of the alloy; $\varphi((T-T_S)/(T_L-T_S))$ – the alloy melting rate expressed in the effective heat capacity using the Heaviside function $\eta(v)$, equal to zero at $v<0$ and set to unity at $v\geq 0$ (here $v=(T-T_S)/T_S$ or $v=(T-T_L)/T_L$, the body volume region $T_S \leq T(t,\vec{r}) \leq T_L$ is explicit, in which the heat is adsorbed during the metal smelting).

To solve these heat conductivity equations, the initial temperature for all the bodies shall be set, which is uniform and equal to the temperature in the shop along with setting the limiting conditions at all the surfaces of the model depending on the time for each zone of material furnace.

The boundary conditions for surfaces heated in the PMP-2 vacuum furnace from thermal heaters or surfaces of other bodies shall be determined by heat radiation and have the form (4):

$$\lambda\frac{\partial T}{\partial n} = \varepsilon\sigma\left[T^4 - T_C^4\right], \text{ at } r = r_{INT} \tag{4}$$

where n – external normal to the considered surface; ε – heat adsorption factor of the surface; σ – Stefan-Boltzmann constant; T_C – temperature of the surface which radiation exchange is effected with;

r_{INT} – body surface point coordinate. The system of equations for each body forms a closed system of equations relative to the temperature of each body.
The primary factors determining the quality of turbine blade manufacturing process are the operating temperature conditions in each of the zones of the casting unit and the speed of movement of the moulding flask train (by changing the induction motor power).

3. Boundary and initial conditions

3.1. Initial conditions

The manufacturing process for single-crystal turbine blades of heat-resistant nickel alloys takes place in the modernized PMP-2 through-type continuous furnace. From the thermal point of view, this installation represents a conduit with the flasks moving in it. On both sides of the flask train, flat heaters are located. A thermal node of the furnace consists of 8 conditional zones, forming 3 regions: heating region (zone I, II); melting region (zone III, IV) and the crystallization region (zones VI-IX). Each zone is characterized by a certain temperature, which is set by the individual power system. At the interface region of zones IV and VI (in a strictly defined point), the mechanical stopper device is activated allowing to open the runner with metal tapping to the form in front of the crystallizer.
The following elements were included in the design region of the turbine blade casting process: 1st stage blade casting model at a minimum wall thickness of 1 mm; a ceramic rod; a seed; a plug; a ceramic mould with the thickness of 10 mm; graphite filler; a stopper; a graphite flask [9, 10, 11].
The following initial process parameters were used:

- Smelting-pouring of ZhS30-VI alloy in the PMP-2 batch furnace;
- Stucco material (of a ceramic mould) is aluminium oxide Al_2O_3;
- Pouring time – 4-6 sec;
- Pouring temperature – 1510 ^{0}C;
- Temperature conditions by PMP-2 furnace zones;

Zone I 1290 +10 ^{0}C; zone VI 1510 +10 ^{0}C;
Zone II 1370 +10 ^{0}C; zone VII 1450 +10 ^{0}C;
Zone III 1400 ±10 ^{0}C; zone VIII 1390 +10 ^{0}C;
Zone IV 1510 +10 ^{0}C; zone IX 1370 +10 ^{0}C.
The temperature fields obtained from the solution of the problem of initial heating of the graphite flask were used as the initial thermal conditions for all elements of the computational region.

3.2 Boundary condition and loading.

Using the User Function module of the ProCast software product enables effecting flexible control of boundary conditions in time and space, i.e. changing the radiation exchange conditions in the heating region initially set at the surface of the graphite flask by the convective heat exchange conditions in the crystallization region and thus simulating the movement of the flask train in the body of the PMP-2 furnace.
The furnace body consisting of 8 zones, was divided into two identical parts, for each of which the user-defined functions were written in the C ++ programming language. The geometric dimensions of the furnace zones were replaced by time values through a constant flask moving speed (0.25 mm/s). To simplify the displaying of the boundaries of the elements of the process system, the adjacent zones of the graphite flask, graphite filler and ceramic form will be considered as a single whole.
In designing of user functions, much attention was paid to heat exchange of the furnace-moulding-flask-filler-metal-stopper process system with the external environment. In this case, heat exchange occurs in a vacuum, therefore, the appropriate values of the emissivity factor and temperature shall be set on the corresponding surfaces of the bodies of the process system (figure 2). In addition, the heat transfer factor was set at the boundary "ceramic mould - seed" along with all the necessary thermophysical properties of the ZhS30-VI alloy.

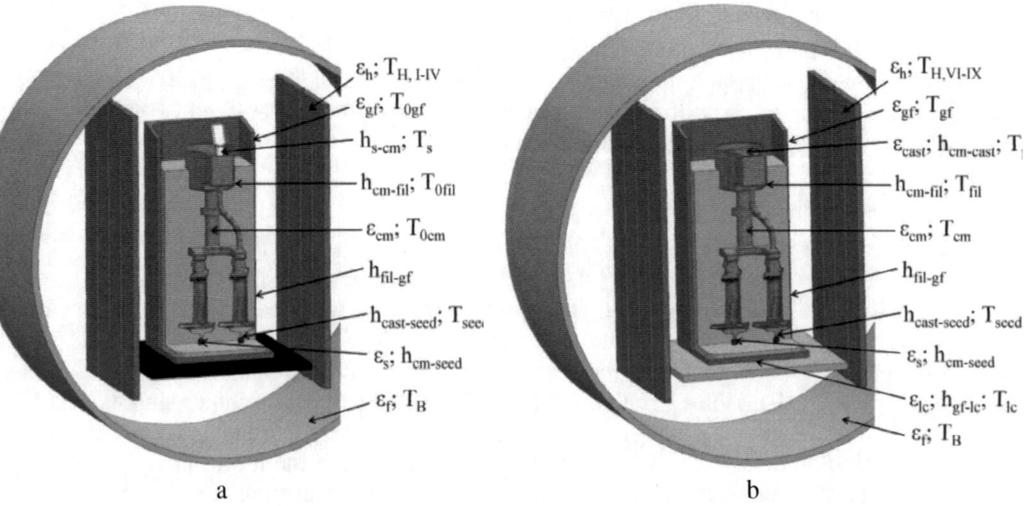

Figure 2. Boundary and initial conditions of the furnace-moulding flask-filler-metal-stopper process system: a – during heating of ceramic mould and graphite flask; b – during pouring and block cooling

ε_h – emissivity factors of heaters; ε_f – emissivity factor of PMP–2 furnace; ε_{gf} – emissivity factor of graphite flask; ε_{cm} – emissivity factor of ceramic mould; ε_s – emissivity factor of seed; ε_{cast} – emissivity factor of the casting; ε_{lc} - emissivity factor of liquid crystallizer; $T_{H,I-IV}$ – temperature of heaters in zones I–IV, ^{0}C; $T_{H,VI-IX}$ – temperature of heaters in zones VI–IX, ^{0}C; T_B – PMP-2 furnace body temperature, ^{0}C; T_{0gf} – initial temperature of graphite flask, ^{0}C; T_{gf} –temperature of graphite flask in crystallization zone, ^{0}C; T_{0fil} – initial temperature of graphite filler ^{0}C; T_{fil} – temperature of graphite filler, ^{0}C; T_{0cm} – initial temperature of ceramic mould, ^{0}C; T_{cm} – temperature of ceramic form, ^{0}C; T_s – stopper temperature, ^{0}C; T_{seed} – seed temperature, ^{0}C; T_p – pouring temperature, ^{0}C; T_{lc} – liquid crystallizer temperature, ^{0}C; v_p – average speed of pouring of molten metal in ceramic mould, mm/s; v_t – flask train movement speed, mm/s; h_{cm-fil} – heat transfer factor at the boundary between the ceramic mould and graphite filler, W/(m^2-K); $h_{cm-seed}$ – heat transfer coefficient at the boundary between the ceramic mould and the seeds, W/(m^2-K); $h_{cm-cast}$ –heat transfer factor at the boundary between the ceramic mould and casting, W/(m^2-K); h_{fil-gf} – heat transfer factor at the boundary between the graphite filler and the graphite flask, W/(m^2-K); $h_{cast-seed}$ – heat transfer factor at the boundary between the casting and seed, W/(m^2-K); h_{s-cm} – heat transfer factor at the boundary between the stopper and the ceramic mould, W/(m^2-K); h_{gf-lc} – the heat transfer factor at the boundary between the graphite flask and the liquid crystallizer, W/(m^2-K).

4. Results and discussion

4.1 Process system heating

The numerical solution of the mathematical model in the three-dimensional setting was carried out with the help of the finite-element method of the ProCast software product, since this method is capable of describing the complex geometry of castings and generators of ceramic forms to a more accurate degree. Figure 3 represents the simulation results for the process system heating.

The process system is heated by flat heaters located in the installation on both sides relative to the flask train. The heating process is uniform, the temperature front is formed on the outer surfaces of the graphite flask with spreading to the centre of the process system, namely to its bottom part, which allows concluding on the directed nature of heating. Along with the heating of the furnace-moulding flask-filler-metal-stopper process system, melting of the charge of the heat-resistant metal occurs.

These two interrelated processes end at the end of the melting region, before the crystallization area approximately in 7 hours after the start of the process cycle. At this point, a mechanical stopper device is triggered; the runner is opened with tapping the molten metal into the ceramic mould.

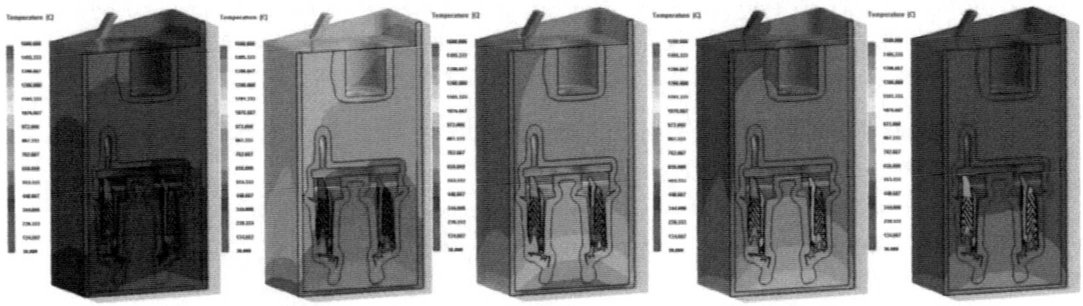

t=9540 sec. t=12260 sec. t=15660 sec. t=191000 sec. t=25000 sec.

Figure 3. The results of numerical simulation of the process system heating.

4.2 Filling a ceramic mould with the molten metal

The results of numerical simulation for heating of the process system were used as initial conditions for simulating the problem of filling a ceramic mould with molten metal (figure 4). The analysis of the results of simulation of the melt flow showed the inefficiency of the runner-feeding system, connected with separation of the metal flow into two parts, as well as the non-simultaneous filling and, as a consequence, crystallization of the castings, which eventually would adversely affect the macrostructure of the blade.

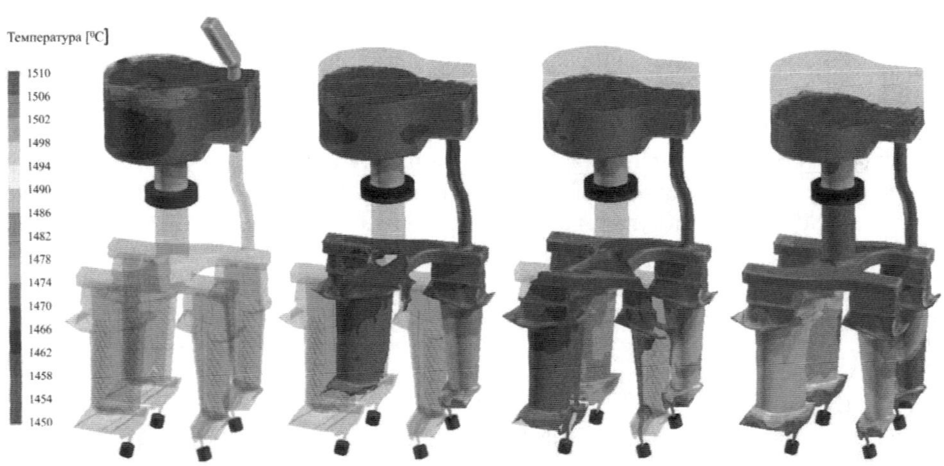

t=4.8 sec. t=12.8 sec. t=20.9 sec. t=27.1 sec.

Figure 4. The results of filling a ceramic mould with molten metal.

4.3 The casting macrostructure simulation results

The computer research of the process of macrostructure formation in the blade casting was performed using the CAFÉ module of the ProCast software product, in which the results of the solved heat problem by finite element method were interpolated for performing the calculations by the cellular automata method. Finally, the result of computer simulation for macrostructure in the blade casting were compared with the structure of a real turbine blade subjected to chemical etching (figure 5). It should be noted that a series of numerical experiments on determining the macrostructure of blade casting were carried out with the selected optimal grain size growth parameters: ΔT_N = 55 ^{0}C, ΔT_σ = 10 ^{0}C and n_{max} =4.7·10^7 m^{-3}.

While analysing the results, it should be noted that the macrostructure obtained by the computer simulation method qualitatively coincides with the macrostructure of the real blank of the turbine blade obtained in the course of chemical etching. Characteristic columnar grains are distinguished in the blade airfoil region. The availability of parasitic grains (figure 5, a) as opposed to the macrostructure of real castings (figure 5, b), may be explained by the use of a complex mathematical model [12] in the CAFÉ calculation module of the ProCast software product and, as a result, a computer simulation error.

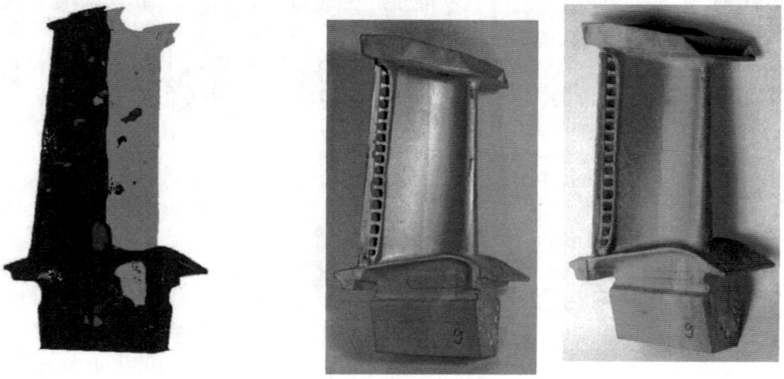

a b

Figure 5. Comparison of the macrostructure of a turbine blade obtained in the course of numerical simulation with real casting: a – macrostructure obtained after computer simulation; b – macrostructure obtained after chemical etching.

5. Conclusion

A cycle of numerical experiments of the process for manufacturing the turbine blade castings (table 1) showed that at the pouring temperature of 1540 ^{0}C, graphite flask pushing cycle of 12 min and the metal running cycle of 4, high quality of blade castings with regard to the macrostructure may be achieved in the PMP-2 discrete continuous unit with regard to the macrostructure at a minimum percentage of defects.

Table 1. The proposed process conditions for the PMP-2 furnace

Zone I, ^{0}C	Zone II, ^{0}C	Zone III, ^{0}C	Zone IV, ^{0}C	Zone VI, ^{0}C	Zone VII, ^{0}C	Zone VIII, ^{0}C	Zone IX, ^{0}C	Defect reduction, %
1150	1330	1460	1460	1450	1440	1380	1310	2
1200	1270	1480	1500	1500	1425	1420	1360	3
1320	1400	1470	1500	1500	1480	1410	1380	4
1200	1265	1470	1510	1510	1420	1400	1340	7
1210	1290	1485	1540	1540	1430	1420	1370	15
1350	1450	1510	1540	1540	1490	1420	1360	30

Employing the developed user-defined functions allowed obtaining the dependence of the migration speed of graphite flasks in the PMP-2 furnace on time based on solving of the temperature problem (figure 6). Using the obtained dependence in the real production conditions at the enterprise allowed to reduce the defects in manufacturing the 1st stage GTE turbine blade blanks by 30%.

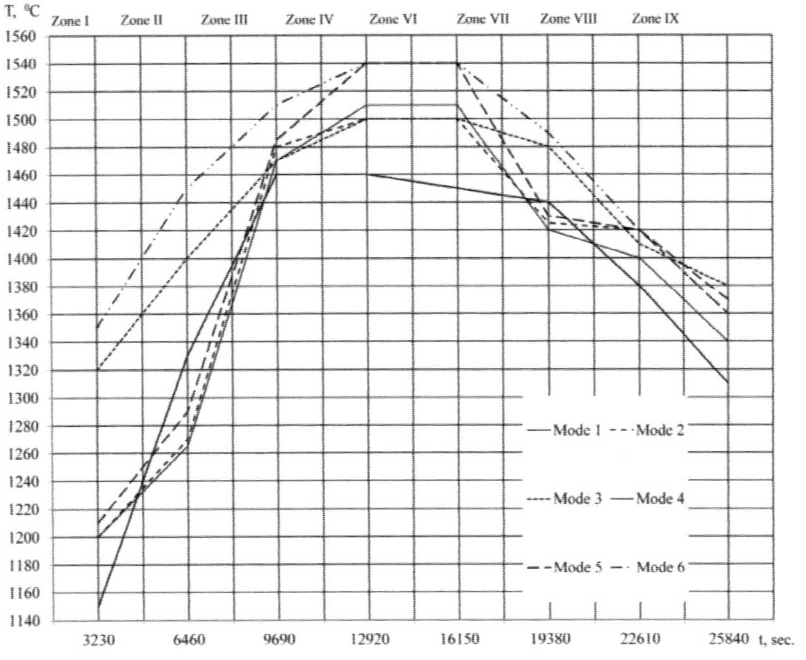

Figure 6. Calculated time-temperature relationships in the PMP-2 unit zones.

References

[1] Szeliga D P, Burbelko A, Motyka M 2014 Modeling of directional solidification of columnar grain structure in CMSX-4 nickel-based superalloy. Materials Engineering and Performance. **23 (3)** 1088–1095

[2] Szeliga D P, Kubiak K, Sieniawski J 2015 The unidirectional crystallization of metals and alloys (turbine blades). The faculty of mechanical engineering and aeronautics department of materials science (Rzeszow University of Technology, Polish)

[3] Smirnov G V, Pronichev N D and Nekhoroshev M V 2017 Effect of technological heredity on the fatigue strength in the manufacture of gas turbine engine blades. IOP Conference Series: Materials Science and Engineering **177 (1)**
Takeshi N, Takeshi O and Kuniyuki I 2014 Development of CMC turbine parts for aero engines. Engineering Review **47 (1)** 29–32

[4] Vdovin R A, Balyakin, A V, Dobryshkina, E M and Alekseev, V P 2018 Rapid Prototyping Technology for Manufacturing GTE Turbine Blades. IOP Conference Series: Materials Science and Engineering **327**

[5] Gell M 1980 The development of single crystal superalloy turbine blades. Superalloys **1** 205–214

[6] Khaimovich A I and Skuratov D L 2017 New damage criteria for machining considering contact friction for FE analysis. Key Engineering Materials **736** 147-152

[7] Vdovin R A and Smelov V G 2016 Research and optimization of the technological process of manufacturing a GTE blades using computer-aided design. IOP Conference Series: Materials Science and Engineering **156 (1)**

[8] Vdovin R A and Smelov V G 2017 Design and optimization of the micro-engine turbine rotor manufacturing using the rapid prototyping technology. IOP Conference Series: Materials Science and Engineering **177 (1)**

[9] Vdovin R A, Smelov V G, Pronichev N D, Bolotov M A 2016 Paths of improving the technological process of manufacture of GTE turbine blades. Innovative Technologies in Engineering. **142** 1–8

[10] Agapovichev A V, Balaykin A V and Smelov V G 2015 Production technology of the internal combustion engine crankcase using additive technologies *Modern Applied Science*. **9** **(4)** 335-343

[11] Kostyshev V A and Khaimovich A I 2015 Analytical modeling of the microstructure evolution of titanium alloys during high speed stamping of the blades of gas turbine engines *Izvestiya VUZ. Tsvetnaya Metallurgiya*. **56** **(2)** 49-56

Ergodic approximation in the theory of binary molecular solutions

Zakharov A.Yu.[1], Schneider A.A[1], Zavorotnev Yu.D.[2], Metlov L.S.[2]

[1] Yaroslav-the-Wise Novgorod State University
173003, Russia, Veliky Novgorod, B. St.-Petersburgskaya st., 41
[2] Donetsk Institute for Physics and Engineering named after A.A.Galkin
83114, Ukraine, Donetsk, Rosa Luxemburg st., 72

E-mail anatoly.zakharov@novsu.ru

Abstract. An exact representation of the classical partition function of a two-component solution with interatomic potentials of a general form via a functional integral is obtained. The ergodic Weyl theorem (ergodic approximation) is used to calculate the functional integral. It is obtained an expression for the Helmholtz free energy via Fourier transforms of interatomic potentials, including renormalization in the case of their singularity. It is shown that, in order to correctly describe the separation of binary solutions into phases, it is nesessary to take into account the ergodic contribution to the Helmholtz free energy. The necessary condition of the phase separation in binary solutions is established.

1. Introduction

Statistical thermodynamics is the theoretical basis for investigating the thermodynamic properties of solutions at the present time. Within statistical thermodynamics the problem of finding thermodynamic functions reduces to calculation of the configuration integral of a system of interacting particles

$$Z_{\mathrm{conf}} = \int_{(V)} \ldots \int \left[\prod_{\alpha=1}^{M} \prod_{s=1}^{N_\alpha} \frac{d\mathbf{R}_s^\alpha}{V} \right] \exp\left[-\beta W\left(\ldots, \mathbf{R}_s^\alpha, \ldots\right) \right], \qquad (1)$$

where $\beta = 1/T$ is reciprocal temperature T in energy units (the Boltzmann constant is equal to 1), $W(\ldots, \mathbf{R}_s^\alpha, \ldots)$ is the energy of interactions of all particles, depending on their coordinates $\mathbf{R}_s^\alpha$, M is the number of components in the system, V is the system volume, index s numbers the particles, Greek indices α, γ are components indexes, N_α is total number of particles of component α. The configuration integral in the formula (1) is normalized to unity, i.e. in the absence of interactions it is equal to 1.

An exact analytic calculation of the configuration integral even for one-component system has by now been performed only for an extremely limited set of model potentials that are very far from "real" potentials [1]. Note that among the exactly solved models there are no three-dimensional models. Estimates of the errors of existing approximate methods for calculating configurational integrals also do not exist.

Most theoretical studies of solutions are based on phenomenological models [2, 3, 4, 5]. In these models some form of dependence of thermodynamic functions on thermodynamic variables (temperature, volume, pressure, composition of the system) is postulated. Interatomic interactions are accounted by means of appropriate empirical parameters.

The functional integration method for calculating the configuration integral of one-component systems with a central interatomic potential admitting the Fourier transform was proposed in papers [6, 7, 8, 9]. This method was further developed in [10, 11]. In particular, the equations of state of one-component systems with model potentials are derived in these papers and the equilibrium between the liquid and gas phases is investigated. A good agreement with the experimental data was obtained.

The purpose of this paper is to generalize the method of functional integration to the case of two-component systems with two-body central interatomic potentials of a general type and the construction of a microscopic theory of binary molecular solutions.

2. Model

We consider a two-component system characterized by two-body interatomic potentials $v_{\alpha\gamma}\left(\mathbf{R}_s^\alpha - \mathbf{R}_{s'}^\gamma\right)$. The expression for the potential energy W of a system of interacting particles is the following

$$W = \frac{1}{2}\sum_{\alpha,\gamma=1}^{2}\left\{\sum_{s=1}^{N_\alpha}\sum_{s'=1}^{N_\gamma} v_{\alpha\gamma}\left(\mathbf{R}_s^\alpha - \mathbf{R}_{s'}^\gamma\right)\right\} - \frac{1}{2}\sum_{\alpha=1}^{2} N_\alpha v_{\alpha\alpha}(0), \tag{2}$$

where the value $-\frac{1}{2}\sum_{\alpha=1}^{2} N_\alpha v_{\alpha\alpha}(0)$ compensates the terms of self-action, i.e. the summands with coincident indices $s = s'$ in the interaction energy of identical atoms ($\alpha = \gamma$).

Suppose that the interatomic potentials $v_{\alpha\gamma}\left(\mathbf{R}_s^\alpha - \mathbf{R}_{s'}^\gamma\right)$ allow a representation in the form of the Fourier integral

$$v_{\alpha\gamma}\left(\mathbf{R}_s^\alpha - \mathbf{R}_{s'}^\gamma\right) = \frac{1}{V}\sum_{\mathbf{k}\in\Omega} \tilde{v}_{\alpha\gamma}(\mathbf{k}) e^{i\mathbf{k}\cdot\mathbf{R}_\mathbf{s}^\alpha} e^{-i\mathbf{k}\cdot\mathbf{R}_{\mathbf{s}'}^\gamma} + \frac{1}{V}\tilde{v}_{\alpha\gamma}(0). \tag{3}$$

The summation over $\mathbf{k}$ in the right side of this relation is fulfilled over the set of all wave vectors Ω except for the vector $\mathbf{k} = 0$: this summand is taken into account by the last term in formula (3).

Let us note that the condition for the existence of a thermodynamic limit imposes certain restrictions on the interatomic potentials [12, 13, 14, 15]. For one-component systems with interatomic potential $v_{\alpha\alpha}(\mathbf{r})$ this condition has the form [16]

$$\tilde{v}_{\alpha\alpha}(\mathbf{k}) \geq 0, \tag{4}$$

where $\tilde{v}_{\alpha\alpha}(\mathbf{k})$ is the Fourier transform of the interatomic potential. For binary systems, the condition of non-negativity of the diagonal elements of the matrix $\tilde{v}_{\alpha\gamma}(\mathbf{k})$ should be supplemented by the following condition:

$$\det \tilde{v}(\mathbf{k}) \geq 0. \tag{5}$$

Interatomic potentials that do not satisfy conditions (4), (5) are called *catastrophic* or *unstable*, since there is no thermodynamic limit $N \to \infty$, $V \to \infty$, $N/V = n = \text{const}$ for them. Therefore, we assume that the Fourier transforms of the interatomic potentials satisfy the conditions of noncatastrophicity (4), (5).

By summation over s and s' in the expression contained in the curly brackets of formula (2), the sums of the following type occur after substituting relation (3)

$$\tilde{\rho}_\alpha(\mathbf{k}) = \sum_{s=1}^{N_\alpha} e^{i\mathbf{k}\cdot\mathbf{R}_s^\alpha} = \sum_{s=1}^{N_\alpha}\left[\cos\left(\mathbf{k}\cdot\mathbf{R}_s^\alpha\right) + i\sin\left(\mathbf{k}\cdot\mathbf{R}_s^\alpha\right)\right] = C_\alpha(\mathbf{k}) + iS_\alpha(\mathbf{k}). \qquad (6)$$

These variables (collective coordinates), $\tilde{\rho}_\alpha(\mathbf{k})$, as well as $C_\alpha(\mathbf{k})$ and $S_\alpha(\mathbf{k})$ were first introduced in [17]. They are related to the microscopic density of the number of particles by the Fourier transform and are often used in statistical mechanics since the 1950s [6, 18, 19, 20]

$$\rho_\alpha(\mathbf{r}) = \sum_{s=1}^{N_\alpha}\delta\left(\mathbf{r} - \mathbf{R}_s^\alpha\right). \qquad (7)$$

We represent the expression in the curly brackets of the formula (2) in terms of the variables $C_\alpha(\mathbf{k})$ and $S_\alpha(\mathbf{k})$:

$$\sum_{s,s'} v_{\alpha\gamma}\left(\mathbf{R}_s^\alpha - \mathbf{R}_{s'}^\gamma\right) = \frac{1}{V}\sum_{\mathbf{k}\in\Omega}\tilde{v}_{\alpha\gamma}(\mathbf{k})\left[C_\alpha(\mathbf{k}) + iS_\alpha(\mathbf{k})\right]\left[C_\gamma(\mathbf{k}) - iS_\gamma(\mathbf{k})\right]$$
$$+\tilde{v}_{\alpha\gamma}(0)\frac{N_\alpha N_\gamma}{V} = \frac{1}{V}\sum_{\mathbf{k}\in\Omega}\tilde{v}_{\alpha\gamma}(\mathbf{k})\left[C_\alpha(\mathbf{k})C_\gamma(\mathbf{k}) + S_\alpha(\mathbf{k})S_\gamma(\mathbf{k})\right] + \tilde{v}_{\alpha\gamma}(0)\frac{N_\alpha N_\gamma}{V}. \qquad (8)$$

Let us note that sums of the type

$$\sum_{\mathbf{k}\in\Omega}\tilde{v}_{\alpha\gamma}(\mathbf{k})C_\gamma(\mathbf{k})S_\gamma(\mathbf{k}) = 0, \qquad (9)$$

by virtue of the fact that

$$\tilde{v}_{\alpha\gamma}(\mathbf{k}) = \tilde{v}_{\alpha\gamma}(-\mathbf{k}); \quad C_\alpha(-\mathbf{k}) = C_\alpha(\mathbf{k}); \quad S_\alpha(-\mathbf{k}) = -S_\alpha(\mathbf{k}). \qquad (10)$$

We also note that the relations (10) imply the linear dependence of both the set of variables $C_\alpha(\mathbf{k})$, and the set of variables $S_\gamma(\mathbf{k})$, defined on the space Ω of all wave vectors $\mathbf{k}\in\Omega$. Therefore, in the sequel, the summation over the wave vectors $\mathbf{k}\in\Omega$ will be replaced by summation over the half-space $\mathbf{k}\in\Omega/2$, obtained by intersection of the space Ω by an arbitrary plane passing through the origin of coordinates $\mathbf{k}=0$:

$$\sum_{\mathbf{k}\in\Omega}\ldots \to 2\sum_{\mathbf{k}\in\Omega/2}\ldots. \qquad (11)$$

As a result, we have the following expression for the potential energy of a two-component system of interacting particles via collective variables $C_\alpha(-\mathbf{k})$ and $S_\alpha(-\mathbf{k})$:

$$W = -\frac{N_1}{2}v_{11}(0) - \frac{N_2}{2}v_{22}(0) + \frac{1}{2V}\sum_{\alpha,\gamma=1}^{2}\tilde{v}_{\alpha\gamma}(0)N_\alpha N_\gamma$$
$$+\frac{1}{V}\sum_{\mathbf{k}\in\Omega/2}\left\{\sum_{\alpha,\gamma=1}^{2}\tilde{v}_{\alpha\gamma}(\mathbf{k})C_\alpha(\mathbf{k})C_\gamma(\mathbf{k}) + \sum_{\alpha,\gamma=1}^{2}\tilde{v}_{\alpha\gamma}(\mathbf{k})S_\alpha(\mathbf{k})S_\gamma(\mathbf{k})\right\}. \qquad (12)$$

3. Representation of the partition function by means of the functional integral

To calculate the partition function, it is necessary to perform averaging of the exponent containing the potential energy of the system

$$Z_{\text{conf}} = \int_{(V\cdots V)}\cdots\int\left[\prod_{\alpha=1}^{2}\left(\prod_{s=1}^{N_\alpha}\frac{d\mathbf{R}_s^\alpha}{V}\right)\right]e^{-\beta W}, \qquad (13)$$

where

$$e^{-\beta W} = \mathrm{A}\prod_{\mathbf{k}\in\Omega/2}\exp\left\{-\frac{\beta}{V}\sum_{\alpha,\gamma=1}^{2}\tilde{v}_{\alpha\gamma}(\mathbf{k})\left[C_\alpha(\mathbf{k})C_\gamma(\mathbf{k}) + S_\alpha(\mathbf{k})S_\gamma(\mathbf{k})\right]\right\}, \qquad (14)$$

$$\mathsf{A} = \exp\left[\beta\left(\frac{N_1}{2}v_{11}(0) + \frac{N_2}{2}v_{22}(0) - \frac{1}{2V}\sum_{\alpha,\gamma=1}^{2}\tilde{v}_{\alpha\gamma}(0)N_\alpha N_\gamma\right)\right]. \tag{15}$$

Atomic coordinates are contained only in the variables $C_\alpha(\mathbf{k})$ and $S_\alpha(\mathbf{k})$. Each of the factors of the product with respect to $\mathbf{k}$ on the right-hand side of formula (14) contains exponents of quadratic forms with respect to variables $C_\alpha(\mathbf{k})$ and $S_\alpha(\mathbf{k})$.

We use the Stratonovich-Hubbard transformation, based on the well-known identity

$$\exp\left[-\frac{1}{2}(\mathbf{b}, A\mathbf{b})\right] = \frac{1}{(\det A)^{1/2}(2\pi)^{n/2}} \times \int_{-\infty}\!\!\dots\int \exp\left[-\frac{1}{2}(\mathbf{x}, A^{-1}\mathbf{x}) - i(\mathbf{b},\mathbf{x})\right][dx], \tag{16}$$

where A is arbitrary positive definite square matrix $n\times n$, $\mathbf{b}$ is column vector consisting of n elements, $[dx] = \prod_{j=1}^{n} dx_j$.

Note that the identity (16) expresses the exponential of the quadratic form $(\mathbf{b}, A\mathbf{b})$ on the left-hand side of this formula in terms of the exponent of the linear form $(\mathbf{b},\mathbf{x})$ on the right-hand side. As a result, we have the following exact representation of the quantity $e^{-\beta W}$ through the functional integral

$$\begin{aligned} e^{-\beta W} = \mathsf{A}\int_{-\infty}\!\!\dots\int\Bigg\{&\left(\prod_{\mathbf{k}\in\Omega/2}\frac{dx_1(\mathbf{k})dx_2(\mathbf{k})dy_1(\mathbf{k})dy_2(\mathbf{k})}{(2\pi)^2}\right)\\ &\times\exp\left[-\frac{1}{2}\sum_{\mathbf{k}\in\Omega/2}\left(\sqrt{\det\tilde{v}(\mathbf{k})}\sum_{\alpha,\gamma=1}^{2}x_\alpha(\mathbf{k})[\tilde{v}_{\alpha\gamma}(\mathbf{k})]^{-1}x_\gamma(\mathbf{k})\right)\right]\\ &\times\exp\left[-\frac{1}{2}\sum_{\mathbf{k}\in\Omega/2}\left(\sqrt{\det\tilde{v}(\mathbf{k})}\sum_{\alpha,\gamma=1}^{2}y_\alpha(\mathbf{k})[\tilde{v}_{\alpha\gamma}(\mathbf{k})]^{-1}y_\gamma(\mathbf{k})\right)\right]\\ &\times\exp\left[-i\sqrt{\frac{2\beta}{V}}\sum_{\mathbf{k}\in\Omega/2}\left[(\det\tilde{v}(\mathbf{k}))^{1/4}\sum_{\alpha=1}^{2}x_\alpha(\mathbf{k})C_\alpha(\mathbf{k})\right]\right]\\ &\times\exp\left[-i\sqrt{\frac{2\beta}{V}}\sum_{\mathbf{k}\in\Omega/2}\left[(\det\tilde{v}(\mathbf{k}))^{1/4}\sum_{\alpha=1}^{2}y_\alpha(\mathbf{k})S_\alpha(\mathbf{k})\right]\right]\Bigg\}. \end{aligned} \tag{17}$$

4. Factorization of the configuration integral with respect to atomic coordinates

To calculate the configuration integral, it is necessary to integrate expression (17) over all atomic coordinates $\mathbf{R}_s^\alpha$. Only the factors in the last two lines of this expression depend on atomic coordinates. Integrating these lines with respect to atomic coordinates, we obtain

$$
\begin{aligned}
&\int\cdots\int\left(\prod_{s_1=1}^{N_1}\frac{d\mathbf{R}_{s_1}}{V}\right)\left(\prod_{s_2=1}^{N_2}\frac{d\mathbf{R}_{s_2}}{V}\right)\\
&\times\exp\left\{-i\sum_{s_1=1}^{N_1}\sum_{\mathbf{k}\in\Omega/2}\sqrt{\frac{2\beta}{V}}(\det\tilde{v}(\mathbf{k}))^{\frac{1}{4}}\left[x_1(\mathbf{k})\cos(\mathbf{k}\mathbf{R}_{s_1})+y_1(\mathbf{k})\sin(\mathbf{k}\mathbf{R}_{s_1})\right]\right\}\\
&\times\exp\left\{-i\sum_{s_2=1}^{N_2}\sum_{\mathbf{k}\in\Omega/2}\sqrt{\frac{2\beta}{V}}(\det\tilde{v}(\mathbf{k}))^{\frac{1}{4}}\left[x_2(\mathbf{k})\cos(\mathbf{k}\mathbf{R}_{s_2})+y_2(\mathbf{k})\sin(\mathbf{k}\mathbf{R}_{s_2})\right]\right\}\\
&=\left\{\int_{(V)}\frac{d\mathbf{R}_1}{V}\exp\left[-i\sum_{\mathbf{k}\in\Omega/2}\sqrt{\frac{2\beta}{V}}(\det\tilde{v}(\mathbf{k}))^{\frac{1}{4}}\left[x_1(\mathbf{k})\cos(\mathbf{k}\mathbf{R}_1)+y_1(\mathbf{k})\sin(\mathbf{k}\mathbf{R}_1)\right]\right]\right\}^{N_1}\\
&\times\left\{\int_{(V)}\frac{d\mathbf{R}_2}{V}\exp\left[-i\sum_{\mathbf{k}\in\Omega/2}\sqrt{\frac{2\beta}{V}}(\det\tilde{v}(\mathbf{k}))^{\frac{1}{4}}\left[x_2(\mathbf{k})\cos(\mathbf{k}\mathbf{R}_2)+y_2(\mathbf{k})\sin(\mathbf{k}\mathbf{R}_2)\right]\right]\right\}^{N_2}.\\
&=(F_1)^{N_1}(F_2)^{N_2},
\end{aligned}
\tag{18}
$$

where

$$F_\alpha=\int_{(V)}\frac{d\mathbf{R}_\alpha}{V}\exp\left[-i\sum_{\mathbf{k}\in\Omega/2}\sqrt{\frac{2\beta}{V}}(\det\tilde{v}(\mathbf{k}))^{\frac{1}{4}}\left[x_\alpha(\mathbf{k})\cos(\mathbf{k}\mathbf{R}_\alpha)+y_\alpha(\mathbf{k})\sin(\mathbf{k}\mathbf{R}_\alpha)\right]\right]. \tag{19}$$

The values F_α depend on the corresponding auxiliary variables $x_\alpha(\mathbf{k})$ and $y_\alpha(\mathbf{k})$, as well as on the Fourier transform of the interatomic potentials $\tilde{v}_{\alpha\gamma}(\mathbf{k})$, the volume V and the inverse temperature β.

We single out the principal terms in the quantities $(F_\alpha)^{N_\alpha}$ in the thermodynamic limit $N_\alpha\to\infty$, $V\to\infty$, $\dfrac{N_\alpha}{V}=n_\alpha=\text{const}_\alpha$.

We note that the functions F_α satisfy the inequality

$$|F_\alpha|\le 1, \tag{20}$$

since the modulus of the exponential in the integrand of (19) is equal to one.

Function

$$\Phi_\alpha(\mathbf{R}_\alpha)=\sum_{\mathbf{k}\in\Omega/2}\sqrt{\frac{2\beta}{V}}(\det\tilde{v}(\mathbf{k}))^{\frac{1}{4}}\left[x_\alpha(\mathbf{k})\cos(\mathbf{k}\mathbf{R}_\alpha)+y_\alpha(\mathbf{k})\sin(\mathbf{k}\mathbf{R}_\alpha)\right], \tag{21}$$

contained in the exponent of formula (19), by virtue of the Riemann-Lebesgue lemma [21], tends to zero at $|\mathbf{R}_\alpha|\to\infty$. Therefore, outside of some finite domain V_0 the exponent in the integrand of the formula (19) tends to one. Hence it follows

$$\lim_{V\to\infty}F_\alpha=1. \tag{22}$$

Thus, the value $(F_\alpha)^{N_\alpha}$ represents indeterminacy of the form 1^∞.

5. Evaluation of one-particle integrals (19)

To calculate the integral (19) we use the Weyl theorem [22].

Theorem. Let us assume that

1. $\psi_\nu(x)=a_\nu+b_\nu x$ $(\nu=1,2,\ldots,M)$, where the numbers a_ν and b_ν is the set of constants;
2. numbers b_ν are rationally independent;

3. function $f(\psi_1,\psi_2,\ldots,\psi_M)$ is periodic in all variables ψ_ν with the period 1 and is integrable in the cube $0 \le \psi_\nu \le 1$.

Then the following equality holds

$$\lim_{x\to\infty}\frac{1}{x}\int_0^x f(\psi_1(x),\psi_2(x),\ldots,\psi_M(x))dx = \int_0^1\int_0^1\cdots\int_0^1 f(\psi_1,\psi_2,\ldots,\psi_M)\prod_{\nu=1}^{M}d\psi_\nu . \quad (23)$$

This theorem has an evident generalization to the case of vectorial variable x.

We carry out the following transformations in the integral (19)

$$\sqrt{\frac{2\beta}{V}}(\det\tilde{v}(\mathbf{k}))^{\frac{1}{4}}[x_\alpha(\mathbf{k})\cos(\mathbf{kR}_\alpha)+y_\alpha(\mathbf{k})\sin(\mathbf{kR}_\alpha)]$$
$$=\sqrt{\frac{2\beta}{V}[x_\alpha^2(\mathbf{k})+y_\alpha^2(\mathbf{k})]}(\det\tilde{v}(\mathbf{k}))^{\frac{1}{4}}\sin[2\pi\psi_\alpha(\mathbf{k})],\quad \psi_\alpha(\mathbf{k})=\frac{1}{2\pi}(\mathbf{kR}_\alpha+\varphi_\alpha(\mathbf{k})). \quad (24)$$

For $V \to \infty$, the integral (19) after applying the Weyl theorem is transformed to the following form

$$F_\alpha = \prod_{\mathbf{k}\in\Omega/2}\int_0^1 \exp\left(-i\sqrt{\frac{2\beta}{V}[x_\alpha^2(\mathbf{k})+y_\alpha^2(\mathbf{k})]}(\det\tilde{v}(\mathbf{k}))^{\frac{1}{4}}\sin[2\pi\psi_\alpha(\mathbf{k})]\right)d\psi_\alpha(\mathbf{k})$$
$$=\prod_{\mathbf{k}\in\Omega/2}J_0\left(\sqrt{\frac{2\beta}{V}[x_\alpha^2(\mathbf{k})+y_\alpha^2(\mathbf{k})]}(\det\tilde{v}(\mathbf{k}))^{\frac{1}{4}}\right), \quad (25)$$

where $J_0(x)$ is the modified Bessel function of zero order. Hence we obtain

$$F_\alpha^{N_\alpha} = \prod_{\mathbf{k}\in\Omega/2}J_0^{N_\alpha}\left(\sqrt{\frac{2\beta}{V}[x_\alpha^2(\mathbf{k})+y_\alpha^2(\mathbf{k})]}(\det\tilde{v}(\mathbf{k}))^{\frac{1}{4}}\right). \quad (26)$$

The largest value of the Bessel function $J_0(x)$ is equal to one and this value is achieved at a single point $x=0$. Using the asymptotics of the Bessel function in the neighborhood of this point, we obtain a value $F_\alpha^{N_\alpha}$ in the thermodynamic limit

$$F_\alpha^{N_\alpha} = \exp\left[-\frac{\beta n_\alpha}{2}\sum_{\mathbf{k}\in\Omega/2}(\det\tilde{v}(\mathbf{k}))^{\frac{1}{2}}\left(x_\alpha^2(\mathbf{k})+y_\alpha^2(\mathbf{k})\right)\right]. \quad (27)$$

Now we turn to the calculation of the partition function and Helmholtz free energy of the solution.

6. The partition function and Helmholtz free energy of the solution

Substituting formulas (27) and (17) in (13), we obtain the following expression for the configuration integral

$$Z_{\text{conf}} = \mathsf{A}\prod_{\mathbf{k}\in\Omega/2}\left(\frac{1}{1+\beta n_1\tilde{v}_{11}(\mathbf{k})+\beta n_2\tilde{v}_{22}(\mathbf{k})+\beta^2 n_1 n_2\det\tilde{v}(\mathbf{k})}\right). \quad (28)$$

This implies the expression for the partition function

$$Z=\frac{\mathsf{A}}{N_1!N_2!}\frac{V^{N_1+N_2}}{\lambda_1^{3N_1}\lambda_2^{3N_2}}\prod_{\mathbf{k}\in\Omega/2}\left(\frac{1}{1+\beta n[c_1\tilde{v}_{11}(\mathbf{k})+c_2\tilde{v}_{22}(\mathbf{k})]+(\beta n)^2c_1c_2\det\tilde{v}(\mathbf{k})}\right), \quad (29)$$

where λ_α is de Broglie thermal wavelength of the α-th component, $N=N_1+N_2$; $c_1=\frac{N_1}{N}$; $c_2=\frac{N_2}{N}=1-c_1$.

The Helmholtz free energy in the thermodynamic limit $N \to \infty, V \to \infty, N/V = n = \text{const}$ has the following form

$$F = -T \ln Z = -\frac{N}{2}(c_1 v_{11}(0) + c_2 v_{22}(0)) + \frac{N^2}{2V}\left[c_1^2 \tilde{v}_{11}(0) + 2c_1 c_2 \tilde{v}_{12}(0) + c_2^2 \tilde{v}_{22}(0)\right] + TN[c_1 \ln c_1 + c_2 \ln c_2] + TN \ln\left[n(\lambda_1)^{3c_1}(\lambda_2)^{3c_2}\right] + \frac{TV}{2}\int_{\mathbf{k}\in\Omega}\frac{d\mathbf{k}}{(2\pi)^3}\ln\left\{1 + n\beta(c_1\tilde{v}_{11}(\mathbf{k}) + c_2\tilde{v}_{22}(\mathbf{k})) + c_1 c_2 (n\beta)^2 \det \tilde{v}(\mathbf{k})\right\} \quad (30)$$

The appearance of the factor $1/2$ in the last line of this formula is due to the replacement of $\mathbf{k} \in \Omega/2$ of the product in (29) by $\mathbf{k} \in \Omega$ in the integral term of formula (30).

7. Isolation of the singular part of the interatomic potentials at $\mathbf{r} = 0$

The expression (30) for the Helmholtz free energy contains the values $v_{11}(0), v_{22}(0)$ — the values of interatomic potentials $v_{11}(\mathbf{r})$ and $v_{22}(\mathbf{r})$ at the origin of coordinates $\mathbf{r} = 0$. As is known, actual interatomic potentials have a singularity at this point and therefore formula (30) should be improved.

We note that the potentials at the origin $v_{\alpha\alpha}(0)$ are related to the Fourier transforms of the potentials by the relation

$$v(0) = \int_{\mathbf{k}\in\Omega}\frac{d\mathbf{k}}{(2\pi)^3}\tilde{v}(\mathbf{k}). \quad (31)$$

The singularity of the interatomic potentials at a point $\mathbf{r} = 0$ means the divergence of this integral at $k = |\mathbf{k}| \to \infty$, i.e. insufficiently rapid decrease of the function $\tilde{v}(\mathbf{k})$ at $k = |\mathbf{k}| \to \infty$. By virtue of the fact that at $k \to \infty$

$$\ln\left\{1 + n\beta(c_1\tilde{v}_{11}(\mathbf{k}) + c_2\tilde{v}_{22}(\mathbf{k})) + c_1 c_2 (n\beta)^2 \det \tilde{v}(\mathbf{k})\right\} \cong n\beta(c_1\tilde{v}_{11}(\mathbf{k}) + c_2\tilde{v}_{22}(\mathbf{k})), \quad (32)$$

then the singularity of the interatomic potential $v(\mathbf{r})$ leads to the divergence of the integral over $\mathbf{k}$ in the formula (30). However, this divergence is completely compensated by the term

$$-\frac{N}{2}(c_1 v_{11}(0) + c_2 v_{22}(0)) = -\frac{TV}{2}\int_{\mathbf{k}\in\Omega}\frac{d\mathbf{k}}{(2\pi)^3} n\beta(c_1\tilde{v}_{11}(\mathbf{k}) + c_2\tilde{v}_{22}(\mathbf{k})) \quad (33)$$

in this formula. Combining this expression with the integral term in (30), we obtain

$$F = TN[c_1 \ln c_1 + c_2 \ln c_2] + TN \ln\left[n(\lambda_1)^{3c_1}(\lambda_2)^{3c_2}\right] + \frac{N^2}{2V}\left[c_1^2 \tilde{v}_{11}(0) + 2c_1 c_2 \tilde{v}_{12}(0) + c_2^2 \tilde{v}_{22}(0)\right] + \frac{TV}{2}\int_{\mathbf{k}\in\Omega}\frac{d\mathbf{k}}{(2\pi)^3}[\ln\left\{1 + n\beta(c_1\tilde{v}_{11}(\mathbf{k}) + c_2\tilde{v}_{22}(\mathbf{k})) + c_1 c_2 (n\beta)^2 \det \tilde{v}(\mathbf{k})\right\} - n\beta(c_1\tilde{v}_{11}(\mathbf{k}) + c_2\tilde{v}_{22}(\mathbf{k}))]. \quad (34)$$

This expression for the Helmholtz free energy is applicable not only for regular interatomic potentials, but also for singular potentials. In other words, the singularity of the interatomic potentials in the neighborhood of the origin of coordinates does not prevent the use of the formula (34) for calculating thermodynamic functions. This agrees with the general result of Zagrebnov and Pastur [23] that thermodynamic potentials have the property of asymptotic independence from the extension of the potential in the singularity neighborhood.

Direct use of formulas (34) and (30) for the study of the thermodynamic functions of solutions is significantly complicated by the presence of a logarithmic function in the integrand. Therefore, it is advisable to exclude the logarithmic function without loss of accuracy and generality. In the case when all the interatomic potentials are central, i.e.

$$v_{\alpha\gamma}(\mathrm{r}) = v_{\alpha\gamma}(r), \quad r = |\mathrm{r}|, \tag{35}$$

the Fourier transformants of all potentials $\tilde{v}_{\alpha\gamma}(\mathbf{k})$ depend only on $k = |\mathbf{k}|$. We first consider the integral term in expression (30). It has the following structure

$$\tilde{W} = \int_{k\in\Omega} \frac{d\mathbf{k}}{(2\pi)^3} \ln[1 + \tilde{w}(k)], \tag{36}$$

where

$$\tilde{w}(k) = n\beta[c_1\tilde{v}_{11}(k) + c_2\tilde{v}_{22}(k)] + c_1c_2(n\beta)^2 \det\tilde{v}(k). \tag{37}$$

Passing to spherical coordinates in the $\mathbf{k}$ – space and integrating by parts, we obtain

$$\tilde{W} = \frac{1}{2\pi^2}\left\{ \ln[1 + \tilde{w}(k)]\frac{k^3}{3}\bigg|_0^\infty - \frac{1}{3}\int_0^\infty \frac{k^3\tilde{w}'(k)}{1 + \tilde{w}(k)}dk \right\}, \tag{38}$$

Similarly, we transform the integral on the right-hand side of the formula (33):

$$\begin{aligned} &-\int_{k\in\Omega} \frac{d\mathbf{k}}{(2\pi)^3}[c_1\tilde{v}_{11}(k) + c_2\tilde{v}_{22}(k)] \\ &= -\frac{1}{2\pi^2}\left\{ [c_1\tilde{v}_{11}(k) + c_2\tilde{v}_{22}(k)]\frac{k^3}{3}\bigg|_0^\infty - \frac{1}{3}\int_0^\infty k^3[c_1\tilde{v}'_{11}(k) + c_2\tilde{v}'_{22}(k)] \right\}. \end{aligned} \tag{39}$$

It should be noted, that the terms outside the integrals in expressions (38) and (39) become zero for non-singular potentials, and are mutually compensated for singular potentials because of the relation (32). Therefore, adding the right-hand sides of formulas (38) and (39) and substituting the result into formula (34), we obtain the final expression for the Helmholtz free energy

$$\begin{aligned} F &= TN[c_1 \ln c_1 + c_2 \ln c_2] + TN \ln\left[n(\lambda_1)^{3c_1}(\lambda_2)^{3c_2}\right] \\ &\quad + \frac{N^2}{2V}\left[c_1^2\tilde{v}_{11}(0) + 2c_1c_2\tilde{v}_{12}(0) + c_2^2\tilde{v}_{22}(0)\right] \\ &\quad - \frac{TV}{12\pi^2}\int_0^\infty k^3\left[\frac{\tilde{w}'(k)}{1 + \tilde{w}(k)} - n\beta[c_1\tilde{v}'_{11}(k) + c_2\tilde{v}'_{22}(k)]\right]dk, \end{aligned} \tag{40}$$

where $\tilde{w}(k)$ is defined by the formula (37).

Thus, we assume that all the interatomic potentials $v_{\alpha\gamma}(\mathbf{r})$ have the corresponding Fourier transforms $\tilde{v}_{\alpha\gamma}(\mathbf{k})$, and the asymptotic properties of Fourier-transformant potentials ensure the convergence of the integral over the wave vectors in formula (40).

8. Conclusion

According to the formula (34), the Helmholtz free energy is represented as the sum of three contributions. The first contribution is the free energy of an ideal gas. The second contribution is the configurational energy in the Gorsky-Bragg-Williams approximation [24, 25, 26, 27, 28]. The third contribution is the integral term, which we shall call the ergodic contribution.

Let us discuss the dependence of each of these contributions on the temperature T and the composition of the solution x ($x = c_1, c_2 = 1 - x$).

1. The free energy of an ideal gas (the first line in (34)) is proportional to temperature and is a *concave* function x on the whole interval $0 \le x \le 1$.

2. The configuration part of the free energy in the Gorsky-Bragg-Williams approximation does not depend on temperature. The conditions of noncatastrophicity of the potentials (5) and (4) imply that values $\tilde{v}_{\alpha\gamma}(\mathbf{k})$ at all values $\mathbf{k}$ (including $\mathbf{k} = 0$) are related by inequalities

$$|\tilde{v}_{12}(\mathbf{k})| \le \sqrt{\tilde{v}_{11}(\mathbf{k})\tilde{v}_{22}(\mathbf{k})} \le \frac{\tilde{v}_{11}(\mathbf{k}) + \tilde{v}_{22}(\mathbf{k})}{2}. \quad (41)$$

Under these restrictions, this contribution is also a *concave* function of the composition x.

3. The ergodic contribution to the free energy of the solution is defined by the integral term in the formula (34). The integrand of this term has the following form

$$\phi(\beta, x, \mathbf{k}) = \frac{\ln\left\{1 + n\beta\left(x\tilde{v}_{11}(\mathbf{k}) + (1-x)\tilde{v}_{22}(\mathbf{k})\right) + x(1-x)(n\beta)^2 \det\tilde{v}(\mathbf{k})\right\}}{\beta} - n\left(x\tilde{v}_{11}(\mathbf{k}) + (1-x)\tilde{v}_{22}(\mathbf{k})\right). \quad (42)$$

The second derivative of this function with respect to the variable x is everywhere negative, therefore, in contrast to the two previous contributions, both the function itself and the ergodic contribution to Helmholtz free energy are a *convex* function of the composition x. Thus, without taking into account the ergodic contribution to the free energy, the phase separation of the solution into phases is impossible. A phase transition in a binary solution occurs only if there is such an interval of compositions $[a;b]$ $(0 < a < b < 1)$, in which the Helmholtz free energy (40) is a *convex* function of the composition. The number of critical points in a binary solution is equal to the number of positive solutions of the equation $\frac{\partial^2 F(x,T)}{\partial x^2} = 0$ with respect to temperature T.

Acknowledgments. The work was carried out with the financial support of the Ministry of Education and Science of the Russian Federation within the framework of the project part of the State order (grant No.3.3572.2017).

References

[1] Baxter R J 1984 Exactly Solved Models in Statistical Mechanics (London: Academic Press)
[2] Guggenheim E A 1952 Mixtures. The theory of the equilibrium properties of some simple classes of mixtures, solutions and alloys (Oxford: Clarendon Press)
[3] Boublk T, Nezbeda I and Hlavatý K 1980 Statistical Thermodynamics of Simple Liquids and Their Mixtures (Amsterdam: Elsevier)
[4] Prigogine I R 1957 The Molecular Theory of Solutions (Amsterdam: North-Holland)
[5] Smirnova N A, Victorov A V 2000 Quasilattice Equations of State for Molecular Fluids Equations of State for Fluids and Fluid Mixtures, ed J V Sengers, R F Kayser, et al (Amsterdam: Elsevier) chapter 7 pp 255–288
[6] Zubarev D N 1954 Evaluation of configuration integrals for a system of particles with Coulomb interaction Doklady 95 757–60
[7] Zakharov A Yu 1990 Exact Calculation Method of Partition Function for One-Component Classical Systems with Two-Body Interactions Phys. Lett. A 147 442–4
[8] Zakharov A Yu and Loktionov I K 1999 Classical statistics of one-component systems with model potentials Theor. and Math. Phys. 119 532–9
[9] Zakharov A Yu 2000 Functional integration and method of factorization in classical statistical mechanics Russ. J. Phys. Chem. 74 40–5
[10] Loktionov I K 2000 Determination of the critical parameters of a classical one-component system with model interaction potential High Temperature 38 494–6
[11] Loktionov I K 2015 Prediction of equilibrium thermodynamic properties of simple liquids in the model with four-parametric oscillating interaction potential Technical Physics 85 317–26
[12] Ruelle D 1963 Classical Statistical Mechanics of a System of Particles Helv. Phys. Acta 36 183–97

[13] Fisher M E 1964 The Free Energy of a Macroscopic System Arch.Rat. Mech. Anal. 17 377–410
[14] Dobrushin R L 1964 Investigation of Conditions for the Asymptotic Existence of the Configuration Integral of Gibbs' Distribution Theory Probab. Appl. 9 566—81
[15] Ruelle D 1969 Statistical Mechanics. Rigorous results (New York: Benjamin)
[16] Baus M and Tejero C F 2008 Equilibrium Statistical Mechanics. Phases of Matter and Phase Transitions (New York: Springer)
[17] Pines D and Bohm D 1952 A Collective Description of Electron Interactions: II. Collective vs Individual Particle Aspects of the Interactions Phys. Rev. 85 338–53
[18] Percus J K and Yevick G J 1958 Analysis of Classical Statistical Mechanics by Means of Collective Coordinates Phys. Rev. 110 1–13
[19] Edwards S F 1959 The statistical thermodynamics of a gas with long and short-range forces. Phil. Mag. 4 1171–82
[20] Edwards S F and Schwartz M 2003 Statistical Mechanics in Collective Coordinates J. Stat. Physics. 110 497–502
[21] Whittaker E T and Watson G N 1927 A course of modern analysis (Cambridge: Cambridge University Press) p 172
[22] Weyl H 1916 Über die Gleichverteilung von Zahlen mod. Eins *Math. Annalen* **77** 313–52
[23] Zagrebnov V A and Pastur L A 1978 Singular interaction potentials in classical statistical mechanics *Theor. Math. Phys.* **36** 784–97
[24] Gorsky W 1928 Röntgenographische Untersuchung von Umwandlungen in der Legierung Cu Au *Z. Physik* **50** 64–81
[25] Bragg W L and Williams E J 1934 The Effect of Thermal Agitation on Atomic Arrangement in Alloys *Proc. R. Soc. Lond.* A **145** 699–730
[26] Bragg W L and Williams E J 1935 The Effect of Thermal Agitation on Atomic Arrangement in Alloys–II. *Proc. R. Soc. Lond.* A. **151** 540–66
[27] Williams E J 1935 The Effect of Thermal Agitation on Atomic Arrangement in Alloys–III. *Proc. R. Soc. Lond.* A. **152** 231–52
[28] Püschl W, Numakura H and Pfeiler W (2007) Point Defects, Atom Jumps, and Diffusion *Alloy Physics: A Comprehensive Reference*, ed. W Pfeiler (Weinheim: Wiley-VCH) chapter 5 pp 173–280

Features of the thermodynamic properties of gaseous and solid methane

A Yu Zakharov[1], A V Leont'eva[2], A Yu Prokhorov[1]

[1]Yaroslav-the-Wise Novgorod State University, Veliky Novgorod, 173003, Russia
[2]Donetsk Physics and Engineering Institute named after A.A. Galkin, 83114, Donetsk, Ukraine
E-mail:anatoly.zakharov@novsu.ru

Abstract. A detailed analysis and calculation of the rotational component of the heat capacity of gaseous and solid methane is carried out. Correction of the generally accepted estimates of the contribution of quantum effects to the rotational heat capacity of methane in both gaseous and solid state is performed.

1. Introduction

In recent years, there has been a sharp increase in interest in solid methane. In particular, it proved to be the best moderator of cold neutrons in pulsed nuclear power sources, it is widely distributed on Saturn, Jupiter and their satellites; it is one of the cryocrystals possessing a wide spectrum of physical properties. The range of properties is so wide that some reference books on the properties of cryocrystals [1] ignored the completely exceptional experimental data on solid methane in temperature range 65-70 K. Some experimental data in this temperature range have not yet been properly interpreted theoretically.

2. Results and discussion

In a remarkable experimental work by Tomita [2], it was found that a sharp change in the nature of the rotations of solid methane molecules occurs at a temperature of about 65 K. This fact is extremely surprising. First of all, the fact of the existence of rotation of methane molecules in the solid phase, in spite of the very likely inhibition of molecular rotations due to the "ears", is surprising. If we start from the hypothesis that the anomalies are associated with the rotation of molecules, then the only feature of the rotational component of the heat capacity is the nonmonotonicity of its temperature dependence. There is a point of maximum heat capacity at a temperature of order

$$T_{rot} = \frac{\hbar^2}{2kT} \tag{1}$$

where k is the Boltzmann constant.

First, we calculate the rotational temperature T_{rot} from the known parameters of the methane molecule using formula (1). The moment of inertia of a spherical top is determined by the relation [3, 4].

$$I = ma^2 = \frac{8}{3}ml^2 \tag{2}$$

where m is the proton mass, a is the distance between the hydrogen atoms in the methane molecule, and $l = 1.093 \cdot 10^{-10}\,\text{m}$ is the length of the bond between the carbon and hydrogen atoms. As a result, we find the moment of inertia of the methane molecule

$$I = 5.324 \cdot 10^{-47}\ \text{kg} \cdot \text{m}^2. \tag{3}$$

This is in agreement with the experimental value of the moment of inertia of methane $5.17 \cdot 10^{-47}\ \text{kg} \cdot \text{m}^2$ -[5]. Substituting (3) into (1), we obtain the value of the rotational temperature of methane

$$T_{\text{rot}} = 7.57\ K. \tag{4}$$

The physical meaning of the rotational temperature is the temperature at which quantum effects begin to manifest themselves clearly in the rotational degrees of freedom of methane molecules.

Note that the effect detected by Tomita occurs at temperatures that are an order of magnitude higher than the rotational temperature of gaseous methane of 7.5 K. In other words, T=60 K lies in the temperature range at which the rotational degrees of freedom of the methane molecules seemingly, must obey classical statistics. Therefore, in this temperature range it was initially difficult to expect the manifestation of quantum effects in rotational degrees of freedom.

Nevertheless, a hypothesis was proposed in [6] that the feature observed by Tomita is related to the manifestation of the quantum nature of the rotational degrees of freedom of methane molecules in the crystalline state. The rotation of molecules in the solid state is possible due to the tunnel effect [7], therefore, in formula (1), instead of the moment of inertia of the methane molecule, the effective moment of inertia of elementary excitations corresponding to the collectivization of rotational degrees of freedom of methane molecules should enter. Since the moment of inertia of elementary excitations (topons) is an order of magnitude much smaller than the moment of inertia of the methane molecule.

In this case, the rotational component of the thermodynamic functions of solid methane is described in the framework of the model of an ideal gas consisting of topon quasiparticles. The thermodynamics of an ideal quantum gas with rotational degrees of freedom of molecules was created in the late 1920s [8]. The rotational component of the heat capacity of an ideal gas consisting of tops is a function of the dimensionless temperature

$$C_{\text{rot}} = f(\tau), \quad \tau = \frac{T}{T_{\text{rot}}}. \tag{5}$$

To calculate the temperature dependence of the rotational heat capacity of methane, it is required to find the corresponding partition function

$$Z = \sum_{n=0}^{\infty} g_n e^{-\frac{n(n+1)}{\tau}}, \tag{6}$$

where g_n is the multiplicity of the degeneracy of the n-th state of the top, τ is the dimensionless temperature determined by the relation (5).

The whole specificity of the thermodynamics of a particular ideal quantum gas in terms of a dimensionless temperature lies in the nature of the dependence of the degeneracy multiplicity of a top on its state number. In contrast to other degrees of freedom, the rotational component of the heat capacity is a nonmonotonic function of temperature: before rushing to zero at $T \to 0$, the rotational heat capacity increases to a certain maximum value at $T=T_0$ and only then rushes to zero. The value of the rotational heat capacity at the maximum point is greater than its classical value, which is reached at high temperatures [9]. It should be noted that the presence of this maximum is the first evidence of quantum nature of the object of research.

Before analyzing the thermodynamic properties of crystalline methane in the framework of the topon model, we will discuss the current state of the theory of ideal methane gas. Note that the theoretical calculations of the rotational component of the heat capacity of methane, carried out by different authors [10, 11, 12], significantly contradict each other. The results of these calculations are partially given in the monographic and educational literature [9, 13].

The type of connection between the degeneracy multiplicity and the state number of the top depends on the symmetry type of the spherical top, and the type of symmetry is predetermined by the internal degrees of freedom of the top.

The multiplicity of degeneracy of polyatomic molecules with large spins of identical nuclei can be calculated from the formula [14]

$$g_n = (2n+1)^2. \tag{7}$$

In this case there are two types of degeneracy, each of which has multiplicity $g = 2n+1$. The first is connected with different directions of the angular momentum with respect to the molecule, and the second - with different directions of the angular momentum in space [14, 15].

However, methane is a spherical top because of its symmetry and also has small spins of identical nuclei. Therefore, methane exists in the form of three different spin-modifications $5A_2, 1E, 3F_1$, each of which is characterized by its rotational partition function [12, 14, 15, 16]. A simple formula of the type (7) for calculating the degeneracy multiplicities does not exist in this case. Only the first few terms in the low-temperature expansion of the statistical sum of methane are taken into account with allowance for all three spin-modifications [9, 15]. Moreover, the transition probabilities between different spin-modifications are extremely small, therefore, the real mixtures of modifications are usually nonequilibrium.

Using the first few terms in the expansion of the partition function of the methane with all three spin modifications, we obtain [1, 11]

$$Z = 5 + 9e^{-\left(\frac{2}{\tau}\right)} + 25e^{-\left(\frac{6}{\tau}\right)} + 77e^{-\left(\frac{12}{\tau}\right)} + 117e^{-\left(\frac{20}{\tau}\right)} + \cdots. \tag{8}$$

In this expansion, terms are represented up to $n=4$. This expression is applicable only for $T \leq T_0$. Outside the low-temperature region (i.e., at $T \geq T_0$), the omitted terms do not decrease as fast as the stored ones, and they can be taken into account in a rather rough approximation (7), which is applicable only for large n values [14]. As a result, we obtain the following expression for the partition function

$$Z = \left[5 + 9e^{-\left(\frac{2}{\tau}\right)} + 25e^{-\left(\frac{6}{\tau}\right)} + 77e^{-\left(\frac{12}{\tau}\right)} + 117e^{-\left(\frac{20}{\tau}\right)}\right] + \sum_{n=5}^{\infty} (2n+1)^2 e^{-\frac{n(n+1)}{\tau}}. \tag{9}$$

The heat capacity of methane is found using the relation

$$\tilde{N} = \beta^2 \frac{\partial^2 (\ln Z)}{\partial \beta^2}, \qquad \beta = \frac{1}{\tau}. \tag{10}$$

The results of numerical calculations of the dependence of the rotational component of the heat capacity of methane on temperature are presented in Figs.1, 2.

In all calculations, a characteristic maximum of the specific heat capacity is observed, but the value of T_0 is very sensitive to the form of the dependence of the degeneracy multiplicity on n. In particular, replacing expression (7) by (8) at the same value of the rotational temperature of the ball top changes

IOP Conf. Series: Materials Science and Engineering **441** (2018) 012060 doi:10.1088/1757-899X/441/1/012060

the value of T_0=0.59 by 3, i.e. approximately 5 times. And the reason for this is only that these two tops differ from one another in symmetry properties.

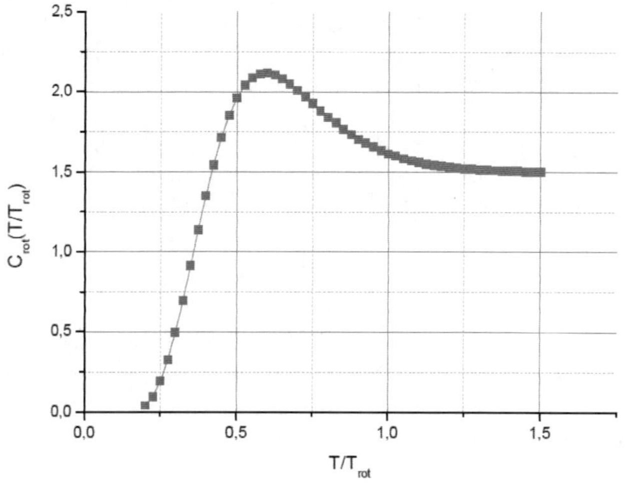

Figure 1. The rotational heat capacity of a spherical top with degeneracy multiplicity $g_n = (2n+1)$.

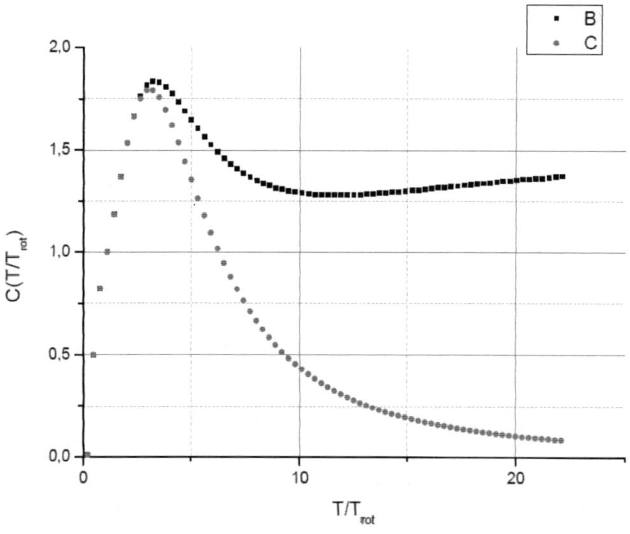

Figure 2. The rotational heat capacity of methane with only the first five terms of the low-temperature expansion (8) of the partition function (curve C) taken into account, and with compensation of the discarded terms in approximation (7) (curve B).

As is known, an arbitrarily small distortion of a spherical top results in a violation of its symmetry and a sudden change in the partition function for the gas [15]. For example, if a methane molecule is transformed into an asymmetric top with a point symmetry $\mathbf{D}_{2h}$ as a result of distortion of the configuration, then the first terms of the low-temperature expansion of the rotational partition function have the form [15]

$$Z = 7 + 27e^{-\left(\frac{2}{\tau}\right)} + 115e^{-\left(\frac{6}{\tau}\right)} + 175e^{-\left(\frac{12}{\tau}\right)} + 351e^{-\left(\frac{20}{\tau}\right)} + \cdots. \qquad (11)$$

We substitute this expression in place of (8) in (9) and find the corresponding temperature dependence of the heat capacity. The results of the numerical calculation are shown in Fig.3.

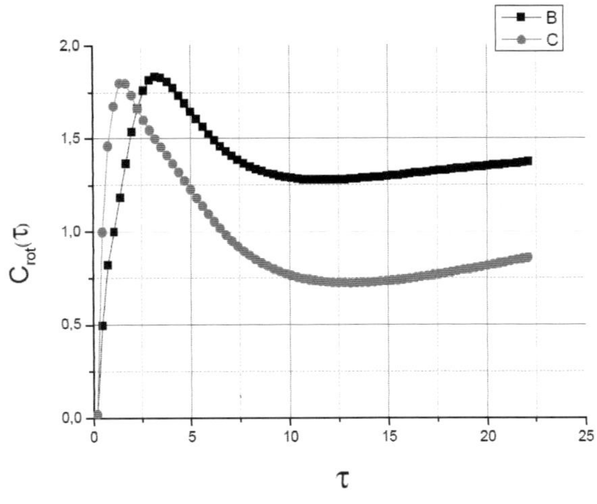

Figure 3. Teat capacity versus temperature for two variants of the low-temperature part of the expansion of the partition function: curve (B) corresponds to formula (8), curve (C) - to formula (11).

Thus, the heat capacity of an ideal gas consisting of spherical-top type molecules as a function of the dimensionless temperature has a maximum point, the position of which depends very substantially on the type of symmetry of the top.

It is significant that the experimental determination of the maximum of the rotational heat capacity of real gaseous methane is very complicated by at least two circumstances.

1. The long duration of the establishment of thermodynamic equilibrium in a mixture of spin modifications of methane because of the extremely small probabilities of the transitions between these modifications [14].

2. Even a negligible distortion of the spherical top results in a sudden change in the partition function. The reasons for the distortion are abound - from inherent intermolecular interactions, to background electromagnetic radiation.

In connection with this, it is unlike to expect any reproducibility of the results of an experimental study of the low-temperature heat capacity of methane in the gaseous state.

Unfortunately, this important conclusion is absolutely absent even in the literature of recent years, including Landau books.

Let us now turn to the thermodynamics of crystalline methane. Tomita [2] interpreted a sharp change in the character of the rotation of solid methane molecules in the temperature range 60-65 K as a result of a 20-fold change in the height of the barrier, which must be overcome by rotating molecules. The hypothesis of the rotation of methane molecules in the solid phase has not only experimental [2], but also theoretical justification [7]. It is pertinent to note here that Pauling [7] showed that in some molecular crystals (including, for example, CH_4, N_2, O_2), with an increase in temperature up to the melting point, the vibrational degrees of freedom of molecules can be transformed into rotational degrees of freedom. This is due, in particular, to the thermal expansion of solids.

In [6] it was assumed that in solid methane, instead of independent rotation of individual molecules characteristic of the gas phase, collective excitations of the rotational degrees of freedom of molecules should be considered because of the intermolecular interaction. Such a representation is in agreement with the results of Pauling [7] and is applicable at temperatures between the phase transition point (about 20 K) from the low-temperature cubic phase to the orientation-disordered phase along the hydrogen atoms of the fcc phase, and the melting point.

The corresponding quasiparticle (topon) is characterized by a single parameter - the effective moment of inertia. In contrast to the methane molecule, the topon is a spherical top not because of its symmetry, and therefore the multiplicity of the degeneracy of the rotational levels of the topon is determined by the formula (7). Then the graph of the temperature dependence of the heat capacity has the form shown in Fig.1.

Identifying the temperature of a sharp change in the character of the rotation of methane molecules with the maximum heat capacity of methane established by Tomita, we find a rough estimate of the rotational temperature of the topon $T_{rot} \cong 100$ K, i.e. the rotational temperature of topons is about 13 times higher than the rotational temperature of methane molecules in the gas phase. This means that solid methane in the entire temperature range of existence is a quantum cryocrystal.

It is interesting to note that solid methane exhibits anomalies in a number of physical properties of crystalline methane in the temperature range 60-65 K: a sharp change in the nature of the rotation of molecules [2]; temperature dependence of the molar volume [1]; coefficient of thermal expansion [1]; longitudinal and transverse ultrasound velocity [17]; low-frequency internal friction [18]; plasticity [18]; adhesion [19], radiolysis of defects at a neutron source [20], etc. These anomalies can be interpreted within the framework of the concept of collectivization of the rotational degrees of freedom of methane molecules in the solid phase [6, 21].

3. Conclusion

The temperature dependence of the heat capacity of methane in the gas phase is calculated and refined. It is shown that, because of the symmetry properties of the methane molecule, the quantum region of the heat capacity of gaseous methane exceeds its rotational temperature by an order of magnitude.

It is shown that solid methane in the entire temperature range of existence is a quantum cryocrystal.

4. Acknowledgments

The authors are grateful to L.P. Mezhov-Deglin, L.P. Pitaevsky, and O.I. Kirichek, for their discussion of the work and valuable advice.

Two of the authors (A. Yu. Zakharov and A. Yu. Prokhorov) are grateful to the Ministry of Education and Science of the Russian Federation for financial support of the work within the framework of the project part of the Government Assignment (Grant No. 3.3572.2017).

References

[1] Manzhelii V G, Prokhvatilov A I, Gavrilko V G and Isakina A P 1998 Structure and thermodynamic properties of cryocrystals: Handbook Begell House Inc. New York Wallingfort U.K

[2] Tomita K 1953 States of Solid Methane as Inferred from Nuclear Magnetic Resonance *Phys. Rev.* **89(2)** 429

[3] Landau L D and Lifshits E M 2007 Mechanics (Amsterdam: Elsevier)

[4] Quick reference book of physical and chemical quantities, ed Ravlel A A, Ponomareva A M 2003 (St. Petersburg: Ivan Fedorov Publishing (in Russian))

[5] Dickinson R G, Dillon R T and Rasetti F 1929 Raman Spectra of Polyatomic Gases *Phys. Rev.* **34** 582

[6] Zakharov A Yu, Leont'eva A V, Prokhorov A Yu and Erenburg A I 2014 Anomalous Properties of Crystalline Methane in the Range of 60–70 K *Phys. of the Solid State* **56** 1501

[7] Pauling L 1931 The Rotational Motion of Molecules in Crystals *Phys. Rev.* **36** 430

[8] Brillouin L 1931 Die Quantenstatistik und ihre Anwendung auf die Elektronentheorie der Metalle. Berlin: Springer
[9] Landau L D and Lifshits E M 1995 Statistical Physics, Part 1 (Oxford: Pergamon Press)
[10] Villars D S and Schultze G 1931 The Low Temperature Rotational Heat Capacities and the Relative Amounts of the Nuclear Singlet, Triplet and Quintet Symmetry Modifications of Methane *Phys. Rev.* **38** 998
[11] MacDougall D P 1931 The Heat Capacity of Methane *Phys. Rev.* **38** 2296
[12] Maue A-W 1937 Die Rotationswarme des Methans *Annalen der Physik.* **422** 555
[13] Fowler R H, Guggenheim E A 1956 Statistical Thermodynamics (Cambridge: Cambridge University Press)
[14] Herzberg G 1991 Molecular Spectra and Molecular Structure II: Infrared and Raman of Polyatomic Molecules (Princeton: Van Nostrand).
[15] Kaplan I G 1975 Symmetry of Many-Electron Systems (New York: Academic Press)
[16] Wilson E B J. 1935 The Statistical Weights of the Rotational Levels of Polyatomic Molecules, Including Methane, Ammonia, Benzene, Cyclopropane and Ethylene *Chem. Phys.* **3** 276
[17] Leont'eva A V, Prokhorov A Yu, Zakharov A Yu and Erenburg A I 2014 Effect of Molecular Rotational Degrees of Freedom on Mechanical and Thermodynamic Properties of Solid Methane at Temperatures above 50 K *J. of Exper. Phys.* **2014** Art.Y.D. 127050
[18] Leont'eva A V, Marinin G A, Prokhorov A Yu and Sukharevskii B Ya 1994 Anomalies of low-frequency internal friction in crystalline methane *Low Temp. Phys.* **20** 815
[19] Kirichek O, Church A J, Thomas M G, Cowdery D, Higgins S D, Dudman M P and Bowden Z A 2012 Adhesion, plasticity and other peculiar properties of solid methane *Cryogenics* **52** 325
[20] Kirichek O I, Lavson C R, Jenkins D M, Ridley C J T and Haunes D J 2017 Solid methane in neutron radiation: Cryogenic moderators and cometary cryo volcanism *Cryogenics* **88** 101
[21] Zakharov A Yu, Leont'eva A V, A. Yu. Prokhorov and Erenburg A I 2016 High-Temperature Quantum–Classical Transition in Solid Methane *J. of Low Temp. Phys.* **185** 140

IOP Conf. Series: Materials Science and Engineering **441** (2018) 012061 doi:10.1088/1757-899X/441/1/012061

Towards non-statistical foundation of thermodynamics

A Yu Zakharov
Yaroslav-the-Wise Novgorod State University, Veliky Novgorod, 173003, Russia

E-mail anatoly.zakharov@novsu.ru; A.Yu.Zakharov@gmail.com

Abstract. It is shown that the irreversible behavior exists even in two-body system. It is due to the retardation of the interactions between the particles. It is established that a two-particle oscillator with a delayed interaction between particles has an infinite spectrum of both stationary and non-stationary oscillations, which depends on the equilibrium distance between the particles.

1. Introduction

It is well known that all the field theories have a property of delay of the interactions. Such are the electromagnetic interaction of charges in vacuum, the interaction of vibrational systems through a medium, the interactions between the particles in a plasma or colloidal systems, etc. The delay of the interactions leads to violation of Newton's third law on the equality of action and reaction [1], and also it is one of the mechanisms for the irreversible behavior of many-body systems [2, 3].

Note that in papers [2, 3] neither probabilistic considerations nor the conditions $N >> 1$ (N is number of degrees of freedom in the system) were used. In this connection, it is of interest to study in detail the mechanisms of the manifestation of retardation in the irreversibility of systems with a small number of degrees of freedom.

We define a two-body oscillator as a system of two particles with identical masses m, the interaction between which **at rest** is described by the potential $W(\mathbf{R}_1 - \mathbf{R}_2)$ with the following properties.

1. The function $W(\mathbf{R}_1 - \mathbf{R}_2)$ has a minimum at $|\mathbf{R}_1 - \mathbf{R}_2| = L$.
2. Near the minimum point, this function can be approximated by quadratic function

$$W(\mathbf{R}_1 - \mathbf{R}_2) = W(L) + \frac{k(|\mathbf{R}_1 - \mathbf{R}_2| - L)^2}{2}. \tag{1}$$

Let us consider the case of one-dimensional **small oscillations** of these particles along a straight line connecting these bodies. In the absence of retardation, the solution of this problem is trivial and describes the oscillations of the particles with circular frequency $\left(\frac{2k}{m}\right)^{1/2}$. The frequency of these oscillations does not depend on either L or $W(L)$. The retardation in the interactions between these two particles leads to a qualitatively different result, including the irreversible behavior of the system as a whole.

2. Oscillator with retardation

We denote the deviations from the equilibrium positions of the particles in the rest state by $x_1(t)$ and $x_2(t)$. Equations of motion for the system in view of the interaction retardation τ have the form:

$$\begin{cases} \ddot{x}_1(t) + \omega_0^2\left[x_1(t) - x_2(t-\tau)\right] = 0; \\ \ddot{x}_2(t) + \omega_0^2\left[x_2(t) - x_1(t-\tau)\right] = 0, \end{cases} \tag{2}$$

where $\omega_0 = \left(\dfrac{k}{m}\right)^{1/2}$.

In general case, τ is not a constant, but a function of the position of the particles, that is, of the solution of the system of equations (2). The solution of such a problem in a general formulation goes far beyond the possibilities of modern mathematics. The condition of smallness of the oscillations is

$$|x_k(t)| = L, \quad (k = 1,2). \tag{3}$$

It substantially simplifies the situation since in this case the principal term of the interaction retardation is reduced to the constant

$$\tau = \frac{L}{v}, \tag{4}$$

where v is the propagation velocity of the interaction. Thus, in the case of small oscillations, the system of equations (2) is linear.

The Euler substitution

$$x_k(t) = C_k e^{i\omega_0 \Lambda t} \tag{5}$$

leads to the following characteristic equation with respect to dimensionless frequency Λ:

$$\Lambda^4 - 2\Lambda^2 + \left[1 - e^{-2i\omega_0\Lambda\tau}\right] = 0. \tag{6}$$

The roots of this transcendental equation depend on τ and in general case are complex:

$$\Lambda = \alpha + i\gamma. \tag{7}$$

The equation (6) with respect to the **complex** value Λ is equivalent to a system of two equations for **real** quantities α and γ:

$$\begin{cases} \alpha^4 - 6\alpha^2\gamma^2 + \gamma^4 - 2\left(\alpha^2 - \gamma^2\right) + \left[1 - e^{2\omega_0\gamma\tau}\cos(2\omega_0\alpha\tau)\right] = 0; \\ 4\alpha\gamma\left(\alpha^2 - \gamma^2 - 1\right) + e^{2\omega_0\gamma\tau}\sin(2\omega_0\alpha\tau) = 0. \end{cases} \tag{8}$$

Note that α determines the oscillation frequency, and γ characterizes the rate of change in the amplitude of the oscillations. Therefore, the condition for stationarity of the oscillations is that $\gamma = 0$, whence we get

$$2\omega_0 \alpha \tau = \pi n, \tag{9}$$

where n is an arbitrary integer.

Substituting τ from (9) and $\gamma = 0$ into equation (8), we obtain the equation for α

$$\alpha^4 - 2\alpha^2 + 2\sin^2\left(\frac{\pi n}{2}\right) = 0. \tag{10}$$

The roots of this equation are real if and only if n is an even number. Under this condition, the solutions are as follows:

$$\alpha_1 = \alpha_2 = 0; \qquad \alpha_{3,4} = \pm\sqrt{2}. \tag{11}$$

Thus, stationary oscillations of a two-particle oscillator with retarded interactions occur only for a discrete set of equilibrium distances L between the particles, determined by the condition (9).

However, there exists a set of values of the parameter L for which both stationary and non-stationary oscillations are possible, i.e. solution of the system of equations (8) with $\gamma = 0$ and $\gamma \neq 0$, respectively. From the immense set of solutions of this system of equations depending on the parameter L, we consider a subset for which the condition (9) is satisfied:

$$L = \frac{\pi n v}{2\omega}. \tag{12}$$

In this case, the system of equations (8) is greatly simplified:

$$\begin{cases} \alpha^4 - 6\alpha^2\gamma^2 + \gamma^4 - 2(\alpha^2 - \gamma^2) + \left[1 - e^{2\omega_0\gamma\tau}(-1)^n\right] = 0; \\ 4\alpha\gamma(\alpha^2 - \gamma^2 - 1) = 0. \end{cases} \tag{13}$$

For $\gamma \neq 0$ we have

$$\gamma = \eta\sqrt{\alpha^2 - 1}, \tag{14}$$

where $\eta = \pm 1$.

Substituting this expression for γ into the first of the equations (13), we find

$$(-1)^n e^{2\omega_0\gamma\tau} = 4\alpha^2(1 - \alpha^2). \tag{15}$$

As follows from the domain of definition of the expression (14) for γ, the right-hand side of the equation (15) is non-positive, so n in this equation is an odd number ($n = 2s + 1$), and the equation itself becomes

$$(2s+1)\eta = \frac{1}{\pi}\frac{\alpha \ln\left[4(\alpha^4 - \alpha^2)\right]}{(\alpha^2 - 1)^{1/2}}, \tag{16}$$

where $\alpha^2 > 1$.

The right-hand side of this equation is a monotonically increasing function on the interval $(1, +\infty)$.

The range of this function fills the entire interval $(-\infty,+\infty)$. Therefore, for each value of $(2s+1)\eta$, the equation (16) has an unique solution $\alpha_s(\eta)$.

The graph of the function contained in the right-hand side of this equation is shown in Fig.1.

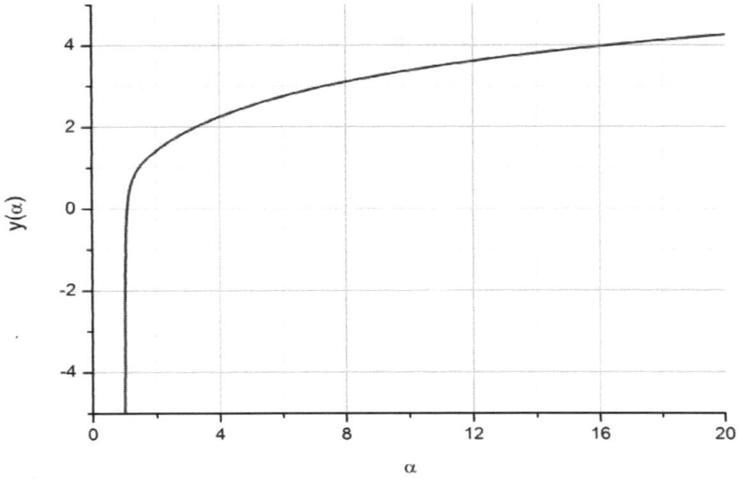

Figure 1. The graph of the function $y(\alpha) = \frac{1}{\pi}\frac{\alpha \ln\left[4\left(\alpha^4 - \alpha^2\right)\right]}{\left(\alpha^2 - 1\right)^{1/2}}$.

3. Discussion

The delay in the interaction between the constituents of a two-body oscillator leads to the following effects:

1. Stationary oscillations exist only for a discrete set of equilibrium distances between the particles, determined by the condition (9). The frequencies of all stationary oscillations ω are the same:

$$\omega = \sqrt{2}\omega_0. \tag{17}$$

Condition (9), with account for the relation (17), can be represented in an equivalent form

$$L = \frac{n}{4}\lambda, \tag{18}$$

where $\lambda = 2\pi v\left(\frac{m}{2k}\right)^{1/2}$ is the wavelength corresponding to the oscillation frequency (17).

2. If the equilibrium distance L between the particles does not satisfy the condition (9), then there are only non-stationary oscillations in the system. Depending on the sign of $\gamma = \mathrm{Im}\Lambda$, the corresponding oscillations either decay or increase. In both of these cases, the energies of the oscillations are not

conserved. For $\gamma > 0$, the corresponding oscillations are damped and the mechanical energy of these oscillations decreases. For $\gamma < 0$, the amplitude of the corresponding oscillations increases exponentially and the assumptions (3), (4) are no longer applicable. In particular, τ is no longer a constant (4), but a function that depends on the solution of the system of equations (2). Thus, the delay of interactions between the particles performs functions similar to those of a thermodynamic reservoir, which, however, is not something external, but is inextricably linked to the particles.

3. Especially it should be noted that even under the condition (9), in addition to stationary oscillations, the non-stationary oscillations with different frequencies corresponding to the complex roots of the characteristic equation may also exist (6):

$$\omega_s(\eta) = \alpha_s(\eta)\omega_0, \quad \gamma_s(\eta) = \eta\omega_0 \left(\alpha_s^2(\eta) - 1\right)^{1/2}. \tag{19}$$

Thus, taking into account the seemingly insignificant but always existing retardation in the interaction between the particles leads to a qualitative change in the dynamics of the mechanical system. In addition to the appearance of an infinite spectrum of oscillations, the phenomenon of the irreversibility of dynamics also takes place. Therefore, the classical theory of systems of particles with retarded interactions can be used for the microscopic probability-free foundation of thermodynamics. This approach has certain advantages over the Gibbsian probabilistic foundation of classical thermodynamics.

Omitting the discussion of the enormous mathematical difficulties associated with the use of Gibbs and related approaches to the practical implementation of statistical mechanics, we will discuss more fundamental problems in the theory of many-body systems.

The microscopic foundation of thermodynamics within the framework of statistical mechanics without establishing of the source of system stochastization is not satisfactory because of the numerous paradoxes generated by the use of probabilistic concepts in combination with deterministic classical mechanics. A certain clarity in the origin of these paradoxes was achieved by Kac [4]. He proposed a dynamic model – namely, a ring model, for which he obtained two solutions. The first solution is an exact dynamic solution that is reversible in time and also periodic in time. The second solution was obtained using the probabilistic assumptions; it has the property of irreversibility. In short time intervals, both solutions are almost identical, and for large time scales, they differ dramatically. Therefore, the probability approach to foundation of thermodynamics is not satisfactory. Exact theorems of classical mechanics are at variance not only with the probabilistic description of macroscopic systems, but also with the thermodynamic behavior of systems. Therefore, the physical cause of irreversibility is beyond the Newtonian classical mechanics.

In this regard, it is appropriate to note two essential points.

- The first point is the amazing work by Synge [5]. He solved "the electromagnetic two-body problem, based on the hypotheses (i) that the bodies are particles, (ii) that the fields are given by the retarded potential, (iii) that the force acting on a particle is the Lorentz ponderomotive force without a radiation term. It is found that energy disappears from the motion, so that the orbital particle slowly spirals in".
- The second point is connected with the outstanding result of Wheeler and Feynman [6, 7]. They proved that classical electrodynamics is equivalent to the theory of direct non-instantaneous interaction between particles. This means that the field concept is not necessary for describing interactions between the particles. This result makes it possible to exclude from consideration the field through which the interaction between the particles occurs. An infinite number of degrees of freedom of the excluded field is manifested in the functional-differential equations of motion of the particles through retardations the interactions.

Thus, there are two ways of describing the evolution of a system of interacting particles.

1. The first way is to use a complete system of equations of motion for both particles and the fields. These equations of motion with the corresponding initial conditions for both particles and fields determine the evolution of both particles and the fields. At present, this is the main way in the theory of many-particle problems.
2. The second way is to use the motion equations of the particles on the basis of the concept of direct retarted interaction between the particles. One of the advantages of this approach is the possibility of describing the evolution of a system of particles interacting via potentials of a general type. In this case, setting the initial conditions only for particles is insufficient to uniquely determine the evolution. The solution of functional-differential equations depends not only on the initial state of the system, but also on its prehistory.This means that under the same initial conditions of a Cauchy-type problem for a particle system, different versions of its history lead to different versions of the evolution of this system. Therefore, the solution of the classical Cauchy problem for a system of particles with retarded interactions is not unique.

Probably for the first time the idea of the non-uniqueness of the Cauchy problem for many-body systems as the cause of irreversibility was proposed in the paper [8]. It is assumed that the non-uniqueness of the Cauchy problem solution is due to certain features of the interatomic potentials. In particular, potentials such as the top of a hill are considered as examples. In this case, there are infinitely many variants to roll down from this hilltop. Hence it was concluded that the non-uniqueness of the solution of the Cauchy problem leads to an indeterministic behavior of the system.

But, as it can be seen from our solution of the two-body oscillator model with delayed interaction, the deterministic and at the same time irreversible behavior of the system is provided by a more complete setting of the initial conditions: it is necessary to take into account the state of the system not only at the initial moment of time. In other words, a dynamical system with retarded interactions has the property of heredity, because of which dynamics depends not only on the actual state of the system at the initial moment of time, but also on the preceding states [9, 10].

Numerous attempts to explain the irreversibility on the basis of the ergodic theory linking probability and thermodynamics have never led to a clear understanding of the nature of irreversibility, at least within the framework of classical mechanics. However, it should be noted the recent papers [11, 12], in which the mechanism of macroscopic irreversibility as a consequence of microscopic irreversibility was developed. In these papers, atoms and molecules are considered as open systems being in continuous interaction with photons flows from their surroundings.

4. Conclusion

The described simple example leads to the following conclusions.

1. The irreversible behavior of systems of interacting particles is a common property for both few-body and many-body classical system, having its origin in the interactions retardations.
2. The unavoidable retardation of interactions is sufficient for the irreversible behavior of the systems. The systems are irreversible in itself — there is no need to use any probabilistic hypotheses or other assumptions to explain the phenomenon of irreversibility in systems of interacting particles.

5. Acknowledgments

I am grateful to Profs. Ya. I. Granovsky and V. V. Uchaikin for useful discussions.
The work was carried out with the financial support of the Ministry of Education and Science of the Russian Federation within the framework of the project part of the State Assignment, project No. 3.33572.2017.

References

[1] Ivlev A V, Bartnick J, Heinen M, Du C K, Nosenko V and Löwen H 2015 *Phys. Rev.* X **5** 011035
[2] Zakharov A Yu and Zakharov M A 2016 *Phys. Lett.* A **380** 365-9
[3] Zakharov A Yu 2017 *Physica A: Statistical Mechanics and its Applications* **473** 72-76
[4] Kac M 1956 *Bull. de l'Académie Royale de Belgique* (*Classe des Sciences*) **42** 356-61
[5] Synge J L 1940 *Proc. Roy. Soc.* A. **177** 118-39
[6] Wheeler J and Feynman R 1945 *Rev. Mod. Phys.* **17** 157-91
[7] Wheeler J and Feynman R 1949 *Rev. Mod. Phys.* **21** 425-33
[8] Kosyakov B P 2008 *Found. Phys.* **38** 76-88
[9] Volterra V 1959 *Theory of Functionals and of Integral and Integro-Differential Equations* (London: Dover Edition)
[10] Uchaikin V V 2013 *Fractional Derivatives for Physicists and Engineers* vol I, II (Berlin: Springer Verlag)
[11] Lucia U 2014 *Scientific Reports* **6** 35796
[12] Lucia U 2018 *Physica A: Statistical Mechanics and its Applications* **492** 962-8

IOP Conf. Series: Materials Science and Engineering **441** (2018) 012062 doi:10.1088/1757-899X/441/1/012062

Calculation of the state diagrams of binary Sn-Pb and Pb-Sb solutions in the framework of the generalized lattice model

M A Zakharov[1]
[1]Novgorod State University, ul. B. St. Petersburgskaya, 41 173003 Veliky Novgorod, Russia
E-mail: Maxim.Zakharov@novsu.ru

Abstract. This paper is devoted to a theoretical study of mono- and non-invariant equilibria in binary Sn-Pb and Pb-Sb solutions. Modeling of Sn-Pb and Pb-Sb solutions was carried out within the framework of the generalized lattice model, corresponding eutectic phase diagrams were obtained.

1. Introduction

Binary lead-based alloys, such as Sn-Pb and Pb-Sb, are widely used in a wide variety of industries. In particular, solid solutions of Sn-Pb and Pb-Sb are traditionally used as solders in microelectronics, including for particularly strong soldering. In this connection, the study of phase equilibria in Sn-Pb and Pb-Sb systems over a wide range of temperatures and concentrations is an important task in materials science. It should be noted that most of the data on phase equilibria in these binary systems is experimental (see, for example, [1]), which makes it difficult to predict the thermal and mechanical properties of Sn-Pb and Pb-Sb alloys as their compositions and / or temperature change. This paper is devoted to the theoretical construction of phase diagrams of Sn-Pb and Pb-Sb solutions within the framework of the generalized lattice model (GLM) (see, for example, [2-4]).

2. The model

It has been experimentally noted [1] that the phase diagrams of Sn-Pb and Pb-Sb solutions refer to the same type of eutectic phase diagrams and are characterized by the limited solubility of both components in the solid state (see Figure 1).

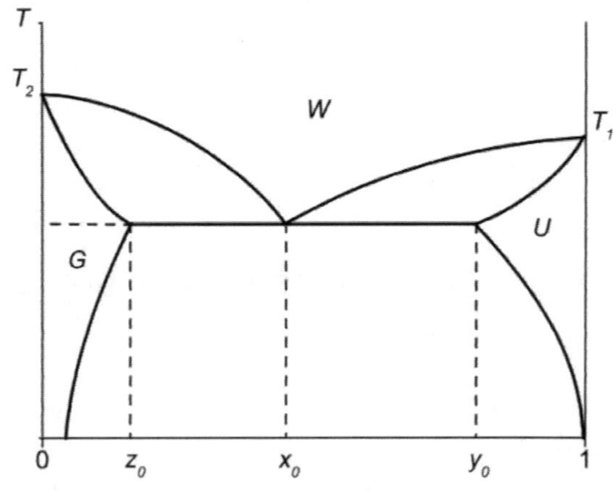

Figure 1. Phase diagram of a binary system of eutectic type in the presence of mutual solubility of components in the solid state.

According to the main provisions of the GLM, the chemical potentials of a binary homogeneous solution (per 1 mole of substance) have the following form [4]

$$\mu_1 = \mu_{10} + RT\ln x + W\lambda\left(\frac{1-x}{x+\lambda(1-x)}\right)^2, \quad \mu_2 = \mu_{20} + RT\ln(1-x) + W\left(\frac{x}{x+\lambda(1-x)}\right)^2, \tag{1}$$

where μ_{i0} is the standard value of the chemical potential of the i component, x is the mole fraction of the first component; T - temperature of the system, reduced to the absolute scale; R is the universal gas constant, W is an analog of the mixing energy in the GLM; $\lambda = \omega_2 / \omega_1$ is the ratio of "owned" atomic volumes of components, here ω_i – the "owned" volume of atoms of the i component.

Let x, y and z be the molar fractions of the first component in the liquid phase, in the solid α-solution and in the solid β-solution, respectively. We also agree to denote the mixing energies of the melt and solid-and-solutions by the letters W, U and G. Then the relations (1) determine the concentration dependences of the chemical potentials of the components in the liquid phase (melt), and analogous relationships corresponding to the solid α-solution and the β-solution look like

$$\mu_1^\alpha = \mu_{10}^S + RT\ln y + U\lambda\left(\frac{1-y}{y+\lambda(1-y)}\right)^2, \quad \mu_2^\alpha = \mu_{20}^S + RT\ln(1-y) + U\left(\frac{y}{y+\lambda(1-y)}\right)^2, \tag{2}$$

and

$$\mu_1^\beta = \mu_{10}^S + RT\ln z + G\lambda\left(\frac{1-z}{z+\lambda(1-z)}\right)^2, \quad \mu_2^\beta = \mu_{20}^S + RT\ln(1-z) + G\left(\frac{z}{z+\lambda(1-z)}\right)^2. \tag{3}$$

For the further calculation of the state diagram, it is necessary to calculate four OPM parameters: W, U, G and λ. To this end, we consider a non-invariant three-phase equilibrium, characterized by the compositions of coexisting phases x_0, y_0 and z_0 at an eutectic temperature T_0. In addition, the conditions of chemical equilibrium in the three-phase system must be met, i.e.

$$\mu_1^L(x_0,T_0) = \mu_1^\alpha(y_0,T_0) = \mu_1^\beta(z_0,T_0), \quad \mu_2^L(x_0,T_0) = \mu_2^\alpha(y_0,T_0) = \mu_2^\beta(z_0,T_0). \tag{4}$$

IOP Conf. Series: Materials Science and Engineering **441** (2018) 012062 doi:10.1088/1757-899X/441/1/012062

Using the expressions for the chemical potentials (1) - (3) and the chemical equilibrium conditions (4), it is not difficult to express the parameters W, U, G and λ through the coordinates of the ends of the eutectic conode, the calculation method of which is described in detail in [5].

A further calculation of the phase diagram reduces to modeling the corresponding two-phase equilibria above (or below) the temperature of the eutectic. Thus, in particular, the problem of calculating the liquidus and solidus curves splits into two:

i. Liquid—solid α-solution equilibrium $\left(x > x_0, y > y_0, T > T_0\right)$:

$$T(x,y) = \frac{q_1 T_1 + W\lambda\left(\dfrac{1-x}{x+\lambda(1-x)}\right)^2 - U\lambda\left(\dfrac{1-y}{y+\lambda(1-y)}\right)^2}{q_1 - \ln\left(\dfrac{x}{y}\right)} = \frac{q_2 T_2 + W\left(\dfrac{x}{x+\lambda(1-x)}\right)^2 - U\left(\dfrac{y}{y+\lambda(1-y)}\right)^2}{q_2 - \ln\left(\dfrac{1-x}{1-y}\right)}. \tag{5}$$

ii. Liquid—solid β-solution equilibrium $\left(x < x_0, z < z_0, T > T_0\right)$:

$$T(x,z) = \frac{q_1 T_1 + W\lambda\left(\dfrac{1-x}{x+\lambda(1-x)}\right)^2 - G\lambda\left(\dfrac{1-z}{z+\lambda(1-z)}\right)^2}{q_1 - \ln\left(\dfrac{x}{z}\right)} = \frac{q_2 T_2 + W\left(\dfrac{x}{x+\lambda(1-x)}\right)^2 - G\left(\dfrac{z}{z+\lambda(1-z)}\right)^2}{q_2 - \ln\left(\dfrac{1-x}{1-z}\right)}. \tag{6}$$

Here $q_i = \Delta H_i / RT_i$, where ΔH_i is the latent heat of the liquid-solid transition in pure components; T_i - melting points of pure components, reduced to an absolute scale.

Finally, for a two-phase equilibrium of a solid α-solution — solid β-solution $\left(y > y_0, z < z_0, T < T_0\right)$, we have

$$T(y,z) = \frac{G\lambda\left(\dfrac{1-z}{z+\lambda(1-z)}\right)^2 - U\lambda\left(\dfrac{1-y}{y+\lambda(1-y)}\right)^2}{\ln\left(\dfrac{y}{z}\right)} = \frac{G\left(\dfrac{z}{z+\lambda(1-z)}\right)^2 - U\left(\dfrac{y}{y+\lambda(1-y)}\right)^2}{\ln\left(\dfrac{1-y}{1-z}\right)}. \tag{7}$$

Thus, using the closed system of equations (5) - (7), it is possible to construct a phase diagram of the eutectic type in the entire range of concentrations and temperatures.

3. Diagram calculations and discussion of the results

Using the method of Ref. 6, the GLM parameters for the Sn-Pb and Pb-Sb systems were obtained, which are given in Table 1.

Table 1. GLM parameters

A-B	q_1	q_2	λ	W/R, K	U/R, K	G/R, K
Sn–Pb	0.955	1.684	0.874	688.4	1036.4	1820.3
Pb–Sb	2.717	0.955	4.395	-572.3	2403.8	6726.5

Figure 2 shows the results of modeling the Sn-Pb and Pb-Sb state diagrams, taking into account the GLM parameters given above.

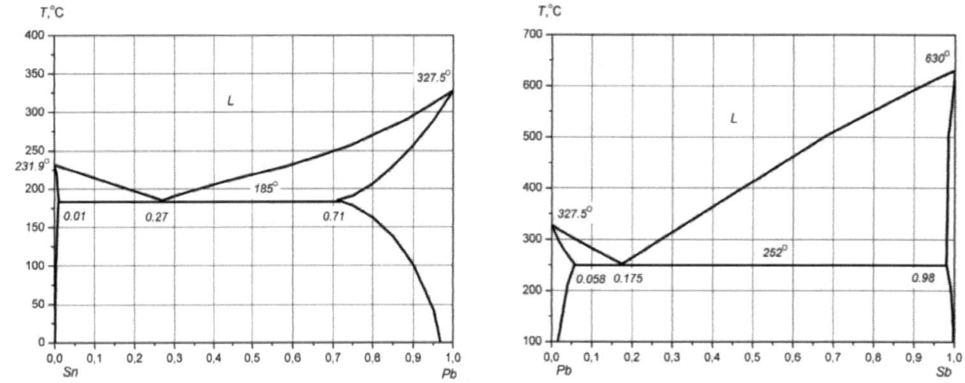

Figure 2. Sn–Pb and Pb–Sb state diagrams

The obtained state diagrams correspond the generalized diagrams of the Sn-Pb and Pb-Sb binary systems presented in [1], which indicates the adequacy of the proposed model.

In conclusion, we note that the GLM can be used to construct phase diagrams of practically any real binary solutions of the eutectic type both in the absence of solubility of the components in the solid state and in the presence of mutual solubility of the components.

Acknowledgments. The work was supported by the Ministry of Education and Science of the Russian Federation within the framework of the project part of the state assignment (grant No. 3.3572.2017).

References

[1] Lyakishev N. P., 1996 *Phase Diagrams of Binary Metallic Systems* (Moscow: Mashinostroenie)
[2] Zakharov A. Y., Zakharov M A and Loginova O V 2004 *Int. J. Quant. Chem.* **100** 435
[3] Zakharov M. A., 2007 *Phys. Solid State* **49** 2312
[4] Korzun E. L. and Terekhov S. V., 1987 Zh. Phys & Chem. **61** 1186
[5] Zakharov M. A., 2016 *NovSU Bulletin* **98** 22

Computer control system for automatic manufacturing Industry 4.0

A V Gurjanov[1], D A Zakoldaev[2], A V Shukalov[2], I O Zharinov[2], O O Zharinov[3]
[1] Director, Stock Company «Experimental Design Bureau «Electroavtomatika» named after P A Yefimov, 40, Marshala Govorova St., Saint Petersburg, 198095, Russia
[2] Department of Computer Design of On-Board Computer Equipment, Saint Petersburg National Research University of Information Technologies, Mechanics and Optics, 49, Kronverksky Av., Saint Petersburg, 197101, Russia
[3] Department of Problem-Oriented Computing Complexes, Saint Petersburg State University of Aerospace Instrumentation, 69, Bolshaya Morskaia str., Saint Petersburg, 190000, Russia
E-mail: mpbva@mail.ru

Abstract. Technical solutions are proposed for the organization of technical support for automatic instrument manufacturing. The basis of the production management system is a three-level computer network. The structure of the automatic production section of surface mounting and individual technological operations is considered in detail.

1. Introduction

According to classical approaches how to organize manufacturing we can distinguish today technological, subject and linear (massive tide, serial tide, continuous tide, intermittent tide, group) ways of technological manufacturing sections organization according to the volume of on-going production and their nomenclature [1-3].

By analyzing advantages and disadvantages of given ways of organization we can see that a small level of automatizing of technological operations is realized normally with linear tide option. And also the tide lines mostly are oriented to prolonged manufacturing of low-nomenclature items according to «quality» documentation and cannot support adaptive self-organizing and adjusting properties of perspective manufacturing which are typical for technically advanced manufactures of the Industry 4.0 [4-6].

In this line the most suitable option to organize production sections of an item designing company of the Industry 4.0 is a «hybrid» solution — so called the flexible manufacture. The hybrid production sections include a set of automatic machines with memories which are destined to realize technological operations of item components creation in a programmed succession. All automatic machines have the same computerized center of control and can cinematically interact with the section transport system — industrial manipulating robot which primary objective is to carry the blanks (PCBs, components and other) between several production places in the limit of a single production line [7-10].

2. Automatic control system of the digital mechanical assembly production

Central robotized control of the mechanical assembly production is an effective tool to develop an item designing company which functions according to the principles of the Industry 4.0. The base for the manufacturing control system is (see figure 1) a three-level informative and measuring and controlling net of the industrial Internet of Things (IoT).

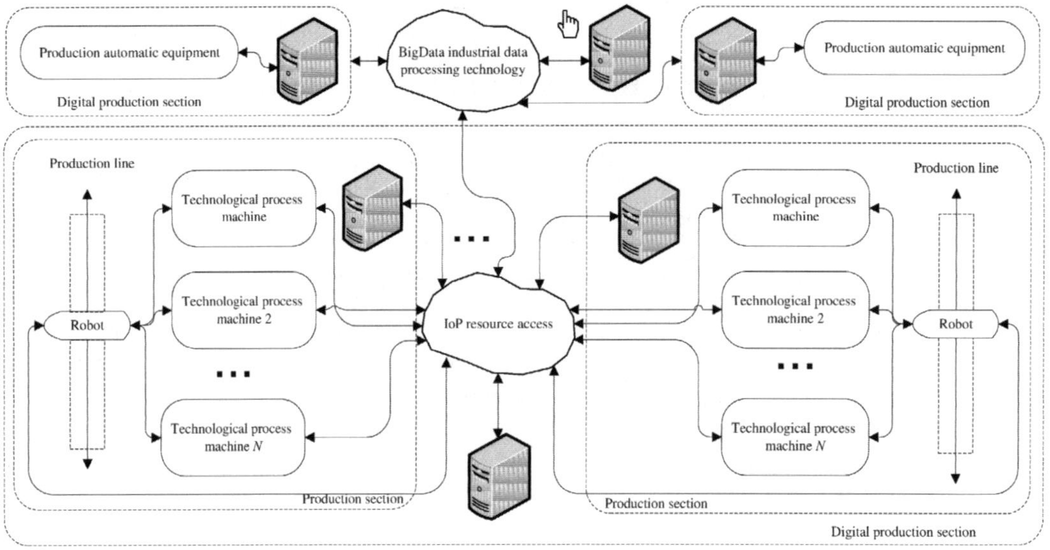

Figure 1. Three-level hierarchy computerized system of mechanical assembly production control of digital item manufacturing company.

Hierarchy low level of the control system is the section technological equipment primary and auxiliary manufacturing systems equipped with controllers and condition detectors of manufacturing processes. The equipment control of production section (workshop) is done by the server which controls the servers of technological lines control.

For a complex production section (for components surface montage) several technological lines can be united in the general automatic inner-section infrastructure.

Technological lines control servers control directly the technological equipment. The upper level of hierarchy system control is the server of mechanical assembly production for information exchange with production sections control servers (workshop).

Computerized system of automatic control of the digital mechanical assembly production include:

- passive integration of digital manufacturing processes of item designing company;
- vertical and horizontal integration technological process (inner production chains) production company, which means those necessary components which creates in the company infrastructure to realize approach the Industry 4.0.

3. Automatic section of surface montage

Principle of organization automatic production of the Industry 4.0 can be studied logically in the example of below. Organization scheme of automatic section of surface montage of digital production is given in figure 2. The given part of scheme of the company main production.

Auxiliary part of production system in the section of surface montage includes:

- air conditioning system in the principle of climate control;
- system of drop and deactivation of water which is used to wash PCBs;
- energized system of powering automatic equipment;
- air filtration system (dust and other materials) in the production room and other.

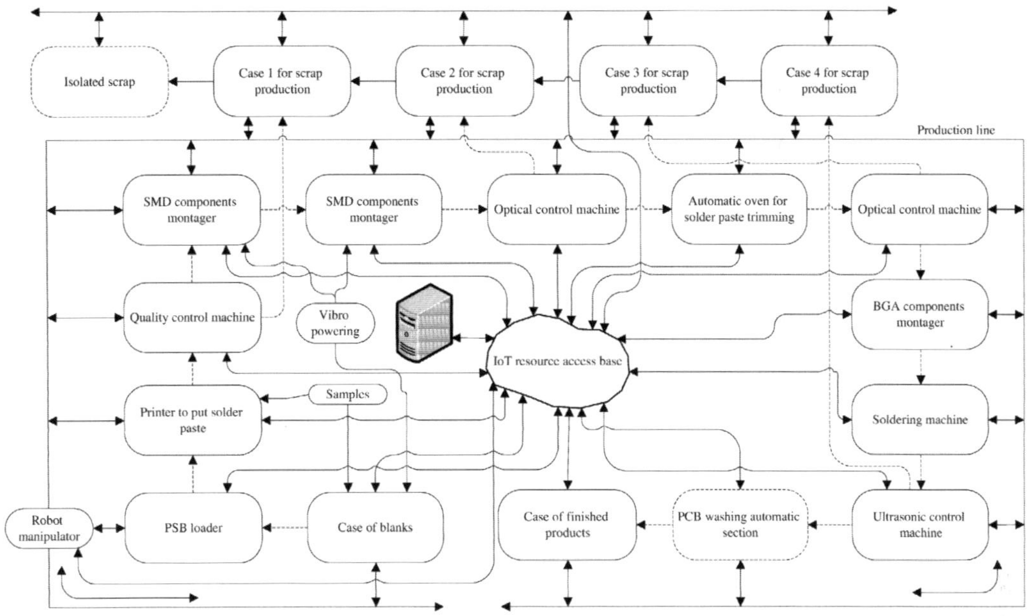

Figure 2. Project solution scheme how to organize automatic section of digital production surface montage.

The main part of section is automatic technological equipment which can be subdivided on the principle of technological operations acceptability in four given groups:
- automatic application of solder paste on a PCB;
- automatic placement of components on a PCB;
- automatic soldering of a PCB;
- automatic washing of a PCB.

Cases of blanks, «finished products» and scrap parts consist a part of production infrastructure of an item designing company which is used to automatically store including dry storage function.
In the figure 2 there is a long option how to place manipulator robot which execute the transport functions in the production line according to the technological equipment. This option is the most suitable in comparison with cross-section and angle placement options.
When there a lot of units of technological equipment and their dimensions are huge it is reasonable to apply ring placement option of technological equipment according to the robot manipulator. The choice how to place transport robot and technological equipment can be made with the criterion of average number of materials connections among them.

4. Automatic application of solder paste

The principle of sample printing is widely known when they organize production sections of serial or low-serial manufacturing of item designing components. In this case the solder paste is applied with ribbon pressure through an aperture in the laser cut made metallic (stainless steel) sample. The volume of applied on a PCB solder material is dependent on the component type and defines with linear sizes of hole and the thickness of sample.
To maintain the technological process of solder paste application it is reasonable to use simple and stepped samples (with different thickness). The printer of sample printing can automatically made:
- PCB application;
- sample application;
- concurrence of sample and PCB;

- solder paste application when the ribbon is being run under pressure near the plate;
- removing from the sample left-over of solder.

Functionality algorithm of technological equipment which maintain the principle of dosing and sample printing and being realized with the downloaded software which is a part of delivery from the manufacturer.

5. Automatic component placement on a PCB

The application of automatic option of components placement on a PCB is in accordance with the principles with an item designing company functioning by the rules of the Industry 4.0 and has some significant advantages:

- significant reduction of mass and dimensions characteristics of item designing components being projected and manufactured;
- significant increase in technological signs of item designing component manufacturing;
- significant reduction of self-price of manufacturing and increase in speed (volume) of issue of item designing components;
- increase in quality of montage of components, increase the signs of reparability of item being manufactured, increase in electrical features of a via «component-printed conduct» and other.

Placement of components on a PCB is being done with special machines of placement. Trimming of solder paste (wave soldering) can be done in a single «run» of convection beam inside the oven (glue is polymerized), or one by one for which purpose the plate with already placed components is being rotated turn by turn.

For montage units which include according to the documentation some components with pin outs is being realized the principle of mixed (mixed-dispersed) montage when the placement of body elements is done on one side only and bodiless components – on both sides. In this case trimming of solder (wave soldering) always must be done in convection oven on both sides being manufactured.

A specific feature which may influence the choice of technological equipment which is a part of production line of surface montage is a step of outputs of electronic components being installed (enlisted in construction documentation) which prove the requirements of high accuracy of work of placement machine.

In case that components on the plate are of different output step it is reasonable to apply the placement option of two stages first in highly productive equipment to install the components with large output step and then in a precision machine to install the components with small step of outputs.

6. Automatic solder of PCBs

Temperature profile of most convection ovens consists of 5 parts in the limits of which with a programmed speed of temperature changing automatically provides:

- heating of plate with installed components (to avoid the heat strike);
- maintaining under some temperature of plate with installed components (removing of moisture, activation of flux and other);
- heating of plate with installed components to the limit acceptable value of temperature (to avoid the heat strike);
- small maintaining of plate with components in conditions of influence of limit acceptable temperature (executes the trimming of solder);
- cooling down of plate with components (to avoid the heat strike).

Solder temperature profile is unique for each type of item (plate with components) and develops in the stage of technological documentation preparation with acceptable modes of functioning (storage) of element base which is used in module.

7. Automatic technological line of PCBs washing

In ultrasonic washing bathes they use the liquid based on alcohol with mixtures which has the increased activity to the solder materials applying in technological equipment of surface montage.

General recommendations when choosing the washing liquid for ultrasonic bathes of Industry 4.0:
- not having chemical activity to the applied materials (components of element base, PCB material and other; we know the case when after PCB washing with installed components plant marking was partially lost);
- low viscosity and fast evaporating;
- compatibility with solder materials which includes natural and synthetic fluxes;
- activity to the left-over of solder paste, flux, solder with flux substance.

Typical project solution for organization of washing line of 4 steps of PCBs applying ultrasonic bathes is given in figure 3.

For PCBs which contains consists special types of electrical and radio items (for example quartz resonator or element components which are sensitive to ultrasonic vibration) the preference of technological equipment choice to wash PCBs it is necessary to give priority to tidal way of washing where it is reasonable to conduct pre-washing of a PCB and washing liquid.

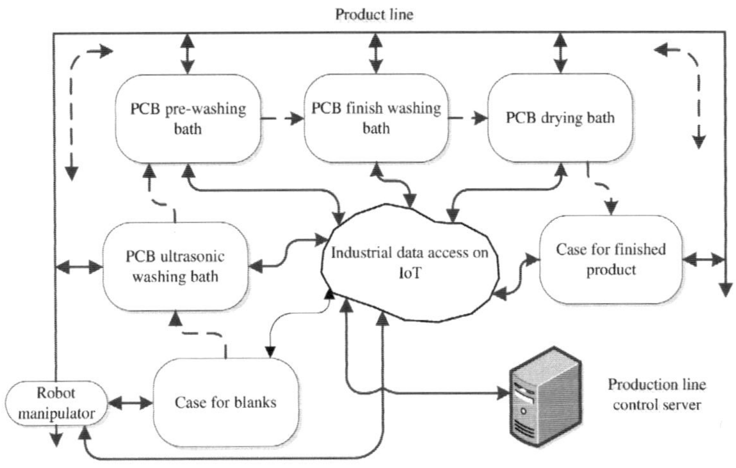

Figure 3. Solution project scheme how to organize digital production section of PCB automatic washing.

8. Conclusion

As unique features of organization new digital production of the Industry 4.0 company we can say the following:
- the presence of automatic computerized equipment equipped with detectors of production process conditions and integration in single informative and measuring and controlled system to prepare the highly qualified management personnel in a company and ability to constantly monitor the condition of manufacturing processes in a company or in other words a transfer to «preventive» maintenance;
- the presence of algorithm and software placed in controllers of equipment and servers of production sections control (workshops, lines) which may give the processing of digital production data applying the technology of BigData cloud technologies, additive technologies and the technology of Internet of Things (data processing include automatic formation processes how to deal with the emergency measure in the production — for example equipment break-down; no more blanks in the case components or raw material which is necessary to make some technological operations and other; emergency in system of auxiliary production, — for example functionality of cleaning facilities, energizing facilities and other).

References

[1] Stef I D, Draghici G, Draghici A 2013 *Procedia technology* **9** 451-462
[2] Almada-Lobo F 2015 *Journal of innovation management* **3** 16-21
[3] Prause G 2015 *Journal of security and sustainability issues* **5** 159-169
[4] Bligh-Wall Sh 2017 *Cyber security: a peer-reviewed journal* **1** 61-68
[5] Lee J, Kao H-A, Yang Sh 2014 *Procedia CIRP* **16** 3-8
[6] Zhong R Y, Xu Ch, Chen Ch and Huang G Q 2017 *International journal of production research* **55** 2610-21
[7] Shariatzadeh N, Lundholm Th, Lindberg L and Sivard G 2016 *Procedia CIRP* **50** 512-517
[8] Gjeldum N, Mladineo M, Veza I 2016 *Procedia CIRP* **54** 158-163
[9] Ruttimann B G, Stockli M T 2016 *Journal of service science and management* **9** 458-500
[10] Kemeny Z, Beregi R J, Erodus G and Nacsa J 2016 *Procedia CIRP* **54** 53-58

IOP Conf. Series: Materials Science and Engineering **441** (2018) 012064 doi:10.1088/1757-899X/441/1/012064

Production control organization system in the Industry 4.0 item designing company

D A Zakoldaev[1], A V Shukalov[1], I O Zharinov[1], V A Nechaev[1]

[1] Department of Computer Design of On-Board Computer Equipment, Saint Petersburg National Research University of Information Technologies, Mechanics and Optics, 49, Kronverksky Av., Saint Petersburg, 197101, Russia

E-mail: mpbva@mail.ru

Abstract. It is about task how to organize an Industry 4.0 item designing company. It is proposed a general algorithm how to manufacture an item designing component and a manufacturing algorithm for an item designing component in the Industry 4.0 automatic production. The article primary attention is put at production control options applied in the company. They defined production control options and life cycle stages where they are realized during the manufacturing.

1. Introduction

Organization of an Industry 4.0 item designing company is one of the most actual directions of industrial companies. This field of research is being maintained in the USA, EU countries, Russia, India, China and some other countries [1, 2].

In Germany the application of Industry 4.0 initiative is directed to [3, 4]:

- development of cyber and physical systems of industrial application including some robotized digital automatic equipment united in smart production complexes;
- development of Internet of Things technology including some protocols for information exchange for industrial Internet of Things;
- development and improvement of technologies like «Machine-to-Machine», «Systems-to-Systems», substituting in production the technology of «Human-to-Machine» which is typical for the Industry 3.0.

The program of Advanced Manufacturing Partnership is being developed in the USA as a part of partnership in government and private levels between departments when they necessarily participate the divisions of the Ministry of Defense, Ministry of Energy, National Science Fund, National Association of Perspective Industrial Technologies, NASA, National Institutes of Standards and Technologies, leading teaching organizations. The teaching sector of the program participants for Advanced Manufacturing Partnership is being represented by Massachusetts Technological Institute, Stanford University, California University, Michigan University and some others. The industrial USA sector participating in the program of Advanced Manufacturing Partnership is being represented by some huge companies like Intel, Ford, United Technologies, Honeywell and others.

The EU countries approach to develop the new digital industrial technologies generally is being represented as the strategy Vision for Manufacturing which was developed as technological platform ARTEMIS and program «Horizon-2020».

In Great Britain to develop the prioritized directions of digital modelling and digital production they created special competition centers Catapult Centers which combine the innovative industrial companies. Competition centers are created under the protection of the national counsel of strategical technologies which functions under the English government to transfer the ideas and technologies received in the academic environment and then transfer them into high-tech production.
The industry of China interested in creation of humanless digital production and development of its own technological base is probing now new technologies created under special national «Internet of Things zone» where participate some hundreds of manufacturing companies and special national structure Internet of Things Center. The primary directions of China industrial policy in the field of how to organize digital manufacturing complexes are oriented to develop the equipment of new generation (for example intellectualized technological equipment for production) and development of company technological processes of digital control system.
In India in 2011 the Ministry of Communication and Information proposed something that is successfully being developed today which is the project of the innovation center of cyber and physical systems. The center primary task is to research for development of intellectualized production environment of company, the most important high-tech operations in the order of their completion by automatic robotized equipment.
Russian approach to modernize the Industry 3.0 companies and to apply the principles of the Industry 4.0 in the digital economy is oriented to create the new types of production companies [5, 6]: digital factory, smart factory, virtual factory.

2. Production control options for the item designing industry

To manufacture the item designing components is a hard manufacturing process which involves to realize different technological operations practically. The general algorithm how to manufacture the item designing components is given in figure 1. The key technological operations being completed in production are [7, 8]:
- to put the components in the item (capacitors, resistors, microchips and other) and their assortment and grouping;
- to prepare blanks and use them to make components which are necessary to assemble the item;
- to create assemble units which will be used to assemble the final item;
- final item assemblage;
- checking, setting and adjustment of the final items;
- testing of ready item and tests have been done (electrical, climatical and other) on a ready item in the sufficient quantity.

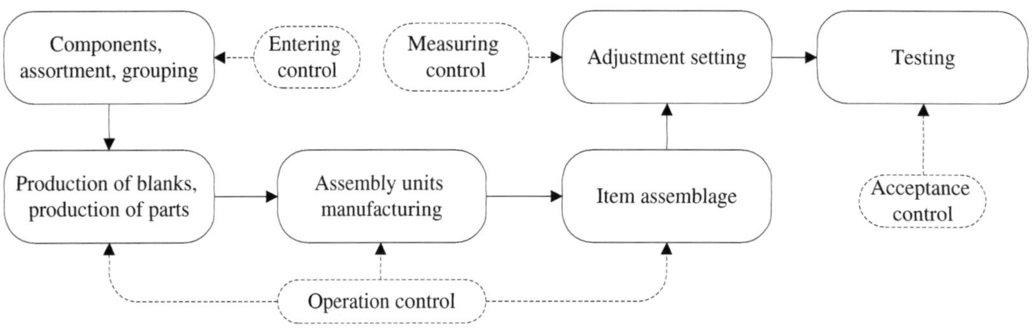

Figure 1. Item designing components manufacturing generalized algorithm.

The completion of technological operations how to manufacture an item in the production can be done automatically (in the Industry 3.0 company) and in automatized mode (in the Industry 4.0 company). In both cases the correspondence of the item to the given quality requirements of technical

documentation that can be valuated after the production control.
Production control in an item designing company includes the following types of control:
- entering control;
- operational control;
- measurement control;
- acceptance control.

Entering control is organized by personnel of an item designing company (designers, production workshop personnel, entering control section personnel of the technological control department). Entering control proves the quality of components (item designing components) which the company receives to manufacture the item.

Operational control is organized by personnel of an item designing company (production workshop personnel, technological department personnel which functions in the workshop). Operational control proves the quality of technological operations made inside the workshop and also its assembly units.

Measurement control is organized by personnel of an item designing company (designers, technological department personnel, metrology department personnel). Measurement control proves the quality of the item functioning as it must be according its technical and tactical requirements given in its technical task.

Acceptance control is organized by personnel of an item designing company (designers, production workshop personnel, adjustment and testing laboratory personnel, technological department personnel). Acceptance control is the final form of production control which will be followed by the item shipping from its manufacturer to its user.

3. Production control realization in the Industry 4.0 company

The figure 2 shows the generalized algorithm of the item designing component manufacturing in an Industry 4.0 automatic production. The item here means a constructive and functioning module which is a module of radio and electronics with its own construction and made with a multi-layer PCB (Printed Circuit Board). The figure 2 includes software; technical requirements; automatized working place.

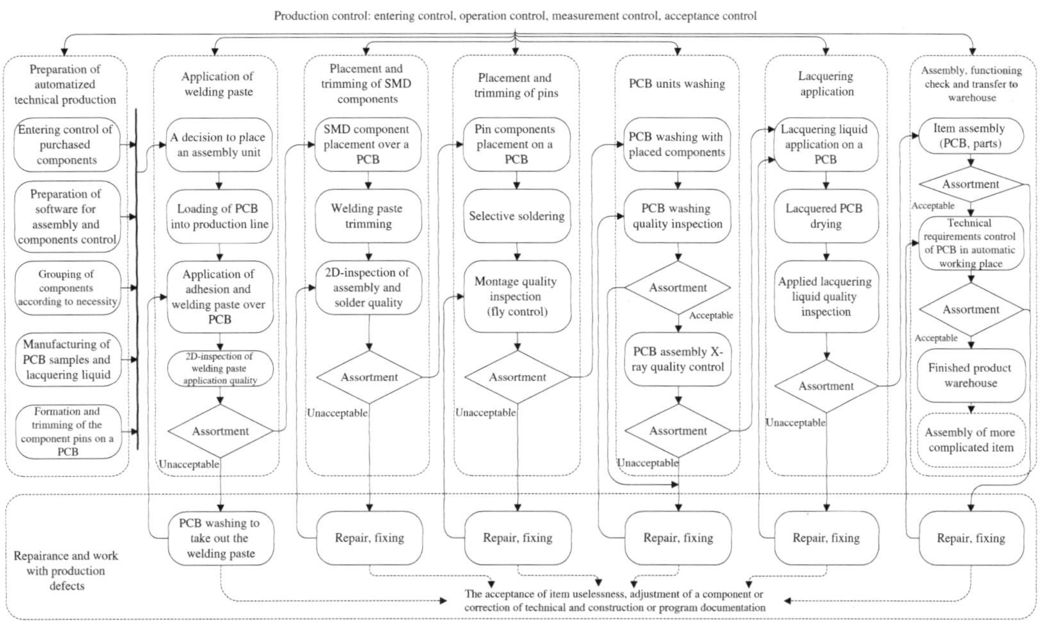

Figure 2. Item designing component manufacturing algorithm in an Industry 4.0

automatic production (SMD – Surface Mounted Devices).

The key technological operations when a radio and electronics module is being manufactured include the following procedures [9, 10]:
- the preparation of automatic technological production;
- the application of welding paste over a PCB;
- surface placement of electronic components on a PCB and solder trimming;
- pin components montage on a PCB and solder trimming;
- PCB washing after the montage;
- the application of anti-moisture lacquer on a PCB;
- item assembly and check of its functioning with the following transfer of the module to the finished product warehouse.

The stage of preparation of automatic technological item designing production includes:
- item assembly processes, grouping and assortment of elemental components which will be used in the production (elemental base has a preliminary processing to form the pins or to make the pins necessarily long and formed which are prescribed in the technical task for the item);
- the process of creation special components for software which will be loaded in production technical equipment engaged in the item manufacturing;
- the manufacturing of samples to apply welding paste on a PCB and preparation of anti-moisture lacquer liquid which protects the item from the environment influence.

In the stage of preparation of automatic technological item designing production the following forms of production control are engaged: entering control, operational control.

The stage of welding paste application includes the transfer of a PCB into special technological equipment where the samples profiles and PCBs are being corresponded and the process of welding paste application. Operational control which is being done during the welding paste application on a PCB supposes the material quality inspection and may include the second time of the paste application if the inspection finds some technological process completion quality violations. In this case it is necessary to wash a PCB from the welding paste being applied badly.

The trimming stage and surface placement of radio and electronics components (capacitors, resistors, microchips and other) include the placement itself of components on a PCB and trimming of welding paste to fixture mechanically and to provide electrical contact of elements with the PCB elements (pads, textolyte). Operational control of its stage includes the corresponding technological operations quality check.

The stage of pin components montage (capacitors, resistors, microchips and other) and solder trimming include the component placement itself, selective soldering (including trimming for placed pin components; the components of surface placement installed before will not be damaged) and the montage quality check process. Operational control of this stage can be done with special technological equipment (the system of fly checks) which includes a check of the electrical contact presence between the components and the PCB itself.

The stage of PCB washing includes taking-out of solder left-overs, flux from the PCB surface and X-ray quality control of the made before elemental base placement technological operations on the PCB surface. Operational control of this stage can be done with special technological equipment which is based on optical and X-ray principles.

Application of the anti-moisture lacquer on the PCB surface includes the application of lacquering liquid (lacquer), the liquid drying and lacquering quality check. Operational control of this stage can be done with special technological equipment which is based on optical principle.

The stage of item assembly and functioning quality check includes mechanical parts placement on a PCB with installed before components and the item check over electrical functionality in several modes. In this stage of production cycle they do measurement control and acceptance control of the made technological operations.

All existing forms of the production control may include the correspondence of technological operations requirements to the technical documentation and also if this correspondence exists at all. If the technological operations requirements do not correspond to the technical documentation the company organizes item reparation works (assembly unit) or the works to take out those found defects. Repairing, not taking-out defects, includes to divide some methods in an item designing company which exclude in the future any re-appearance of those defects.

4. Conclusion

The structure itself of the Industry 4.0 item designing company includes a transfer from the standard ways of components manufacturing which are done in automatized mode to the ways of automatic mode. It is necessary to create for that purpose the technological equipment of new generation which is cyber and physical systems. Cyber and physical systems function with the application of technology of industrial Internet of Things, cloud technologies, sensorics technologies and other. This kind of robotized equipment will be a base for industrial company automatic production line which will be used to manufacture item designing components without involvement of operators.

Existing now forms how to control quality of manufactured item designing components can be applied in the Industry 4.0 company in the corresponding parts of manufacturing cycle.

References

[1] Theorin A, Bengtsson K, Provost J, Lieder M, Johnsson Ch and Lundholm Th 2017 *International journal of production research* **55** 1297-11

[2] Liao Y, Deschamps S, Loures E F and Ramos L F 2017 *International journal of production research* **55** 3609-29

[3] Jung K, Choi S S, Kulvatunyou B, Cho H and Morris K S 2017 *Production planning & control* **28** 108-122

[4] Radziwon A, Bilberg A, Bogers M and Madsen E S 2014 *Procedia engineering* **69** 1184-90

[5] Zakoldaev D A, Shukalov A V, Zharinov I O and Zharinov O O 2018 *Journal of Physics: Conference Series* **1015** 052033

[6] Gurjanov A V, Zakoldaev D A, Shukalov A V and Zharinov I O 2018 *Journal of Physics: Conference Series* **1015** 052034

[7] Poonpakdee P, Koiwanit J, Yuangyai C 2017 *Procedia computer science* **110** 46-53

[8] Hwang G, Lee J, Park J and Chang T-W 2017 *International journal of production research* **55** 2590-02

[9] Qu T, Thurer M, Wang J, Wang Z, Fu H and Li C 2017 *International journal of production research* **55** 2622-49

[10] Vogel-Heuser B, Rosch S, Fischer J, Simon Th, Ulewicz S and Folmer J 2016 *Journal of software engineering and applications* **9** 1-43

IOP Conf. Series: Materials Science and Engineering **441** (2018) 012065 doi:10.1088/1757-899X/441/1/012065

Improvements in the performance of the n^+ cap-layer GaN in the formation of transistor structures based on AlGaN / GaN

A V Zhelannov[1], B I Seleznev[2]

[1]OAO OKB-Planeta
Russia, 173004, Veliky Novgorod, ul. Fyodorovskiy ruchei, 2/13,
e-mail: ZhelannovAV@okbplaneta.ru
[2] Novgorod State University, ul. B. St. Petersburgskaya, 41
173003 Veliky Novgorod, Russia

Abstract. The use of heavily doped protective layers of gallium nitride in AlGaN/GaN heterostructures makes it possible to improve the characteristics of transistors. In this paper, using structures with a protective cap layer and the process of selective etching, structures of HEMT transistors are fabricated. Using the n^+-GaN layer, remove the contact resistance from 1 Ω·mm to 0.74 Ω·mm. The transistors show a maximum drain current of 380mA/mm and a steepness of 140mS/mm with an output power of 5W/mm.

1. Introduction
Highly efficient gallium nitride transistors are successfully used in microwave power devices and circuits, thanks to a unique combination of semiconductor material properties, namely, a large band gap, a high electric breakdown voltage and a high saturation speed of carriers.

When developing instrument structures, the correct choice of the parameters of the initial material largely determines the achievement of the ultimate characteristics of the finished device. To obtain the ultimate characteristics of AlGaN / GaN-based transistor structures, it is important to use protective layers that can act as thin dielectric films (Si_3N_4, SiO_2, Al_2O_3) or thin semiconductor layers of gallium nitride (doped or undoped). However, due to the low mobility and wide bandgap of undoped AlGaN layers, conventional approaches to using AlGaN as a surface protective layer limit the performance of devices, especially for microwave power applications. To reduce the contact resistance, unalloyed or slightly doped thin GaN layers are introduced, while the characteristics of the devices are not improved because of the increase in the distance from the gate to the channel and the reduction of the induced piezoelectric charges. Therefore, the addition of a heavily doped n + -GaN layer over the traditional HEMT structure of AlGaN / GaN can improve instrument performance. At the same time, this doped layer must be removed before the formation of Schottky gate metallization to reduce the leakage current of the gate.

2. Instrument structure of AlGaN / GaN
The structures used in the work were grown on sapphire substrates by the method of chemical deposition from the gas phase using organometallic compounds (MOCVD). The structure of the layers is shown in Figure 1. The mobility and concentration of charge carriers in the structure with the n + layer were 1820 cm^2 / V · s and 1.04 × 1013 cm^{-2}. For structures without a protective layer, these parameters were 1675 cm^2 / V · s and 1.21 × 1013 $cm^{-2.}$ The decrease in carrier concentration is caused by the reverse direction of the piezoelectric polarization between the n + -GaN layer and the AlGaN barrier layer, where piezoelectric charges are the main charge carriers in the HEMT AlGaN-GaN [1]. In this case, a decrease in the concentration of sheet carriers always leads to a higher mobility of a two-dimensional electron gas. After removing the protective layer under the gate metallization, the reverse direction of the piezoelectric effect can be neglected.

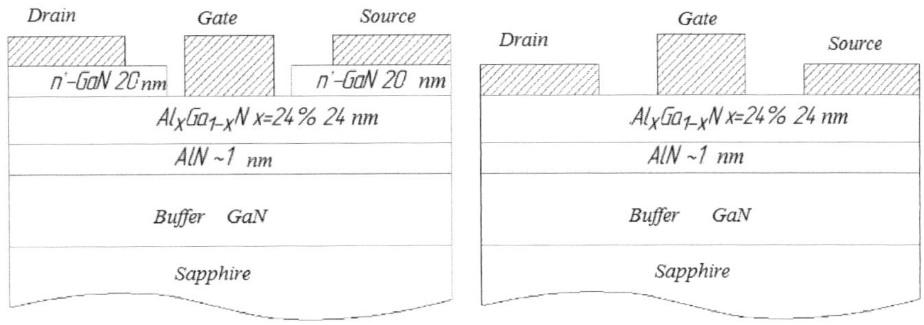

Figure 1. The cross section of the structures used in the work.

3. Manufacturing technology of instrument structures

The fabrication of instrument structures began with the formation of mesa-isolation. Under the formation of mesic, it is meant to create a structure on a heteroepitaxial plate with conductive (active) and nonconducting layers by etching the heterostructure surface. The purpose of the operation is to separate the active regions on which the devices are formed from each other in order to avoid electrical contact between the instruments. The limitations of liquid etching methods for nitride semiconductors lead to the need to introduce plasma etching techniques. The most effective at present is the technology of nitride semiconductors etching in plasma using sources on an inductive high-frequency discharge. Such sources allow to create a denser plasma with high uniformity and controllability. In this work, etching in a mixture of $Cl_2/BCl_3/Ar$ using inductively coupled plasma (ICP) at a rate of 800 nm / min was used to form the mesa-isolation [2].

The ohmic contacts of the drain and the source were formed with the help of an explosive photolithography process using a two-layer photoresist system LOR 10A and S1813. The use of a two-layer system based on LOR resist (Figure 2) allows to form a negative slope of the walls, facilitating the process of metallization removal and providing an even edge of the contact pads. As the metallization of ohmic contacts, a Ti /Al/Ni/Au system was used, formed by electron beam deposition. The annealing of the metallization was carried out at a temperature of 820 °C for 30 seconds under a nitrogen atmosphere.

The contact resistance was evaluated by the long-line method [3] with a set of square contact pads 100 × 100 μm at a distance of 5, 10, 15, 20, 25 and 30 μm. The calculation of the contact resistance gives a value at the level of 0.8 Ohm · mm for a structure with n^+-GaN protective layer and about 1.0 Ω · mm for structures without a protective layer.

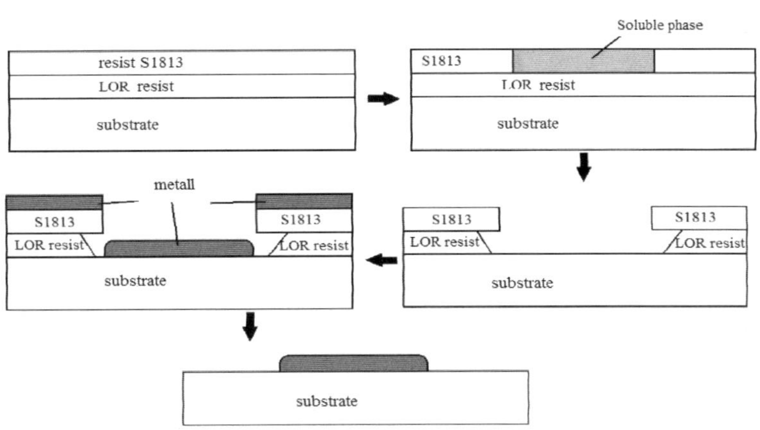

Figure 2. Stages of forming a resistive mask for "Explosive" photolithography.

Measurements of the resistance versus distance by the long-line method and the current-voltage characteristics of contacts with a distance of 30 μm for the structures under consideration are shown in Figures 3 and 4.

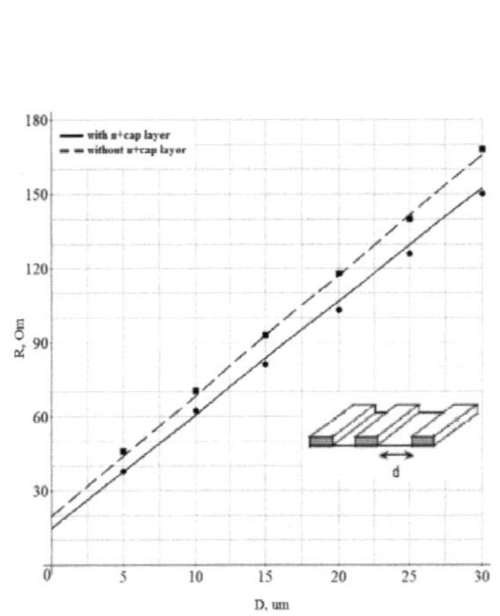

Figure 3. Dependence of the resistance on the distance between the pads. The inset shows a fragment of the test structure.

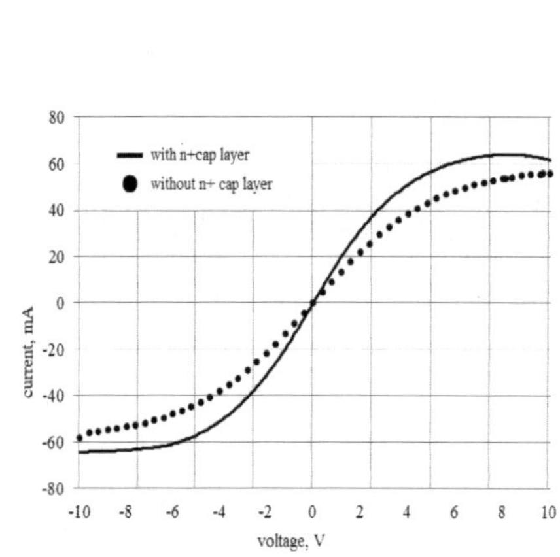

Figure 4. Volt-ampere characteristics of drain-source on different structures.

Before barrier metallization formation, it is necessary to remove the conductive cap layer of GaN. To carry out this operation, a process of selective etching of GaN with respect to AlGaN was developed, at which a sharp decrease in velocity occurs when the AlGaN layer is reached. Such a process is realized by adding an oxygen additive to the chlorine-containing mixture Cl_2/Ar, which, reacting with aluminum, forms an Al_2O_3 compound, which limits further etching. At the same time, the selectivity of GaN/AlGaN etching is determined not only by the amount of oxygen, but also by a number of parameters. This is the power of the inductively-coupled plasma source, high-frequency power and pressure in the chamber. As a result of the work, the optimum etching regimes were determined from the point of view of obtaining the required etching rate of gallium nitride and the selectivity. The dependence of GaN and AlGaN etching rate and etching selectivity on these parameters is shown in Figure 5.

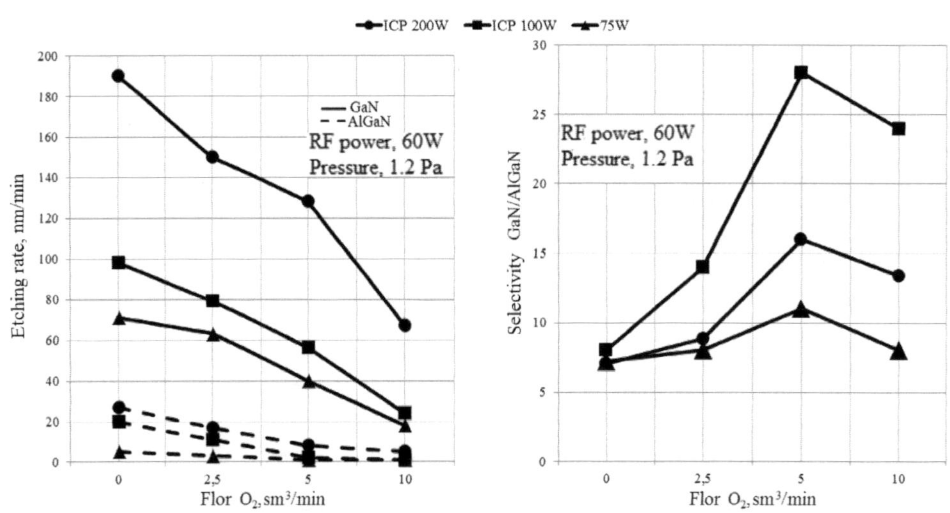

Figure 5. Dependence of the rate and selectivity of etching on regimes.

In accordance with the presented dependences, the etching mode was chosen: $Cl_2/Ar/O_2$ 60/10/10 cm^3/min: ISP power – 200W, RF power – 30W and pressure - 1.2Pa, permitting etching selectivity about 30:1.

After removal of the n^+-GaN layer, barrier metallization was formed. The gates of transistor structures with a length of 1.0 µm and a width of 100µm were manufactured by successively performing electron-beam deposition of the Ni/Au system and explosive photolithography.

IOP Conf. Series: Materials Science and Engineering **441** (2018) 012065 doi:10.1088/1757-899X/441/1/012065

After the formation of the transistor structure contact systems, the devices were passivated by a Si_3N_4 film obtained by plasma-chemical deposition at a temperature of 250 °C. The dielectric film provides stabilization of the surface of the instrument structure.

4. Volt-ampere and voltage-voltage characteristics of transistor structures

As a result of the technological process, models of transistor structures were obtained. The study of characteristics began with the determination of the parameters of barrier metallization using the test structures of Schottky diodes. The height of the barrier and the nonideality coefficient calculated from the current-voltage characteristics of Schottky test diodes were determined. The results of the calculation are given in Table 1.

Table 1 – Calculation of barrier metallization

	Coefficient of nonideality	Barrier height, eV
n^+-layer	0.62	1.77
without n^+-layer	0.66	1.73

The close values of the barrier metallization parameters show that the developed process of selective etching does not introduce significant defects into the barrier layer of AlGaN.

The current-voltage and transient characteristics of the fabricated structures are shown in Figures 6 and 7.

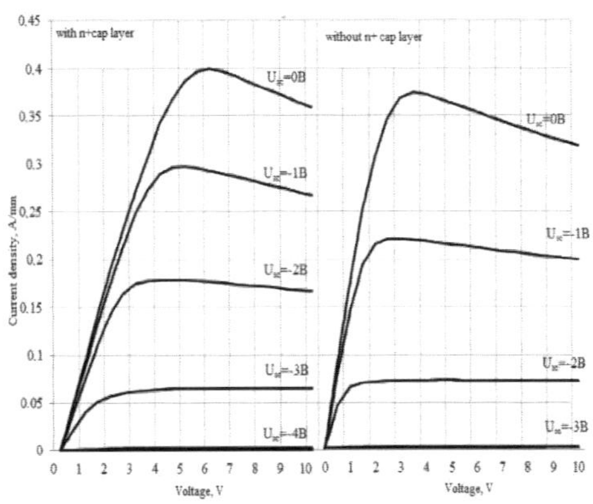

Figure 6. Output volt-ampere characteristics of transistor structures.

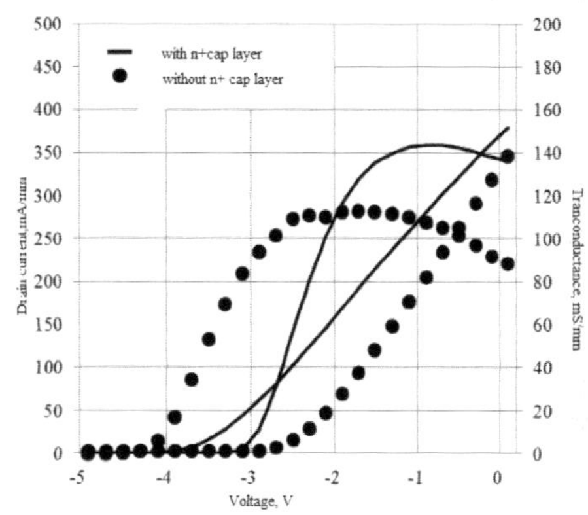

Figure 7. Transient characteristics of transistor structures.

Note that the addition of a conductive layer provides a low resistivity of the conductive channel. This leads to a low threshold voltage, greater steepness and current density. For structures with n^+-GaN layer, the maximum steepness and current density were 380 mA/mm and 143 mS/mm, while for structures without a protective layer they were 345 mA/mm and 112 mS /mm.

Another important parameter of transistors is the size of the gate-drain capacitance. The capacitance value has a key influence on the frequency characteristics of transistors. Dependencies of the gate-drain capacitance on the gate voltage for the structures under consideration are shown in Figure 8. At a gate voltage of 0 V, the capacitance values are 0.66 pF for structures with a protective layer and 1.12 pF for structures without an n^+-layer.

IOP Conf. Series: Materials Science and Engineering **441** (2018) 012065 doi:10.1088/1757-899X/441/1/012065

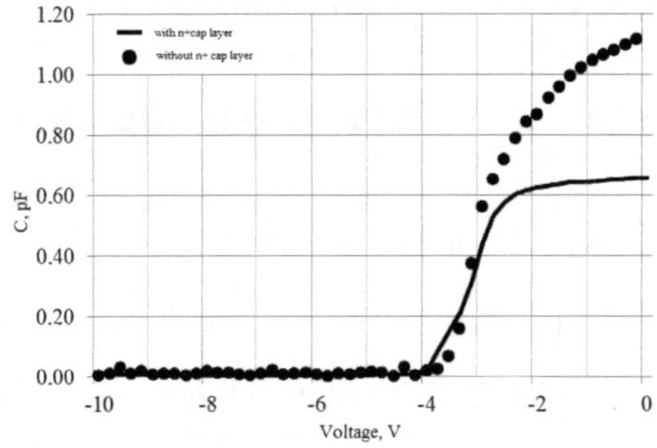

Figure 8. Gate-source capacitance voltage on the gate.

5. Conclusion

In this paper the characteristics of transistor structures with a protective heavily doped layer of gallium nitride and without it were compared. Improved performance is shown when using the n^+-layer due to the reduction of contact resistance and channel resistance. The current density is 380mA/mm, the steepness is 143mS/mm and the output power density is 5W/mm.

The publication has been prepared with the financial support of the Ministry of Education and Science of the Russian Federation within the framework of the project part of the state task, project No. 3.33572.2017.

References

[1] Heikman S, Keller S 2003 *J. Appl. Phys.* **93** 10114.
[2] Zhelannov AV, Ionov AS, Petrov AV, Seleznev BI 2017 Nano- and microsystem technology 19 399.
[3] Reeves G K, Harrison H B 1982 IEEE Electr. Dev. Leter. 5 111.

AUTHOR INDEX

AUTHOR INDEX